T0245138

LAGRANGIAN ANALYSIS AND PREDICTION OF COASTAL AND OCEAN DYNAMICS

Written by a group of international experts in their field, this book is a review of Lagrangian observation, analysis and assimilation methods in physical and biological oceanography. In recent years a large number of floating and drifting research buoys have been deployed in the global oceans to study the state of the ocean and its variation in terms of water mass properties, circulation and heat transport. Lagrangian techniques are required to analyze the data from these buoys.

This multidisciplinary text contains observations, theory, numerical simulations, and analysis techniques. It presents new results on nonlinear analysis of Lagrangian dynamics, the prediction of particle trajectories, and Lagrangian stochastic models. It includes chapters on floats and drifters, Lagrangian-based analysis methods and models in marine biology, the statistics of particle trajectories in the ocean, numerical simulations and their relationship with classical turbulence results, and nonlinear Lagrangian-based theory for studying ocean transport and particle trajectories. The book contains historical information, up-to-date developments, and speculation on future developments in Lagrangian-based observations, analysis, and modeling of physical and biological systems.

Containing contributions from experimentalists, theoreticians, and modelers in the fields of physical oceanography, marine biology, mathematics, and meteorology this book will be of great interest to researchers and graduate students looking for both practical applications and information on the theory of transport and dispersion in physical systems, biological modeling, and data assimilation.

Cover illustration: The cover depicts the abrupt breakup of a large ocean eddy in the Gulf of Mexico. Eddy Fourchon was tracked by assimilating satellite data into the University of Colorado version of the Princeton Ocean Model (developed by L. H. Kantha). Lagrangian analysis by researchers at the University of Delaware (led by A. D. Kirwan, Jr.) and the City University of New York (A. C. Poje) produced the time sequence of marked particles in the middle of the Gulf between July 28 and August 17, 1998. On July 28, Fourchon appears to be a typical large elliptical ocean eddy. Over the next two and a half weeks, interactions with nearby mesoscale features split the core in half. The larger colored region is determined by computing the Lagrangian boundaries of the eddy on the initial day with red/yellow assigned to those particles within the eddy which eventually split to the north/south respectively. The contrasting inscribed circles show the stirring inherent in each sub-region during the evolution. Figure design by Patrick Fagan.

LAGRANGIAN ANALYSIS AND PREDICTION OF COASTAL AND OCEAN DYNAMICS

Edited by

ANNALISA GRIFFA
Rosenstiel School of Marine and Atmospheric Science
University of Miami
Istituto di Scienze Marine, Consiglio Nazionale Ricerche, La Spezia, Italy

A. D. KIRWAN, JR.
University of Delaware

ARTHUR J. MARIANO
Rosenstiel School of Marine and Atmospheric Science
University of Miami

TAMAY M. ÖZGÖKMEN
Rosenstiel School of Marine and Atmospheric Science
University of Miami

THOMAS ROSSBY
Graduate School of Oceanography
University of Rhode Island

CAMBRIDGE
UNIVERSITY PRESS

University Printing House, Cambridge CB2 8BS, United Kingdom

One Liberty Plaza, 20th Floor, New York, NY 10006, USA

477 Williamstown Road, Port Melbourne, VIC 3207, Australia

314-321, 3rd Floor, Plot 3, Splendor Forum, Jasola District Centre, New Delhi - 110025, India

79 Anson Road, #06-04/06, Singapore 079906

Cambridge University Press is part of the University of Cambridge.

It furthers the University's mission by disseminating knowledge in the pursuit of education, learning and research at the highest international levels of excellence.

www.cambridge.org
Information on this title: www.cambridge.org/9781108460552

First published 2007
First paperback edition 2018

A catalogue record for this publication is available from the British Library

ISBN 978-0-521-87018-4 Hardback
ISBN 978-1-108-46055-2 Paperback

Contents

The color plates are situated between pages 228 and 229.

Contributors

Amy S Bower
Department of Physical
Oceanography
Woods Hole Oceanographic Institute
Woods Hole, MA 02543
USA

Annalisa Bracco
Department of Physical
Oceanography
Woods Hole Oceanographic
Institution
Woods Hole, MA 02543
USA

Giuseppe Buffoni
ENEA
Santa Teresa – Lerici
La Spezia I-19100
Italy

Jim Carton
University of Maryland
Stadium Drive
College Park, MD 20742-0001
USA

Luca R Centurioni
Scripps Institute of Oceanography
9500 Gilman Drive
La Jolla, CA 92093-0213
USA

Toshio Chin
RSMAS/MPO
University of Miami
4600 Rickenbacker Causeway
Miami, FL 33149
USA

Curtis A Collins
Code Oc/Co
Department of Oceanography
Naval Postgraduate School
833 Dyer Road
Monterey, CA 93943-5122
USA

Robert K Cowen
RSMAS/MBF
University of Miami
4600 Rickenbacker Causeway
Miami, FL 33149
USA

Heather Furey
Department of Physical Oceanography
Woods Hole Oceanographic Institute
Woods Hole, MA 02543
USA

Newell Garfield
San Francisco State University
Geosciences Dept.
3152 Paradise Drive
Tiburon, CA 94920
USA

Annalisa Griffa
RSMAS/MPO
University of Miami
4600 Rickenbacker Causeway
Miami, FL 33149
USA
ISMAR/CNR
Forte Santa Teresa
La Spezia I-19036
Italy

Semyon Grodsky
Department of Meteorology
University of Maryland
College Park, MD 20742
USA

Gary L Hitchcock
RSMAS/MBF
University of Miami
4600 Rickenbacker Causeway
Miami, FL 33149
USA

Kayo Ide
Institute of Geophysics & Planetary Physics
UCLA
Los Angeles, CA 90095-1567
USA

Christopher Jones
University of North Carolina at Chapel Hill
CB #32350
UNC-CH
Chapel Hill, NC 27599
USA

YooYin Kim
Scripps Institute of Oceanography
9500 Gilman Drive
La Jolla, CA 92093-0213
USA

Vassiliki Kourafalou
RSMAS/MPO
University of Miami
4600 Rickenbacker Causeway
Miami, FL 33149
USA

Leonid Kuznetsov
Applied Mathematics
University of North Carolina at Chapel Hill
(Phillips Hall 362)
Chapel Hill, NC 27599
USA

Matthias Lankhorst
Leibniz-Institut fur
Meereswissenschaften
(IFM-GEOMAR)
Dusternbrooker Weg 20
Kiel D-24105
Germany

Dong-Kyu Lee
Department of Marine Sciences
Busan National University
Busan 609-735
South Korea

Thomas N Lee
RSMAS/MPO
University of Miami
4600 Rickenbacker Causeway
Miami, FL 33149
USA

Rick Lumpkin
Atlantic Oceanographic &
Meteorological Lab
NOAA/AOML/PhOD
4301 Rickenbacker Causeway
Miami, FL 33149
USA

Svend-Aage Malmberg
Marine Research Institute
1 Hafrannsoknansofnunin
PO Box 1390
Skulgata 4
Reykjavik 121
Iceland

Arthur J Mariano
RSMAS/MPO
University of Miami
4600 Rickenbacker Causeway
Miami, FL 33149
USA

Maria Grazia Mazzocchi
Stazione Zoologica A. Dohrn
Villa Communale
Napoli I-80121
Italy

Anne Molcard
LSEET
University of Toulon
Forte Santa Teresa
La Spezia I-19036
Italy

Pearn P Niiler
Scripps Institute of Oceanography
9500 Gilman Drive
La Jolla, CA 92093-0213
USA

Donald B Olson
RSMAS/MPO
University of Miami
4600 Rickenbacker Causeway
Miami, FL 33149
USA

Tamay Özgökmen
RSMAS/MPO
University of Miami
4600 Rickenbacker Causeway
Miami, FL 33149
USA

Nathan Paldor
Hebrew University of Jerusalem
Institute of Earth Sciences
Edmund Safra Campus, Givat Ram
Jerusalem 92509
Israel

Sara Pasquali
CNR-IMATI
via Bassini, 15
Milano I-20133
Italy

Claudia Pasquero
Earth System Science Dept.
University of California
3224 Croul Hall
Irvine, CA 92697-3100
USA

Mayra C Pazos
Atlantic Oceanographic &
Meteorological Lab
NOAA/AOML/PhOD
4301 Rickenbacker Causeway
Miami, FL 33149
USA

Leonid Piterbarg
University of Southern California
Kaprielian Hall, Room 108
3620 Vermont Avenue
Los Angeles, CA 90089-2532
USA

Pierre-Marie Poulain
Istituto Nazionale di Oceanografia e
di Geofisica
Sperimentale (OGS)
Borgo Grotta Gigante 42/c
Trieste I-34010
Italy

Antonello Provenzale
Institute of Atmospheric Sciences &
Climate
CNR
Corso Fiume, 4
Torino I-10133
Italy

Thomas A Rago
Department of Oceanography
Naval Postgraduate School
833 Dyer Road, Rm 328
Monterey, CA 93943
USA

Thomas Rossby
University of Rhode Island
Graduate School of Oceanography
Kingston, RI 02881
USA

Volfango Rupolo
ENEA
via Anguillarese, 301
Roma I-00060
Italy

Edward H Ryan
RSMAS/MPO
University of Miami
4600 Rickenbacker Causeway
Miami, FL 33149
USA

Vitalii A Sheremet
Graduate School of Oceanography
University of Rhode Island
Narranganset, RI 02882
USA

Hedinn Valdimarsson
Marine Research Institute
1 Hafrannsoknansofnunin
PO Box 1390
Skulagata 4
121 Reykjavik
Iceland

Jeffrey B Weiss
Dept. of Atmospheric & Oceanic
Science
PAOS
University of Colorado
Boulder, CO 80309-0311
USA

Elizabeth Williams
RSMAS/MPO
University of Miami
4600 Rickenbacker Causeway
Miami, FL 33149
USA

Walter Zenk
Leibniz-Institut fur
Meereswissenschaften
Ozeanographie
Dusternbrooker Weg 20
D-24105 Kiel
Germany

Preface

This book has been motivated by the recent surge in the density and availability of Lagrangian measurements in the ocean, recent mathematical and methodological developments in the analysis of such data to improve forecasts and transport characteristics of ocean general circulation models, and numerous applications to dispersion of biological species. Another source of motivation has been the Lagrangian Analysis and Prediction of Coastal and Ocean Dynamics (LAPCOD) workshops (www.rsmas.miami.edu/LAPCOD/meetings.html).

The main purpose of this book is to conduct a review of Lagrangian observations, analysis and assimilation methods in physical and biological oceanography, and to present new methodologies on Lagrangian analysis and data assimilation, and new applications of Lagrangian stochastic models from biological dispersion studies. Some of the chapters included in this volume were presented at LAPCOD workshops, while others have been specifically written for this collection. Given the size of the Lagrangian field, the present work cannot be considered as an exhaustive effort, but one which is aimed to cover many of the central research topics. It was our intent to maintain a good balance between historical and state-of-the-art developments in Lagrangian-based observations, theory, numerical modeling and analysis techniques.

This book seems to be a first of its kind because the central theme is the Lagrangian viewpoint for studying the transport phenomena in oceanic flows. Another unique and timely aspect of this book is its multidisciplinary nature with contributions from experimentalists, theoreticians, and modelers from diverse fields such as physical oceanography, marine biology, mathematics, and meteorology.

The book starts with a historical perspective of the development and application of Lagrangian methods, while more recent measurements and results are presented in Chapter 2. Some striking examples of Lagrangian trajectories are depicted by a collection of authors in Chapter 3. A number of new theoretical approaches to understand and describe particle motion are outlined in Chapters 4, 5, 6, and 9. New methods for assimilating Lagrangian data in ocean models to improve their forecast are described in Chapters 7 and 8. A suite of applications of Lagrangian techniques to transport of biological species are given in Chapters 10 to 12. Finally, we close with an extensive observational and theoretical review of Lagrangian techniques that were presented in the three LAPCOD workshops held in 2000, 2002, and 2005.

We would like to express special thanks to Dr. Manuel Fiedeiro from the US Office of Naval Research (ONR) for sponsoring much of the research presented in this book, while fostering collaboration between many groups of researchers and initiating LAPCOD workshops. We also thank Dr. Jerry Miller from ONR-London for supporting some of the LAPCOD workshops. Special thanks are also due to Edward Ryan, who has dedicated countless days to help organize this book. We also thank anonymous reviewers for many useful suggestions to help improve the chapters, and for maintaining a quality standard of scientific work. Finally, we thank all the scientists who have played important roles in the advancement of Lagrangian observations and analysis, but are not directly represented in this book.

Annalisa Griffa, Denny Kirwan, Arthur Mariano,
Tamay Özgökmen and Tom Rossby

1

Evolution of Lagrangian methods in oceanography

T. ROSSBY

Graduate School of Oceanography, University of Rhode Island, Kingston, Rhode Island, USA

1.1 Introduction

A complete description of a dynamical system must include information about two things: its state and its kinetics. The first part defines its condition or state at some instant in time, but nothing about its motion. The latter does the opposite, it tells us how the system is evolving, but nothing about its state. Thus, for a full understanding of a dynamical system, we need information on both. If we consider the ocean as such a system, its state would be determined by the distribution of mass while the kinetics of the system would be given by the distribution of currents. Since the birth of modern oceanography, we have developed an increasingly accurate picture of the state of the ocean, more specifically the distribution of heat and salt: the two properties that determine the mass field and hence the internal forces acting on it. Progress has been much slower – and more recent – with respect to a corresponding description of the kinetics of the ocean. Indeed, our view of the ocean circulation is still incomplete and depends to a significant extent upon assumptions about its internal dynamics in order to estimate ocean currents from the observed mass field. We have employed this methodology out of convenience and necessity because for a very long time we did not have the tools to observe the ocean in motion directly.

Fluid motion can be specified in two ways. The first, generally known as the Eulerian method, specifies the velocity field as a function of location and time. The other – Lagrangian – method specifies the position of labeled fluid parcels as a function of time. The two methods have strengths and weaknesses. Virtually all theoretical and numerical research of fluid dynamics uses the Eulerian specification because of the clear separation of the independent variables, space and time. In the Lagrangian frame, the spatial information enters through the initial position of each and every particle. The dependent

Lagrangian Analysis and Prediction of Coastal and Ocean Dynamics, ed. A. Griffa, D. Kirwan, A. Mariano, T. Özgökmen, and T. Rossby. Published by Cambridge University Press.

variables are the particles' subsequent positions as a function of time. The equations that describe this system can be solved only for certain, very simple flows, and as a result are hardly ever used. However, from an observational point of view the Lagrangian method has a major advantage in that it tells us precisely how fluid parcels move about in space. This may seem like a self-evident if not tautological statement, but it assumes special significance when one considers how difficult it is to accurately determine the horizontal structure of ocean currents, particularly at depth.

How does one observe the spatial structure of currents and how well can this be done? In the Eulerian frame resolution is set by the number of observation points. In the Lagrangian frame it is set by the number of markers or tagged parcels that are released. For synoptic applications such as tracking weather, the Eulerian specification is the method of choice. The synoptic approach can be very effective in the ocean too but, for a wide range of questions, the Lagrangian approach provides a more natural and cost-effective match. The reason is that continuous tracking of Lagrangian markers allows us to map out the horizontal structure of the flow field in extraordinary detail. Even a single trajectory can be quite revelatory about the underlying fluid dynamics. To illustrate this point, consider a steady flow such as the orbital motion in an eddy. A Eulerian current meter will register a steady flow in a certain direction whereas a Lagrangian marker will immediately paint out the circular structure of the eddy. The technology to study fluid motion in the ocean with Lagrangian techniques has expanded enormously since its birth a half a century ago. In this chapter we review the development of Lagrangian methods for observing ocean currents and how these observations have helped to expand our knowledge of the dynamics of ocean currents.

The development of Lagrangian techniques in the ocean has depended upon two physical properties of the ocean: the existence of an acoustic waveguide and the very low absorption of sound. The waveguide traps sound so it spreads only in the horizontal, cylindrically instead of spherically (acoustic energy decreases as r^{-1} instead of r^{-2}), and the low absorption lets it spread out over great distances. Although it defies the imagination, one can, at low frequencies with special equipment, detect a one-watt sound source at a distance of 1000 km! It is this property of the ocean that permits us to use underwater sound to track Lagrangian markers or floats over wide areas. But this is easy to say now. In fact, the development of the float technology, from the 1950s Swallow float to today's RAFOS float, has depended crucially on the ability to generate low frequency signals to locate and track floats out to distances of 1000s of km. The first section of this retrospective of Lagrangian methods describes the evolution of the acoustic float technology; the second section discusses a range of platform-based *in situ* measurements that have added significant value to the Lagrangian trajectory data. This is followed by a brief review of the sound source technology and acoustic navigation techniques. We then attempt a summary of lessons learned from the use of Lagrangian methods, and the brief last section speculates on likely developments in the near future. Common to the entire discussion given here is the acoustic transparency of the ocean, the property that allows us to generate and detect acoustic signals at great distances for the purposes of location and tracking.

1.2 History of floats

The first neutrally buoyant float designed to track water movements at depth was developed by John Swallow, a British oceanographer. It consisted of two aluminum pipes strapped together with a battery and timer circuit that would excite a magnetostrictive transducer, a "pinger," hanging underneath (Swallow, 1955), Figure 1.1. The signals could be heard from a ship overhead.

Figure 1.1 This widely reproduced photo shows John Swallow on deck preparing his float for launch. The float consisted of two aluminum tubes strapped together. The toroidal pinger (barely visible to Swallow's right) hangs underneath. Note the ship's cat paying close attention!

Using acoustic triangulation the ship could determine the float's position and depth. The weight of the float was carefully trimmed so that it would float at a desired depth. It is also essential that the float be less compressible than seawater. Imagine that for some reason the float is displaced downward a bit. Since it won't compress as much as the surrounding water, it will be lighter than the displaced water and thus return to its equilibrium depth.

The new Swallow float enjoyed great success with the discovery of a south-flowing deep western boundary current (Swallow and Worthington, 1961) that had been predicted by Stommel (1957). The direct observation of southward motion of several floats was far more convincing than estimating currents from geostrophy given knowledge of the density field, also known as the dynamic method which yields only relative and not absolute velocity information. Steele *et al.* (1962) used Swallow floats to confirm and estimate the deep transport out of the Norwegian Sea south of Iceland. The Swallow float was also used to study deep flow west of Bermuda. The premise had been that the currents would be rather weak revealing the interior broad scale abyssal circulation required to balance the rapidly flowing deep western boundary flow. Instead, very energetic mesoscale motions on time scales of a few weeks were observed, precluding any accurate estimate of the mean circulation at depth (Crease, 1962). It was perhaps this study more than any other that alerted oceanographers to the existence of an energetic deep ocean. But it also made clear the prohibitive cost of tracking floats for any length of time on the high seas from surface vessels. Swallow (1971) reflects on those early observations.

It had been known for some time that sound can propagate great distances through the ocean and that this acoustic "transparency" of the ocean could be used to track subsurface drifters. Stommel, in a letter to the editor in Deep Sea Research in 1955, proposed to deploy subsurface drifters that would be located from time to time by generating explosive signals that would be picked up at distant hydrophones (underwater microphones) using the acoustic wave guide, usually known as the deep sound or SOFAR (SOund Fixing And Ranging) channel (see next paragraph). Stommel had in mind a float that would release a small explosive device that would detonate when it had reached a certain depth or pressure. At the same time it would release a compensating buoyant element so that the float from which these were released would remain at the same depth as it continued to drift. Actually, he touched on the idea of using floats in an earlier paper (Stommel, 1949) where he suggested using the SOFAR channel to study eddy diffusion in the ocean – the dispersion of fluid parcels – on the gyre scale. Nothing came of these ideas at the time, but they did highlight the potential for using the SOFAR channel for tracking subsurface drifters over

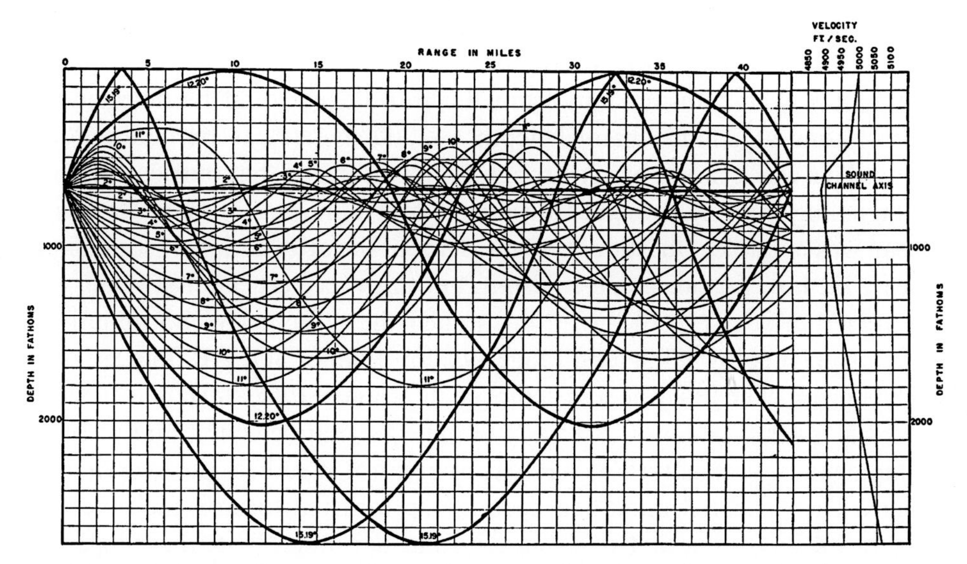

Figure 1.2 Ray diagram of sound propagation through a stratified ocean showing both refracted and surface reflected rays (from Ewing and Worzel, 1948).

great distances for extended periods of time. These early ideas remained on his mind, as we shall see.

The SOFAR channel was discovered during World War II at the Woods Hole Oceanographic Institution as part of its research into the acoustics of the ocean, a very important aspect of antisubmarine warfare (Ewing and Worzel, 1948). The SOFAR channel is an acoustic waveguide due to a sound velocity minimum below the warm upper ocean. The speed of sound is a positive, nearly linear function of both temperature and pressure such that the speed of sound first decreases with increasing depth due to the rapid decrease in temperature. Beyond a certain depth below which temperature only slowly decreases, the pressure effect dominates such that the speed of sound increases with further increase in depth. This minimum in sound speed, typically about 1490 ms^{-1} around 1000–1300 m below the surface in tropical and subtropical waters, gives rise to a sound channel which tends to trap sound near these depths, in short an acoustic waveguide. Figure 1.2 (widely reproduced from the 1948 Ewing and Worzel paper) illustrates the trapping of sound in the sound channel.

In 1965–1966, at Stommel's initiative, M. J. Tucker, on sabbatical leave from England at the Massachusetts Institute of Technology, and Douglas Webb at the Woods Hole Oceanographic Institution (WHOI), used a low frequency piezoelectric transducer, small enough for neutrally buoyant float applications, to evaluate its use for acoustic signaling in the SOFAR channel (Webb and Tucker, 1970). It was tested from the RV Chain in early 1966, Figure 1.3. The transmissions from the transducer could be detected at a hydrophone at sound channel depth (~1200 meters) at Bermuda 270 km away. This was a major breakthrough for it confirmed that purposeful acoustic signaling from lightweight transducers over hundreds of kilometers in the ocean was possible. Questions remained about the attenuation of sound in the ocean, i.e. how far one could transmit, but this was a crucial first step.

Later that year (1966) the author joined the Stommel/Webb initiative to develop and deploy a first long-range Swallow float or SOFAR float as it came to be known. The first one was deployed in January 1968 and could clearly be heard at 846 km distance. Specially designed signal detection and recording equipment using existing SOFAR hydrophones at Eleuthera (in the Bahamas), Puerto Rico and Bermuda had been set up to listen to and track the float. Figure 1.4 shows the first signals received at Bermuda. Note the roughly five second spread in the arrival of each 1.4 second long signal reflecting the early arrivals of the off-axis rays first (the path is longer but the average speed of sound is greater) and the more solid last arrivals due to the concentration of rays closer to the sound channel axis. Unfortunately, the float failed after only one week. A second deployment a few months later lasted only 2 days. After a substantial redesign effort, a third float was deployed in October 1969 and was tracked for four months before it failed. One possibility for the first two failures may have been that the cables, which hung loosely between the aluminum sphere and the transducer, were vulnerable to biological attack (fish bite). Although this was only a guess, the float was redesigned to make it more rugged, Figure 1.5. The transducer sits inside an oil-filled white polypropylene boot only the top part of which is visible in the photograph (due to the protective frame used for deployment).

In parallel with the SOFAR float work, the author explored other approaches to estimating currents from 2-point displacement vectors, i.e. from knowing the start and end points only. These would give us mean flow patterns at much lower cost albeit without any spatial or temporal detail along the trajectories. Ensembles of mean displacement vectors would nonetheless provide independent information on the mean circulation at depths where the dynamic method gives little insight. Building upon Stommel's original idea of using small explosives to study currents, but instead of a single float carrying many charges, we used many small floats, each one with a single pressure-actuated explosive. The charge was a standard SOFAR signaling device secured inside a small glass flotation sphere. The sphere float had a small port to allow it to be flooded at a certain time. This would cause it to fill until

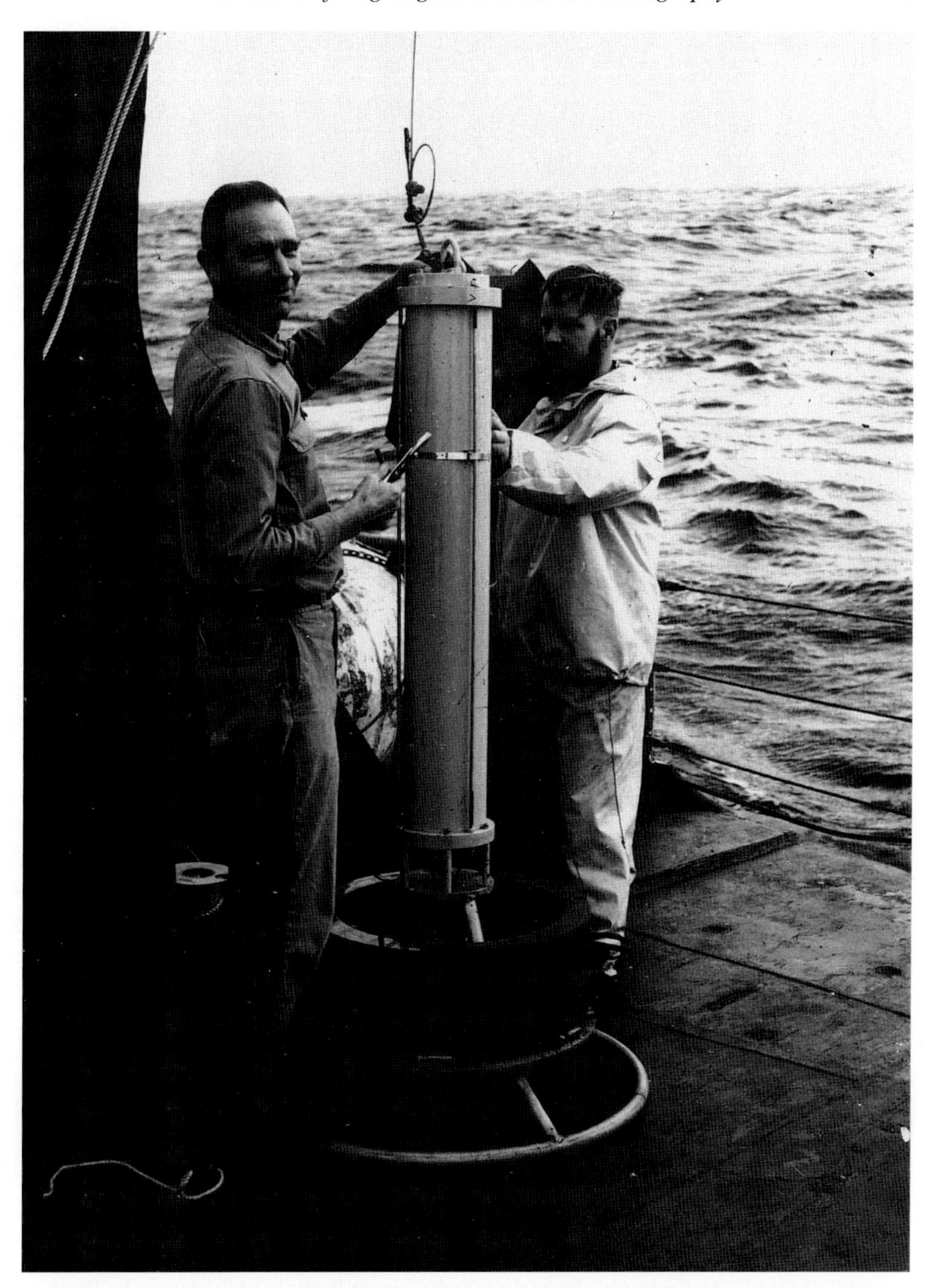

Figure 1.3 Gordon Volkmann (left) with acoustic transducer about to be lowered to sound channel depth to test long-range transmission through the ocean. The ring transducer visible below the shiny electronics module operated at 778 Hz.

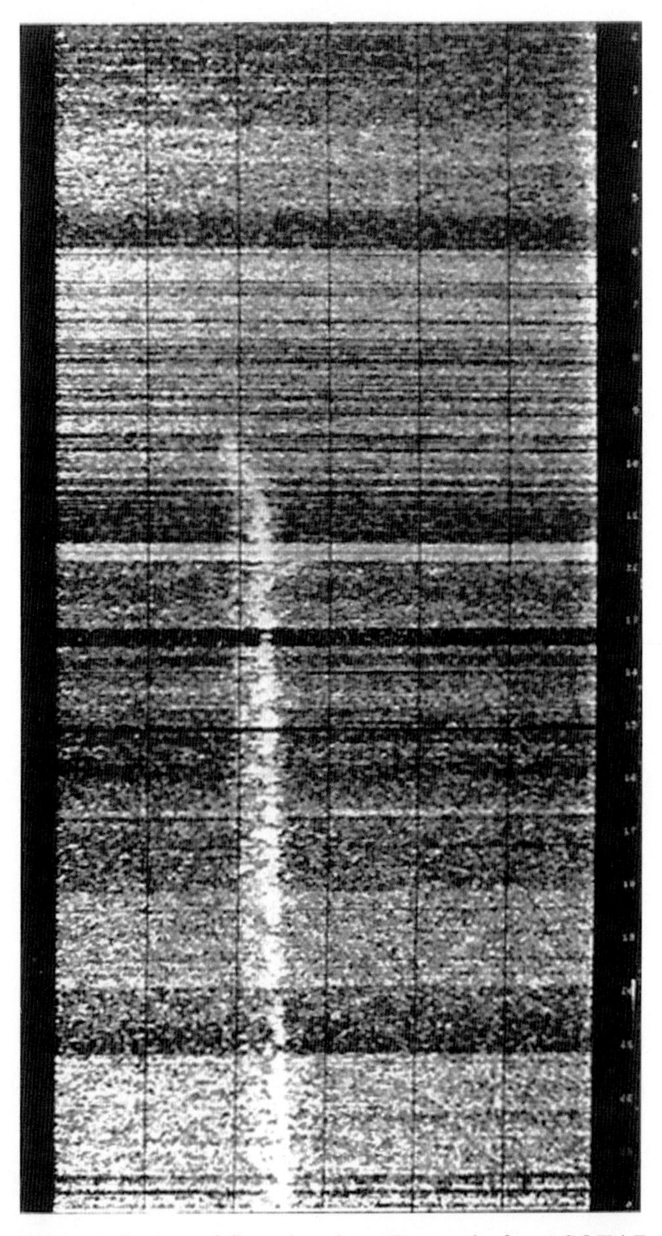

Figure 1.4 Visual display of first signals at Bermuda from SOFAR float #1 January 20, 1968. The scales are 1 minute across and 0–24 hours top to bottom. The bright line shows the one per minute transmissions starting at 0930 GMT. At first only the early off-axis arrivals show up because the float has just been deployed and has not yet reached sound channel depth (Rossby and Webb, 1970).

Figure 1.5 The aluminum sphere SOFAR float with the transducer underneath (mostly concealed by launch frame). See Plate 1 for color version.

the pressure reached the point where the charge was hydrostatically armed and subsequently triggered. An exploratory experiment to test the concept took place with a dozen floats set to drift for three weeks. Unfortunately, due to a leak in the flooding mechanism only one displacement vector was obtained. While the test showed that the concept did work and could have been improved upon, working with explosives, although small and safe, held little appeal for further use.

1.2.1 The SOFAR float

Regardless of the reasons for the short life lengths of the first generation SOFAR floats, it became clear that the high cost of fabrication of the machined aluminum flotation sphere and the piezoelectric transducer hanging underneath would preclude their use in large numbers. To address this, Doug Webb suggested using a resonating tube instead. It would be excited by means of a thin piezoelectric bender plate at the closed end of the tube, the length of which would be carefully trimmed to achieve resonance at the desired operating frequency. Instead of using spheres, flotation would be provided by commercially available extruded aluminum tubes cut to the appropriate length. True, the floats were much larger than the compact spheres used earlier, but this could be countered by the use of suitable handling tools. The first design, Figure 1.6, had two resonator tubes, one to each side of the center flotation tube. Later, a single tube was mounted end-to-end to the main tube, Figure 1.7. These instruments proved enormously successful. They enjoyed their first major use in the Mid-Ocean Dynamics Experiment (MODE) in 1973. Twenty floats were deployed at a depth of 1500 meters. They could be called to the surface acoustically by sending a command to drop a small ballast weight. This proved valuable for many of the floats had to be recovered and repaired due to small leaks to the oil-filled bender plates. But the repairs could be completed at sea, and the floats performed admirably thereafter. Numerous studies have been written using these data. The data are described in detail in a technical report (Dow *et al.*, 1977) and two papers (Rossby *et al.*, 1975; Freeland *et al.*, 1975) provide a first synthesis of the observations. This was the first study to use coherent arrays of floats to study sub-mesoscale dynamics.

1.2.2 The mini-MODE float

Another very clever Lagrangian contribution to the MODE consisted of the mini-MODE program (Swallow *et al.*, 1974). These floats were transponding

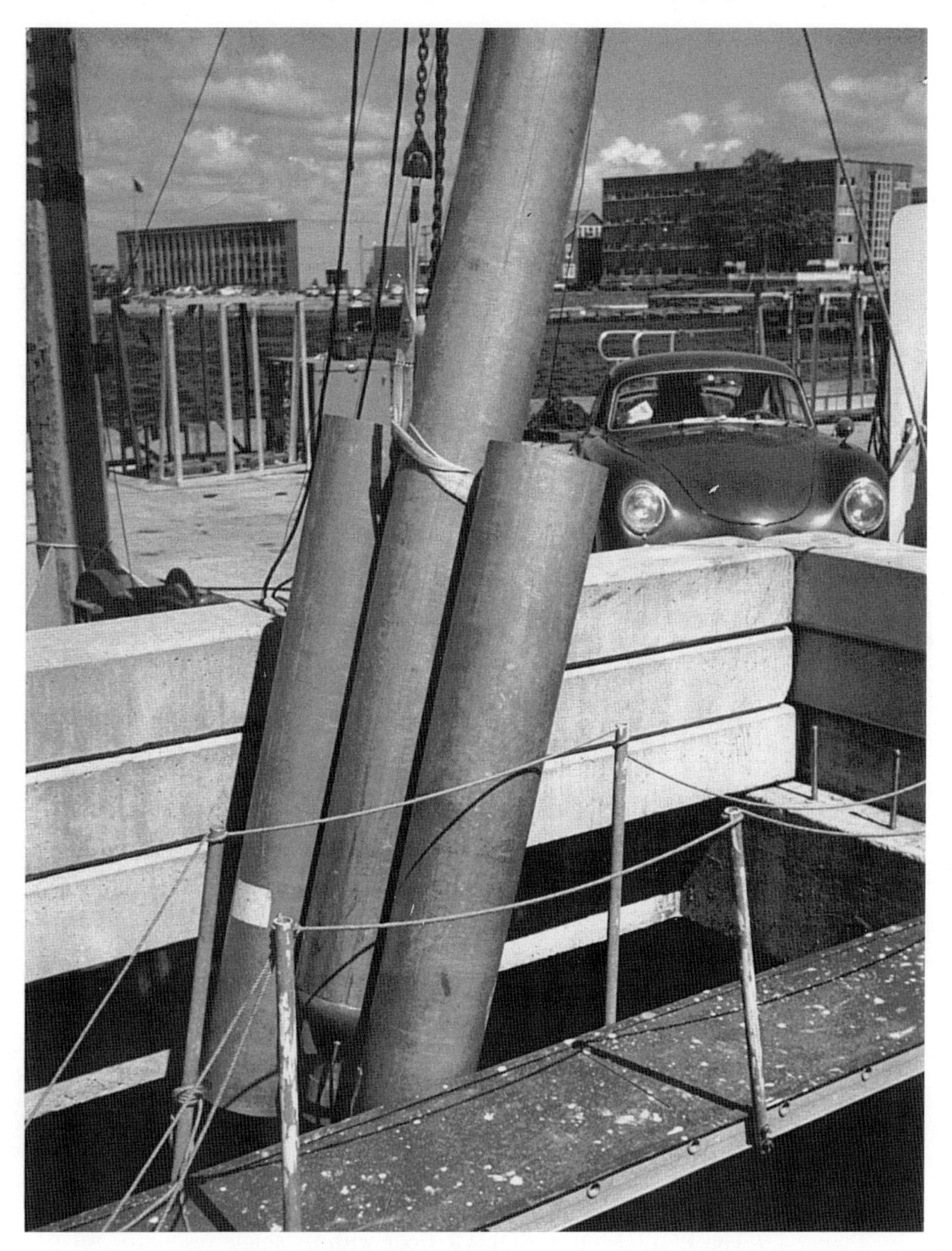

Figure 1.6 The MODE SOFAR float with two resonator tubes mounted on opposite sides of the flotation tube. See Plate 2 for color version.

Figure 1.7 The POLYMODE SOFAR float with its single resonator tube mounted end-to-end. See Plate 3 for color version.

devices that would transmit a reply whenever then received an interrogation pulse. Up to 18 floats out to 70 km distance could be tracked simultaneously from a ship. An acoustic interrogator was lowered with the conductivity-temperature-depth (CTD) profiler. Due to the wide range in depths of the floats and the complexity of sound propagation, the system was designed to transmit the interrogation signals at depths such that the floats would be likely to hear the signals (see the ray propagation diagram in Figure 1.2). It is somewhat of a mystery that this method never became more widely used. The ability to track a large number of floats and conduct hydrography simultaneously made it a very powerful combination for hydrodynamical studies of mesoscale phenomena.

1.2.3 The POLYMODE float

Immediately following the successful MODE, Doug Webb suggested switching to a different and more efficient acoustic transmission system. Instead of repeatedly sending short continuous wave (CW) pings, the float could transmit a single long frequency modulated (FM) signal. The purpose was to take advantage of the structure of the signal at the receiver by searching for signals with a given phase pattern. While the received signal varies strongly in amplitude due to the superposition of the many paths along which sound travels through the ocean to the receiver, the phase relationship between these at the receiver changes only little during the reception of the 80 second long signal. This is because the fastest time scale of the thermal field, upon which the sound velocity structure depends, is set by the internal wave activity, which has a much longer O(10–20) minute period in the main thermocline. Figure 1.8 shows the nature of the FM signaling system. The top line shows an ideal signal while the other lines show examples of received signals. The signal detection algorithms (a cross-correlation) can be implemented very effectively using digital techniques. This signaling system has been so successful it continues to be used today. In 1978 40 floats of this new design were employed in a US–USSR cooperative study called the POLYMODE Local Dynamics Experiment (LDE). The LDE was designed to study mean flow–eddy interactions in a region west of Bermuda near where the original Swallow float study of the deep ocean circulation took place (Crease, 1962). Half of the floats were deployed at 700 and the other half at 1300 meters. A technical report by Spain *et al.* (1980) gives details on the entire operation and all data. Much of the results from the LDE and the POLYMODE program as a whole can be found in “Eddies in Marine Science”, A. R. Robinson, editor (1983) and a special issue of the Journal of Physical Oceanography (March 1986).

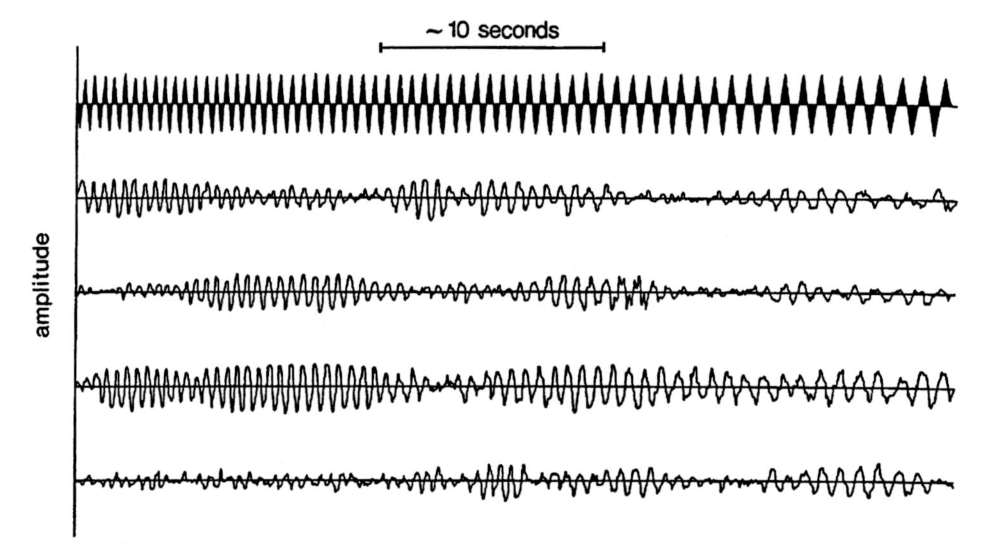

Figure 1.8 Examples of received FM transmissions. Make a transparent copy of the ideal signal (top line) and slide it over one of the signals. You will notice that there is a position at which the signal seems to overlap or agree the best. This best fit determines the arrival time of the signal.

1.2.4 The autonomous listening station (ALS)

In the early days of SOFAR float work up through and including the LDE, the floats were tracked using a network of hydrophones that had been installed around the western North Atlantic in the 1950s as part of a system to determine the splashdown point of missiles fired from Cape Canaveral (known then as the Missile Impact Location System or MILS). The missiles would release a small explosive device that would arm and detonate once it reached the depth of the SOFAR channel. From the arrival time of this signal at three or more hydrophones one could determine the target accuracy of the missile. The availability of these hydrophones greatly facilitated our work by permitting us to focus on the development of the float technology. Later, once confidence with the floats and the acoustic tracking became well established, impatience with the regional coverage of the hydrophones stimulated interest in a 'portable' tracking network that could be deployed anywhere. In the late 1970s Dr. Albert Bradley (1978) at WHOI developed an autonomous listening system (ALS) to listen for and log all acoustic transmissions from SOFAR floats. Figure 1.9 shows conceptually the arrangement between the moored ALS and the SOFAR float and earlier shore-based listening systems. This very successful development opened up a host of research programs with SOFAR floats all over the Atlantic Ocean such as tagging Mediterranean eddies a.k.a.

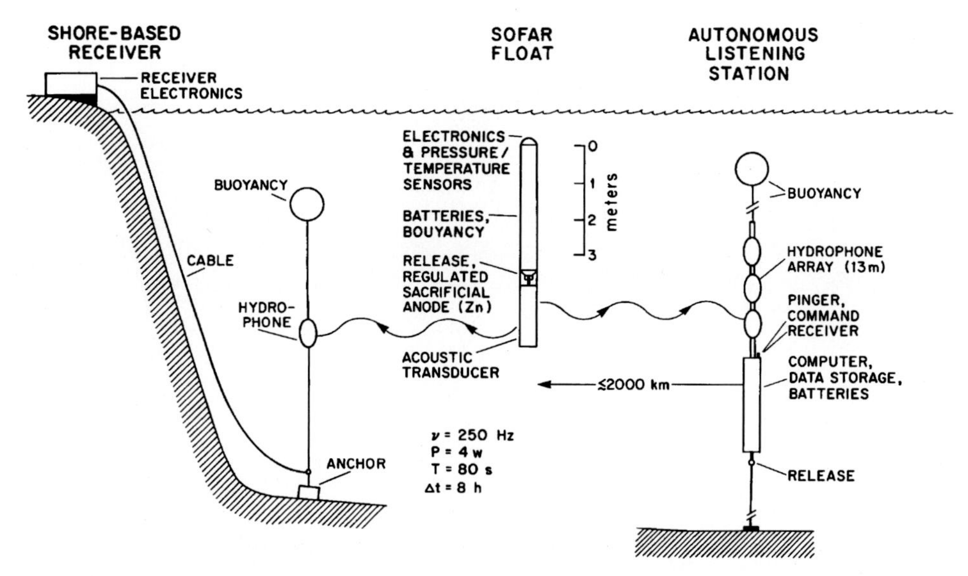

Figure 1.9 Schematic diagram of SOFAR float and listening systems.

Meddies (Richardson *et al.*, 1989) and cross-equatorial flows (Richardson and Schmitz, 1993). Owens (1991) gives a comprehensive discussion of the mean circulation and eddy variability based on all SOFAR data from the 1970s and 1980s.

While the SOFAR float was a clear technical and scientific success, it was not an inexpensive instrument to deploy and use. Further, the substantial investment required just to establish a basic acoustic tracking system meant that the technology was not scalable. One needed at least three ALS moorings and preferably one more for redundancy to provide for reliable tracking of all floats. Such an investment was not defensible unless the study called for a large number of floats to be deployed. Thus all SOFAR float programs in the 1970s and 1980s typically involved many tens of floats at a time.

1.2.5 The RAFOS float

The increasing capability of microprocessor technology that led to the ALS set the stage a few years later for the development of the RAFOS float. In particular, the advent of CMOS-based microprocessors with their extremely low energy requirements meant that the electronics required to detect and record the arrival time of acoustic signals could be reduced to a small package requiring little space and little energy. This became the stimulus for the development of the RAFOS float. Conceptually the RAFOS float is identical

Figure 1.10 Phil Richardson (WHOI) holding a RAFOS float. The drop weight (not shown) is attached just before launch. See Plate 4 for color version.

to the SOFAR float except that instead of transmitting, it listens to moored SOFAR floats, now serving as an underwater navigation system.

The RAFOS float consists of a 1.5 to 2.2 meter long glass pipe housing the hydrophone and signal processing circuits, microprocessor and battery. They weigh about 10–15 kg, depending upon length. Jim Fontaine and Don Dorson at the University of Rhode Island were instrumental in the development of the RAFOS float concept. The name RAFOS (SOFAR spelled backwards) indicates the opposite direction of acoustic signaling. These floats weigh only a few percent of the SOFAR float reducing the materials cost substantially as well as making them much easier to handle (Rossby *et al.*, 1986). Figure 1.10 shows a float just about to be launched with another waiting nearby. After deployment the float drifts with the waters for a preset period of time. During this time it listens for and records the arrival times of the acoustic transmissions according to a prearranged schedule. The duration of drift depends upon the objectives of the study. In the early studies of the Gulf Stream the floats surfaced after a few weeks whereas in other programs they have been programmed to remain submerged for several years (Hogg and Owens, 1999). At the end of their missions the floats drop a ballast weight, return to the surface and telemeter their data to a satellite system, Service ARGOS, which can collect data from small transmitters anywhere around the globe.

1.2.6 The ALACE float

The autonomous Lagrangian circulation explorer or ALACE float was developed to map out ocean currents without the need for acoustic tracking (Davis *et al.*, 1992). The floats drift like all floats at some particular depth. Every so often, typically once a week or every ten days, the float surfaces so that its position can be determined by satellite. At the same time it reports the temperature profile obtained during ascent. The advantage is of course that there is no need to deploy and maintain the moored sound sources required for acoustic navigation. The disadvantage is that one loses considerable information about the mesoscale structure of the ocean. In regions of strong currents and fronts this is a significant shortcoming, but in general the ALACE float has contributed in a major way to our knowledge of the large-scale circulation. The article by Davis and Zenk (2001) gives a concise overview of the extensive ALACE activities during WOCE. Interestingly, the focus of the ALACE float activities has shifted from its Lagrangian roots towards the profiling of temperature and salinity during its excursions to the surface. The expanding global array of profiling ALACE (PALACE) floats will in the coming years total some 3000 floats in all major ocean basins. It is planned that they should provide a broad view of the thermocline and upper ocean density field and variability on time scales of weeks and longer. Notwithstanding the shift in emphasis towards profiling, the floats' displacement vectors will continue to provide fundamental information on the large-scale circulation of the ocean at the parking depth of the floats.

1.2.7 The ALFOS and MARVOR floats

If the RAFOS float was a significant technical and cost improvement over the earlier SOFAR float there was a major drawback, one had to wait until the underwater mission was over before getting any data back. At least with the shore-based listening system we could track the floats in real-time. Two groups developed a partial solution to the problem based on the ALACE concept. By combining a RAFOS receiver with an ALACE float one can retain the advantages of high resolution Lagrangian tracking without having to wait until the end of the mission; each time a float surfaces it telemeters all data collected since the float last surfaced. This is exactly what Breck Owens (WHOI, personal communication) did. The WHOI version, dubbed the ALFOS float (ALACE + RAFOS), has been used in a number of studies, and has been particularly useful as a monitor of the moored sound sources, to make sure they are performing properly while a particular study is underway. In France a similar development has taken place, but in this case the float was designed from the outset as a multi-surfacing RAFOS float with a design life of five years (Ollitrault *et al.*, 1994). This design, known as the MARVOR float, has established itself as a highly reliable multi-year instrument and is used in large numbers. The floats are programmed to drift at depth for two to three months at a time, interrupted by visits to the surface for data telemetry. The floats do not require precise ballasting beforehand because they actively control their volume to remain at a prescribed pressure.

1.2.8 Isopycnal operation

The glass pipe housing of the RAFOS float deserves special mention because it enables the float to track fluid motion vertically as well, more specifically to stay on the same isopycnal surface. Borosilicate glass has a very small coefficient of thermal expansion, about 1/20th that of water (in volume units). Such a small coefficient means that the float's volume all but remains unaffected by temperature. The ocean is almost always thermally stratified with warmer waters above colder waters. Now imagine that for some reason a float bobs up into warmer water. Warmer water is less dense, but the float, being made of glass that can't expand, will find itself heavier than the displaced water and thus sink back to its equilibrium temperature. Similarly, the float is completely impervious to salinity variations. The result is that the float will always stay at the same layer of water regardless of temperature or salinity since neither affect the volume (density) of the float. But there is one other requirement.

1.2.9 The compressee

The density of water also depends upon pressure, so as water up- or downwells, it will expand or compress. For a float to follow a fluid parcel as it moves vertically it must have the same compressibility as the ambient waters (thus canceling out the pressure restoring force used by Swallow for his constant depth floats as mentioned earlier). This can be arranged by adding a compressible element consisting of a spring-backed piston that moves in or out in proportion to the ambient pressure (Rossby *et al.*, 1985). Figure 1.11 shows the spring-piston arrangement: By carefully adjusting the diameter of

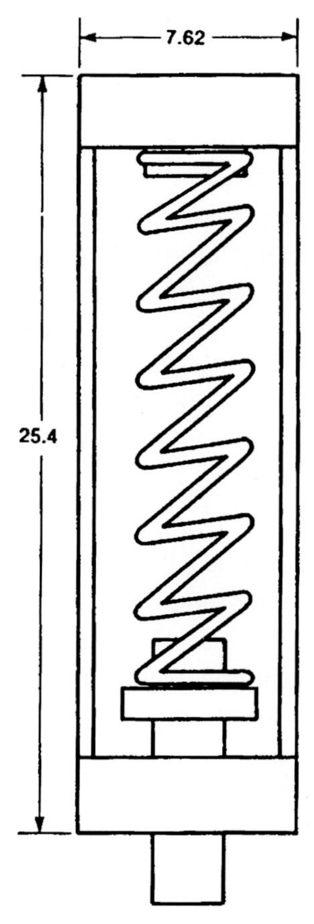

Figure 1.11 Sketch of compressee. The spring constant and diameter of the piston, the position of which depends upon pressure, are chosen to add an additional calibrated volume loss to the glass pipe (from Rossby *et al.*, 1985).

the piston – for a given spring constant – one can match the compressibility of the float package to that of seawater to within about $\pm 1\%$. The use of borosilicate glass such that the float feels neither thermal nor salinity variations, and the compressee to give the float the same compressibility as water, results in a package that inherently remains with the water parcels around it as they drift through space, horizontally and vertically. It is a natural water parcel follower where the waters are stratified. Swift and Riser (1994) give a detailed discussion on how to design glass pipe floats for isobaric, isopycnal, and neutral surfaces.

1.2.10 The COOL float

In upwelling systems where a fluid parcel's density decreases through downward mixing of buoyancy, a standard compressee-equipped float cannot follow the upwelling flow, but will remain behind on the same isopycnal surface. Conceptually, a solution to this is to measure the vertical flow past it and then adjust its volume actively so that on average there is no vertical flow past it, i.e. the float adds its own buoyancy to match that of the surrounding waters. For example, off the Oregon coast in summer, fluid parcels upwell from depth along the sloping bottom and flow back out to sea near the surface. As the waters shoal they become lighter due to downward mixing. Whereas the standard isopycnal float will remain at the same isopycnal, a float equipped with tilted vanes and a magnetic compass can detect vertical motion relative to the float. With the addition of a small pump and appropriate control logic, a float can then adjust its volume, i.e. buoyancy, to ensure that this rotation does not take place. In other words the float will rise with the waters so that – on average – there is no vertical velocity relative to the float. This added buoyancy would then equal exactly the buoyancy that is mixed down from the surface. While a complete servo-controlled float has yet to be attempted, vane-and-compass-equipped floats have been used extensively in studies of upwelling in fronts, particularly along the New England shelfbreak front, Figure 1.12. These studies have shown that isopycnal floats in stratified waters track the vertical movements of an isopycnal surface closely, even on the short time-scales of the buoyancy or Brunt–Väisälä frequency. To show how well such floats actually track a density surface, the top panel in Figure 1.13 shows a three hour record of O(15) m vertical movements of internal waves (the pressure record) with only a gradual change in temperature at the float. The bottom panel shows the corresponding velocities: the vertical velocity from pressure shows $> \pm 5$ cm/s variations while the relative displacement (measured from rotation) remains well under 1 mm/s. The added drag provided by the vanes no doubt helped the float to track the isopycnal more closely on fast time scales (Hebert *et al.*, 1997).

1.2.11 Bottom-following floats

Almost all water in the deep ocean originates in one of two principal outflows: The Filchner Ice Shelf at the southern end of the Weddell Sea in the Antarctic, or the Nordic Seas (Stommel and Arons, 1960). In all cases the outflow appears to take place in the form of a thin layer of dense water flowing obliquely along the bottom gradually increasing in depth. What does this flow look like, and to what extent does it interact and mix with the ambient waters? The nature and degree of mixing determines the depth at which these dense waters settle out, i.e. whether at intermediate or greater depths. To investigate these processes, another class of Swallow floats is under

Figure 1.12 The COOL float. The float has a compressee underneath so that the float rides on an isopycnal surface. But to the extent that the float does cross isopycnal surfaces, the slanted vanes will force it to rotate indicating how much relative motion there is. The bright spot at the top is the flasher. Photo courtesy Dave Hebert. See Plate 5 for color version.

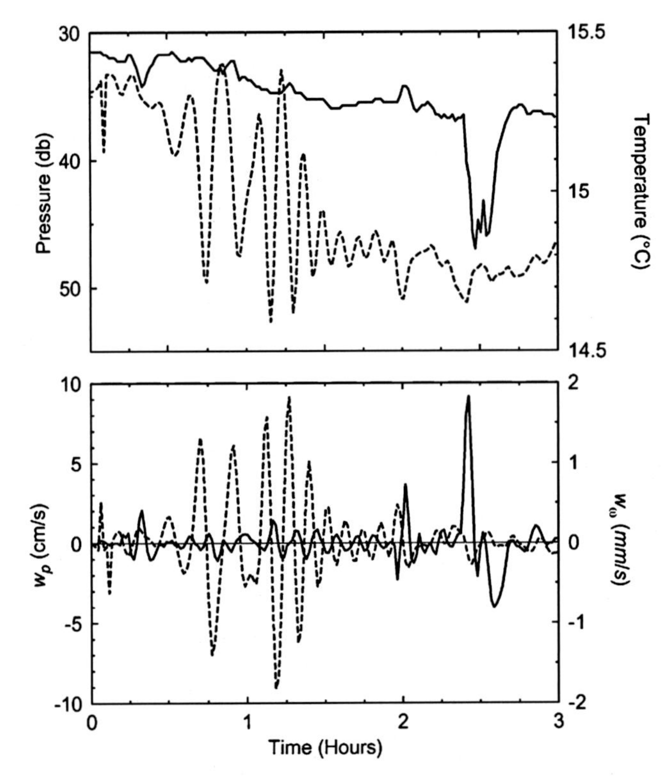

Figure 1.13 Top panel: three hour in-situ time series of pressure (dashed line) and temperature (solid). Bottom panel: the dashed line shows the vertical velocity of the isopycnal surface (left scale in cm/s) while the solid line shows the vertical velocity of the float relative to the isopycnal (right scale in mm/s). Figure provided by D. Hebert.

development. Here, we require that a float follow the waters along but without touching the bottom. The first generation bottom-following float used to study Nordic Sea overflow through the Faroe Bank Channel consisted of a standard RAFOS float that dragged a very light line along the bottom keeping it at a fixed elevation, Figure 1.14. As the waters flowed downslope the float remained within that bottom layer. This approach has been successfully demonstrated in the northern North Atlantic where waters from the Nordic Seas spill into the deep North Atlantic (Prater and Rossby, 2004). Unfortunately, despite efforts to minimize the wire's contact with the bottom sooner or later the line would snag on the bottom arresting the float's drift. The obvious solution to this will be to replace the wire with an acoustic altimeter, a device that measures the distance to the bottom, and a pump to adjust its buoyancy to stay within a certain distance. We anticipate that such a float will be of great use for studies of downslope currents. For example, how

Figure 1.14 Deployment of the bottom-following float. The red 'sails' enhance the float's ability to drift with the waters in the presence of drag forces from the bottom-following line (faintly visible) below the float. Photo courtesy Mark Prater. See Plate 6 for color version.

stable are such flows, when and where do they become supercritical, and how is criticality affected by tidal motions which might be quite strong over continental slopes. In addition to overflow systems, bottom-following floats should be quite effective for benthic circulation studies in shallow seas such as the Gulf of Maine and the Barents Sea.

1.2.12 Convecting floats

The above examples have in common that the waters are stratified, or nearly so. Lagrangian studies of waters subject to intense mixing require a different approach. For example, intense heat loss at the surface can lead to deep vigorously mixed layers. Here a float must execute the same large and rapid vertical excursions as the fluid itself. An approach that meets this requirement very well uses some of the elements mentioned above. The first is to give the float the same compressibility as water at the temperature it will be used. The

Figure 1.15 The '3-D' Lagrangian float. The large 'skirt' adds considerable drag so that the float follows vertical movements as closely as possible. Note the ribbing of the pressure case to give it compressibility without collapsing at high pressure. Photo courtesy Eric D'Asaro. See Plate 7 for color version.

second requirement is to maximize the vertical drag coefficient so that any remaining mismatch in float-water compressibility and hence buoyancy-driven relative motion will be minimized. An elegant design to meet these requirements consists of an aluminum tube, which has been machined to obtain the right amount of compressibility. However, a thin-walled tube will collapse even at modest pressure. The solution is to leave machined ribs or ridges around the tube to prevent buckling instabilities while reducing the wall thickness in-between. This combination gives the tube the ability to withstand high hydrostatic pressure yet flex elastically to give it the same compressibility as the surrounding waters. A circular plate mounted perpendicular to the vertical tube increases the drag coefficient in the vertical enormously. Figure 1.15 shows such a float (D'Asaro *et al.*, 1996; D'Asaro, 2003). This float has been used very effectively to study deep convection in the Labrador Sea and mixed layer dynamics. The match in compressibility is so close that the float tracks vertical motions to better than 1 cm/s. The floats can be RAFOS equipped for long-range acoustic tracking. Even in the Labrador Sea in winter when the thermocline is absent and all sound must reflect against the surface, useful acoustic tracking at distances in excess of 400 km were possible (Prater, 2002).

1.3 Acoustic navigation

As mentioned earlier, the SOFAR floats were tracked first using permanently installed hydrophones, and later using autonomous listening stations (ALS). The RAFOS floats used moored SOFAR floats to provide the acoustic navigation. The use of SOFAR floats for many years is clear testimony to their success. In the early 1990s there was interest in providing a wider area of coverage in the open ocean. This led to the development of the Sparton source (Rossby *et al.*, 1993). The source itself consists of an open pipe, like the SOFAR float, but instead of exciting it with a bender plate at one end, a piezo-electric ring transducer is prestressed into compression onto the outside of the pipe. As the transducer is excited, the pipe walls vibrate such that resonant standing waves are induced inside the pipe. This resonance provides efficient conversion of electric into acoustic energy. Deployed at sound channel depth (1200 m) in the open ocean (well away from continental boundaries that limit the sound to certain directions), and with an estimated source level of 195 dB re 1 μPa at 1 m, these sources provide good navigation out to distances well beyond 2000 km. More recently, a new source has been developed, which is also an open pipe, but whose resonance is excited by means of a monopole transducer (radial oscillations) suspended inside the pipe. The pipe itself is now completely passive. Although the source cannot put out very much power (about $\sim$180 dB re 1 μPa at 1 m), it is also quite efficient allowing for a rather small battery pack.

Accurate navigation depends upon accurate timing and knowledge of the average speed of sound between sound source and receiver. By timing we mean the time required for a signal to reach the receiver. This is turn depends upon clock accuracy at both ends. Any unaccounted error in either the sound source clock or the float clock will limit the positioning accuracy of the float. The float clock error is generally constrained by checking the clock just prior to deployment, and by knowing the scheduled radio transmission times after the float surfaces. Since each message received at the satellite is given a time stamp, we can readily determine the float clock error at the end of the mission.

But what about clock error during the mission? The default approach is to assume that the error accumulates linearly. This would be necessary if only two sound sources can be heard. However, with three or more sound sources one can use hyperbolic navigation, i.e. position is determined from the time of arrival differences between pairs of arrivals. Any clock error cancels out. We then find out how much the float clock has to be adjusted for the hyperbolic and range-range positions to agree. This gives us the float clock error at various times in a mission. Generally speaking the float clock errors exhibit a linear growth with time.

In addition to travel time we also need to know the average speed of sound. In principle one can use acoustical models of the ocean, but in most cases that the author is aware of the average speed of sound is determined empirically. We know where a float is deployed and where it surfaces. By equating the first and last subsurface acoustic positions with these surface positions (with a possible correction for drift since they are not obtained at exactly the same time), we know the distance, and hence an average speed of sound since we know the travel times. This technique also allows us to determine sound source clock errors at the end of an experiment. Using floats that surface near a sound source (such that any uncertainty in speed of sound matters less), and after their clock errors have been determined, we can estimate from the last signal arrival time when that transmission took place. Repeating this process for a number of floats constrains the sound source clock error to within a few seconds at most.

1.4 Float-based *in-situ* observations

Increasingly, floats are serving as platforms for *in-situ* studies. Except for some early generation Swallow floats most SOFAR and all RAFOS floats measure pressure and temperature. Pressure in particular gives helpful feedback on the accuracy with which a float settles at the desired depth. Gradual changes in temperature give information on thermocline depth along the trajectory while high frequency variations provide information on internal wave activity. But many other aspects of *in-situ* activity can be monitored from floats. In this section we give a brief survey on the use of floats as platforms. It will be evident that the variety of applications is virtually limitless. Indeed, one could argue that RAFOS is merely the navigational support to whatever *in-situ* activity one has in mind.

1.4.1 Vertical velocity

While we tend to think of the ocean as stratified with virtually no diapycnal activity, this does not imply an absence of vertical motion; on the contrary, vertical movements and internal waves are present everywhere over a wide range of space and time scales. But these velocities are very small compared to the horizontal velocity field and thus more difficult to observe. Floats become ideal platforms for their drift naturally removes the horizontal component of motion. Webb and Worthington (1968) showed how a float equipped with slanted vanes could be used to measure vertical motion. The rotation induced by the vanes was measured with a magnetic compass. The same float technology was used to study convective motions in the Mediterranean during cold air outbreaks in winter, the so-called Mistral. Stommel *et al.* (1971) reported on strong sustained vertical motions consistent with convective cooling. This was a very impressive application of the vertical current meter concept.

With the advent of isopycnal float technology, one can let the float ride with the waves, i.e. float on a density surface as it rises and falls. As noted earlier, Hebert *et al.* (1997) have shown how isopycnal floats quite accurately can track internal wave activity even at Brunt-Väisälä frequencies. Since the measurement of pressure and temperature requires little energy, one could readily measure and record these at a fast rate. Unfortunately, limitations in the volume of data that can be returned reliably through Service Argos at the end of a RAFOS float mission preclude sampling pressure or temperature often enough to resolve internal wave activity. But pressure can, even when sampled less frequently, give useful information on the internal tide, for example. One can use a technique called complex demodulation (essentially a running least squares fit of a sine wave) to examine its amplitude and phase variability over time. Figure 1.16 shows the amplitude of the semi-diurnal internal tide signal from the pressure record of float 475 in the southern Irminger Sea (Anderson-Fontana *et al.*, 2001). Note the conspicuous fortnightly modulation of the internal tide for the 75–125 day period. During this time the float is drifting north just over the western slope of the Reykjanes Ridge, so the large tide and its modulation most likely were excited by the local topographic relief. Earlier, Rossby (1988) found evidence for semi-diurnal internal tides in the eastern Atlantic using isobaric floats (these floats had no acoustic tracking so we don't know their trajectories), which also respond to density changes but with less amplitude. One of the floats recorded a fortnightly modulation that peaked a few days after a total solar eclipse.

1.4.2 Relative motion

The first float study to investigate relative motion and dispersion at depth in the ocean appears to be that by Pochapsky (1963) who used pairs of floats to measure their radial separation acoustically. The floats were ballasted for the same pressure and they typically settled out within O(50) dBars of each other. The subsequent relative motion typically had two components, a gradual increase

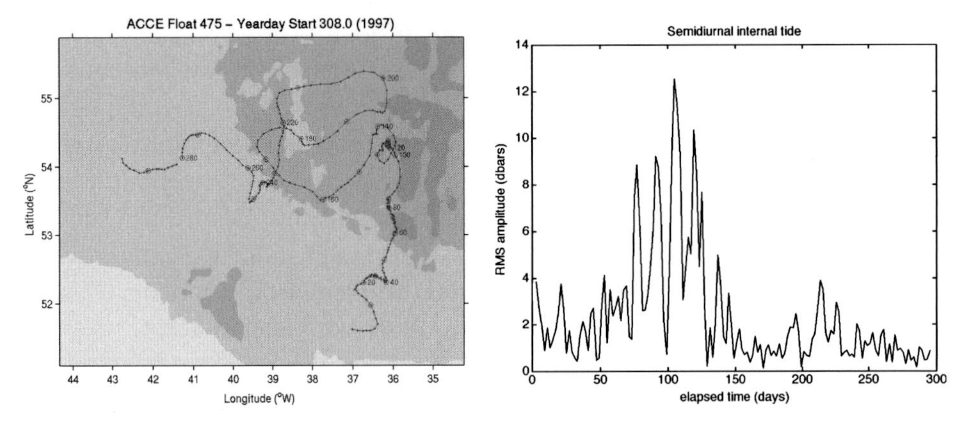

Figure 1.16 Left panel: trajectory of float 475 just west of the mid-Atlantic Ridge (depth contours at 2, 3, and 4 km). Right panel: the semi-diurnal pressure fluctuations along the track of isopycnal float #475. Note the large amplitude fortnightly variations while the float is over the western slope of the Reykjanes Ridge and only there. See Plate 8 for color version.

in separation and an oscillatory component. The larger the mismatch in depth between the two floats the more rapid their rate of separation, almost certainly due to a background vertical shear. The oscillatory component may have been dominated by inertial motion but the short records precluded an accurate estimation. The first attempt to study relative dispersion on isopycnal surfaces took place in the NAC program where several floats were deployed in pairs. In all cases it took at least a couple of weeks for the floats to reach an O(30) km distance beyond which they rapidly separated. One pair of floats in NAC followed the same track for more than two months (Rossby, 1996). Relative dispersion has become an area of intensive research (e.g. LaCasce and Bower, 2000; Mariano & Ryan, 2006, this volume).

1.4.3 Vertical movements in fronts

The very first application of isopycnal RAFOS floats took place in the Gulf Stream. Their trajectories clarified the nature of lateral motion in the meandering current (Bower and Rossby, 1989). On shallow isopycnals floats tended to stay in the current as they were rapidly swept downstream whereas on deeper surfaces, they could cross the entire current with relative ease. It was these observations that showed how a front could be a barrier to lateral exchange at shallow depths and enhancer of lateral exchange at depth, the “barrier or blender” concept (Bower *et al.*, 1985). The changes in depth give direct measurements of vertical velocities, frequently as large as 1–3 $\mathrm{mm\,s^{-1}}$ in large amplitude meanders (Song *et al.*, 1995).

1.4.4 Static stability or f/h

The isopycnal technology can readily be extended to measure stratification. By sequentially expanding and contracting the float's volume a small amount, the float will first rise and then sink to the corresponding isopycnal surfaces above and below its parking surface. The difference in pressure measures the layer thickness. This technique has been used to study layer thickness (or stretching vorticity) variations in fronts and eddies (Rossby *et al.*, 1994) and subduction (Price, 1996). Floats passing through sharp meander crests and troughs may exhibit the expected decrease and increase in layer thickness associated with the conservation of potential vorticity. Similarly, floats caught in coherent cyclonic eddies may experience O(25) % increases in layer thickness. But because these cycling floats sample only O(100) m of the full water column, they are sensitive to high-order vertical structure even if it is of small amplitude compared to the lowest internal modes. As a result, the records from these "f/h-floats" can be difficult to interpret (Rossby and Prater, 2004). On a more speculative note, one might also be able to estimate local stratification by tracking the internal wave sea state at the float. It is known that the IW spectrum drops rapidly above the Brunt-Väisälä frequency, $N = (-g/\rho_o \partial < \rho > /\partial z)^{1/2}$. If one could track the IW spectrum over time (weeks to months), one might be able discern gradual changes in stratification. Of course, such measurements would give much information on the variability of the internal sea state as a function of time including the internal tide discussed earlier.

1.4.5 Estimating salinity

Given the equation of state of a float, i.e. its compressibility and coefficient of thermal expansion and the isopycnal upon which it floats, one can deduce the salinity of the water with sufficient precision to track changes in water mass composition over the life of a float trajectory. A thorough analysis of this method has been done by Boebel *et al.* (1995). They show that a precision of 0.02 PSU is achievable, limited only by possible drift in the pressure gauge and material creep of the glass pipe float housing. Accuracy of the salinity depends upon the ballasting accuracy but they suggest that if a hydrographic profile is taken at the time of deployment an accuracy of 0.02 PSU might be possible. There is no requirement that the float be isopycnal, one just needs to know the float's equation of state, i.e. its compressibility and coefficient of thermal expansion.

1.4.6 Oxygen

In a recent study Lazarevich *et al.* (2004) used floats equipped with oxygen sensors to study changes in oxygen along isopycnal surfaces as these upwell and outcrop in winter and subduct the following spring. The floats were equipped with commercial Clark-type sensors operated in the "pulsed" mode pioneered by Dr. Chris Langdon (1986). The sensors are activated only briefly once a day giving them a useful life to match that of the float. From a Lagrangian viewpoint, interpreting the variability of dissolved oxygen presents the challenge of distinguishing between biological and physical processes. However, at certain times the nature of the variability clearly indicates that one process dominates over the other. In winter the surface waters may become supersaturated due to bubble formation. During the springtime phytoplankton blooms significant, if temporary, drawdown can occur, most likely due to concentration of detrital material on shallow newly stratified density surfaces. Lagrangian oxygen history might provide helpful insight into complex biophysical processes. The drawdown of oxygen continues throughout the summer due to bacterial remineralization of additional material settling through the water column from the surface. Clusters of oxygen-equipped isopycnal floats can quantify the rate of drawdown in recent and perhaps even older subducted waters. With accurate oxygen sensors (the best sensors today have 2% absolute accuracy and long-term stability) and cluster averaging one might be able to estimate the 'a' in $dO_2/dt = -a$ even for waters several years old. How does 'a' vary in space and evolve with time; does it have a seasonal component reflecting detrital fallout from the biologically active surface waters?

1.4.7 The barotropic component

It is also possible to determine the vertically averaged horizontal or barotropic velocity from a float using the Earth's magnetic field. The reason is that since salt water is a good electric conductor, small electric fields are generated as water in motion cuts across the earth's magnetic field. This has been demonstrated very elegantly with the Electric Field Float (EFF). The EFF is a standard constant pressure RAFOS float (no compressee) to which have been added two pairs of electrodes and a compass. Slanted vanes on the electrode arms, Figure 1.17, cause the float to rotate in the presence of internal wave activity allowing the float to resolve the amplitude and direction of the electric field, which is induced by both the movement of the water and the float. Given independent knowledge of the Earth's magnetic field and the motion of the float from the acoustic tracking one can back out water column average

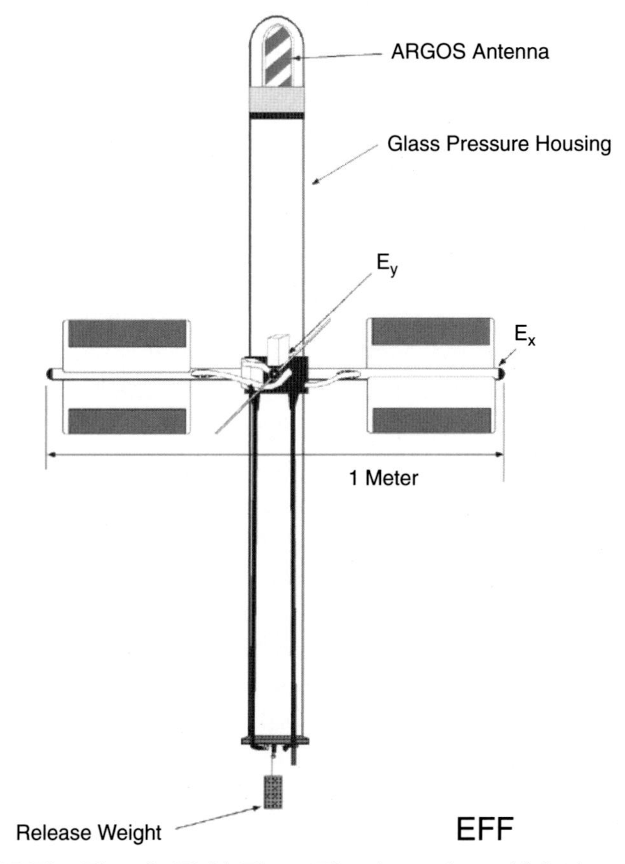

Figure 1.17 The Electric Field Float. The electrodes, which sit at the outer end of each arm, measure the *in-situ* electric field. The slanted vanes cause the float to rotate (induced by the internal wave field) so the electric field measurements can be correlated with direction thereby removing ("chopping" to use an electrical engineering term) errors due to electrochemical potentials. Drawing courtesy Tom Sanford. See Plate 9 for color version.

velocity (Sanford *et al.*, 1995). The added value of obtaining the barotropic component of motion is enormous, particularly considering the paucity of information about currents and their variability in the deep ocean. One can't but help think that it is only a matter of time before the EFF concept will become widely adapted, especially on the global array of profilers.

1.5 Some lessons learned

The advantage of the Lagrangian specification of fluid motion is that it gives unambiguous information on the fate of fluid parcels. The measurement is position, not velocity, or to put it differently, an integral of a fluid parcel's velocity through space. This explains why the temporal development of trajectories has a much smoother character than a time series of velocity at a point. Floats, of course, are not fluid parcels, but they can be given properties so that they can track fluid parcels quite closely for both stratified and homogeneous fluids. We know this for the stratified case from a number of cases where floats have been deployed in pairs. The closer they are deployed and settle out on the same isopycnal surface, the longer they will remain in close proximity, where longer is measured in weeks to months. Since two fluid parcels will also separate over time, it seems plausible to assume that the rate of separation of two floats represents that of two parcels similarly separated – at least when the initial separation is larger than the scale of the float. We suggest therefore that floats have characteristics quite representative of fluid parcel behavior on scales large compared to that of the float itself. This is not a trivial statement for fluid dynamical concepts have, at heart, a Lagrangian formulation: conservation of momentum, mass, heat, potential vorticity, etc. In the absence of forcing or dissipation these properties are conserved along their trajectories. The substantial time derivative is sometimes known as the Lagrangian derivative. For example, a linear segment of a trajectory gives immediately the geostrophic balance since the Coriolis term is known: $-\rho_0 f\mathbf{v} = -\partial p/\partial n$. Any deviation from a linear trajectory, whether change in speed or direction, will give us insight into the nature of the accelerations causing a departure from geostrophy, i.e. the $d\mathbf{v}/dt$ term in the momentum equation.

Similarly, change in temperature, i.e. veronicity or spiciness, as a float drifts along an isopycnal, gives us information on lateral mixing processes. A couple of NAC floats drifted from the North Atlantic Current some 10° of latitude south into the Mediterranean Salt Tongue, where they registered an increase in temperature indicating a warmer, saltier water mass. Change in local *in-situ* properties, whether temperature or layer thickness, can also occur quite suddenly such as during entrainment into fronts or coherent mesoscale eddies and lenses. It is easy to say that lateral shear allows floats to "jump" from one water type to another (as will fluid parcels), but what is actually taking place is less clear, and this leads us back to questions about what happens at scales comparable to the float itself. Someday, perhaps, a float might actually consist of a platform of sensors arrayed to study such things as intrusions and straining of internal waves. Floats have given us much diagnostic information about lateral (isopycnal) mixing, and almost certainly more can be discerned about small-scale stirring, interleaving and mixing processes using Lagrangian platforms of observation and analysis.

Floats have given us much insight into the mesoscale structure of fronts and coherent vortices. We have learned that major fronts such as the Gulf Stream are quite stiff in their lateral scale regardless of position and direction of flow (Bower and Rossby, 1989; Song *et al.*, 1995). Floats trapped in coherent eddies can remain there for extended periods of time (weeks to months) mapping out the radial structure of the eddy as the floats orbit closer and farther from the center. Even single instruments can provide enormous richness of information and detail (e.g. Richardson *et al.*, 1989).

Implausible as it may seem, one can also use arrays of RAFOS floats to quantify the transport patterns of a region. This involves two steps. The first step is to determine the absolute velocity field for the surface the floats are drifting on, for the case considered here an isopycnal surface. To do this the velocities and depths along the float trajectories are binned into small regions; for each of these bins one can estimate the mean velocity and depth and their standard deviations. This converts the Lagrangian information into a gridded Eulerian vector field. These vectors are then used to construct maps of the mean circulation using standard objective analysis techniques. The second step projects the mean isopycnal depth and velocity information onto the rest of the water column using the gravest empirical mode (GEM) technique. This technique recognizes the fact that depth variations of the density profile tend to be highly correlated in the vertical which means that we can, from a synthesis of all hydrography in the vicinity of that site, determine with considerable accuracy the full density profile if we know the depth of one isopycnal. Since we already know the depth of the isopycnal from the gridding of the float data, we effectively have an estimate of the mean three-dimensional distribution of density field everywhere. The combination of density or dynamic height and an absolute velocity on one surface gives us the mean absolute circulation of the region (Perez-Brunius *et al.*, 2004).

Topography plays a large role in constraining fluid motion in a rotating system. We have seen this in the polarization of eddy kinetic energy in regions of slope topographic relief and in the tendency for float trajectories to align themselves along continental slopes. A more subtle but no less striking observation from the recent Atlantic Climate Change Experiment (a component of WOCE) concerns the warm water pathways across the mid-Atlantic ridge. A large number of RAFOS floats were deployed west of the ridge on the 27.5 sigma-t surface. The floats revealed a distinct tendency to converge towards the fracture zones before crossing into the European Basin. The major route, by far, was via the Subpolar Front that crosses the ridge at the Charlie-Gibbs Fracture Zone (~52°N). But a subset also converged on the Faraday Fracture Zone at ~50°N and an even smaller set (three floats) crossed the ridge near the

Maxwell Fracture Zone at 48°N. The fracture zones act as 'virtual funnels' in the sense that openings in the ridge are felt up through the water column even though the ridge crest is well below the main thermocline (Bower *et al.*, 2002). This was a beautiful demonstration of how effectively Lagrangian techniques can capture subtle features in the spatial structure of the ocean in motion.

1.6 Future trends

As the survey above indicates, the acoustic float technology is evolving in many different directions, a clear indication of the flexibility of Lagrangian methods to address a wide range of questions. To date almost all work has taken place in the open ocean, initially in subtropical water with its well-defined sound channel to ensure good tracking over large distances. But experience taught us that even in far less optimal conditions one can still obtain quite useful listening ranges, for example 1000 km for floats at 4 km depth in the Brazil Basin (Hogg and Owens, 1999) and >400 km in the Labrador Sea in winter (Prater, 2002). Given this growing confidence in acoustic navigation in significantly less than optimal regions, one imagines that the next step will be to adapt the technology to studies of circulation in shallow seas and continental shelves such as the Gulf of Maine and the North Sea. To do so floats will have to operate in shallow waters with seasonally varying density. This will require floats to actively position themselves at a certain depth or distance above bottom. Second, a denser network of sound sources will be required due to the rapid absorption of sound at reflection in the bottom sediments. Neither of these requirements is "difficult", but once available the technology will open up new and societally important areas of research, such as the effect of tidal currents on the dispersal of effluents, fate of pollutants, currents and fisheries management, etc.

A significant effort is presently underway to develop an acoustic RAFOS receiver on an integrated circuit chip powered by a tiny battery such that it could be used to study the movements and migratory habits of various types of fish. Such a device, generically known a data storage tag, would listen for and log arrival times of signals from standard RAFOS sound sources. If and when the fish is caught, the data storage tag would be sent to a return address for data retrieval and processing. A wide variety of data storage tags are used to study fish behavior. One class of tags measures incoming light, and from the time of sunrise and sunset one obtains longitude and from the amount of light (after correcting for attenuation by knowing the depth, i.e. pressure, of the fish) one obtains latitude. Needless to say an acoustic tag, which would work at great depths as well, would be a major improvement.

The same "fishchip" could be used in a new generation of very small and inexpensive floats, particularly if they are used in shallow seas or estuaries where less costly forms of telecommunication than the satellite-based Service Argos could be used for data collection. The same acoustic navigation system could be used to map the currents and their variability on the one hand, and the movement of particular stocks of fish on the other, permitting one to explore questions such as the importance of the dynamics of the environment to the behavior of fish.

Finally, it seems inevitable that acoustic float techniques will come into systematic use to study currents in Arctic and Antarctic waters. Eulerian methods using current meters suspended from the ice work well, but one is always curious to know about the pattern of circulation, spatial scales of eddy activity, and the constraints of topography, all of which demand high resolution in the horizontal of the flow patterns. As we noted earlier, float trajectories have the potential to yield highly cost-effective information on the structure of currents in regions where stratification is weak and thus the constraint of topography is likely to be felt quite strongly throughout the water column.

Acknowledgments

As should be evident from this chapter, many engineers and scientists have contributed to this continuing evolution of Lagrangian methods in Oceanography. But without question Henry Stommel's efforts to get a SOFAR float program going were central to everything that followed. He brought Tommy Tucker (IOS) and Doug Webb (WHOI) together for the first test of the feasibility of using the SOFAR channel for long-range float tracking. Webb's continued development of floats was central to the SOFAR float evolution through MODE and POLYMODE, and set the stage for the RAFOS float. The development of the RAFOS float, which began in the early 1980s by Don Dorson and Jim Fontaine, continues to evolve as new needs and applications arise. RAFOS floats can be bought commercially from SeaScan in Falmouth, MA. Mr. Pierre Tillier, has significantly redesigned the floats, which now have excellent time keeping, to within about one second/year. Mr. Martin Menzel, with help and encouragement from Dr. Olaf Boebel, rewrote the tracking software into a modern user-friendly package that runs under MATLAB. This greatly simplified the efforts of new users to become conversant with the RAFOS tracking system. In my work I continue to benefit enormously from my colleagues, students and friends. Drs. Amy Bower, Dave Hebert and Mark Prater deserve special acknowledgment. The reviewers' comments are much appreciated. In the early years of SOFAR development the Office of Naval Reseach played a central role. The MODE, POLYMODE and later programs such as WOCE were primarily supported by the National Science Foundation.

References

Anderson-Fontana, S., P. Lazarevich, P. Perez-Brunius, M. Prater, T. Rossby, and H-M. Zhang, 2001. *RAFOS float data report of the Atlantic Climate Change Experiment (ACCE) 1997–2000*. Tech. Rept. Ref. 01–4. Graduate School of Oceanography, University of Rhode Island. 112 pp.

Boebel, O., K. L. Schultz Tokos, and W. Zenk, 1995. Calculation of salinity from neutrally buoyant RAFOS floats. *J. Atmos. Oceanic Tech.*, **12**, 923–34.

Bower, A., H. T. Rossby, and J. Lillibridge, 1985. The Gulf Stream barrier or blender? *J. Phys. Oceanogr.*, **15**, 24–32.

Bower, A. S. and T. Rossby, 1989. Evidence of cross-frontal exchange processes in the Gulf Stream based on isopycnal RAFOS float data. *J. Phys. Oceanogr.*, **19**, 1177–90.

Bower, A. S., B. Le Cann, T. Rossby, W. Zenk, J. Gould, K. Speer, P. Richardson, M. D. Prater, and H. M. Zhang, 2002. Directly-measured mid-depth circulation in the northeastern North Atlantic Ocean. *Nature*, **419**, 603–7.

Bradley, A., 1978. Autonomous listening stations. *Polymode News* (4). Unpublished manuscript.

Crease, J., 1962. Velocity measurements in the deep water of the western North Atlantic. *J. Geophys. Res.*, **67**, 3173–6.

Davis, R. E., D. C. Webb, L. A. Regier, and J. Dufour, 1992. The autonomous Lagrangian circulation explorer (ALACE). *J. Atmos. Ocean. Tech.*, **9**, 264–85.

Davis, R. E. and W. Zenk, 2001. Subsurface Lagrangian observations during the 1990s. In *Ocean Circulation and Climate: Observing and Modeling the Global Ocean*, ed. J. Church, G. Siedler, and J. Gould. San Diego: Academic Press, Chapter 3.2.

D'Asaro, E. A., D. M. Farmer. J. T. Osse, and G. T. Dairiki, 1996. A Lagrangian float. *J. Atmos. Ocean. Tech.*, **13**, 1230–46.

D'Asaro, E. A., 2003. Performance of autonomous Lagrangian floats. *J. Atmos. Ocean. Tech.*, **20**, 896–911.

Dow, D. L., H. T. Rossby, and S. R. Signorini, 1977. *SOFAR Floats in MODE*. Technical Report. Ref. No. 77–3. Graduate School of Oceanography, University of Rhode Island. 108 pp.

Ewing, M. and J. L. Worzel, 1948. Long-range sound transmission. *Geol. Soc. America*, **27**.

Freeland, H. J., P. Rhines, and H. T. Rossby, 1975. Statistical observations of the trajectories of neutrally buoyant floats in the North Atlantic. *J. Mar. Res.*, **33**, 383–404.

Hebert, D., M. Prater, J. Fontaine, and T. Rossby, 1997. *Results from the test deployments of the Coastal Ocean Lagrangian (COOL) float*, GSO Tech. Rpt. 97–2, University of Rhode Island, 27pp.

Hogg, N. G. and W. B. Owens, 1999. Direct measurement of the deep circulation within the Brazil Basin. *Deep-Sea Res.* II, **46**, 335–53.

LaCasce, J. H. and A. Bower, 2000. Relative dispersion in the subsurface North Atlantic. *J. Mar. Res.*, **58**, 863–94.
Langdon, C., 1986. Pulsing technique for improved performance of oxygen sensors. Proceedings of the Oceans 86 Conference. Washington D.C.
Lazarevich, P., T. Rossby, and C. McNeil, 2004. Oxygen variability in the near-surface waters of the northern North Atlantic: observations and a model. *J. Mar. Res.*, **62**, 663–83.
Ollitrault, M., G. Loaëc, and C. Dumortier, 1994. MARVOR: a multi-cycle RAFOS float. *Sea Technology*, **35**, 43–4.
Owens, W. B., 1991. A statistical description of the mean circulation and eddy variability in the northwestern Atlantic using SOFAR floats. *Prog. Oceanography*, **28**, 257–303.
Perez-Brunius, P., T. Rossby, and R. Watts, 2004. A method to obtain the mean transports of ocean currents by combining isopycnal float data with historical hydrography. *J. Atmos. Ocean. Tech.*, **21**, 298–316.
Pochapsky, T. E., 1963. Measurement of small-scale oceanic motions with neutrally buoyant floats. *Tellus*, **15**, 352–62.
Prater, M. D., 2002. Eddies in the Labrador Sea observed by profiling RAFOS floats and remote sensing. *J. Phys. Oceanogr.*, **32**, 411–27.
Prater, M. D. and T. Rossby, 2004. Observations of the Faroe Bank Channel overflow using bottom-following RAFOS floats. *Deep-Sea Res.* Accepted.
Price, J. F., 1996. Bobber floats measure currents' vertical component in the Subduction Experiment. *Oceanus*, **39**, 26.
Richardson, P. L., D. Walsh, L. Armi, M. Schröder, and J. F. Price, 1989. Tracking three meddies with SOFAR floats. *J. Phys. Oceanogr.*, **19**, 371–83.
Richardson, P. L. and W. J. Schmitz, Jr., 1993. Deep-cross-equatorial flow in the Atlantic measured with SOFAR floats. *J. Geophys. Res.*, **98**, 8371–87.
Robinson, A. R. (ed.), 1983. *Eddies in Marine Science*. Heidelberg: Springer-Verlag.
Rossby, T. and D. Webb, 1970. Observing abyssal motions by tracking Swallow floats in the SOFAR Channel. *Deep-Sea Res.*, **17**, 359–65.
Rossby, H. T., A. Voorhis, and D. Webb, 1975. A quasi-Lagrangian study of mid-ocean variability using long-range SOFAR floats. *J. Mar. Res.*, **33**, 355–82.
Rossby, T., E. R. Levine, and D. N. Conners, 1985. The isopycnal Swallow float – A simple device for tracking water parcels in the ocean. In Essays in Oceanography: a tribute to John Swallow. *Prog. Oceanogr.*, **14**, 511–25.
Rossby, T., D. Dorson, and J. Fontaine, 1986. The RAFOS system. *J. Atmos. Ocean. Tech.*, **3**, 672–9.
Rossby, T., 1988. Five drifters in a Mediterranean salt lens. *Deep Sea Res.*, **35**, 1653–63.
Rossby, T., J. Ellis, and D. C. Webb, 1993. An efficient sound source for wide area RAFOS navigation. *J. Atmos. Ocean. Tech.*, **10**, 397–403.
Rossby, T., J. Fontaine, and E. C. Carter, Jr., 1994. The f/H float – measuring stretching vorticity directly. *Deep-Sea Res.*, **41**, 975–92.
Rossby, T., 1996. The North Atlantic Current and surrounding waters: At the crossroads. *Rev. Geophysics*, **34**, 463–81.
Rossby, T. and M. D. Prater, 2004. On variations in static stability along Lagrangian trajectories. *Deep-Sea Res.* Accepted.
Sanford, T. B., R. G. Drever, and J. H. Dunlap, 1995. Barotropic ocean velocity observations from an electric field float, a modified RAFOS float. In *Proceedings of the IEEE Fifth Working Conference on Current Measurement*, ed. S. Anderson,

G. Appell, and A. J. Williams Taunton, MA: William S. Sullwood Publishing, 24–9.

Song, T., T. Rossby, and E. Carter, Jr., 1995. Lagrangian studies of fluid exchange between the Gulf Stream and surrounding waters. *J. Phys. Oceanogr.*, **25**, 46–63.

Spain, D. L., R. M. O'Gara, and H. T. Rossby, 1980. *SOFAR Float Data Report of the POLYMODE Local Dynamics Experiment*. Technical Report 80–1. Graduate School of Oceanography, University of Rhode Island. 200 pp.

Steele, J. H., J. R. Barrett, and L. V. Worthington, 1962. Deep currents south of Iceland. *Deep-Sea Res.*, **9**, 465–74.

Stommel, H., 1949. Horizontal diffusion due to oceanic turbulence. *J. Mar. Res.*, **8**, 199–225.

Stommel, H, 1955. Direct measurements of sub-surface currents. *Deep-Sea Res.*, **2**, 284–5.

Stommel, H., 1957. A survey of ocean current theory. *Deep-Sea Res.*, **4**, 149–84.

Stommel, H. and A. B. Arons, 1960. On the abyssal circulation of the world ocean – II. An idealized model of the circulation pattern and amplitude in oceanic basins. *Deep-Sea Res.*, **6**, 217–33.

Stommel, H., A. Voorhis, and D. Webb, 1971. Submarine clouds in the deep ocean. *American Scientist*, **59**, 716–22.

Swallow, John, 1955. A neutral-buoyancy float for measuring deep currents. *Deep-Sea Res.*, **3**, 74–81.

Swallow, J. C. and L. V. Worthington, 1961. An observation of a deep countercurrent in the western North Atlantic. *Deep-Sea Res.*, **8**, 1–19.

Swallow, J. C., 1971. The *Aries* measurements in the western North Atlantic. *Phil. Trans. Royal Society of London*, **A 270**, 451–60.

Swallow, J. C., B. S. McCartney, and N. W. Millard, 1974. The minimode tracking system. *Deep-Sea Res.*, **21**, 573–95.

Swift, D. D. and S. C. Riser, 1994. RAFOS Floats: Defining and targeting surfaces of neutral buoyancy. *J. Atmos. Ocean. Tech.*, **11**, 1079–92.

Webb, D. C. and M. J. Tucker, 1970. Transmission characteristics of the SOFAR channel. *J. Acoust. Soc. Am.*, **48**, 767–9.

2

Measuring surface currents with Surface Velocity Program drifters: the instrument, its data, and some recent results

RICK LUMPKIN AND MAYRA PAZOS

National Oceanographic and Atmospheric Administration, Atlantic Oceanographic and Meteorological Laboratory, Miami, Florida, USA

2.1 Introduction

For centuries, our knowledge of the oceans' surface circulation was inferred from the drift of floating objects. Dramatic examples include wrecked Chinese junks and Japanese glass fishing balls which have washed ashore on the US west coast (Sverdrup *et al.*, 1942). Such observations could only provide crude ideas of gyre-scale currents, as there was no way to tell the exact beginning (in time or space) of the drifter's journey, or the trajectory it had taken. Currents can be more accurately inferred from ship drift measurements. A ship's motion relative to the surrounding water is measured by the ship log; its absolute motion is estimated from navigational fixes. In the absence of wind and the "sailing" force of flow around the hull and keel, the difference between absolute and relative motion is the velocity of the water (the current). However, due to relatively large navigational errors in the mostly pre-GPS data set of ship drifts, such current estimates can have errors of O(20 cm/s) (Richardson and McKee, 1984). In addition, a drifting ship is exposed to both currents and wind, making the relative role of the two forces difficult to separate. Comparison of ship drifts with less windage-prone measurements have revealed significant differences in the tropical Pacific (Reverdin *et al.*, 1994) and Atlantic (Richardson and Walsh, 1986; Lumpkin and Garzoli, 2005). To reduce the wind force, investigators in the early 1800s began using drift-bottles to map surface currents. These bottles were typically weighed down so that they were almost entirely submersed, and typically carried a note indicating their launch location and time (Sverdrup *et al.*, 1942). Bottles have been used to map currents in regions such as the North Sea (Fulton, 1897; Tait, 1930) and northwestern Pacific Ocean (Uda, 1935).

A major step towards collecting true Lagrangian time series of velocity was made by attaching a sea anchor, or "drogue", to a buoyant object that would

Lagrangian Analysis and Prediction of Coastal and Ocean Dynamics, ed. A. Griffa, D. Kirwan, A. Mariano, T. Özgökmen, and T. Rossby. Published by Cambridge University Press.

not extend far above the surface but could be tracked by triangulation from a fixed point (such as an anchored ship). Observations of this nature were collected off the US east coast as early as the mid-1700s (Franklin, 1785; Davis, 1991) and were collected worldwide in the famous 1872–76 *Challenger* oceanographic survey at most of the 354 hydrographic stations (Thomson, 1877; Niiler, 2001). With the advent of radio, drifter positions could be transmitted from small, low-drag antennae and triangulated from shore (Davis, 1991). Drifters of this type are still manufactured today, often inspired by the cruciform-shaped design used in the Coastal Ocean Dynamics Experiment (CODE). In CODE, 164 drifters were used to map currents and their variability and to calculate Lagrangian integral scales and dispersion off the California coast (Davis, 1985).

The early 1970s saw the introduction of positioning via satellite observations of an earthbound transmitter's Doppler shift. This was first done using the NIMBUS satellites and later the more accurate Argos Data Collection and Location System carried aboard the National Oceanic and Atmospheric Administration (NOAA) TIROS-N polar-orbiting satellites. Several independent groups promptly developed and deployed satellite-tracked surface drifting buoys. One of the earliest such deployments was in 1975 as part of the North Pacific Experiment (NORPAX). NORPAX drifters were 3 m long, 38 cm-diameter fiberglass cylinders, drogued at 30 m with a 9 m parachute (McNally *et al.*, 1983). An array of 35 drifters, drogued with 200 m of polypropylene line and either a 25 kg weight or a windowshade sail, was deployed in the Gulf Stream region in 1976–78 (Richardson, 1980). These drifters included a tether strain sensor to indicate drogue status, but this sensor frequently failed shortly after deployment. A large array of over 300 Argos-tracked drifters was deployed as part of the Global Atmosphere Research Program (GARP) First Global GARP Experiment (FGGE) in 1979–80 in the Southern Ocean (Garrett, 1980). FGGE drifters were configured in various designs, with the "regular" variety having a tall (3.4 m) surface float and 100 m line acting as a drogue, weighted at the end with 29.5 kg of chain. Other FGGE drifters had large vanes for measuring wind direction. From 1981–84, 113 HERMES-type drifters, drogued with a windowshade sail at 100 m depth, were deployed in the eastern and northern North Atlantic (Krauss and Böning, 1987). In 1983–85, 53 TIROS and mini-TIROS drifters were deployed in the tropical Atlantic as part of the SEQUAL (Seasonal Response of the Equatorial Atlantic) and FOCAL (Programme Francaise Océan-Climat en Atlantique Equatorial) programs. The TIROS design had a windowshade drogue of area 20 m^2 centered at 20 m depth, while the mini-TIROS had a 2.2 m^2 windowshade centered at 5 m depth (Richardson and Reverdin, 1987).

The drifters did not have sensors to indicate drogue presence, although two TIROS drifters recovered after 217 days at sea had drogues in "excellent condition" (Richardson and Reverdin, 1987).

2.2 The SVP drifter

In 1982 the World Climate Research Program (WCRP) recognized that a global array of drifters would be invaluable for oceanographic and climate research, but uncertainties and large variations in the water following properties of various drifter designs posed a major challenge, along with the high costs and excessive weight of some drifter types (World Climate Research Program, 1988; Niiler, 2001). The WCRP declared that a standardized, low-cost, lightweight, easily-deployed surface drifter should be developed, with a semi-rigid drogue which would maintain its shape in high-shear flows. This development took place under the Surface Velocity Program (SVP) of the Tropical Ocean Global Atmosphere (TOGA) experiment and the World Ocean Circulation Experiment (WOCE). Initial funding was provided by the US Office of Naval Research, with subsequent support from NOAA and the National Science Foundation. Competing designs were submitted by NOAA's Atlantic Oceanographic and Meteorological Laboratory (AOML), MIT's Draper Laboratory, and Scripps Institution of Oceanography (SIO) (Niiler, 2003).

During the 1980s these designs continued to evolve, and in 1985–89 they were rigorously evaluated on a number of criteria including their water-following characteristics, quantified by attaching vector-measuring current meters to the tops and bottom of the drogues (Niiler *et al.*, 1987, 1995). Several drogue designs were examined and various problems identified. For example, windowshade drogues could twist and sail across a current; parachute drogues could collapse and subsequently provide very little drag; the line-and-chain FGGE drogue provided too little area compared to the surface float, resulting in significant slip with respect to the currents at the drogue depth (Niiler *et al.*, 1987, 1995; Niiler and Paduan, 1995; Pazan and Niiler, 2001). Other factors were also considered; for example the rigid three-dimensional tristar drogue was found to have somewhat better water-following characteristics than the holey-sock drogue developed at AOML, but the increased manufacturing and shipping costs and more difficult deployments for the tristar outweighed this benefit. By 1993 a clear design for the SVP drifter had emerged which combined the holey-sock drogue of the AOML drifters with reinforced tether ends and surface float designs from SIO. This design (Sybrandy and Niiler, 1992) became the foundation for future SVP drifter development.

The modern data set of "SVP" drifters includes all drifters deployed during the 1979–93 development period that had a holey-sock drogue centered at 15 m depth. AOML spar-type drifters with holey-sock drogues were first deployed in February 1979 as part of the TOGA/Equatorial Pacific Ocean Circulation Experiment (EPOCS). Large-scale deployments of the first modern SVP drifters took place in 1988 (World Climate Research Program, 1988) with the goal of mapping the tropical Pacific Ocean's surface circulation. This effort was expanded to global scale as part of WOCE and the Atlantic Climate Change Program (ACCP), in which the array of SVP drifters was extended to cover the Pacific and North Atlantic Oceans by 1992 and the Southern and Indian Oceans by 1994 (Niiler, 2001). The array spanned the tropical and South Atlantic Ocean by 2004 (Lumpkin and Garzoli, 2005).

Today the array of SVP drifters is known collectively as the Global Drifter Program (GDP), a component of NOAA's Global Ocean Observing System (GOOS) and Global Climate Observing System (GCOS) and a scientific project of the Data Buoy Cooperation Panel (DBCP) of the World Meteorological Organization and International Oceanographic Commission. The scientific objectives of the GDP are to provide operational, near-real-time surface velocity, sea surface temperature (SST) and sea level pressure observations for numerical weather forecasting, research, and *in-situ* calibration/verification of satellite observations. The GDP is managed in close cooperation between NOAA/AOML in Miami, Florida, NOAA/SIO's Joint Institute of Marine Observations (JIMO) in La Jolla, California, and three private US drifter manufacturers (Clearwater, Pacific Gyre and Technocean). AOML arranges and conducts drifter deployments, processes the data, maintains files which describe each drifter, and hosts the GDP website (www.aoml.noaa.gov/phod/dac/gdp.html). JIMO supervises the industry, acquires the drifters from the various manufacturers, upgrades the technology, develops new sensors, and creates enhanced data sets (Pazan and Niiler, 2004) for the research community. The manufacturers build the SVP drifters according to closely monitored specifications.

2.2.1 Design

At present, there are two basic sizes of SVP drifters: the original, relatively large SVP drifter and the new "mini" version (Fig. 2.1). The original design is extremely robust but is relatively expensive and heavy. A more gracile redesign was proposed in DBCP Specification Rev. 1.2 in December 2002. This "mini" drifter is produced alongside SVP drifters of original dimensions by several manufacturers.

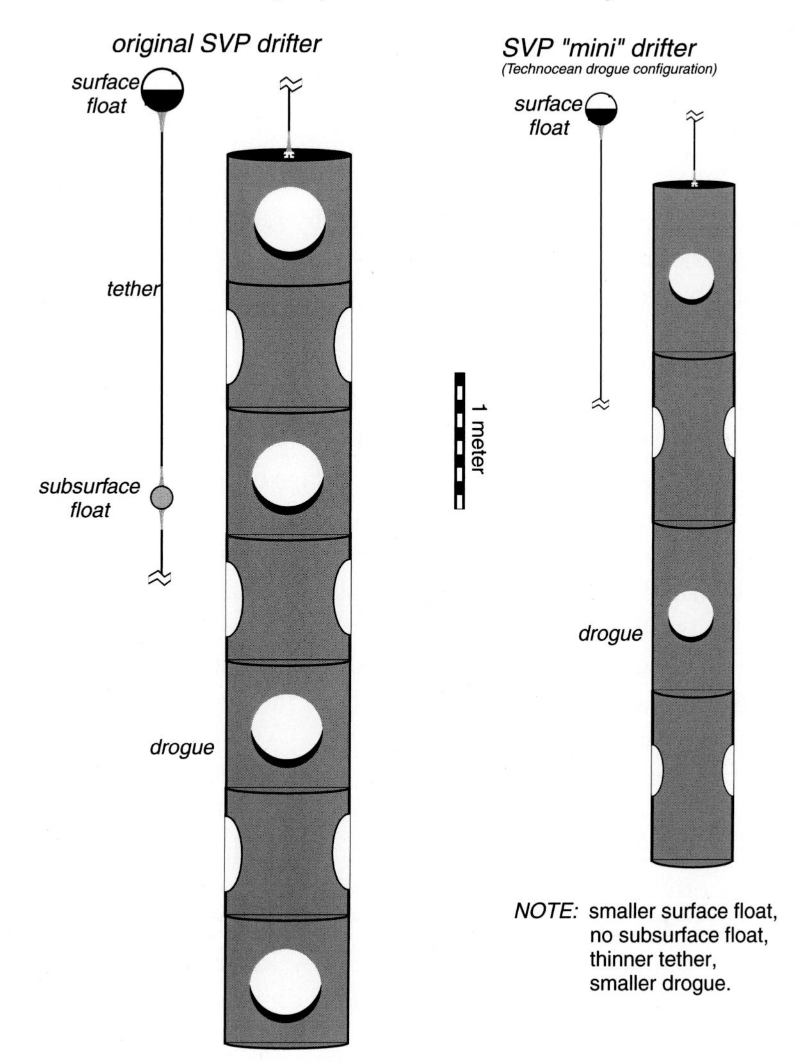

Figure 2.1 Schematic view of two SVP drifter types. All components are shown to scale; much of the tether length has been excluded. The drogues are centered at a depth of 15 m.

Present manufacturers of SVP drifters include Clearwater Instrumentation (Watertown, MA USA; www.clearwater-inst.com), Marlin-Yug (Sevastopol, Ukraine; marlin.stel.sebastopol.ua), Metocean Data Systems (Dartmouth, Nova Scotia, Canada; www.metocean.com), Pacific Gyre (Oceanside, CA USA; www.pacificgyre.com), and Technocean (Cape Coral, FL USA; www.technocean.com).

The surface float of an SVP drifter ranges from 30.5 cm (the smallest mini) to 40 cm in diameter. Originally, the surface float hull was made of

0.3–0.4 cm-thick fiberglass (thicker at the tether protrusion; cf. Sybrandy and Niiler 1991, Fig. 3). Most manufacturers have now switched to less expensive ABS (Acrylonitrile-Butadiene-Styrene) plastic for surface float hull construction. The surface float contains: batteries in diode-protected packs, typically 4–5 packs each with 7–9 alkaline D-cell batteries; a satellite transmitter (401.650 MHz, ±10 kHz) typically activated by removing a magnet from the float hull; a thermistor for sub-skin sea surface temperature, located at the base of the float to avoid direct radiative heating; and possibly other instruments measuring barometric pressure, wind speed and direction, salinity, or ocean color. Most surface floats also include a submergence sensor, consisting of two screws extruding from the hull. This sensor is used to identify the presence of the drogue, which frequently drags the surface float beneath the sea surface – an abrupt drop in the percentage of submerged time indicates that the drogue has been lost. One manufacturer (Clearwater) has replaced the submergence sensor with a tether strain sensor, which more directly measures drogue presence. Most manufacturers apply cuprous oxide paint to the bottom half of the surface float to reduce biofouling. A polypropylene-impregnated wire rope tether connects the surface float to a subsurface float (original design; 5.6 mm diameter tether) or directly to the drogue (mini drifter, with a 3.2 mm diameter tether).

An SVP drifter has its drogue centered at 15 m depth to measure mixed layer currents; other options (such as 100 m) have been made available to individual researchers. The outer surface of the drogue is Cordura nylon cloth. In the original design, it is composed of 7 sections, each 92 cm long and 92 cm in diameter for a total length of 6.44 m. Mini drogues are not yet standardized among the manufacturers: they are composed of 4 (Pacific Gyre) or 5 (Marlin-Yug) sections of original dimensions, or 4 (Clearwater) or 5 (Technocean) redesigned sections of diameter 61 cm, length 1.22 m/section. Throughout the drogue, PVC or polypropylene pipe rings with wire rope spokes provide support, maintaining the drogue's cylindrical shape. The top ring is filled with polyurethane foam to provide some positive buoyancy, reducing accordion-type oscillations when tether strain is low (Sybrandy and Niiler, 1992). Lead weights (in some drifters, sand ballast in a polypropylene pipe) sewn into the base of the drogue insure that it hangs nearly vertically. The drogue is a "holey-sock," i.e. each drogue section contains two 46 cm (mini: 30 cm) diameter opposing holes, which are rotated 90° from one section to the next (see Fig. 2.1). These holes act like the dimples of a golf ball by disrupting the otherwise laminar flow which would generate organized lee vortices. As a consequence, the drogue does not experience an abrupt change in drag coefficient across a critical

Reynolds number which would be associated with vortex shedding (Nath, 1977).

While the sizes of the surface float and drogue vary, the manufacturers all aim for a specific nondimensional goal: a drag area ratio of ~40. This ratio is the drag area (drag coefficient times cross-sectional area) of the drogue, divided by the drag area of all other components. At a drag area ratio of 40, the resulting downwind slip (see Section 2.4.1) is 0.7 cm/s in 10 m/s winds; for comparison, a standard FGGE-type drifter had a drag area ratio of 10–12 and a downwind slip of 8 cm/s in 10 m/s winds (Niiler and Paduan, 1995; Pazan and Niiler, 2001). In practice, the manufacturer-provided drag area ratios range from 37.5–45.9. Some modified drifters include substantial additional components which greatly reduce the drag area ratio. An example is Marlin-Yug's SVP-BTC drifter which has a thermistor chain extending to 60 m depth, reducing the drag area ratio to 6.8.

2.2.2 Deployment

Original-sized SVP drifters are packaged two per cardboard box of dimension 1.07 m (3′6″) cubed. Each drifter weights approximately 45 kg (100 lbs) and can be deployed by a single person, although in heavy seas it is recommended that two people deploy an original SVP drifter. Mini drifters are packaged five per cardboard box of the same dimensions, or two in a box of size $1.17 \times 0.89 \times 0.56$ m. The mini weighs approximately 20 kg (44 lbs) and can easily be deployed by one person. The drogue and tether are bound with paper tape which dissolves in the water, and the tether is sometimes wrapped around a water-soluble cardboard tube to protect it from kinking. The drifter is deployed by throwing it from the stern of a vessel, preferably from the lowest deck and within 10 m of the sea surface. The ship may be underway; successful deployments have been conducted from cargo ships steaming at up to 25 knots. After deployment, it may take approximately an hour for the drogue to become fully soaked, allowing the paper tape to dissolve and trapped air bubbles to be released, before the drogue sinks.

Drifters have also been air-deployed out of Lockheed C-130 Hercules, operated by the Air Force Reserve "Hurricane Hunters" (53rd Weather Reconnaissance Squadron, 403rd Wing, Keesler Air Force Base), and by the Naval Oceanographic Office (NAVOCEANO) which conducts surveys supporting naval operations primarily in the northern hemisphere. Deployments have also been conducted from a C-141 Starlifter. Air deployment requires extensive rigging of the drifter package, including adding the parachute, at a cost approximately equal to the drifter itself (P. Niiler, pers. comm.).

During the one-year period September 2003–August 2004, a total of 658 drifters were deployed in NOAA's contribution to the Global Drifter Program. Of these, 440 were deployed off research vessels, 201 off volunteer observation ships (typically cargo vessels), and 17 were air-deployed.

2.2.3 Data transmission

Drifter sensor data (including SST and battery voltage) are typically sampled at intervals of 90 s. Averages are calculated over an observation cycle of seven to ten samples and transmitted. Submergence or strain data transmission varies by manufacturer. For example, Metocean drifters sample the submergence four times every 90 s, and sum the total number of underwater samples over a 30 minute averaging period to determine the percentage of time submerged. The data are formatted for transmission with checksum entries provided in a 32-bit Argos (see below) message. Older drifters often used a duty cycle of 1/3 (typically the transmitter spent one day on, then two days off), which could lead to significant biasing of high latitude inertial motion (Bograd *et al.*, 1999). Nearly all drifters since 2001 operate on a 100% duty cycle. Recently developed drifters exploit multiplexing techniques to increase data transmission. During one satellite pass, 6–7 data frames can be sent every 90 s, or twice as many with a 45 s repetition rate. Each transmitter is assigned a Platform Terminal Transmitter (PTT) code by Argos, often referred to as the drifter ID number.

Argos (www.argosinc.com) is an American–French satellite-based system for collecting, processing and distributing data. It is operated by Collecte Localisation Satellites (CLS) in Toulouse, France, with a subsidiary (Service Argos, Inc.) in Largo, Maryland USA. Since 1978, the Argos system has been carried on the US National Oceanographic and Atmospheric Administration (NOAA) Polar Orbiting Environmental Satellites to obtain global coverage. In addition, a second-generation Argos system was carried aboard the Japanese Advanced Earth Observing Satellite II (ADEOS-II), launched in December 2002. This joint Argos/ADEOS-II program ("Argos Next") was declared operational on May 5, 2003; unfortunately, the satellite failed on October 25, 2003. Future launches with next-generation Argos systems are planned aboard the European METOP satellites, beginning in the last quarter of 2005.

The position of a drifter is inferred from the Doppler shift of its transmission. The position-deducing algorithm can be summarized as follows (Argos, 1996). As the satellite approaches, passes and recedes from the latitude of a drifter, the satellite's speed (7.4 km/s) Doppler shifts the signal. The timing of the swing from blue to red (but not exactly the latitude of the shift, due to drifter motion; see below) gives the drifter's latitude, and the rapidness of the swing gives the off-track direction (the closer the satellite pass is to the drifter, the more step-like the swing). The absolute motion of the drifter introduces an additional Doppler shift: at 20°N, a fixed point on the Earth's surface travels 437 m/s westward. Thus, if the drifter is east of the satellite pass, an additional blue shift is superimposed which reaches its maximum as the satellite crosses the drifter's latitude. This Doppler shift decays with increased distance from the drifter at a rate dependent upon the minimum satellite/drifter distance (greater distance equals slower decay). The sign of this shift is estimated from least-squares fitting and the previous history of the drifter, and gives the off-track direction.

Argos estimates the accuracy of position fixes at 150–1000 m. The largest errors occur when a satellite pass is close to, but not directly over, a drifter. In this situation the Doppler shift from the absolute drifter motion is a relatively brief spike which can be difficult to resolve – possibly causing the Argos position algorithm to "place" the drifter on the wrong side of the pass. When this happens, there are relatively large errors in longitude, with much smaller latitudinal errors. Additional errors can arise due to satellite instrumental noise and inaccuracies in orbit and time coding (Hansen and Poulain, 1996). From a 70-day time series of position fixes for a grounded drifter in Honolulu, Hawaii, we have calculated the root-mean-square error in a fix to be 630 m zonally, 270 m meridionally. This drifter's transmitter was not held ridgidly vertical, possibly introducing a bias to some of the satellite fixes (J. Wingenroth, pers. comm.) that would not be experienced by a drogued drifter in the water. The 70-day median of the Argos fixes was not significantly different from a GPS fix.

2.2.4 Drifter lifetime

The manufacturers' estimate for an original SVP drifter lifetime is ~400 days. In order to examine the accuracy of this claim, we have calculated the half life for SVP drifters deployed in 1998–2003. This was calculated for all drifters which had not run aground or been picked up by boaters. The remaining drifters were sorted by the year in which they had been deployed. For each year's batch of drifters, we calculated how many days of observations were obtained, and how many days of drogued observations were obtained. Drifters still transmitting (and still drogued, for the drogue half life) were assigned a large placeholder value (9999 days). The histogram of these lifetimes was used to identify the amount of time after which one-half of the

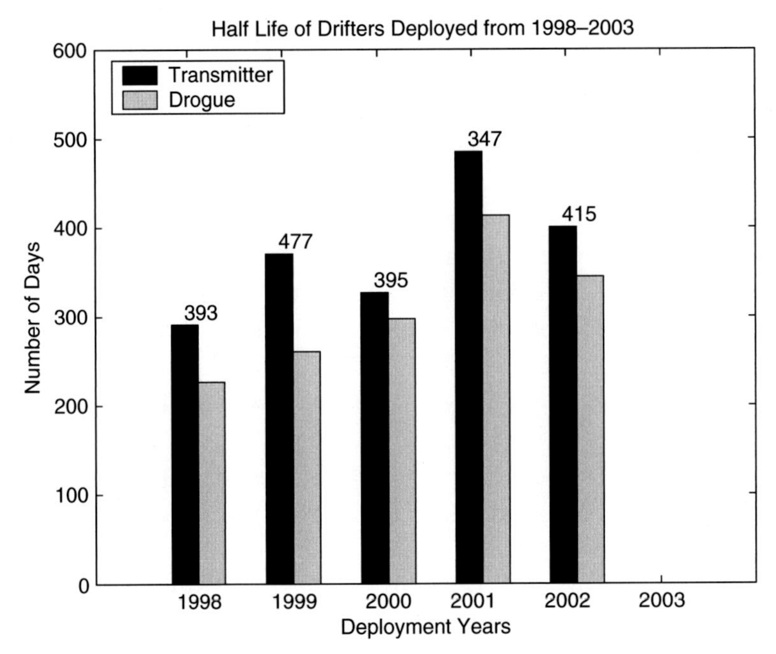

Figure 2.2 Half lives (days) of drifters deployed in 1998–2003. Black bar: half life of transmitter (e.g., velocity observations). Light grey bar: half life of drogued position observations. Numbers above the bars indicate how many drifters went into each calculation, e.g. how many drifters were deployed that year which did not run aground or were picked up. The 2003 half lives could not be determined because more than half are still alive and drogued.

drifters were no longer alive, and after which one-half were no longer providing drogued velocity observations. The half lives are shown in Fig. 2.2. Transmitter lifetimes generally increased during 1998–2002, from 291 days in 1998 to 400 days in 2002, with a peak of 485 days in 2001.

These calculations do not include drifters which failed to transmit upon deployment: 12% of drifters in 1998, 5% in 2001 and 3% in 2003. More than half the drifters deployed in 2003 were still alive and drogued at the time of these calculations (November 2004), making their half life impossible to determine; given the temporal distribution of deployments through 2003 and the histogram of lifetimes so far, it is likely to exceed 450 days.

Thus, while some individual batches of drifters have had a high failure rate due to causes ranging from defects during manufacturing to deployment in extremely harsh conditions (e.g., air-deployed in the path of a hurricane), the general tendency across recent years is encouraging. In addition to the development of the mini drifter described earlier, incremental improvements in drogue reinforcement, tether/drogue attachment, and transmitter design are successfully increasing the value of the SVP drifter. At the same time, the cost of a drifter has steadily dropped, from USD 5475 in 1993 (adjusted to 2003 US dollars; Niiler, 2003) to USD 1750 today.

2.3 Drifter data: quality control, interpolation and coverage

Before December 2004, position fixes acquired by two of the satellites were processed by Argos. For a typical drifter near the equator, 6–9 fixes per day were acquired. At higher latitudes the polar-orbiting satellite passes were closer, yielding 8–10 fixes per day at 20° latitude. The theoretical maximum was 28 fixes per day at the poles (Argos, 1996). In late 2004 the Joint Tariff Agreement (see www.ogp.noaa.gov/argos/overview.htm) negotiated the use of the full satellite constellation (up to four at present). Since mid-December 2004, this "multi-satellite service" has been yielding 16–20 fixes per day for equatorial drifters. AOML's Drifter Data Assembly Center (DAC; www.aoml.noaa.gov/phod/dac/dacdata.html) assembles these raw data, applies quality control procedures, and interpolates them to regular 1/4-day intervals. The raw observations and processed data are archived at AOML's DAC and at Canada's Marine Environmental Data Service (MEDS; www.meds-sdmm.dfo-mpo.gc.ca).

2.3.1 Quality control

The DAC first visually examines drifter data for evidence that the data were transmitted while on the deck of a ship, the drifter was aground, or the drifter has been picked up by a boater. These observations are usually apparent from the trajectories, and can be supported by submergence values and the diurnal variations in temperature. These observations are removed from the data set.

Next, the DAC identifies drifters which have lost their drogues. This is done using the submergence or tether strain observations. The drogue lost dates are compiled in a "directory file" which includes each drifter's deployment time and location, ending time and location, and the type of death (e.g., picked up, ran aground, stopped transmitting). These dates are stored using a modified Julian day convention in which day 1 is January 1, 1979. For a drifter that never lost its drogue, the directory file holds the placeholder value 0 for drogue off time while it is still alive (still transmitting good data), or the date of its final reliable transmission if it has died.

To eliminate the more egregious errors in raw Argos fixes, the DAC applies a two-step quality control scheme (Hansen and Poulain, 1996). In this methodology, the velocity is calculated via finite differencing of the raw fixes both forward and backward in time. A fix is flagged as "bad" if it produces a velocity greater than four standard deviations from the mean velocity in both forward and backward passes. Two-way differencing is used because a forward-only calculation may fail to identify a bad fix if it comes immediately after a gap in data acquisition.

2.3.2 Interpolation

The raw fixes are interpolated to uniform 6-h intervals using an optimal interpolation procedure known as kriging (Hansen and Herman, 1989; Hansen and Poulain, 1996). Latitude and longitude (and SST) are interpolated independently. Kriging assumes that the observations consist of a "true" signal contaminated by noise, the latter assumed to be white and unbiased (zero mean). Interpolated values are constructed by a linear combination of the five observations preceding and five following the interpolation point (if available), using a set of weights constructed such that the root-mean-square difference between the true value and the concurrent interpolated value is minimized.

In the kriging procedure, the (true) position is described by a structure function. For this, Hansen and Poulain (1996) used a fractional Brownian model that can describe motion ranging from uncorrelated Brownian diffusion to perfectly correlated linear advection. They estimated the parameters for the model using tropical Pacific Ocean drifter observations for the period 1988–93. The resulting parameters yielded a model which was a blend of advection and diffusion, with advection more significant for zonal displacements. This result is consistent with studies which have found strong anisotropy in the zonal vs. meridional scales of dispersion in the tropics, with much longer zonal integral length scales (e.g., Lumpkin and Flament, 2001). This model is probably not the most appropriate choice at higher latitudes away from strong zonal currents.

Along with the interpolated positions, AOML provides formal error bars on the positions. These can be invaluable in identifying large gaps – as long as two weeks – across which the data have been interpolated, as these will have large error bars. Researchers working with the drifter data should approach these data with caution; they may be useful for calculating mean advective pathways, but should not be included in calculations of eddy kinetic energy or dispersion (see Fig. 2.3).

Following interpolation, the zonal and meridional components of velocity are calculated via centered finite differencing over 1/2 day displacements. Many investigators interested in subinertial motion (e.g., Ralph and Niiler,

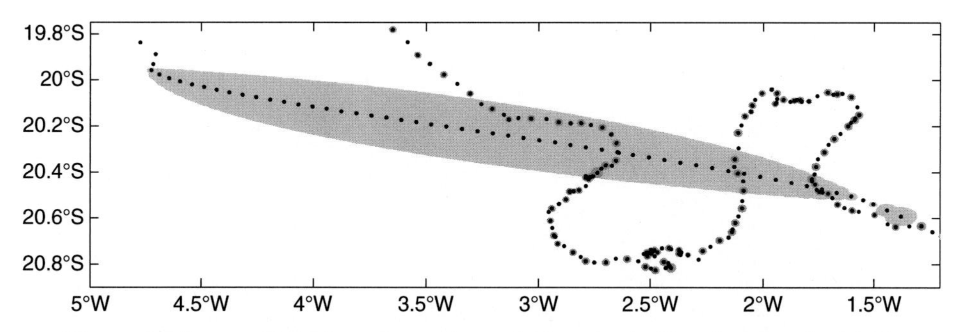

Figure 2.3 Interpolated locations of two drifters in the South Atlantic (black points), with shaded error bars from the kriging. For one of the drifters, 8–10 satellite fixes per day yields narrow error estimates (smaller than the black points for some quarter-day values). Transmissions from the other drifter were not picked up by the satellites for 12 days; interpolation through this gap yields a smooth trajectory and large location error estimates.

1999; Fratantoni, 2001; Lumpkin and Garzoli, 2005) apply a lowpass filter to these velocities before proceeding with their analyses.

2.3.3 Data coverage

SVP drifter observations now cover most areas of the world's oceans at sufficient density to map mean currents at 1° resolution (Fig. 2.4).

The recent growth of the drifter array is shown in Fig. 2.5. The number of drifters in the global array has increased tremendously due to the efforts of many individual investigators and international partnerships contributing to the Global Ocean Observing System (GOOS). The GOOS goal of maintaining a 5° by 5° network of drifters requires an array of at least 1250 drifters. The drifter array reached this size on September 18, 2005 (see www.ogp.noaa.gov/events/20050918_globaldrifter/).

From 1998 to 2003, drifter coverage has increased in all basins shown in Fig. 2.5 except the North Pacific. Recent air deployments by NAVOCEANO south of the Aleutian Islands, along with future deployments from volunteer observation ships running the great circle route between Japan and California, are addressing this gap.

2.4 Velocity observations

Fig. 2.4 shows time-mean surface currents (middle) and eddy kinetic energy (bottom) for the world's oceans, calculated on a 1° by 1° grid via Gauss–Markov decomposition (Lumpkin, 2003). Annual and semiannual amplitudes

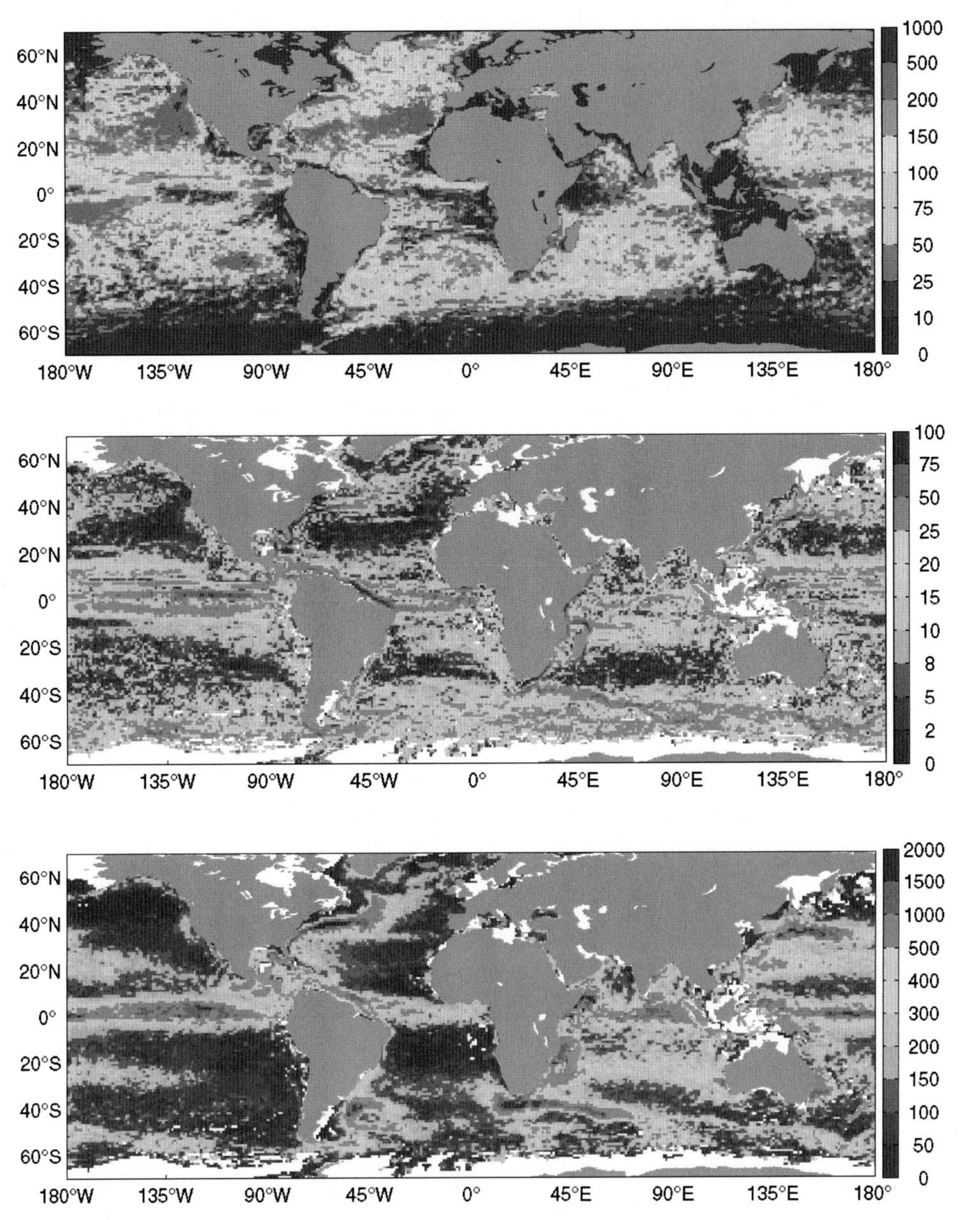

Figure 2.4 Fields calculated from the most recent decade (to October 31, 2004) of quality-controlled SVP drifter observations. Top: density of observations (drifter days per square degree). Middle: mean current speed (cm/s). Bottom: mean eddy kinetic energy (cm^2/s^2). See Plate 10 for color version.

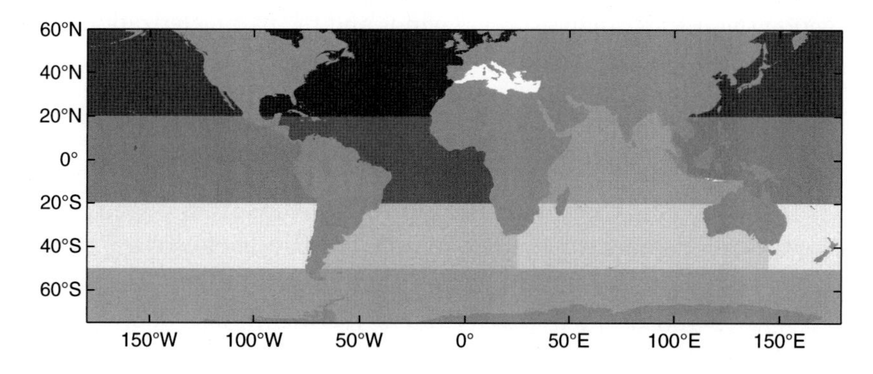

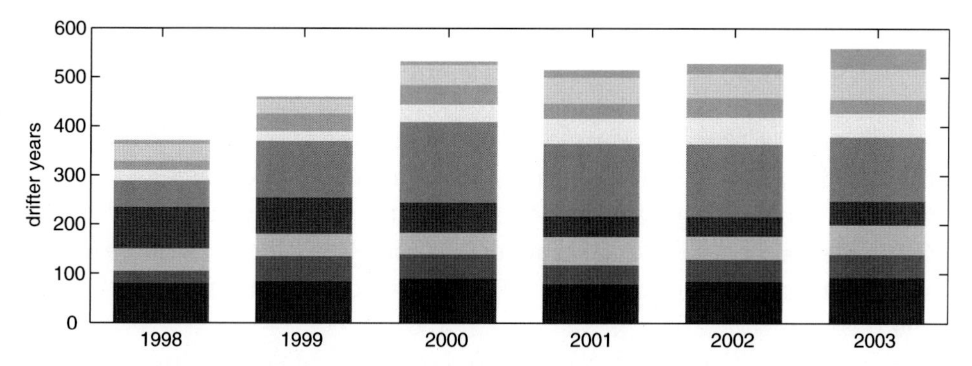

Figure 2.5 Top: regions (colored) used to subdivide data in bottom panel. Bottom: number of observations (drifter-years) per year in the different regions. See Plate 11 for color version.

and phases (not shown) were calculated at the same resolution. All of the major western boundary currents (Gulf Stream, Philippines/Kuroshio, Brazil, North Brazil, East Australian, Mozambique/Agulhas and Somali) are clearly seen as maxima in both mean current speed and eddy energy. The time-mean zonal structure of tropical currents such as the northern South Equatorial Current and North Equatorial Countercurrent are prominent features of Fig. 2.4, as are the monsoon-driven semiannual currents of the equatorial Indian Ocean.

SVP drifters provide imperfect pseudo-Lagrangian observations at the drogue depth – "pseudo" because water parcels can upwell or downwell while the drifter stays at the surface, and imperfect because of slip (see below). The resulting observations of mixed layer velocity are thus a combination of slip, plus the directly wind-driven flow in the upper mixed layer, plus the 15 m-deep signature of the underlying eddy and gyre-scale currents. Time-mean plots of these components, derived from National Centers for Environmental

Prediction (NCEP) Reanalysis Ver. 2 winds and the parameterizations described below, are shown in Fig. 2.6.

2.4.1 Slip, with and without a drogue

Slip is the horizontal motion of a drifter that differs from the lateral motion of currents averaged over the drogue depth. Slip is caused by direct wind forcing on the surface float, drag on the float and tether induced by wind-driven shear, and surface gravity wave rectification (Niiler *et al.*, 1987; Geyer, 1989). In order to reduce wave rectification, the surface float is spherical (Niiler *et al.*, 1987, 1995). The original SVP design included a 20 cm diameter subsurface float between the surface float and drogue, intended to decouple their motion and to provide additional buoyancy off-setting the weighted drogue. The subsurface float has been omitted in the recent mini drifter redesign. The most important design characteristics that minimize slip are low tension between the surface buoy and drogue, which avoids aliasing wave motion, and a large drag area ratio (Niiler *et al.*, 1987). As long as the drogue remains attached to the drifter, the downwind slip is estimated at 0.7 cm/s per 10 m/s of wind speed (Niiler and Paduan, 1995).

Pazan and Niiler (2001) examined a data set of 2334 SVP drifters, including 1845 which had lost their drogues while continuing to transmit location. They examined residual (undrogued minus drogued) drifter motion using a multiple regression model with local surface wave height (from the Fleet Numerical Meteorology and Oceanography Center, FNMOC) and winds (FNMOC, NCEP, and the European Center for Medium range Weather Forecasting, ECMWF). The residual motion did not have a significant relationship with waves, either because wave drift is negligible or because the relationship is obscured by errors in or resolution of the wave data set. Residual motion was aligned on average directly downwind, and was linearly dependent upon wind velocity with a global mean value of 7.9 ± 0.7 cm/s per 10 m/s of wind. This result and those of Niiler and Paduan (1995) indicate that an undrogued SVP drifter experiences a slip of 8.6 ± 0.7 cm/s per 10 m/s wind.

2.4.2 Ekman drift

Niiler and Paduan (1995) analyzed the Lagrangian velocity time series of 8 drogued SVP drifters in the northeast Pacific during October 1989–April 1990. They found that velocity and local wind (using 6-hourly ECMWF operational

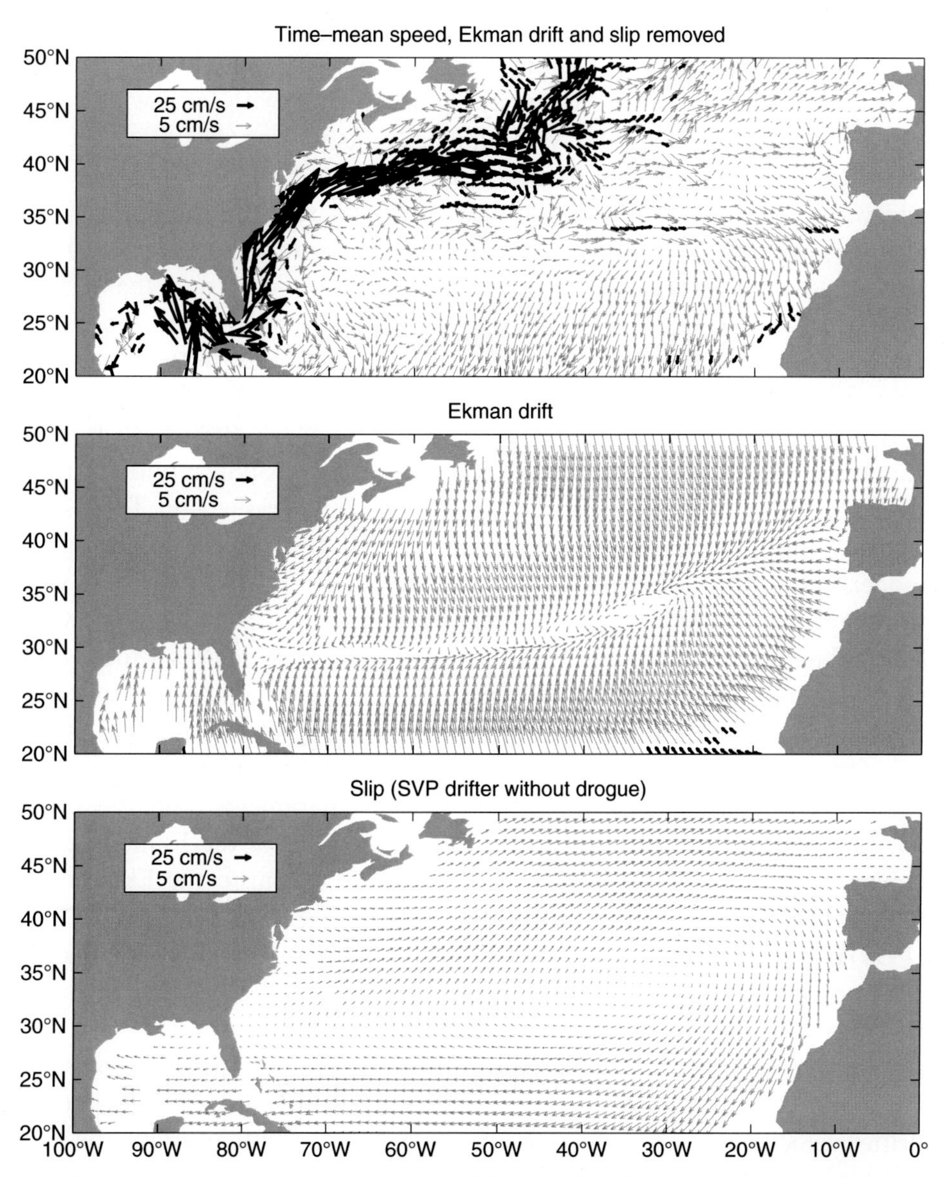

Figure 2.6 Top: time-mean speed of all SVP drifters in the subtropical North Atlantic, October 1989 to April 2004, with Ekman drift and slip removed. A separate scale (bold arrows) is used for speeds exceeding 10 cm/s. Middle: time-mean Ekman drift (Ralph and Niiler, 1999; Niiler, 2001), showing the wind-driven convergence that forces the subtropical gyre (also see Fig. 15 of Rio and Hernandez, 2003). Bottom: time-mean slip of undrogued SVP drifters, using the parameterizations of Niiler and Paduan (1995) and Pazan and Niiler (2001). Drogued drifters experience a slip which is smaller by an order of magnitude. The total time-mean drift of undrogued drifters is given by the sum of these three panels.

winds) was highly coherent at subinertial periods of 5–20 days. At shorter periods inertial oscillations dominated the Lagrangian velocity spectra, while at longer periods mesoscale variations (uncorrelated with local wind) dominated the spectra. The 5–20 day bandpassed velocity was 60–100° to the right of the wind, consistent with Ekman dynamics. For the bandpassed velocities and winds, Niiler and Paduan derived the linear regression model $\boldsymbol{U}=b\boldsymbol{\tau}_o$, where τ_o is the wind stress at the ocean surface, $\boldsymbol{U}=u+\mathrm{i}v$, (u, v) are the zonal and meridional components of velocity, $\mathrm{i}=\sqrt{-1}$ and the complex coefficient b has amplitude $|b|=0.28\,\mathrm{m/(s\ Pa)}$ and angle 77° to the right of the wind. This model accounted for 35% of the variance in the bandpassed drifter velocities.

In a study of 1503 SVP drifters in the tropical Pacific 20°S–20°N, 1988–96, Ralph and Niiler (1999) removed the time-mean geostrophic motion using the hydrographic climatologies of Levitus (1982) and Kessler and Taft (1987). They averaged the residual velocities in 2° by 5° bins and examined various models regressing this velocity onto concurrent, interpolated 6-hourly winds from operational ECMWF. Niiler (2001) repeated this calculation with a larger data set (30°S–30°N, 1988–99) using NCEP reanalysis winds. These studies found that 49% of the variance across the bins could be accounted for by the model $U=Au^*/f^{1/2}$, where U is the magnitude of the residual (Ekman) velocity, $u^*=\sqrt{|\tau_o|/\rho}$ is the friction velocity due to the time-mean surface wind stress τ_o, ρ is density, f is the Coriolis parameter, and the best-fit coefficient $A=(0.081\pm0.013)\ \mathrm{s}^{-1/2}$. The orientation of the Ekman velocity depended upon the ratio of the drogue depth $d=15\,\mathrm{m}$ to the Ekman layer depth scale $D=u^*/A$, which varied from bin to bin. When Ralph and Niiler (1999) plotted the off-wind angle as a function of d/D they found a clockwise-rotating (in the northern hemisphere) spiral consistent with the theory of Ekman (1905).

More recently Rio and Hernandez (2003) subtracted time-mean geostrophic currents and altimetry-derived geostrophic current anomalies from concurrent drifter velocities over the globe for the period 1993–99, to examine the wind-driven ageostrophic motion. They found that the time-varying geostrophic component was significant at periods greater than 10 days in the latitude band 15–30° (north and south), and at periods greater than four days at 30–90°. At higher frequencies drifter velocities are predominantly ageostrophic, with clear peaks in the anticyclonic spectra at inertial and tidal frequencies. The ageostrophic motion is significantly coherent with the wind in the band between 20 days and the inertial/tidal periods. By separately examining summer and winter months, Rio and Hernandez (2003) showed that the Ekman depth (parameterized as in Ralph and Niiler, 1999) varies with the seasonal change of wind speed over much of the world's subtropical and subpolar basins.

2.5 Other observations

The focus of this chapter is on the velocity observations made by SVP drifters. However, this platform is capable of collecting many other types of observations, some of which can be combined with velocity to estimate heat advection, ground-truth satellite-based products, and many other uses. Thus, a brief overview of these observations is warranted.

Sea surface temperature (SST)

All standard SVP drifters measure temperature 15–30 cm beneath the sea surface. These data are disseminated on the Global Telecommunication System (GTS) by Argos within two hours of reception for use in numerical weather forecasting and operational SST analysis, and for calibrating SST products (c.f. Reynolds *et al.*, 2002).

Barometric pressure

Many drifters, known as SVP-Bs, have been outfitted with a barometer to measure air pressure. Large-scale experimental deployments began in 1994; operational barometric observations have been collected since 1997. These data are particularly valuable in numerical weather prediction at high latitudes, where few *in-situ* observations are available if a storm develops outside the major shipping lanes. The barometer port extends ~20 cm above the top of the surface float to minimize spuriously high spikes in the pressure record associated with submergence.

Wind

Some drifters include a hydrophone for noise level, which can be converted to wind speed and precipitation estimates, and a 25 cm-by-20 cm wind vane mounted to the barometer port of the surface float (with accompanying two-axis tilt sensor in the float, and swivel connection for the tether) to measure wind direction. SVP drifters of this type are known as Minimets (Milliff *et al.*, 2003). The WOTAN hydrophone is typically mounted either on the tether, at a depth of 11 m, or between the tether and drogue top. Recent air deployments of these drifters in the path of Category 4 Hurricane Fabian have demonstrated the ability to measure the wind direction to within $\pm 10^\circ$ (Niiler *et al.*, 2004b), mapping the circulation of the hurricane more clearly than in QuickSCAT data, although the hydrophone became saturated at such extreme wind speeds.

Ocean color

Some drifters have included an upwelling radiance sensor mounted on the surface float just beneath the sea surface, along with a downwelling irradiance sensor. Their observations have been used to study chlorophyll variations in remote regions such as the Southern Ocean (Letelier *et al.*, 1996).

Salinity

The first salinity-measuring drifters were developed at Scripps Institution of Oceanography (SIO) by attaching a SeaBird SeaCat (thermistor and conductivity) to the top of the drogue (11 m depth). In 1992–3, 72 of these drifters were deployed in the tropical Pacific and provided observations which compared favorably to the TAO mooring data (Kennan *et al.*, 1998). Four of these drifters were recovered after 310 days at sea, with post-calibration revealing a maximum offset of 0.02 psu.

More recently, drifters have been developed which can measure surface salinity. At SIO, SeaBird Microcats have been mounted to the base of the surface float. Parallel development is being conducted at Woods Hole Oceanographic Institution. Thirty of the SIO drifters were deployed in the East China Sea in the period 2000–4, to help evaluate the effects of the decreasing Yangzte flow on the ecology of South Korea. Two were recovered several weeks after deployment, and showed no detectable offset when post-calibrated (P. Niiler, pers. comm.).

Biofouling presents the major challenge to obtaining extended observations of surface salinity. Ongoing experiments are varying the antifouling paint and the pumping systems for the SeaBird Microcats. Current Global Drifter Program plans include the deployment of SVP-Microcat pairs in the Bay of Biscay west of France, each pair consisting of one drifter with pumping and one without, with sequential recoveries to evaluate the success and necessity of pumping. In the future, these observations will provide calibration and validation for satellite-derived sea surface salinity products.

Subsurface temperature

Several drifter manufacturers are developing drifters with thermistor chains to measure temperature profiles of the ocean's upper O(100 m). These observations would be invaluable for measuring mixed layer heat content variability, which can be poorly correlated with SST changes (Kelly, 2004). An array of drifters including eight with thermistor chains were air-deployed ahead of Hurricane Rita in September 2005, successfully providing upper ocean heat measurements in the Gulf of Mexico prior to the storm's landfall near the Texas/Louisiana border.

2.6 Recent drifter-based studies: an overview

A large number of studies have used drifter observations to map currents and their variability in various regions. Recent examples (2000–present) have focused on the North Atlantic (Flatau *et al.*, 2003), Labrador Sea (Cuny *et al.*, 2002), Caribbean Sea (Centurioni and Niiler, 2003; Richardson, 2004), subtropical eastern Atlantic (Zhou *et al.*, 2000), tropical Atlantic (Grodsky and Carton, 2002; Lumpkin and Garzoli, 2005), eastern South Atlantic (Largier and Boyd, 2001), tropical Pacific (Grodsky and Carton, 2001b; Johnson, 2001; Yaremchuk and Qu, 2004), central subtropical Pacific (Lumpkin and Flament, 2001; Flament *et al.*, 2001), western Pacific (Niiler *et al.*, 2003), North Pacific (Rabinovich and Thomson, 2001), tropical Indian (Grodsky and Carton, 2001a) and Southern (Zhou *et al.*, 2002) oceans. Several recent studies have also focused on marginal seas including the Japan/East (Lee *et al.*, 2000), South China (Centurioni *et al.*, 2004) and Black (Poulain *et al.*, 2004; Zhurbas *et al.*, 2004) seas.

Because drifters follow the two-dimensional surface flow, they are ideal for studying the dispersion of surface particles such as fish larvae and other plankton and buoyant pollutants such as oil spills. Their observations of dispersion can also be used to quantify the effect of mesoscale variability upon mean transports (c.f. Davis, 1991). Early studies of oceanic dispersion from drifter observations (Richardson, 1983; Davis, 1985; Krauss and Böning, 1987) addressed the relevance of Taylor's (1921) classical dispersion theory. Taylor demonstrated that the mean square distance that a particle will diffuse can be characterized in a homogeneous, stationary field of eddies by time and length scales known as the Lagrangian integral scales. These are in principle different from their Eulerian counterparts, although some mixing length-based studies have mapped eddy diffusivities using observations of Eulerian length scales (e.g., Stammer, 1998). Lumpkin *et al.* (2002) compared the Lagrangian and Eulerian length scales from surface drifters and altimetry, and found that they were proportional only in the most energetic parts of the ocean (such as the Gulf Stream) where Lagrangian particles are advected around eddies at time scales much shorter than the Eulerian time scale. In the tropical Pacific, Bauer *et al.* (2002) separated mean and eddy drifter velocities using optimized bicubic splines and found that eddy diffusivity in this region is strongly anisotropic: zonal diffusion is up to seven times larger than meridional diffusion, due to parcel trapping in coherent features such as equatorial and tropical instability waves. More recently, Zhurbas and Oh (2004) mapped the global surface diffusivity using a methodology designed to avoid contaminating eddy statistics by time-mean flow shear. Their maps of apparent diffusivity revealed enhanced values in zonal bands at $\sim 30^\circ$ in both southern and northern hemispheres. They attributed this to meandering, eddy-rich eastward currents in the North Atlantic (Azores Current) and North Pacific (North Subtropical Countercurrent), and to the westward drift of Agulhas retroflection eddies in the South Atlantic. In the South Pacific, they argued that the distribution indicated the presence of a South Subtropical Countercurrent.

Eddy statistics from SVP drifter observations can be compared to those of simulated drifters in general circulation models, in order to assess how well the eddy-resolving models simulate near-surface turbulence and associated Lagrangian scales. This type of analysis has shown that the $1/12^\circ$-resolution Miami Isopycnic Coordinate Ocean Model (MICOM) underestimates EKE in the Gulf Stream extension and ocean interior, perhaps due to the climatological forcing employed or the lack of vertical shear in the bulk mixed layer (Garraffo *et al.*, 2001). Similarly, the $1/10^\circ$-resolution Parallel Ocean Program (POP) model does not reproduce the observed Lagrangian scales, due to the model being too hydrodynamically stable in the surface layer (McClean *et al.*, 2002).

Drifter observations of current and SST are also extremely valuable when evaluating the role of oceanic processes in SST variations. These data have been used to show that tropical instability waves, zonal advection, storage and entrainment all play a major role in the heat budget of the Pacific (Hansen and Paul, 1987; Swenson and Hansen, 1999) and Atlantic (Foltz *et al.*, 2003) equatorial cold tongues.

A number of methodological advances have been published recently, adding value to the SVP drifter data set. For example, as noted earlier, many observations have been collected by non-SVP drifters, including the dense data set collected by FGGE drifters in the Southern Ocean. Recently, Pazan and Niiler (2004) announced a data set of ocean currents derived by combining these data after applying a drifter-dependent slip correction model (Pazan and Niiler, 2001). The types of drifters in this data set include SVP (both with and without drogue), FGGE (including later FGGE-type drifters deployed by the International Ice Patrol), and Navy-AN/WSQ-6.

Many oceanic regions have been sampled very heavily in some times of the year and lightly at other times, due to infrequent batch deployments from research vessels or ships of opportunity. In regions with strong seasonal fluctuations, such as the tropical Atlantic, this will produce biased estimates of the mean flow when the data are averaged in bins. Lumpkin (2003) addressed this problem by dividing the Lagrangian time series into spatial bins and, within each bin, decomposing the time series into time-mean, annual and semiannual harmonics, and a residual. This decomposition was performed using a Gauss–Markov technique familiar in the oceanographic literature for its use in box inverse models (e.g. Ganachaud and Wunsch, 2000). This approach can resolve the amplitudes and phases of seasonal fluctuations throughout most of the tropical Atlantic basin with the present density of SVP drifter observations (Lumpkin and Garraffo, 2005).

An extremely exciting recent development is the synthesis of SVP drifter observations and satellite altimetry. Altimetry is excellent at capturing the variations of sea level height associated with geostrophic velocity anomalies, but cannot yet map absolute sea level with sufficient accuracy to resolve time-mean currents. Furthermore, sea level anomaly fails to account for significant ageostrophic components of velocity variation, such as centrifugal effects that may account for differences in drifter-derived and altimeter-derived eddy kinetic energy (EKE) on either side of the Gulf Stream front (Fratantoni, 2001). In addition, gridded altimetric products smooth the observations and thus tend to systematically underestimate EKE. Niiler *et al.* (2003) describe a methodology to synthesize Ekman-removed drifter speeds with gridded altimetric velocity anomalies in regions where they are significantly correlated. This methodology uses concurrent drifter and altimetry velocities in order to calibrate the altimetry, making its amplitude consistent with the *in-situ* observations, and uses the continuous time series from altimetry to correct for biased drifter sampling of mesoscale to interannual variations. By applying this approach to the global data set of drifter observations, Niiler *et al.* (2004a) have produced a map of the absolute sea level of the global ocean for the period 1992–2002. In a similar study, Rio and Hernandez (2004) synthesized a geoid model, operational winds, and observations from drifters, altimetry, and hydrography to produce an alternative global mean dynamic topography (Rio and Hernandez, 2004). Future work will improve upon these efforts by using results from the Gravity Recovery and Climate Experiment (GRACE) and the rapidly growing data sets of Argo float profiles and drifter trajectories in remote regions such as the Southern Ocean.

2.7 The future

The design of the SVP drifter has clearly not ceased to evolve – exemplified by the recent introduction of the mini drifter – while its qualitative characteristics and water-following properties have remained relatively steady since the earliest deployments. Incremental improvements in design and manufacturing will continue to increase drifter lifetime, and alternative methods for detecting drogue presence (such as tether strain) are being considered. New methodologies for drifter data analysis will continue to be developed, aided by increasing information from the ever-growing drifter array and from other sources of complementary observations. Dense deployments in eddy-rich, frontal regions will help us improve our understanding of eddy fluxes and their role in modifying air-sea heat fluxes and water mass formation.

The quality of drifter data will also improve with updated interpolation schemes. As noted by Hansen and Poulain (1996), the structure functions used in the kriging interpolation are derived from tropical Pacific observations and assume uncorrelated zonal and meridional speeds. At high latitudes, these velocity components are often correlated due to looping motion in mesoscale eddies and inertial oscillations. Future interpolation schemes might exploit Lagrangian stochastic models with spin (e.g., Veneziani *et al.*, 2005) which can include this behavior. Hansen and Poulain (1996) considered a structure function for velocity which was consistent with diffusion, but found that the total (as opposed to eddy, or turbulent) velocity was not robustly interpolated. At the present density of drifter observations, seasonal climatologies of near-surface currents can be mapped at sufficient density to ameliorate this problem. Revised interpolation approaches will provide better error estimates of the interpolated positions and velocity, and may allow improved estimates of the Lagrangian acceleration $\mathrm{D}\boldsymbol{u} / \mathrm{D}t$.

In September 2005 the surface drifting buoy array of NOAA's Global Ocean Observing System and Global Climate Observing System consisted of 1250 SVP drifters at a nominal global resolution of $5^\circ \times 5^\circ$. The major challenge facing AOML's Drifter Operations Center, which coordinates drifter deployments, is to arrange deployments in regions of surface divergence and areas infrequently visited by research or volunteer observation vessels. This logistical challenge is being addressed by air deployments, increased international cooperation, and the development of tools to predict global drifter array coverage based on its present distribution and historical advection/dispersion. As the array grows, it provides invaluable observations of ocean dynamics, meteorological conditions and climate variations, and offers a platform to test experimental sensors measuring surface conductivity, rain rates, biochemical concentrations, and air-sea fluxes throughout the world's oceans.

Acknowledgments

The authors would like to thank the worldwide group of Global Drifter Program operational partners, whose contributions have established a truly global array of surface drifters. We would like to acknowledge Craig Engler, Jessica Redman and Erik Valdes for their frequent assistance. Discussions with Jeff Wingenroth and Sergey Motyzhev were extremely valuable. We thank Peter Niiler for many enlightening conversations, and for his comments on an early draft of this chapter which helped improve it tremendously. Additional comments by Claudia Schmid, Sang-Ki Lee and two anonymous reviewers were also valuable. NCEP Reanalysis 2 data are provided by the NOAA-CIRES Climate Diagnostics Center, Boulder, Colorado from their Web site at www.cdc.noaa.gov. This work was funded by NOAA's Office of Global Programs and the Atlantic Oceanographic and Meteorological Laboratory.

References

Argos, 1996. *Argos user manual 1.0*. Technical report, CLS, Toulouse, France.

Bauer, S., M. Swenson, and A. Griffa, 2002. Eddy-mean flow decomposition and eddy diffusivity estimates in the tropical Pacific Ocean. 2. Results. *J. Geophys. Res.*, **107**, 3154–71.

Bograd, S. J., A. B. Rabinovich, R. E. Thomson, and A. J. Eert, 1999. On sampling strategies and interpolation schemes for satellite-tracked drifters. *J. Atmos. Ocean. Tech.*, **16**, 893–904.

Centurioni, L. R. and P. P. Niiler, 2003. On the surface currents of the Caribbean Sea. *Geophys. Res. Lett.*, **30**, 1279, doi:10.1029/2002GL016231.

Centurioni, L. R., P. P. Niiler, and D. K. Lee, 2004. Observations of inflow of Philippine Sea water into the South China Sea through the Luzon Strait. *J. Phys. Oceanogr.*, **34**, 113–21.

Cuny, J., P. B. Rhines, P. P. Niiler, and S. Bacon, 2002. Labrador Sea boundary currents and the fate of the Irminger Sea Water. *J. Phys. Oceanogr.*, **32**, 627–47.

Davis, R., 1985. Drifter observations of coastal surface currents during CODE: The statistical and dynamical views. *J. Geophys. Res.*, **90**, 4756–72.

Davis, R., 1991. Observing the general circulation with floats. *Deep Sea Res.*, **38**, S531–71.

Ekman, V. W., 1905. On the influence of the Earth's rotation on ocean currents. *Ark. Mat. Astron. Fys.*, **2**, 1–52.

Flament, P. J., R. Lumpkin, J. Tournadre, R. Kloosterziel, and L. Armi, 2001. Vortex pairing in an unstable anticyclonic shear flow: discrete subharmonics of one pendulum day. *J. Fluid Mech.*, **440**, 401–10.

Flatau, M. K., L. Talley, and P. P. Niiler, 2003. The North Atlantic Oscillation, surface current velocities, and SST changes in the subpolar North Atlantic. *J. Climate*, **16**, 2355–69.

Foltz, G. R., S. A. Grodsky, J. A. Carton, and M. J. McPhaden, 2003. Seasonal mixed layer heat budget of the tropical Atlantic Ocean. *J. Geophys. Res.*, **108**, 3146, doi:10.1029/2002JC001584.

Franklin, B., 1785. Sundry marine observations. *Trans. Am. Philos. Soc., Ser. 1*, **2**, 294–329.

Fratantoni, D. M., 2001. North Atlantic surface circulation during the 1990s observed with satellite-tracked drifters. *J. Geophys. Res.*, **106**, 22067–93.

Fulton, T. W., 1897. The surface currents of the North Sea. *Scottish Geographical Mag.*, **13**, 636–45.
Ganachaud, A. S. and C. Wunsch, 2000. Improved estimates of global ocean circulation, heat transport and mixing from hydrographic data. *Nature*, **408**, 453–7.
Garraffo, Z., A. J. Mariano, A. Griffa, C. Veneziani, and E. Chassignet, 2001. Lagrangian data in a high resolution numerical simulation of the North Atlantic. I: Comparison with *in-situ* drifter data. *J. Mar. Sys.*, **29**, 157–76.
Garrett, J. F., 1980. The availability of the FGGE drifting buoy system data set. *Deep Sea Res.*, **27**, 1083–6.
Geyer, W. R., 1989. Field calibration of mixed-layer drifters. *J. Atmos. Ocean. Tech.*, **6**, 333–42.
Grodsky, S. A. and J. A. Carton, 2001a. Anomalous surface currents in the tropical Indian Ocean. *Geophys. Res. Letters*, **28**, 4207–10.
Grodsky, S. A. and J. A. Carton, 2001b. Intense surface currents in the Tropical Pacific during 1996–1998. *J. Geophys. Res.*, **106**, 16,673–84.
Grodsky, S. A. and J. A. Carton, 2002. Surface drifter pathways originating in the equatorial Atlantic cold tongue. *Geophys. Res. Lett.*, **29**, 2147, doi:10.1029/2002GL015788.
Hansen, D. and A. Herman, 1989. Temporal sampling requirements for surface drifting buoys in the tropical Pacific. *J. Atmos. Ocean. Tech.*, **6**, 599–607.
Hansen, D. and P.-M. Poulain, 1996. Quality control and interpolations of WOCE-TOGA drifter data. *J. Atmos. Ocean. Tech.*, **13**, 900–9.
Hansen, D. V. and C. A. Paul, 1987. Vertical motion in the eastern equatorial Pacific inferred from drifting buoys. *Oceanol. Acta*, **6**, 27–32.
Johnson, G. C., 2001. The Pacific Ocean subtropical cell surface limb. *Geophys. Res. Letters*, **28**, 1771–4.
Kelly, K. A., 2004. The relationship between oceanic heat transport and surface fluxes in the western North Pacific. *J. Climate*, **17**, 573–88.
Kennan, S., P. P. Niiler, and A. Sybrandy, 1998. Advances in drifting buoy technology. *International WOCE Newsletter*, **30**, 7–10.
Kessler, W. S., and B. A. Taft, 1987. Dynamic heights and zonal geostrophic transports in the central tropical Pacific during 1979–1984. *J. Phys. Oceanogr.*, **17**, 97–122.
Krauss, W. and C. Böning, 1987. Lagrangian properties of eddy fields in the northern North Atlantic as deduced from satellite-tracked buoys. *J. Mar. Res.*, **45**, 259–91.
Largier, J. and A. J. Boyd, 2001. Drifter observations of surface water transport in the Benguela Current during winter 1999. *S. Afr. J. Sci.*, **97**, 223–9.
Lee, D.-K., P. P. Niiler, S.-R. Lee, K. Kim, and H.-J. Lie, 2000. Energetics of the surface circulation of the Japan/East Sea. *J. Geophys. Res.*, **105**, 19561–73.
Letelier, R., M. Abbott, and D. Karl, 1996. Southern Ocean optical drifter experiment. *Antarct. J. US*, **30**, 108–10.
Levitus, S., 1982. Climatological atlas of the world ocean. *NOAA Professional Paper. U.S. Dept of Commerce National Oceanic and Atmospheric Administration*, **13**, 173.
Lumpkin, R., 2003. Decomposition of surface drifter observations in the Atlantic Ocean. *Geophys. Res. Lett.*, **30**, 1753, 10.1029/2003GL017519.
Lumpkin, R. and P. Flament, 2001. Lagrangian statistics in the central North Pacific. *J. Mar. Sys.*, **29**, 141–55.
Lumpkin, R. and Z. Garraffo, 2005. Evaluating the decomposition of Tropical Atlantic drifter observations. *J. Atmos. Oceanic Technol.*, **22**, 1403–15.

Lumpkin, R. and S. L. Garzoli, 2005. Near-surface circulation in the tropical Atlantic Ocean. *Deep Sea Res., Part I*, **52**, 495–518.

Lumpkin, R., A.-M. Treguier, and K. Speer, 2002. Lagrangian eddy scales in the northern Atlantic Ocean. *J. Phys. Oceanogr.*, **32**, 2425–40.

McClean, J. L., P.-M. Poulain, J. W. Pelton, and M. E. Maltrud, 2002. Eulerian and Lagrangian statistics from surface drifters and a high-resolution POP simulation of the North Atlantic. *J. Phys. Oceanogr.*, **32**, 2472–91.

McNally, G. J., W. C. Patzert, J. A. D. Kirwan, and A. C. Vastano, 1983. The near-surface circulation of the North Pacific using satellite tracked drifting buoys. *J. Geophys. Res.*, **88**, 7634–40.

Milliff, R. F., P. P. Niiler, J. Morzel, A. E. Sybrandy, D. Nychka, and W. G. Large, 2003. Mesoscale correlation length scales from NSCAT and Minimet surface wind retrievals in the Labrador Sea. *J. Atmos. Oceanic Technol.*, **20**, 513–33.

Nath, J. H., 1977. *Wind tunnel tests on drogues*. Technical report, NOAA Data Buoy Office.

Niiler, P., 2001. The world ocean surface circulation. In *Ocean Circulation and Climate*, ed. G. Siedler, J. Church, and J. Gould. San Diego: Academic Press, 193–204.

Niiler, P. P., 2003. A brief history of drifter technology. In *Autonomous and Lagrangian Platforms and Sensors Workshop*. La Jolla, CA: Scripps Institution of Oceanography.

Niiler, P. P., R. Davis, and H. White, 1987. Water-following characteristics of a mixed-layer drifter. *Deep Sea Res.*, **34**, 1867–82.

Niiler, P. P., N. A. Maximenko, and J. C. McWilliams, 2004a. Dynamically balanced absolute sea level of the global ocean derived from near-surface velocity observations. *Geophys. Res. Lett.*, **30**, 2164, doi:10.1029/2003GL018628.

Niiler, P. P., N. A. Maximenko, G. G. Panteleev, T. Yamagata, and D. B. Olson, 2003. Near-surface dynamical structure of the Kuroshio Extension. *J. Geophys. Res.*, **108**, 3193, doi:10.1029/2002JC001461.

Niiler, P. P. and J. D. Paduan, 1995. Wind-driven motions in the northeast Pacific as measured by Lagrangian drifters. *J. Phys. Oceanogr.*, **25**, 2819–30.

Niiler, P. P., W. Scuba, and D.-K. Lee, 2004b. Performance of Minimet wind drifters in Hurricane Fabian. *The Sea, J. Korean Soc. Oceanog.*, **9**, 130–6.

Niiler, P. P., A. Sybrandy, K. Bi, P. Poulain, and D. Bitterman, 1995. Measurements of the water-following capability of holey-sock and TRISTAR drifters. *Deep Sea Res.*, **42**, 1951–64.

Pazan, S. E. and P. P. Niiler, 2001. Recovery of near-surface velocity from undrogued drifters. *J. Atmos. Ocean. Tech.*, **18**, 476–89.

Pazan, S. E. and P. P. Niiler, 2004. New global drifter data set available. *EOS*, **85**, 17.

Poulain, P.-M., R. Barbanti, S. Motyzhev, and A. Zatsepin, 2004. Statistical description of the Black Sea near-surface circulation using drifters in 1999–2003. *Deep Sea Res.*, (submitted).

Rabinovich, A. B. and R. E. Thomson, 2001. Evidence of diurnal shelf waves in satellite-tracked drifter trajectories off the Kuril Islands. *J. Phys. Oceanogr.*, **31**, 2650–68.

Ralph, E. A. and P. P. Niiler, 1999. Wind-driven currents in the Tropical Pacific. *J. Phys. Oceanogr.*, **29**, 2121–29.

Reverdin, G., C. Frankignoul, E. Kestenare, and M. J. McPhaden, 1994. Seasonal variability in the surface currents of the equatorial Pacific. *J. Geophys. Res.*, **99**, 20323–44.

Reynolds, R. W., N. A. Rayner, T. M. Smith, D. C. Stokes, and W. Wang, 2002. An improved in situ and satellite SST analysis for climate. *J. Climate*, **15**, 1609–25.
Richardson, P., 1980. Gulf Stream ring trajectories. *J. Phys. Oceanogr.*, **10**, 90–104.
Richardson, P., 1983. Eddy kinetic energy in the North Atlantic from surface drifters. *J. Geophys. Res.*, **88**, 4355–67.
Richardson, P., 2004. Caribbean current and eddies as observed by surface drifters. *Deep Sea Res.*, (in press).
Richardson, P. and G. Reverdin, 1987. Seasonal cycle of velocity in the Atlantic North Equatorial Countercurrent as measured by surface drifters, current meters, and ship drifts. *J. Geophys. Res.*, **92**, 3691–708.
Richardson, P. and D. Walsh, 1986. Mapping climatological seasonal variations of surface currents in the Tropical Atlantic using ship drifts. *J. Geophys. Res.*, **91**, 10537–50.
Richardson, P. L. and T. K. McKee, 1984. Average seasonal variation of the Atlantic equatorial currents from historical ship drift. *J. Phys. Oceanogr.*, **14**, 1226–38.
Rio, M.-H. and F. Hernandez, 2003. High-frequency response of wind-driven currents measured by drifting buoys and altimetry over the world ocean. *J. Geophys. Res.*, **108**, 3283, doi:10.1029/2002JC001655.
Rio, M.-H. and F. Hernandez, 2004. A mean dynamic topography computed over the world ocean from altimetry, in situ measurements, and a geoid model. *J. Geophys. Res.*, **109**, doi:10.1029/2003JC002226.
Stammer, D., 1998. On eddy characteristics, eddy transports, and mean flow properties. *J. Phys. Oceanogr.*, **28**, 727–39.
Sverdrup, H. U., M. W. Johnson, and R. H. Fleming, 1942. *The Oceans: their Physics, Chemistry and General Biology*. Englewood Cliffs, NJ: Prentice-Hall.
Swenson, M. and D. V. Hansen, 1999. Tropical Pacific Ocean mixed layer heat budget: the Pacific Cold Tongue. *J. Phys. Oceanogr.*, **29**, 69–82.
Sybrandy, A. L. and P. P. Niiler: 1992, *WOCE/TOGA Lagrangian drifter construction manual.* WOCE Rep. 63, SIO Ref. 91/6. Scripps Inst. of Oceanogr., 58 pp.
Tait, J. B., 1930. *The water drift in the northern and middle area of the North Sea and in the Faroe-Shetland Channel.* Technical Report Scientific Investigation no. 4, Fishery Board for Scotland, Edinburgh.
Taylor, G., 1921. Diffusion by continuous movements. *Proc. Lond. Math. Soc.*, **20**, 196–212.
Thomson, C. W., 1877. *A Preliminary Account of the General Results of the Voyage of the HMS* Challenger. London: MacMillan.
Uda, M., 1935. The results of simultaneous oceanographical investigations in the North Pacific Ocean adjacent to Japan made in August, 1933. *Imperial Fisheries Exper. Sta. Journal*, **6**.
Veneziani, M., A. Griffa, Z. D. Garraffo, and E. P. Chassignet, 2005. Lagrangian spin parameter and coherent structures from trajectories released in a high-resolution ocean model. *J. Mar. Res.*, **63**, 753–88.
World Climate Research Program, 1988. *SVP-1 and TOGA pan-Pacific surface current study.* WOCE surface velocity program planning committee report of first meeting WMO/TD-No.323, WCRP-26, World Meteorological Organization, Wormley, 33 pp.
Yaremchuk, M. and T. Qu, 2004. Seasonal variability of the large-scale currents near the coast of the Philippines. *J. Phys. Oceanogr.*, **34**, 844–55.
Zhou, M., P. P. Niiler, and J.-H. Hu, 2002. Surface currents in the Bransfield and Gerlache Straits, Antarctica. *Deep Sea Res., Part I*, **49**, 267–80.

Zhou, M., J. D. Paduan, and P. P. Niiler, 2000. Surface currents in the Canary Basin from drifter observations. *J. Geophys. Res.*, **105**, 21893–911.

Zhurbas, V. and I. S. Oh, 2004. Drifter-derived maps of lateral diffusivity in the Pacific and Atlantic Oceans in relation to surface circulation patterns. *J. Geophys. Res.*, **109**, C05015, doi:10.1029/2003JC002241.

Zhurbas, V., A. G. Zatsepin, V. Yu, V. Grigoreva, V. N. Eremeev, V. V. Kremenetsky, S. V. Motyzhev, S. G. Poyarkov, P.-M. Poulain, S. V. Stanichny, and D. M. Soloviev, 2004. Water circulation and characteristics of currents of different scales in the upper layer of the Black Sea from drifter data. *Oceanology*, **44**, 30–43.

3
Favorite trajectories

In this chapter, a collection of "favorite trajectories" from various authors are presented.

While Lagrangian data analysis uses an extensive array of sophisticated tools, including classical statistics, dynamical system theory, stochastic modelling, assimilation techniques, and many others, visual inspection of individual trajectories still plays an important role, providing the first and often fundamental glimpse of the underlying dynamics. Often, for Lagrangian investigators, looking at trajectories gives the first intuition, then leading to the use of sophisticated and appropriate analysis. Trajectories tell the story of the journey of drifters and floats, and these stories are often complex and fascinating.

In the following sections, a number of investigators take us in the various world oceans, including Atlantic, Pacific and regional Seas, from the Poles to the Tropics, telling us the stories of their favorite trajectories and giving us their intuition and physical insights.

3.1 Mesoscale eddies in the Red Sea outflow region

AMY BOWER AND HEATHER FUREY

Department of Physical Oceanography, Woods Hole Oceanographic Institute, Woods Hole, Massachusetts, USA

In 2001–2002, 50 RAFOS floats were released at the core depth (~650 m) of Red Sea Outflow Water (RSOW) in the Gulf of Aden (northwestern Indian Ocean) as part of the Red Sea Outflow Experiment (REDSOX). The objective was to determine how warm, saline RSOW spreads from its source at the southern end of Bab al Mandeb Strait to the open Indian Ocean. Our hypothesis was that either boundary undercurrents or submesoscale coherent vortices (SCVs like Meddies, but here called "Reddies") were the main transport mechanisms for RSOW. Float releases were from two hydrographic cruises

Lagrangian Analysis and Prediction of Coastal and Ocean Dynamics, ed. A. Griffa, D. Kirwan, A. Mariano, T. Özgökmen, and T. Rossby. Published by Cambridge University Press.

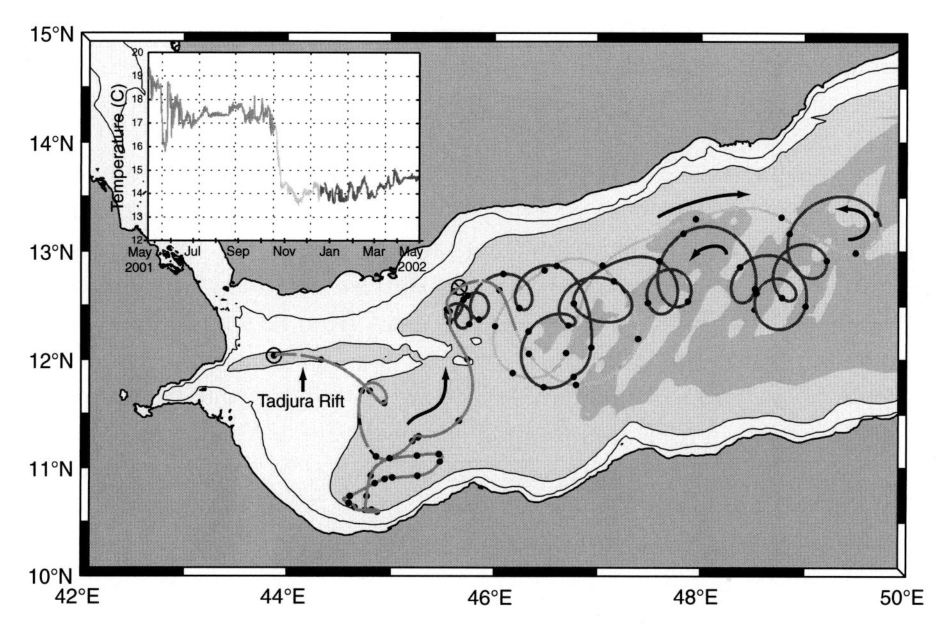

Figure 3.1 Track and temperature record of one RAFOS float in the Gulf of Aden. The multi-colored track reveals the rotation and translation of one cyclonic eddy from the mouth of the gulf westward. Colors allow the reader to match different segments of the track with the temperature record. Black dots indicate position every 10 days. Isobaths are 200 and every 1000 m. See Plate 12 for color version.

(February/March and August/September 2001) and from the sea floor using "float parks." Each float was tracked for one year using an array of five 780 Hz sound sources (see Furey *et al.* (2005) for more technical details).

The main result to emerge from REDSOX is that energetic, deep-reaching, mesoscale eddies in the Gulf of Aden, and not SCVs or boundary undercurrents, are likely responsible for most of the eastward spreading of RSOW through vigorous stirring and mixing. The multi-colored trajectory in Figure 3.1 illustrates the westward translation of one such cyclonic eddy.

The float was released from the sea floor in May 2001 in the far western gulf. Warm temperatures (>18 °C) indicate that it was initially embedded in recently injected outflow water from the previous winter. Temperature decreased somewhat as the float drifted into the southwestern corner of the gulf, where it slowly meandered around for several months. Around mid-October, the float made one large slow cyclonic loop, and temperature dropped by several degrees. Temperature remained relatively low as the float drifted eastward to the mouth of the gulf. Around the end of 2001, the float began looping rapidly in a coherent cyclonic eddy, with average looping period of 9 days and azimuthal speed of 20–30 cm/s. The eddy translated back toward the west over the next five months. Hydrographic and direct velocity observations in the eddies indicate that peak azimuthal speeds are ~50 cm/s near the surface and that the eddies extend nearly to the sea floor (Bower *et al.*, 2002). The smaller loops near the end of the trajectory (near 12° 30′ N, 46°E) suggest that the cyclonic eddy may have cleaved into multiple smaller eddies when it encountered the high seamounts at the eastern end of the Tadjura Rift. Other floats show similar cyclonic and anticyclonic eddies. We suggest that RSOW, injected into the open ocean at a western boundary where mesoscale eddy energy is typically elevated, spreads away from its source mainly by the stirring action of the eddy field, whereas Mediterranean Water, which is injected at an eastern boundary with very low background eddy energy, spreads (at least in part) in SCVs and narrow boundary undercurrents. Even if SCVs and boundary undercurrents carrying RSOW formed in the Gulf of Aden, they would likely not survive long in the presence of the vigorous eddy field in the Gulf of Aden and western Indian Ocean.

References

Bower, A. S., D. M. Fratantoni, W. E. Johns, and H. Peters, 2002. Gulf of Aden eddies and their impact on Red Sea Water. *Geophys. Res. Lett.*, **29**(21), 2025, doi:10.1029/2002GL015342.

Furey, H., A. Bower, and D. Fratantoni, 2005. *Red Sea Outflow Experiment (REDSOX): DLD2 RAFOS Float Data Report, February 2001–March 2003.* Woods Hole Oceanographic Institution Technical Report, WHOI-05-01.

3.2 Conservation of potential vorticity in the Gulf Stream–Deep Western Boundary Current crossover region

AMY BOWER AND HEATHER FUREY

Department of Physical Oceanography, Woods Hole Oceanographic Institute, Woods Hole, Massachusetts, USA

In 1994–95, 30 RAFOS floats were deployed in the Deep Western Boundary Current (DWBC) between the Grand Banks of Newfoundland and Cape Hatteras, North Carolina, split between two depths (~800 m and 3000 m). The objective of these deployments was to observe the spreading pathways and rates of water masses transported by the DWBC, particularly where the DWBC crosses under the Gulf Stream. The floats were tracked for two years using an array of 260 Hz sound sources.

One striking result from this study was the observation of potential vorticity conservation as the 3000-m floats crossed under the Gulf Stream. This is illustrated in Figure 3.2, which shows a segment of one deep float track,

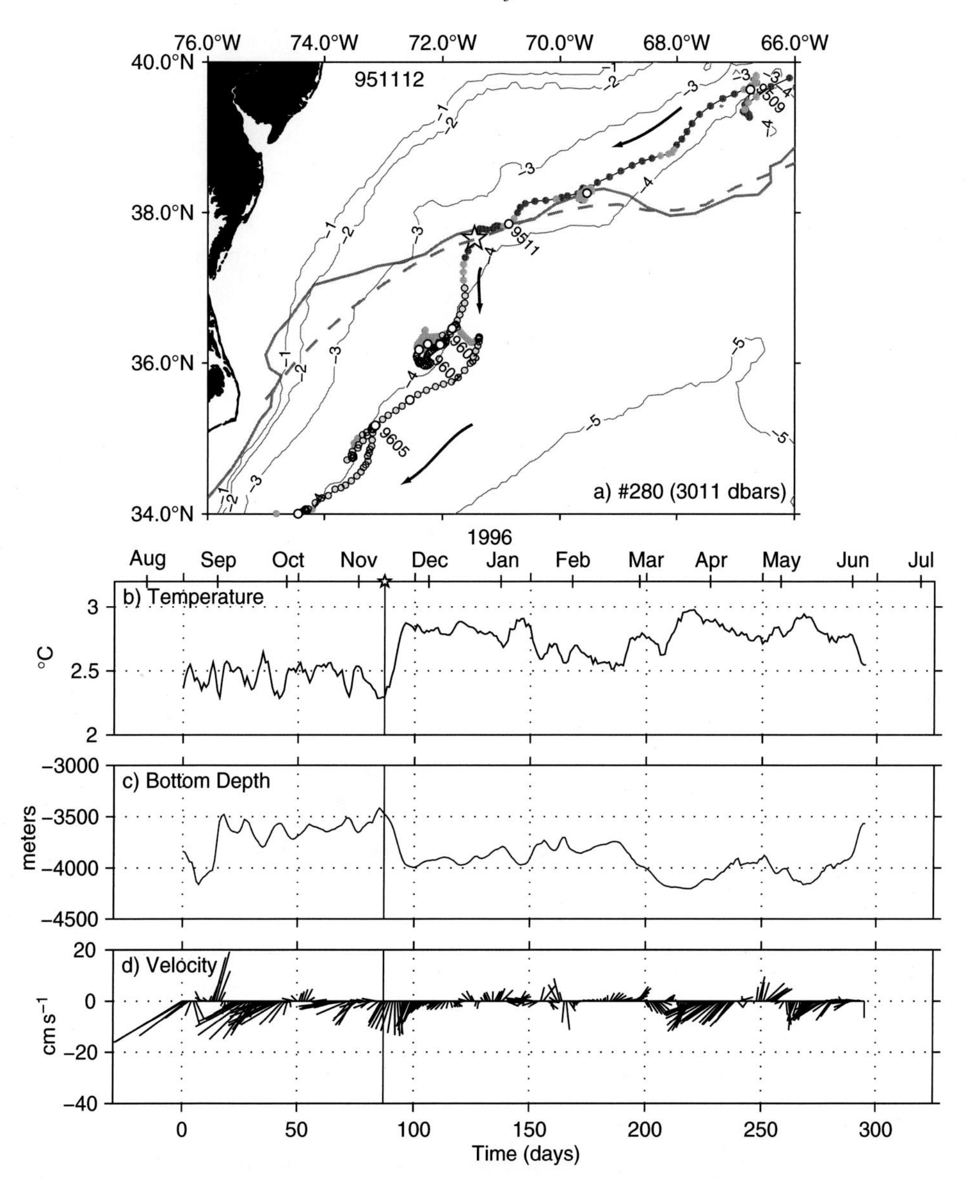

Figure 3.2 Trajectory segment, temperature, bottom depth and velocity of float #280, which was deployed at 3000 m in June 1995 over the continental slope south of Nova Scotia. Colored dots indicate float temperature in 0.5 °C intervals (e.g., blue T ≤ 2.5 °C, green 2.5 < T ≤ 3.0 °C, etc). Dashed red line shows long-term mean position of Gulf Stream north wall as observed with satellite AVHRR, the solid red line shows the instantaneous location of the north wall on November 12, 1995, and the star shows the float position on the same date. Isobaths are every 1000 m. Bottom depth at each float position was determined from the digital bathymetric data base ETOPO5. See Plate 13 for color version.

along with float temperature, bottom depth along the float track, and float velocity time series. Colored dots along the float track indicate temperature at daily positions.

North of the Gulf Stream, the float drifted southwestward between the 3000 and 4000 m isobaths (panel a), and measured relatively cold temperatures (<2.5 °C). On November 12, 1995, the float began to cross under the north wall of the Gulf Stream. This is apparent from the comparison of the float track with AVHRR imagery (panel a), and from the float's temperature record (panel b), which shows an abrupt increase in temperature (isotherms slope downward across the Gulf Stream at all depths). As the float crossed under the Gulf Stream, it moved offshore such that bottom depth under the float increased by ~500 m (panel c). This is approximately the same as the change in depth of the main thermocline across the Gulf Stream, indicating that the thickness of the layer below the main thermocline was approximately conserved as the float (and the deep water column it was following) crossed under the Gulf Stream. This is direct observational confirmation of the conceptual model put forward by Hogg and Stommel (1985), which indicated that DWBC fluid parcels in the lower layer of a two-layer system must migrate into deeper water in order to cross under the Gulf Stream. Other floats in this study showed similar behavior, and Bower and Hunt (2000) went on to show quantitatively the conservation of potential vorticity along the RAFOS float tracks.

References

Bower, A. S. and H. D. Hunt, 2000. Lagrangian observations of the Deep Western Boundary Current in the North Atlantic Ocean. Part II: The Gulf Stream–Deep Western Boundary Current crossover. *J. Phys. Oceanogr.*, **30**(5), 784–804.

Hogg, N. G. and H. Stommel, 1985. On the relation between the deep circulation and the Gulf Stream. *Deep-Sea Res.*, **32**, 1181–93.

3.3 Are there closed surface pathways in the tropical Atlantic?

SENYA GRODSKY AND JIM CARTON

University of Maryland, College Park, Maryland, USA

We find the drifter trajectory shown in Fig. 3.3 to be interesting in many ways. The drifter was deployed in late June 1997 in the central tropical north Atlantic. In the following three months it was transported by the North Equatorial Counter Current (NECC) into the Gulf of Guinea. Its further history is amazing. During the next 17 months the drifter was trapped in the eastern Gulf of Guinea. This is evident in the trajectory shape that displays

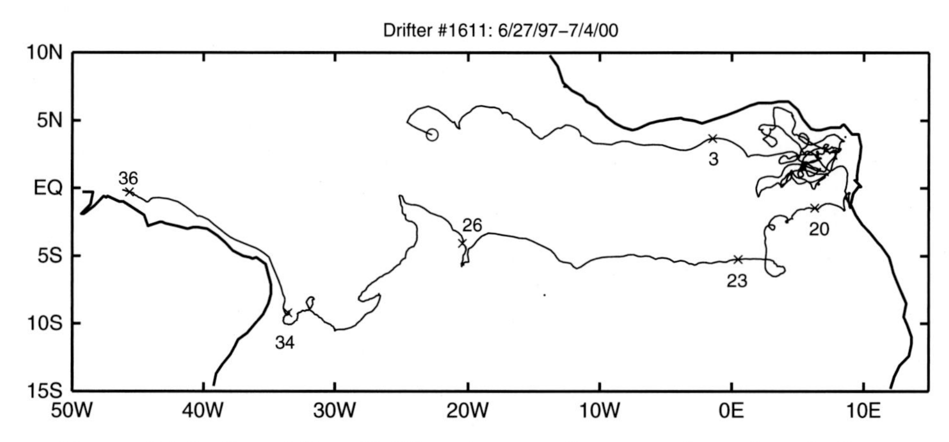

Figure 3.3 Surface drifter #1611 trajectory. Crosses and numbers indicate drifter location and time (in months) since deployment.

a stochastic spaghetti-like behavior indicative of slow currents. By February, 1999 (month #20) the drifter exited the Gulf of Guinea by crossing the equator and entering the southern branch of the South Equatorial Current. This suggests southward cross-equatorial currents in the eastern Gulf of Guinea in late boreal winter. During the long period (almost a year between months #20 and #34) the drifter crossed the south tropical Atlantic and entered the western boundary area at 10°S. During the next three months it was transported northwestward by the North Brazil Current (NBC) along the South America coast. Finally it expired near the Amazon at the end of the third year of service (probably due to time-life of power supply).

But one could easily imagine that if it had survived it might complete a closed loop by further transport in the NBC, then entering the NECC through the NBC retroflection area. A closed tropical Atlantic circulation cell could be hypothesized based on this unique trajectory.

3.4 Near-surface dispersion of particles in the South China Sea

LUCA R. CENTURIONI, PEARN P. NIILER, YOO YIN KIM
Scripps Institution of Oceanography, La Jolla, California, USA

DONG-KYU LEE
Busan National University, Busan, South Korea

VITALII A. SHEREMET
University of Rhode Island, Kingston, Rhode Island, USA

On May 7, 2005, 25 Surface Velocity Program drifters (Niiler, 2001) drogued at 15 m depth were deployed in the South China Sea (SCS), west of the Luzon

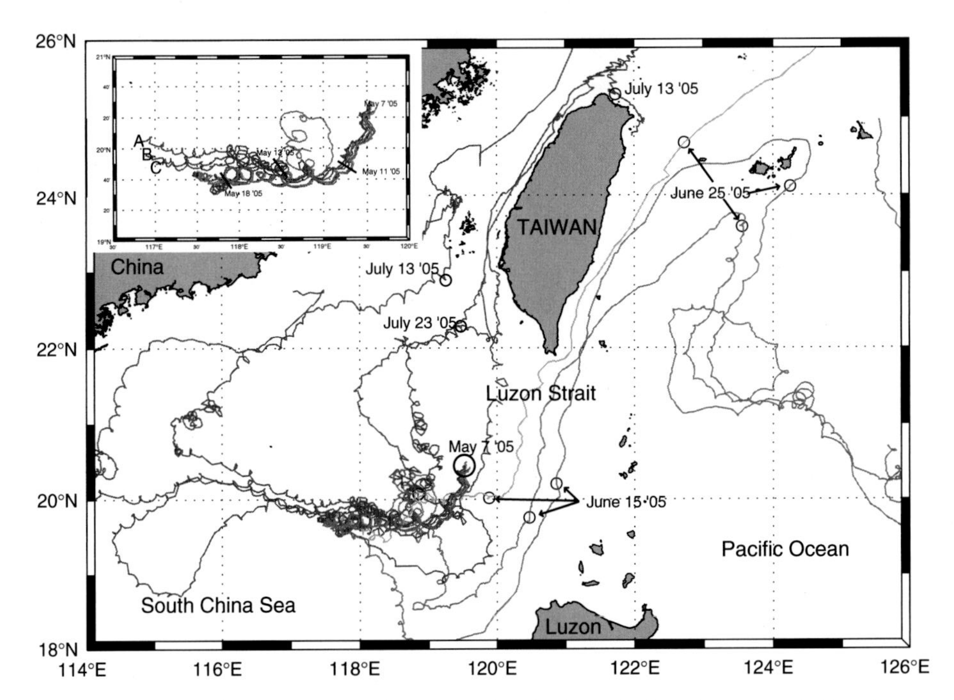

Figure 3.4 Trajectories of 8 drifters launched in May 2005 in the South China Sea west of the Luzon Strait. Blue (red and green) trajectories identify drifters that exited through the Taiwan Strait (Luzon Strait). See Plate 14 for color version.

Strait in a 5×5, 1.5 nm spacing, square array centered at 20° 27′ N, 119° 33′ E. All drifters mounted an Argos transmitter and a GPS receiver. The objective of this experiment was to measure the surface expression of the internal tide and non-linear internal waves that reach their maximum amplitude during spring tides (Ramp *et al.*, 2004).

The purpose of this note is to show a remarkable dispersion of near-surface particles that can occur in the SCS at the onset of the southwest monsoon. Eight selected trajectories (Figure 3.4) illustrate very different processes that occur in the SCS and at its boundaries, such as strong tides, vigorous stirring by the mesoscale eddy field and interaction of mesoscale structures with the Kuroshio (Centurioni *et al.*, 2004).

The drifters, deployed within 19 km of each other, moved coherently to the south-west/west for about 6 days until they started to separate into 3 groups. The southwest monsoon in 2005 was established by June 15 and the associated Ekman currents then push the drifters north-eastward. All drifters exited the SCS following very different routes. The ones that exited through the Taiwan Strait are shown in blue, while the red and the green trajectories are the drifters that exited through the Luzon Strait.

The inset in Figure 3.4 shows the details of the trajectories for 17 days after deployment. Cycloidal paths occur for the first 10 days and have the period of the diurnal tide. The drifters slowly separate into 2 groups as early as May 11, 4 days after deployment. On May 13 one drifter (red) re-circulates northward while the 3 other northernmost drifters (blue, labeled with A, B and C on the inset) slowly move apart from the rest of the array traveling westward. The remaining drifters describe a large tidal loop on May 18 (but smaller loops are present at later times) and then move very slowly. Absolute sea level plots (not shown) computed adding together the absolute mean sea level (Niiler *et al.*, 2003) and Ssalto/Duacs mean sea level anomalies suggest that the mesoscale eddy field has spatial scales of O(50–100 km) and time scales longer than 1 week, thus effectively stirring the drifters when their separation is comparable to the eddy scale. A saddle point on the sea level along the drifter paths starts to form on May 18 and evolves through May 25, dictating the different behavior and fate of the drifters labeled with A, B and C in the inset and the others.

The red drifters cross the Luzon Strait around June 15 (it takes longer – between 2 to 6 weeks – for the others to reach the Taiwan Strait). Their trajectories and the sea level maps suggest that a northward current northwest of Luzon interacts with the Kuroshio and exports SCS water across the Luzon Strait. Once in the Pacific Ocean the red drifters are caught in a persistent cyclonic north/south elongated eddy that occurs east of Taiwan and is connected to the Kuroshio. The red drifters rapidly quit the boundary current and make a deep excursion, in excess of 500 km, to the south. The strong eddy and its genesis are clearly visible in sea level maps from the beginning of June 2005 through mid-August 2005.

References

Centurioni, L. R., P. P. Niiler, and D. K. Lee, 2004. Observations of inflow of Philippine Sea surface water into the South China Sea through the Luzon Strait. *J. Phys. Oceanogr.*, **34**, 113–21.

Niiler, P. P., 2001. The world ocean surface circulation. In *Ocean Circulation and Climate: Observing and Modeling the Global Ocean*, ed. J. Church, G. Siedler, and J. Gould. San Diego: Academic Press, 193–204.

Niiler, P. P., N. A. Maximenko, and J. C. McWilliams, 2003. Dynamically balanced absolute sea level of the global ocean derived from near-surface velocity observations. *Geophys. Res. Lett.*, **30**, 22, 2164, doi:10.1029/2003GL018628.

Ramp, S. R., T.-T. Tang, T. Duda, J. Lynch, A. Liu, C. S. Chiu, F. Bahr, H. R. Kim, and Y. J. Yang, 2004. Internal solitons in the Northeastern South China Sea. Part I: Sources and deep water propagation. *IEEE J. Ocean Eng.*, **29**, 4, 1157–81.

3.5 The Naval Postgraduate School RAFOS Study

NEWELL GARFIELD
San Francisco State University, Tiburon, California, USA
CURTIS A. COLLINS
Department of Oceanography, Naval Postgraduate School, Monterey, California, USA
THOMAS A. RAGO
Department of Oceanography, Naval Postgraduate School, Monterey, California, USA

Figure 3.5 shows the first triad of RAFOS floats launched west of San Francisco, California, by the Naval Postgraduate School's Lagrangian study of the California Undercurrent (Collins *et al.*, 1996 and 2004, Garfield *et al.*, 1999 and 2001). These are our favorite trajectories because this first triad deployment clearly displays the major results of the study: (1) steady poleward flow in the Undercurrent (float NPS5), (2) reversing parallel flow that remained near the continental margin (float NPS3), and (3) westward migration, usually in an anticyclonic submesoscale eddy (SCV), in the ocean interior (float NPS7). These floats were launched July 7, 1993, and surfaced September 5, 1993. In the figure, daily fixes are superimposed on a NOAA POES AVHRR image for August 27, 1993. The daily fixes are open circles, the filled circles are the locations of the floats on the day of the satellite image. On that day NPS3 was at the southern extreme of its trajectory, just starting the return to the north. NPS5 was in an anticyclonic eddy just south of Cape Blanco, where there is a surface signature of cold and warm water deformed anticyclonically. NPS7 was in a warm anticyclonic feature west of San Francisco.

The primary launch site was west of San Francisco, California, at 37° 47′ N, 123° 30′ W, where float triads were launched in a triangular pattern eight kilometers on a side. Two floats were launched above the 1000 m isobath, while the third was launched at the offshore apex of the equilateral triangle. The target pressure surface for this triad was 350 dbar which was the surface attained by floats 3 and 7. Float 5, the float transported in the California Undercurrent, settled to 135 dbar.

The program has documented the flow of the California Undercurrent. The unexpected result is the prevalence of the submesoscale coherent eddies which the data suggest are created in and transported by the Undercurrent. In adding to the "eddy zoo" of eastern boundary formed eddies, we suggested that the Undercurrent SCVs be referred to as "cuddies" (California Undercurrent eddies). Once formed and separated from the Undercurrent, cuddy translation is markedly different than the surface flow. Upon detachment from the Undercurrent, all cuddies moved primarily westward. The

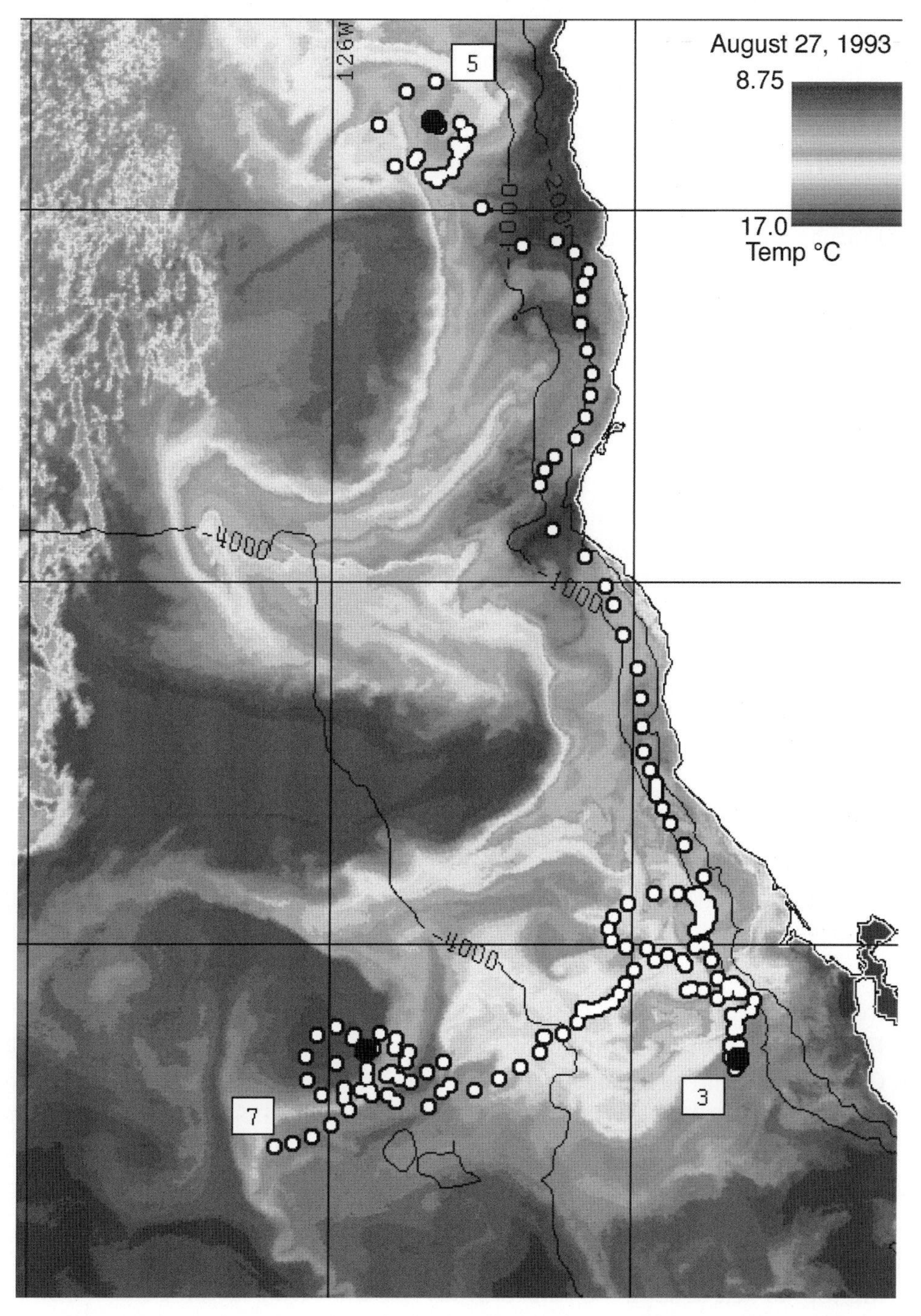

Figure 3.5 Trajectories of a triad of RAFOS floats (NPS5, NPS3, NPS7) launched west of S. Francisco, California in July 1993. See Plate 15 for color version.

westward migration and long duration may explain the offshore subsurface variability in water types observed between the 26 and 27 kg/m^3 density anomaly surfaces. In the California Current region off central California, this corresponds to a depth range of about 150–600 m. The variability may exist because the Undercurrent anticyclonic eddies allow the equatorial Pacific water signature to remain intact for the lifetime of the eddies, providing a mechanism for delivering the equatorial Pacific water offshore.

References

Collins, C. A., N. Garfield, R. G. Paquette, and E. Carter, 1996. Lagrangian measurements of subsurface poleward flow between 38° N and 43° N along the west coast of the United States during summer, 1993. *Geophy. Res. Lett.*, **23**(18), 2461–4.

Collins, C. A., L. M. Ivanov, O. V. Melnichenko, and N. Garfield, 2004. California Undercurrent variability and eddy transport estimated from RAFOS float observations. *J. Geophy. Res.*, in press.

Garfield, N., C. A. Collins, R. G. Paquette, and E. Carter, 1999. Lagrangian exploration of the California Undercurrent, 1992–1995. *J. Phys. Oceanogr.*, **29**(4), 560–83.

Garfield, N., M. E. Maltrud, C. A. Collins, T. A. Rago, and R. G. Paquette, 2001. Lagrangian flow in the California Undercurrent, an observation and model comparison. *J. Mar. Sys.*, **29**, 201–20.

3.6 Favorite drifter trajectories deployed from the western shelf of Florida and the coastal waters of the Florida Keys

VASSILIKI KOURAFALOU, ELIZABETH WILLIAMS, AND THOMAS LEE
Rosenstiel School of Marine and Atmospheric Science, University of Miami, Miami, Florida, USA

From 1995 to 2003, forty seven shallow water, satellite-tracked surface drifters were deployed bimonthly in the Shark River plume at the southwestern tip of Florida just north of Cape Sable. The drifter study is part of the Florida Circulation and Exchange Project of the South Florida Ecosystem Restoration Program. The Technocean ARGO satellite-tracked drifters were chosen for the study due to their shallow water design and relative low cost per unit. They consist of a Telonics ARGOS transmitter, transmitting every 90 seconds at the standard ARGOS frequency of 982.5 mHz (mounted inside the hull of the buoy with a battery pack) and a 2 foot antenna (mounted on top of the buoy). At the latitude of the study area, polar orbiting satellites fitted with ARGOS receivers pass over approximately six to eight times per day resulting in approximately the same number of position fixes per day.

The drifters released near the Shark River mouth (a wide shallow discharge through the Florida Everglades) are a good indicator of the brackish water pathways and the near surface regional circulation at large. The favorite trajectories presented here highlight the regional circulation, which is seasonally dependent, and the connectivity of the west Florida shelf circulation to the larger scale Gulf of Mexico and Straits of Florida currents. As a brief background on the circulation (see also Lee *et al.*, 2002), we note the following. On the wide southwest Florida shelf, the sub-tidal circulation is mainly wind-driven. Near the western boundary of Florida Bay (between Cape Sable and the Florida Keys), drifters are subject to a mean southward, slope-driven current that is due to the presence of the Florida Keys chain (sea level response to wind forcing is opposite for each side of the Keys) and to the overall higher sea level standing in the Gulf of Mexico, as compared to the Atlantic. Further west, the circulation in the eastern Gulf of Mexico is controlled by the presence of the Loop Current (LC), and the cold, cyclonic eddies (diameters about 100–200 km) that travel around the LC front and enter the Straits of Florida in the vicinity of the Dry Tortugas. The drifters may be retained in the Tortugas area, depending on the presence and location of the Tortugas gyre. Then, they travel with the Florida Current (FC) and, depending on the interaction of the FC with the Florida Keys Atlantic shelf, they either continue northward or return southward, if they enter a coastal wind-driven current along the Florida Keys.

Drifters 1 and 2 (Fig. 3.6a and b respectively) were released in winter and they were initially influenced by the prevailing northerly to northwesterly wind component that is characteristic of the atmospheric frontal passages that are frequent in this season. Drifter 1 traveled westward until it was caught up in the LC and the Tortugas gyre. Drifter 2 traveled southward, along the western boundary of Florida Bay, then it crossed the Keys passages to the Atlantic shelf into the coastal circulation along the Florida Keys until it reached the Dry Tortugas area where it reversed direction following the FC through the Florida Straits.

Drifter 3 (Fig. 3.6c) was a summer release and it followed a characteristic pathway for this season: a slow northwestern drift from the deployment location until it reached about 27 N, then entrainment in the Gulf of Mexico circulation through a southwesterly shelf break current and the Loop Current – Florida Current system.

Drifters 4 and 5 (Fig. 3.6d and e respectively) were both released in the fall, but exhibited very different patterns. They both started with westerly to southwesterly flow on the southern part of the west Florida shelf, but only Drifter 4 continued on a seasonally “typical” pathway that after reaching the Tortugas

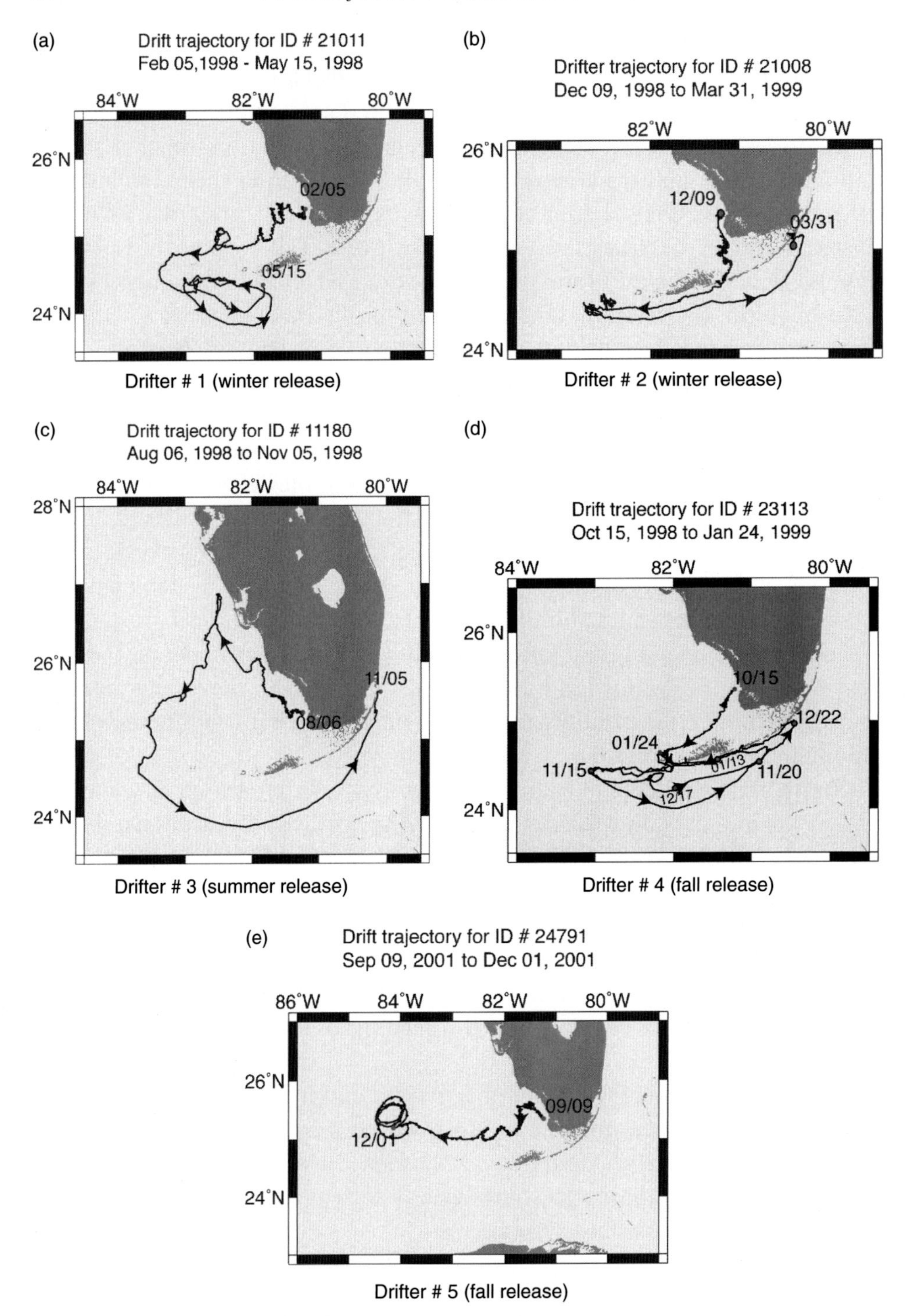

Figure 3.6 Trajectories of drifters launched from the western shelf of Florida in the framework of the Florida Circulation and Exchange Project (1995–2003). Drifters 1 and 2 have been launched during winter, drifter 3 during summer and drifters 4 and 5 during fall. See Plate 16 for color version.

turned cyclonically around the Tortugas eddy, then traveled with the FC (that had been displaced southward) in the Florida Straits. Still, Drifter 4 is quite particular because it gave clear evidence of the intense interactions between shelf and oceanic flows in the Florida Straits. The drifter was first removed from the FC (presumably through eddy to shelf interaction) in the middle Keys and traveled back to the Tortugas through the coastal southwestward current. The whole pattern (Tortugas–FC–southward coastal current) was repeated; the second time around (about a month later) the FC had moved closer to the Keys and the drifter approached the upper Keys before reversing course. Drifter 5 is unique in its continuous southwestward displacement and direct entrainment into an LC eddy.

Acknowledgments

Support for this study was provided by NOAA/CIMAS through the South Florida Program SFP2004, Contract NA17RJ1226.

References

Lee, T. N., E. Williams, E. Johns, D. Wilson, and N. Smith, 2002. Transport processes linking South Florida coastal ecosystems. In *The Everglades, Florida Bay and Coral Reefs of the Florida Keys: An Ecosystem Source Book*, ed. J. Porter and K. Porter. Boca Raton, FL: CRC Press, 309–341.

3.7 On the Intermediate Circulation in the Iceland Basin

MATTHIAS LANKHORST AND WALTER ZENK
Leibniz-Institut für Meereswissenschaften (IFM-GEOMAR), Kiel, Germany

Examples of eddy-resolving RAFOS tracks in the Iceland Basin are shown in Fig. 3.7. These floats were launched in context with the Kiel SFB-460[1] project at the level of the low-saline Labrador Sea Water (LSW), i.e. 1600 m. The latter represents a water mass that is formed convectively in the Labrador Sea each winter and enters the eastern basins mainly through Charlie-Gibbs Fracture Zone (CGFZ). Farther to the north, LSW encounters Iceland Scotland Overflow Water (ISOW) invading the Iceland Basin from the Faroe Bank Channel (FBC). The center of the basin is characterized by enhanced eddy activities.

[1] Sonderforschungsbereich (SFB, collaborative research center) is a type of long-term research initiative funded by the Deutsche Forschungsgemeinschaft, Bonn, Germany.

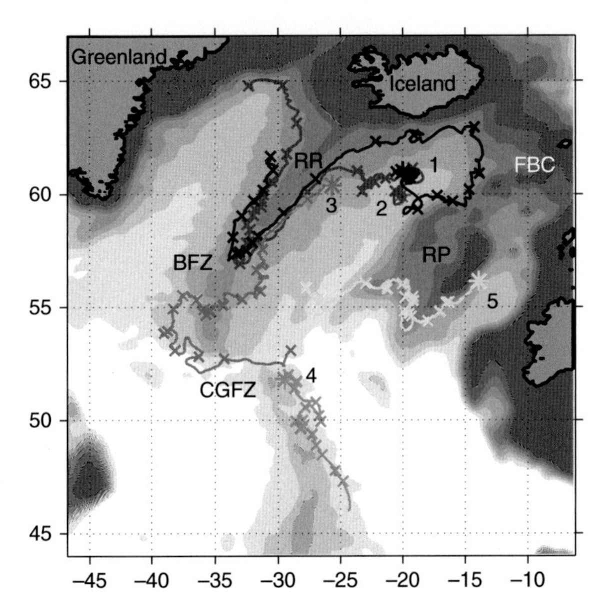

Figure 3.7 Trajectories of floats launched in the Iceland basin in the context of the Kiel SFB-460 project. See Plate 17 for color version.

In total, there were 68 floats launched. Their trajectories can be found at: www.ifm-geomar.de/index.php?id=999&L=1

All trajectories (Figure 3.7) start at the asterisk symbols. They are synthesized from daily positions. Additional ticks are 30 days apart. Observation periods lasted from one to two years. Bottom contours are given by shading every 500 m.

Findings:

1 (Blue): After leaving the central Iceland Basin (high eddy kinetic energy), this float is caught by the ISOW jet just south of Iceland. Following the topography of the Reykjanes Ridge (RR) with high speeds (>10 cm/s), it approaches Bight Fracture Zone (BFZ) and enters the intermediate level of the Irminger Sea where it is advected farther northward towards Denmark Strait.

2 (Red): This float reproduces the pathway of the blue float, except that it is entrained directly by the contour current at the flank of the Reykjanes Ridge. It stresses the fact that BFZ apparently is permeable for some ISOW depending on its earlier entrainment history and, hence, density level of this modified ISOW. Note that the canonical exit for ISOW as found in the literature concentrates on the CGFZ, some 400 km farther south.

3 (Green): This float represents two totally different pathways. In its early "life" it verifies the pathway of entrained ISOW along the ridge including its

export in the newly observed region well north of CGFZ. Instead of being advected northeastward on the western side of the ridge, however, it then is entrained by LSW and returns to the eastern basin via CGFZ.

4 (Magenta): In the center region of the Atlantic along the mid oceanic ridge system, the existence of a deep western boundary current east of the ridge has been postulated. This float shows its path between CGFZ and the Azores Plateau.

5 (Cyan): Some of the ISOW is spilt into Rockall Trough, as opposed to following FBC. This trajectory shows this pathway as it continues counterclockwise around Rockall Plateau (RP).

References

Lankhorst, M. and W. Zenk, 2006. Lagrangian observations of the middepth and deep velocity fields of the Northeastern Atlantic Ocean. *J. of Phys. Oceanogr.*, **36**(1), 43–63.

Machín, F., U. Send, and W. Zenk, 2006. Intercomparing drifts from RAFOS and profiling floats in the deep western boundary current along the Mid-Atlantic Ridge. *Scientia Marina*, **70**(1), 1–8.

3.8 Where is the diffusivity?

ARTHUR J. MARIANO AND EDWARD H. RYAN

Rosenstiel School of Marine and Atmospheric Science, University of Miami, Miami, Florida, USA

Tom Rossby deployed SOFAR floats 64, 67, and 89 in the main thermocline of the POLYMODE Local Dynamics Experiment region at a nominal depth of 700 m during the summer of 1978. The top panel of Figure 3.8 shows the trajectories of the three floats from August 3 to October 17, 1978 with arrows every ten days. There is a general displacement of the floats to the east-southeast, presumably due to advection by the Northwest Atlantic Subtropical Counter Current (NASCC). The arithmetic average float velocity for all these floats is (10.1, −1.3) cm/s. At the time of deployment of these floats, most oceanographers would have predicted a mean west-southwest displacement due to flow of the Gulf Stream recirculation. This cluster and other float trajectories (Rossby *et al.*, 1983), as well as Reid's (1978) analysis of hydrographic data, indicate a net eastward motion embedded in the large-scale westward flow of the North Atlantic subtropical gyre. The confirmation of Reid's indirect geostrophic velocity calculation by direct current measurements forced us to rethink our simple views of ocean circulation.

Another striking feature of this float cluster is the lack of relative dispersion; after two months, the floats are about the same distance from each other as

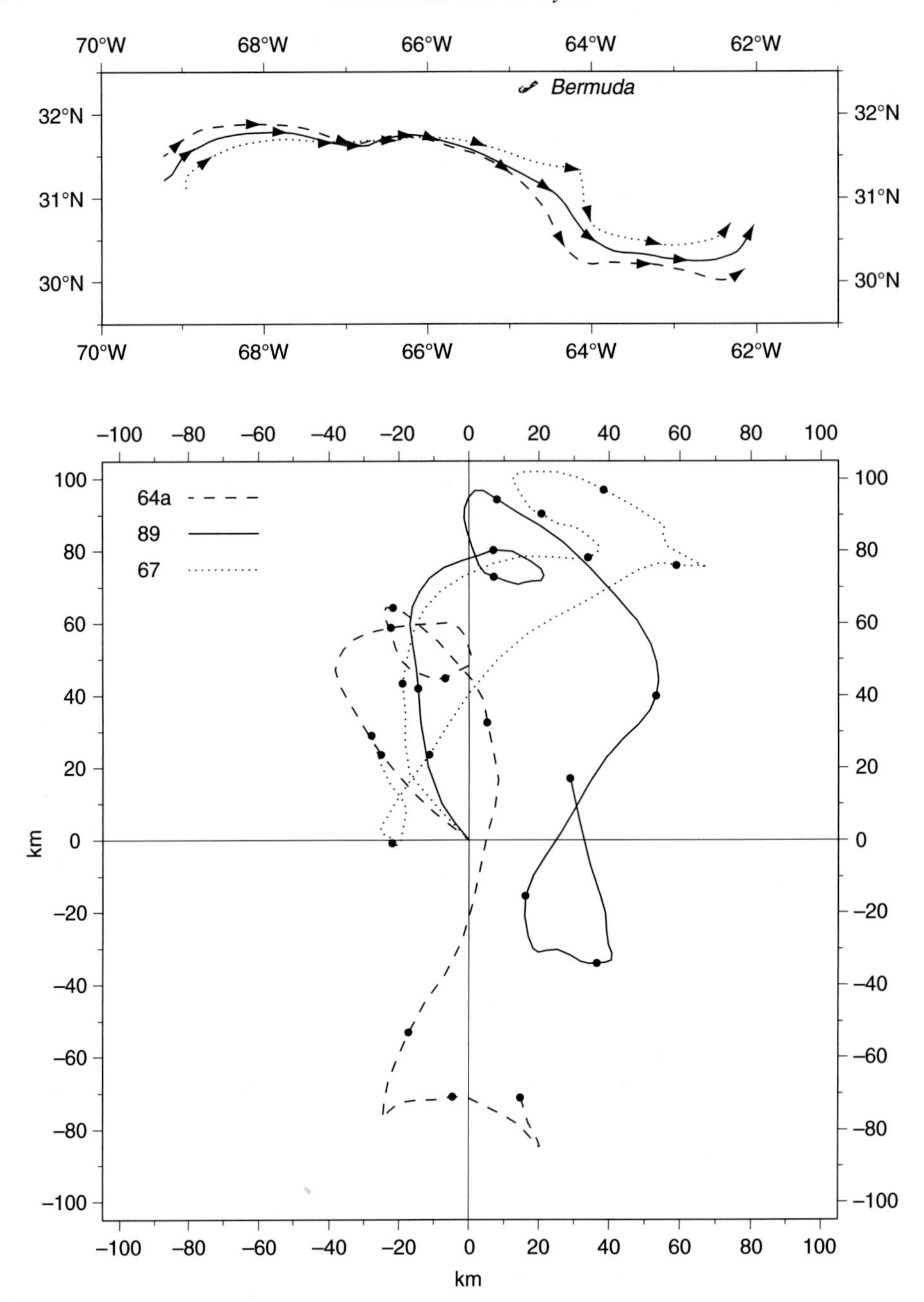

Figure 3.8 Upper panel: trajectories of 3 SOFAR floats (64, 67, 89) launched in the POLYMODE region in summer 1971. Lower panel: same but with mean translational velocity removed.

when they were launched. Given the lack of looping motion in the trajectories, particle trapping in a coherent vortex is probably not the reason for the low dispersion. A plot of the residual motion (lower panel Fig. 3.8) about the mean translational velocity (10.1, −1.3 cm/s) from the float's initial location (lower figure) reveals no dominant sense of rotation and no dominant scale. Another possible explanation is that the floats are in the core of the NASCC and there is little particle exchange across the subtropical front. The plot of the relative motion does not support this view. One can speculate that the floats are in another type of flow feature and that the identification and modeling of this flow feature will require new analysis tools, possibly in a Lagrangian-based coordinate system, such as those discussed in Bennett (2005) and Paldor (Chapter 5).

References

Bennett, A. F., 2005. *Lagrangian Fluid Dynamics*. Cambridge: Cambridge University Press.
Reid, J. L., 1978. On the mid-depth circulation and salinity field in the North Atlantic Ocean. *J. Geophys. Res.*, **83**, 5063–67.
Rossby, H. T., S. C. Riser, and A. J. Mariano, 1983. The Western North Atlantic – a Lagrangian viewpoint. In *Eddies in Marine Science*, ed. A. R. Robinson. Heidelberg: Springer-Verlag, 66–91.

3.9 Opposing trajectories in the Mediterranean!

PIERRE-MARIE POULAIN
Istituto Nazionale di Oceanografia e di Geofisica Sperimentale (OGS), Trieste, Italy

Figure 3.9 shows the partial trajectories of two CODE surface drifters. The first drifter (dotted line) was located south of the southern tip of Sicily on September 1996 (∗ star), moved northeastward in the northern Ionian, entered the Adriatic Sea on the eastern flank of the Strait of Otranto and ended up in the central Adriatic (near longitude 17°E) on December 4, 1996. The second (dashed line) was in the central Adriatic on 9 September 1997 (∗ star), moved to the southeast with the Western Adriatic Current, escaped into the Ionian on the western side of the Strait of Otranto, proceeded generally to the southwest while being entrapped in several eddy structures, before reaching the region south of the southern tip of Sicily on January 19, 1998.

In three to four months, these two drifters have connected the central Adriatic to the Strait of Sicily regions, and vice versa. These generally opposing trajectories are mainly the result of a drastic change in the mean circulation patterns in the northern Ionian Sea. This change occurred in 1997 and is related to the interannual variability of the whole Mediterranean Sea.

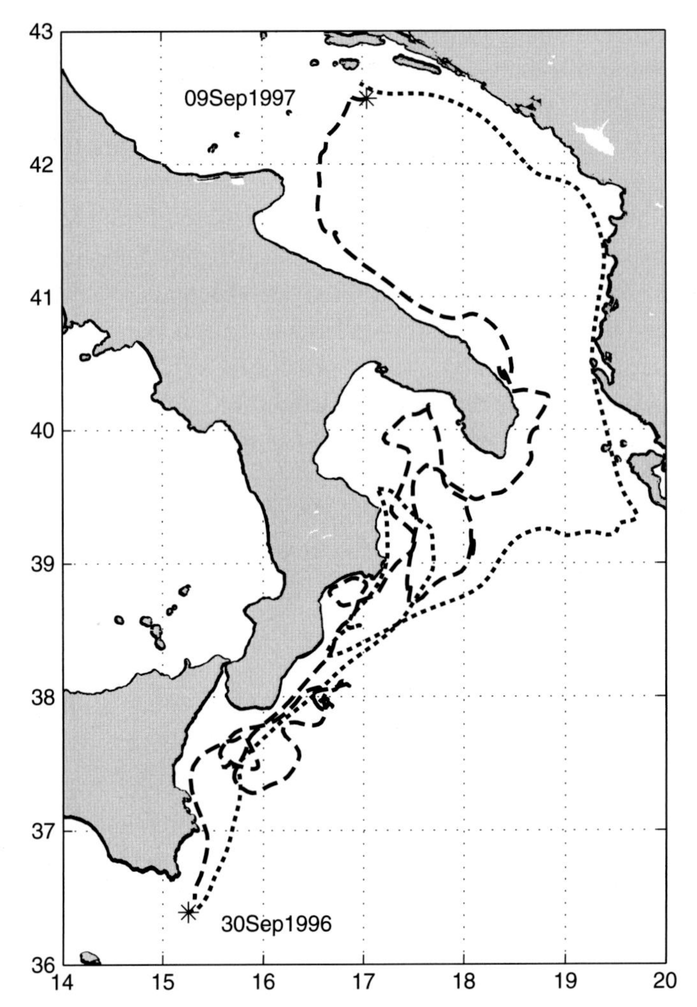

Figure 3.9 Trajectories of two surface CODE drifters launched in the Mediterranean Sea east of Italy in fall 1996 (dotted line) and in fall 1997 (dashed line).

3.10 Tracking the sub-polar gyre

HEDINN VALDIMARSSON AND SVEND-AAGE MALMBERG
Marine Research Institute, Reykjavik, Iceland

Drifter 23508 was deployed in Faxafloi off the west coast of Iceland (see Figure 3.10). This drifter was one of 120 drifters deployed in a SVP WOCE project that was a cooperation between Peter Niiler at the Scripps Institute of Oceanography and the Marine Research Institute in Reykjavík (MRI). Like so many of the drifters deployed at the southwest banks of

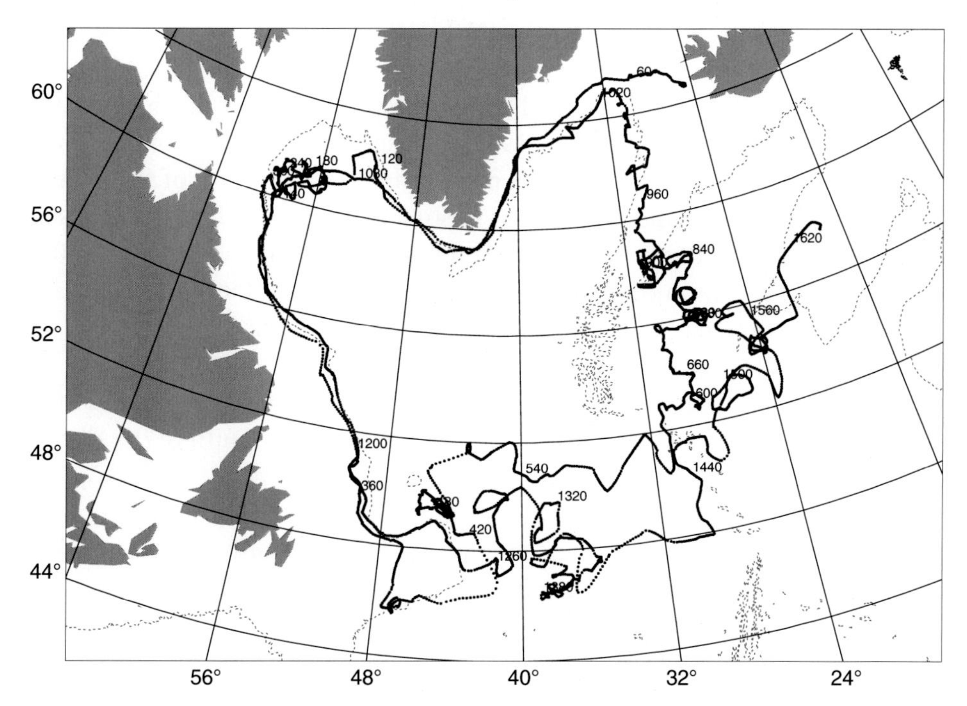

Figure 3.10 Track of drifter 23508. Number indicates drifting days and depth contours of 2000 m depth.

Iceland it drifted west over to Greenland (Valdimarsson and Malmberg, 1999). This drifter became after a while a favorite and the focus of the attention of our group as it sailed the seas for an unusually long time and was followed from day to day like an explorer. The travel of this drifter showing such stamina was described in Malmberg and Valdimarsson (1999) and will be reviewed and updated here. Velocities and temperatures mentioned are means from data processed by the Drifting Buoy Data Assembly Center (Hansen and Poulain, 1996).

From deployment day, August 30, 1995, it drifted slowly northwards over the Icelandic shelf for about 40 days when it quickly changed speed and direction in the southern part of the Denmark Strait (DS) and turned southwest along the East Greenland Current (EGC). After the turn the six hourly mean velocity of the drifter went from 10–20 cm/s up to 30–50 cm/s and the water temperature dropped from 9 °C to 5 °C.

The trip from the Denmark Strait to south of Cape Farwell took about 40 days. Thereafter it drifted in the West Greenland Current and experienced the lowest temperatures on its track (0.2 °C). After spending over six months (July 1996) in the northern Labrador Sea, north of 60°N, the route followed the Labrador Current all the way to south of 46°N. On this part of the trip the mean velocities ranged from 2 to 100 cm/s, with highest velocities around 55°N, over a period of about 3 months. From the Labrador Current (September 1996) the drifter travelled around the Flemish Cap, got stuck in eddies north of the Cap and then at the end of January 1997 it headed eastwards with the North Atlantic Current. It drifted over the mid-Atlantic Ridge south of the Charlie-Gibbs Fracture Zone and was caught in some persistent eddies in the western part of the Iceland Basin for several months especially between 55 and 60°N. The drifter then continued northwards (March 1998) over the Reykjanes Ridge and into the Irminger Sea to complete the trip around the Subpolar Gyre in a little under 3 years in June 1998.

The drifter continued another circle around the Subpolar Gyre and followed closely the former track but in a shorter time until it was south of the Flemish Cap for the second time (January 1999). It turned east north of the Cap and drifted east along 48°N over to the Iceland Basin further south than before. This drifter finished its "lifetime" then on the Hatton Bank in February 2000 after patiently following the currents for four and a half years.

References

Hansen, D. V. and P.-Marie Poulain, 1996. Quality control and interpolations of WOCE/TOGA drifter data. *J. Atmos. Oceanic Tec.*, **13**, 900–9.

Malmberg, S.-A and H. Valdimarsson, 1999. Satellite tracked surface drifters and "Great Salinity Anomalies" in the subpolar gyre and the Norwegian Sea. *Intl. WOCE Newsletter*, **37**, 31–3.

Valdimarsson, H. and S.-A. Malmberg, 1999. Near-surface circulation in Icelandic waters derived from satellite tracked drifters. *Rit Fiskideildar*, **16**, 23–39.

4

Particle motion in a sea of eddies

CLAUDIA PASQUERO
ESS, University of California, Irvine, California, USA

ANNALISA BRACCO
Physical Oceanography Dept., Woods Hole Oceanographic Institute, Woods Hole, Massachusetts, USA

ANTONELLO PROVENZALE
ISAC-CNR, Torino, CIMA, Savona, Italy

AND

JEFFREY B. WEISS
PAOS, University of Colorado, Boulder, Colorado, USA

Abstract

As more high-resolution observations become available, our view of ocean mesoscale turbulence more closely becomes that of a "sea of eddies." The presence of the coherent vortices significantly affects the dynamics and the statistical properties of mesoscale flows, with important consequences on tracer dispersion and ocean stirring and mixing processes. Here we review some of the properties of particle transport in vortex-dominated flows, concentrating on the statistical properties induced by the presence of an ensemble of vortices. We discuss a possible parameterization of particle dispersion in vortex-dominated flows, adopting the view that ocean mesoscale turbulence is a two-component fluid which includes intense, localized vortical structures with non-local effects immersed in a Kolmogorovian, low-energy turbulent background which has mostly local effects. Finally, we report on some recent results regarding the role of coherent mesoscale eddies in marine ecosystem functioning, which is related to the effects that vortices have on nutrient supply.

4.1 Introduction

The ocean transports heat, salt, momentum and vorticity, nutrients and pollutants, and many other material and dynamical quantities across its vast spaces. Some of these transport processes are at the heart of the mechanisms of climate variability and of marine ecosystem functioning. In addition, a large portion of the available data on ocean dynamics are in the form of float and drifter trajectories. These provide a Lagrangian view of the ocean circulation which is not always easy to disentangle.

Lagrangian Analysis and Prediction of Coastal and Ocean Dynamics, ed. A. Griffa, D. Kirwan, A. Mariano, T. Özgökmen, and T. Rossby. Published by Cambridge University Press.

A proper consideration of the issues mentioned above, from climate variability to the reconstruction of the ocean circulation, requires the understanding, modeling and parameterization of oceanic transport processes. However, this task can be difficult, as ocean motions include a full spectrum of different scales, from the displacement of individual molecules to the general circulation at basin and planetary scales. At each scale, structured motions appear, and "eddies" populate the flow.[1] Of course, eddies at different scales behave differently, some of them are coherent objects,[2] such as mesoscale vortices, others are random eddies that live for short times, such as the whirls of three-dimensional turbulence. While short-lived eddies can sometimes be described by approaches based on turbulent cascades and random-Fourier-phase approximations, the long-lived coherent vortices usually lead to a break-up of the random-phase approximation.

Since a complete description of ocean motions at all scales is not feasible, a common approach consists in identifying a scale range of interest where we resolve the dynamics in the best possible way. Fluctuations on smaller scales are described as turbulence (of course, once we define something as "turbulence" this does not mean we have proceeded much in its understanding). Fluctuations at scales larger than those of interest are considered as boundary conditions, or large-scale forcings. Therefore, depending on the detail of knowledge of the mean flow and on the scale we work with, a given dynamical feature can either be included in the mean flow, or accounted for as one component of the turbulent behavior, or included in the boundary conditions. In this chapter, we shall focus on the dynamics of approximately two-dimensional mesoscale eddies, and on their effect on particle transport. Consistently, the term eddy will be used to refer to dynamical features characterized by strong potential vorticity anomalies that survive for several turnover times (typically, eddies of this kind survive for months to years), and which have a size of at least a few kilometers. In this sense, we are describing coherent eddies.

One of the reasons why we focus on these dynamical structures is that mesoscale eddies are widespread in the world oceans, and they account for a large portion of the ocean turbulent kinetic energy. Observations reveal the presence of rings in the Gulf Current and in the Agulhas Current (Olson and

[1] Note that the term eddy refers to a wide variety of dynamical features in the ocean. It can stand for the low-frequency mesoscale variability (i.e. the medium-size fluctuations in the general circulation), for the mesoscale and sub-mesoscale coherent vortices (vortical motions at scales smaller than the internal Rossby radius of deformation; McWilliams, 1985), or for a generic complicated motion in the presence of turbulence.

[2] Here, by coherent we mean a dynamical structure that lives for times that are much longer than a suitably defined local turbulence time.

Evans, 1986; Garzoli *et al.*, 1999; Arhan *et al.*, 1999), Meddies in the North Atlantic subtropical gyre (Richardson *et al.*, 2000; Bower *et al.*, 1997), abyssal vortices in the Brazil Basin (Richardson *et al.*, 1994; Richardson and Fratantoni, 1999; Hogg and Owens, 1999; Weatherly *et al.*, 2002) and in the Red Sea (Shapiro and Meschanov, 1991), and small vortices close to the regions of deep water formation (Gulf of Lyon: Testor and Gascard, 2003; Labrador Sea: Pickart *et al.*, 1996). Mesoscale and sub-mesoscale eddies have radii that range from 5 km in the Gulf of Lyon (Testor and Gascard, 2003) to more than 100 km for Agulhas Current Rings (Garzoli *et al.*, 1999; Arhan *et al.*, 1999).

The presence of eddies is also revealed by Lagrangian data, in particular by floats undergoing looping trajectories. Depending on the geographical area where the floats are deployed, the fraction of floats undergoing looping trajectories (loopers), range from a few percent to almost fifty percent (Richardson, 1993).[3] It is to be noted that the number of observed looping trajectories remains low despite the effort of launching instruments in the core of a ring. For this reason, a direct census of the vortex population is difficult and the quantitative estimate of their statistical properties is questionable. Nevertheless, the available observations suggest that some oceanic regions are densely packed with vortices. Surface structures can be observed by satellite using infrared images (Hooker *et al.*, 1995), or altimetric maps (Goni *et al.*, 1997; Stammer, 1997; Isern-Fontanet *et al.*, 2003). Subsurface structures are more difficult to measure, and their presence can be inferred thanks to the swirling motion of Lagrangian floats, or by measuring lenses of water with different biological, chemical, and/or physical characteristics with respect to the surrounding water (such as the salty Meddies exiting from the Mediterranean at the Gibraltar Strait). With the assumption of homogeneity of the vortex distribution, Richardson (1993) has estimated the vortex population in the North Atlantic to be composed of about a thousand eddies. Recently, a method for estimating some of the statistical properties of the vortex population from the analysis of a few velocity time series has been proposed by Pasquero *et al.* (2002). This method is quite efficient in homogeneous conditions, but its extension to the inhomogeneous distributions typical of the ocean environment has still to be pursued.

A final comment concerns the fact that mesoscale structures affect the population dynamics of phyto- and zooplankton (for a general introduction

[3] A looper is a float that makes at least two consecutive loops. Note that only a part of the loopers are effectively located inside vortex cores: many of them are simply swirling around a vortex a few times, moving elsewhere afterwards.

to the interaction between ocean motions and marine ecosystem dynamics see the book by Mann and Lazier, 1996). Coherent vortices are associated with secondary currents responsible for both horizontal and vertical fluxes of nutrients (e.g., Falkowski *et al.*, 1991; McGillicuddy and Robinson, 1997; McGillicuddy *et al.*, 1998; Siegel *et al.*, 1999; Martin and Richards, 2001; Lévy, Klein and Tréguier, 2001; Martin *et al.*, 2002; Lévy, 2003). The fact that the nutrient fluxes have a fine spatial and temporal detail, generated by the eddy field, has important consequences on primary productivity (Martin *et al.*, 2002; Pasquero *et al.*, 2004, 2005; Pasquero, 2005) and the horizontal velocity field induced by the eddies has been suggested to play an important role in determining plankton patchiness (e.g., Abraham, 1998; Mahadevan and Campbell, 2002; Martin, 2003). Owing to their trapping properties, vortices can also act as shelters for temporarily less-favored planktonic species (Bracco *et al.*, 2000c). Overall, these considerations indicate that mesoscale and sub-mesoscale eddies are important agents in marine ecosystem dynamics.

4.2 The two-component view of mesoscale turbulence

In this contribution we shall advocate the view that ocean mesoscale turbulence can be pictured as a two-component fluid: a sea of coherent vortices immersed into a background turbulence that is quite Kolmogorovian.[4] This two-component viewpoint forms the basis of how we interpret Lagrangian (and Eulerian) measurements and how we infer flow properties from them. This issue has two facets: a direct problem, where we derive the properties of Lagrangian transport from the knowledge of a two-component Eulerian flow, and an inverse problem, where we try to obtain the structure and dynamics of the advecting flow from the knowledge of an ensemble of Lagrangian measurements. The latter is probably the most important of the two from an oceanographic perspective, but we cannot do the inverse problem without first understanding the direct problem. An important issue is how we identify the two components. Twenty years of two-dimensional and three-dimensional quasigeostrophic turbulence research has shown, at least, that spectral analysis is not enough to identify the two components. The spectral viewpoint is local in Fourier space and global in physical space, while coherent vortices are characterized by vorticity distributions that are local in physical space, and thus they are non-local in Fourier space. In addition, the spectral representation of a coherent vortex does not lead to random Fourier phases, and thus the presence of coherent vortices generates significant phase correlations in the

[4] The statistical properties of the background turbulence have been discussed for example by Siegel and Weiss (1997), using the separation procedure provided by their vortex census method.

Fourier spectrum of the field. Any approach that assumes random Fourier phases is thus doomed to failure when applied to vortex-dominated flows.

Until now, the best way to identify vortices has found to be the direct identification by some vortex census algorithm based on the analysis of local vorticity patches in physical space. A variety of such methods exists (Benzi *et al.*, 1987; Farge and Rabreau, 1988; McWilliams, 1990; Siegel and Weiss, 1997; McWilliams *et al.*, 1999; Farge *et al.*, 1999); all of them require the knowledge of the full vorticity field. A simplified version of a vortex census, which requires the knowledge of just a few Eulerian time series and provides the gross features of the vortex statistics such as the vortex density and the average vortex size, has also been proposed (Pasquero *et al.*, 2002).

Although coherent vortices are local vorticity concentrations, their effects are non-local: the velocity field generated by a coherent vortex is non-local as it extends to large distances from the vortex center, well beyond the region where vorticity is significant. The range where the effect of the vortex on the velocity field is significant depends on the vortex shape and on the degree of baroclinicity: barotropic vortices extend their influence to far distances, while baroclinic lenses (such as Meddies) have a shorter range of influence.[5] In terms of the velocity field (and particle dispersion), the two-component view of mesoscale turbulence should not be seen as a purely spatial decomposition of space into separate vortex and non-vortex areas, but rather as the superposition of two dynamical components which can simultaneously act at the same spatial position. Thus, even Lagrangian floats that are always outside vortices could be significantly influenced by vortex dynamics.

In the following, we explore some of the Lagrangian implications of the two-component nature of mesoscale turbulence in the ocean. We discuss some of the properties of the vortex components, such as the role that vortices play as transport barriers and the properties of the velocity field that they induce. After this, we shall exploit the two-component nature of these flows to construct a stochastic parameterization of Lagrangian dispersion in mesoscale turbulence.

4.3 Equations of motion

In this contribution we shall mainly be concerned with the behavior of Lagrangian tracers in simple dynamical models of ocean mesoscale

[5] The Green's function associated with a barotropic (point) vortex decreases proportionally to $\log(r)$, where r is the distance from the core of the vortex. For a baroclinic (point) vortex, the Green's function goes as $1/r$. Therefore barotropic vortices extend their influence to far distance, while baroclinic vortices have a shorter range of influence (Bracco *et al.*, 2004).

turbulence. The description of the oceanic turbulent dynamics can be simplified by considering that diapycnal motion is inhibited by the stable density stratification present in most regions of the ocean. We can think of a stratified, hydrostatic fluid as a stack of infinitesimally thin layers. The dynamics in each layer can be described by the quasi-geostrophic (QG) approximation (Pedlosky, 1987; Salmon, 1998), which refers to a slowly evolving layer of homogeneous, incompressible fluid in a rotating environment. With the further assumption that the fluid is inviscid, the equations of motion correspond to the conservation of potential vorticity, which in the QG approximation can be written as the sum of relative vorticity, planetary vorticity, and the contribution to potential vorticity due to variations in the layer thickness (vortex-tube stretching). The interesting characteristic of this model, from a mathematical point of view, is that potential vorticity is now a single scalar dynamical field that fully describes the dynamics. The model allows for the development of coherent structures but does not capture several features such as, for instance, the surfacing of isopycnals (Flierl, 1987). Barotropic models are a further simplification, that can be obtained from the QG approximation by discarding the effects of the stratification. Two-dimensional turbulence results from neglecting the effects of vortex-tube stretching. For a detailed description of these approximations see, e.g., Pedlosky (1987) and Salmon (1998). Note, also, that although barotropic and baroclinic QG turbulence have different Eulerian characteristics, they nevertheless lead to very similar particle dispersion processes (Bracco *et al.*, 2004).

The Lagrangian equation of motion for an individual fluid particle moving in a two-dimensional flow is

$$\frac{\mathrm{d}\mathbf{X}_i}{\mathrm{d}t} = \mathbf{U}_i(t) = \mathbf{u}(\mathbf{X}_i(t), t) \tag{4.1}$$

where $\mathbf{X}_i(t)$ and $\mathbf{U}_i(t)$ are the position and velocity of the i-th particle, and $\mathbf{u}$ is the Eulerian velocity at the particle position. Note that in this equation we do not equate force to mass times particle acceleration, but rather particle velocity to the velocity of the flow. This happens because the particle is assumed to have negligible size and vanishing inertia with respect to the advecting fluid, i.e., to be a fluid element. When particles have finite size and/or non-vanishing inertia, the equations of motion become more complicated, see e.g. Provenzale (1999) and Babiano *et al.* (2000) for a discussion of the dynamics of inertial and finite-size particles in vortex-dominated flows. In the following, we shall only consider the dynamics of fluid particles for which Equation (4.1) holds.

4.4 Mesoscale vortices as transport barriers

Lagrangian observations indicate that floats deployed inside a mesoscale vortex stay inside the eddy for a long time, undergoing a looping trajectory (Olson, 1991; Richardson, 1993; Garzoli *et al.*, 1999; Richardson *et al.*, 2000). Numerical simulations of barotropic and of baroclinic (stratified) quasi-geostrophic turbulence and of point-vortex systems confirm that the cores of coherent vortices are associated with islands of regular (non chaotic) Lagrangian motion that trap particles for times comparable with the vortex lifetime (Babiano *et al.*, 1994), and that vortices are characterized by a strong impermeability to inward and outward particle fluxes, see e.g. Provenzale (1999) for a review. Particles can have more complex behavior and can eventually migrate from inside to outside of a vortex or vice versa only when highly (and relatively rare) dissipative events take place, such as the deformation of a vortex due to the interaction with a nearby vortex, or the formation of a filament.[6] For this reason, an initially inhomogeneous particle distribution becomes homogeneous only on a very long time scale, which is determined by the typical lifetime of the vortices rather than by the typical eddy turnover time of the individual vortices.

The trapping behavior of coherent vortices can be rationalized in terms of potential vorticity conservation. The most general definition of potential vorticity (PV) is $\Pi = (\underline{\omega_\mathbf{a}} \cdot \underline{\nabla}\eta)/\rho$, where $\underline{\omega_\mathbf{a}} = \underline{\omega} + 2\underline{\Omega}$ is absolute vorticity, given by the sum of relative and planetary vorticity, ρ is the density of the fluid, and η is entropy. Neglecting dissipation, fluid particles move conserving PV, i.e., we can write

$$\frac{\mathrm{D}\Pi}{\mathrm{D}t} = 0 \tag{4.2}$$

where $\mathrm{D}/\mathrm{D}t$ is the material (Lagrangian) derivative, that represents the temporal variation following a fluid element.[7]

If surfaces of constant potential vorticity are known, then much is known about the motion of Lagrangian tracers. For this reason potential vorticity, despite the fact that its definition is not always intuitive and it cannot easily be

[6] An attempt to rationalize a disturbed vortex in terms of a wave-like perturbation superimposed on a regular steady object has recently showed that regular islands of motion in an otherwise chaotic sea characterize the Lagrangian behavior inside the perturbed vortex (Beron-Vera *et al.*, 2004).

[7] We have now three time derivatives. The partial (Eulerian) derivative, $\partial/\partial t$, is the local time derivative of an Eulerian field, function of space and time. The material, or Lagrangian, derivative is the time derivative following the trajectory of the fluid element, $\mathrm{D}/\mathrm{D}t = \partial/\partial t + \underline{\mathbf{u}} \cdot \underline{\nabla}$. The underline symbol, __, is introduced to discriminate two-dimensional vectors (no underline) from three-dimensional vectors. The total time derivative, such as that used on the left hand side of Equation (4.1), is used when the dynamical variable depends only on time, as happens for the particle position.

measured in the ocean, is a quantity that can be of help in the understanding of Lagrangian motions. For oceanographic application, a useful approximation to potential vorticity is $\Pi = \frac{\omega_a}{\rho}\frac{d\rho}{dz}$ (Pedlosky, 1987), where the scalar ω_a is the vertical component of absolute vorticity. In this case, from the knowledge of the horizontal velocity and the vertical density profile, a map of potential vorticity can be drawn (Fratantoni *et al.*, 1995).

A first consequence of potential vorticity conservation is that regions of strong PV gradients can act as transport barriers (McIntyre, 1989). For an ideal fluid with irrotational external forcing PV is conserved. When some little dissipation and/or rotational forcing is acting on the fluid, as it usually happens, PV is not conserved. If the PV-changing effects are small, PV is quasi-conserved. This means that in regions where PV changes slightly, the particles will be able to shift from one PV surface to another. However, strong PV gradients are much more difficult to overcome, as the change in PV that the particle should achieve to climb (or descend) the gradient may be too large compared to the effect of the non-irrotational forcings and dissipation present in the system. As a result, strong PV gradients can act as transport barriers.

This is the main physical reason why intense jets, associated with strong PV gradients, can act as efficient barriers to transport. The same happens for isolated vortices (which can be thought of as jets wrapped on themselves): vortex edges act as barriers to transport because vortices are regions of anomalous potential vorticity, usually embedded in a background where PV oscillates with low variance around some reference value. The vortex edges are therefore characterized by a large potential vorticity gradient, which fluid particles can rarely cross. Entrainment or detrainment of fluid particles from vortices more likely occurs during highly dissipative events (when potential vorticity is not conserved), and when vortices undergo structural changes and disruption, such as in vortex merging or during the ejection of filaments from the vortex core (see, e.g., de Steur *et al.*, 2004 for the study of tracer leakage from modeled Agulhas rings).

4.5 Estimate of Lagrangian statistics

In studying Lagrangian data, we should distinguish between time averages taken along a particle trajectory and ensemble averages performed over a set of different particles. The two averages, in principle, can give rather different results. One of the reasons for this behavior is that Lagrangian particles can have a very long memory when coherent structures, whose lifetime is long compared to other time scales in the problem, are present. For instance, if a Lagrangian particle is initially released in the background turbulence outside vortex cores, it will move around without entering any of the vortex cores present in the turbulent flow, until, in a quite rare event such as the formation of a new vortex, the particle will get trapped inside a newly forming vortical structure. From that moment on, the particle will stay inside the vortex for times comparable with the vortex lifetime.

The above example indicates that the temporal convergence of the statistical properties of a set of Lagrangian trajectories can take place on rather long timescales, related to the lifetime of the coherent structures. Of course, ensemble averages over a large number of homogeneously distributed Lagrangian particles do not suffer from this problem and they usually give a more complete picture of the flow. This illustrates the fact that ergodicity (i.e., equivalence of time and ensemble averages) is reached only on very long times, if ever, for Lagrangian statistics of particles moving in vortex-dominated flows, as discussed by Weiss *et al.* (1998) for point vortices and by Provenzale (1999) and Pasquero *et al.* (2002) for the vortices of two-dimensional turbulence. In the case of ocean floats, usually one does not have access to a full ensemble of simultaneously launched Lagrangian floats, and often a mixture of ensemble and time averages has to be employed. An interesting question, then, concerns the trade-off between the number of particles required to provide a meaningful picture of the flow (i.e., a correct estimate of the statistical properties of the flow) and the length of the trajectories. The larger the number of Lagrangian particles, the shorter the minimum length of the trajectories needed to describe the flow, but the exact balance depends on many factors, including the level of turbulence, the density and intensity of coherent structures, and the location where the Lagrangian probes are released. In such cases, the possible lack of ergodicity should be carefully evaluated when computing Lagrangian statistics. This issue has been discussed in some detail by Pasquero *et al.* (2002), together with the comparison between Lagrangian and Eulerian second-order statistics (i.e., spectra and decorrelation times).

Another delicate aspect related to the estimate of Lagrangian statistics is due to the (spatial and temporal) inhomogeneity of most ocean flows. A standard pre-analysis procedure is the spatial and temporal binning of the data, motivated by the need to group together and average only those Lagrangian data which refer to similar values of the mean and eddy kinetic energy. In fact, analysing data coming from different statistical distributions, and/or very different values of the mean and eddy kinetic energy can lead to spurious results. In theory, the binning operation is a well-defined procedure, and the bin size should in principle be reduced until the statistical properties in each bin converge (Poulain, 2001). This procedure assumes that the number of data is large enough to guarantee statistically significant results for small bin size.

Unfortunately, in most cases the Lagrangian field data sets are not large enough. A compromise has to be achieved between the need for homogeneity and the requirement of statistical significance in each bin. If not properly performed, the binning procedure adopted to eliminate spatial and temporal inhomogeneities can lead to spurious results; see the discussion in Bracco *et al.* (2003).

4.6 Velocity statistics

Barotropic vortices influence the velocity field at large distances compared to their size. This influence is seen in the probability distribution function (PDF) of the velocity. At high Reynolds numbers, when vortices are intense and have sharp profiles, velocity PDFs in barotropic turbulence have non-Gaussian tails indicating that high velocities are more probable than would be the case for a Gaussian field (Bracco *et al.*, 2000a; Pasquero *et al.*, 2001).[8]

This non-Gaussianity has been previously discussed in the context of point vortices, which can be thought of as a simplified model of vortex-dominated flows at very large Reynolds numbers (Min *et al.*, 1996; Jiménez, 1996; Weiss *et al.*, 1998). In this context it has been shown that small velocities have a Gaussian distribution but the PDF has a non-Gaussian tail which is related to the slow decay with distance of the velocity induced by a single vortex. Convergence to a Gaussian PDF is obtained only in systems with an extremely large number of vortices, orders of magnitude more than exist in the ocean (Weiss *et al.*, 1998).

Float trajectories in the North Atlantic (Bracco *et al.*, 2000b) and in the Adriatic Sea (Falco *et al.*, 2000; Maurizi *et al.*, 2004) indicate that velocity PDFs are non-Gaussian, see Figure 4.1. Typically, they have larger kurtosis than a normal distribution: they have a Gaussian-like core and non-Gaussian tails for high velocities. Similar results have been found (Bracco *et al.*, 2003) from midlatitude fluid particle trajectories along isobaric surfaces in a simulation of the Atlantic Ocean dynamics at high resolution (MICOM model, 1/12 of a degree). This similarity in velocity PDFs between float data, ocean general circulation models, simplified turbulence models, and point vortex systems suggests that the non-Gaussian nature of the velocity PDFs is due to the vortex component of the mesoscale turbulence.

Velocity distribution functions alone do not fully determine the characteristics of Lagrangian motion as they provide no information about velocity

[8] We are here referring to either Eulerian or Lagrangian velocity PDFs under the assumption that Lagrangian particles sample the whole domain. In this case, in fact, Lagrangian velocity PDFs must converge to the Eulerian ones. See also the well-mixed condition in Thomson (1987).

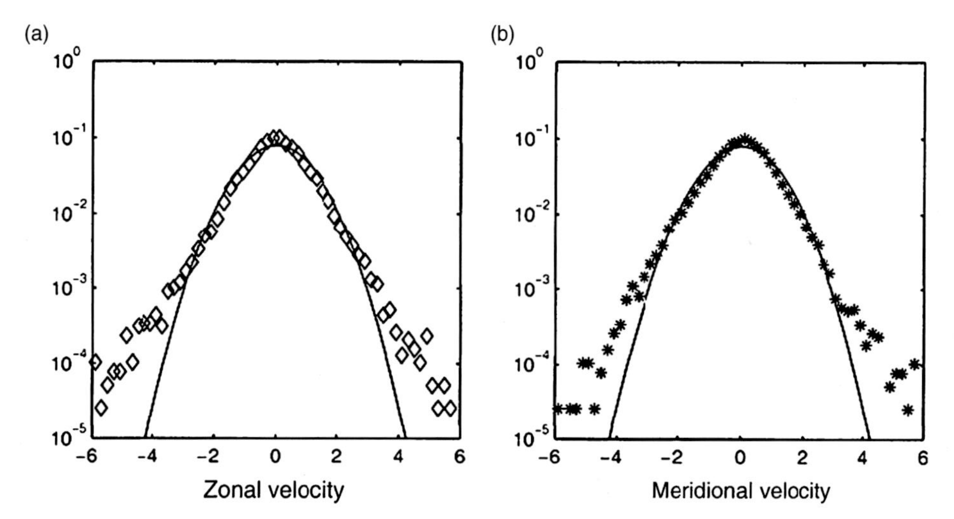

Figure 4.1 Velocity probability density function from floats in the deep Eastern North-Atlantic: (a) zonal component, (b) meridional component [from Bracco *et al.*, 2000b].

correlations. One important correlation is the memory a fluid parcel has of its past velocity, which is often characterized by the velocity autocorrelation function.

The single trajectory autocorrelation function for an individual particle (labeled by the index i) is

$$R_i(\tau) = \frac{\overline{(\mathbf{U}_i(t) - \overline{\mathbf{U}}_i) \cdot (\mathbf{U}_i(t+\tau) - \overline{\mathbf{U}}_i)}}{\sigma_i^2}, \tag{4.3}$$

where $\mathbf{U}_i(t)$ is the velocity of the i-th particle at time t, $\overline{\mathbf{U}}_i$ and σ_i^2 are the mean and variance of the velocity of the i-th trajectory, and the overbar indicates an average over time t. Hence, $R_i(0) = 1$ and $R_i(\tau)$ goes to zero for large τ, when the particle velocity loses memory of its initial value. The functions $R_i(\tau)$ can be extremely different for different particles, depending on the dynamical characteristics of the region in which tracers move.

A particle trapped inside a vortex spins around the center of the vortex, which is itself advected by the flow. In this case, R_i is a decaying oscillatory function (see Figure 4.2a). In other cases R_i can slowly decay with no oscillations, as in the case of a particle moving within a jet (Berloff *et al.*, 2002). Point vortex systems show that long time correlations associated with high velocities occur in the vicinity of vortex pairs (Weiss *et al.*, 1998), and there is indication that this behavior occurs in the ocean as well (Weatherly *et al.*, 2002).

The flow field as a whole is characterized by the ensemble-averaged velocity autocorrelation function (Fig. 4.2b), defined by averaging over all trajectories.

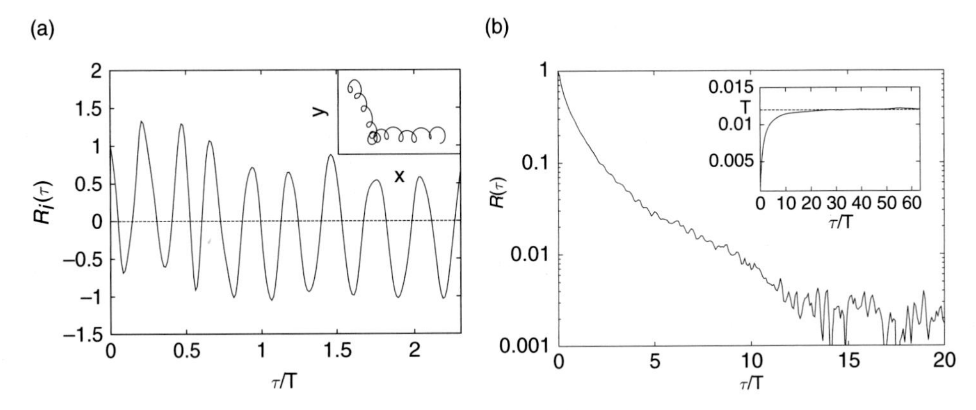

Figure 4.2 (a) Autocorrelation function, $R_i(\tau)$, for an individual particle moving within a vortex in a simulation of two-dimensional turbulence. The particle trajectory is shown in the inset. (b) Ensemble-averaged autocorrelation function, $R(\tau)$, for several thousand particles advected by two-dimensional turbulence. The inset shows the estimate of the Lagrangian integral time, T, as a function of the length of the trajectory [adapted from Pasquero *et al.*, 2001].

Because of the diversity of Lagrangian histories which floats can experience, the estimate of ensemble-averaged autocorrelation functions is particularly sensitive to the typically small number of trajectories available in the ocean. Depending on the characteristics of coherent structures, the ensemble-averaged autocorrelation functions can also have oscillatory behavior (see Veneziani *et al.*, 2004 for an example on float data).

One simple measure of the memory of Lagrangian particles is the Lagrangian integral time, defined as

$$T = \int_0^\infty R(\tau)\mathrm{d}\tau. \tag{4.4}$$

When the autocorrelation function is exponential, the Lagrangian integral time is the decay time of the exponential, $R(\tau) = \mathrm{e}^{-\tau/T}$. Typical values for T in the ocean are of the order of few days (Griffa *et al.*, 1995; Garraffo *et al.*, 2001; Zhang *et al.*, 2001; Bauer *et al.*, 2002).

4.7 Particle dispersion

Lagrangian particles in a Gaussian, homogeneous, stationary and uncorrelated velocity field undergo a Brownian random walk. Under such conditions, the second-order moment of the distribution of particle displacements grows linearly with time:

$$A^2(\tau; t_0) \equiv <(\mathbf{X}_i(t_0 + \tau) - \mathbf{X}_i(t_0))^2 >= 2K\tau \tag{4.5}$$

where K is the dispersion (or diffusion) coefficient. Here, $\mathbf{X}_i(t)$ is the position of the i-th particle at time t, and the angular brackets denote an ensemble average over all particles. The function $A^2(\tau, t_0)$ measures the absolute (or single-particle) dispersion. For a statistically stationary flow, the absolute dispersion A^2 does not depend on the starting time t_0.

Relaxing any of the above assumptions (Gaussianity, homogeneity, stationarity, lack of temporal and spatial correlations) can significantly alter the dispersion law described above. On short time scales, in particular, the Brownian dispersion law is modified by the presence of spatial and temporal correlations in the advecting flow, which induce Lagrangian velocity correlations over a substantial time range. Over times much shorter than the Lagrangian integral time, the velocity is almost constant and one observes a ballistic dispersion phase,

$$A^2(\tau) = 2E\tau^2 \tag{4.6}$$

where E is the mean kinetic energy of the advecting flow.

A standard way of representing absolute dispersion is to define a time-dependent dispersion coefficient, $K(\tau)$, as

$$K(\tau) = \frac{A^2(\tau)}{2\tau}. \tag{4.7}$$

In the ballistic phase, $K(\tau) \to E\tau$ as $\tau \to 0$, while in the Brownian dispersion phase $K(\tau) \to K$ for $\tau \to \infty$. The ballistic regime is sometimes visible in the dispersion curves computed from surface drifter data (e.g. Colin de Verdiere, 1983). Subsurface float trajectories are often characterized by a well-defined ballistic regime, associated with very steep Lagrangian spectra at small times (Rupolo *et al.*, 1996).

At long times, the presence of boundaries defines an upper saturation scale and it can destroy the Brownian behavior (Artale *et al.*, 1997). Well before this, the presence of large-scale dynamical and geographical inhomogeneities and the temporal non-stationarity associated with different oceanic processes can also mask the Brownian dispersion phase. For example, meridional dispersion in the tropical Pacific Ocean has been found to depend on the strong wave activity typical of equatorial regions, which induces strong correlations in the acceleration (Bauer *et al.*, 2002), and in the Adriatic Sea the diffusion coefficient computed from floats does not converge to a Brownian regime (Falco *et al.*, 2000).

The analysis of float and drifter trajectories has also revealed the presence of anomalous dispersion regimes with $K(\tau) \propto \tau^{\alpha}$ and $\alpha \neq 0$ (Osborne *et al.*, 1986, 1989; Sanderson and Booth, 1991; Provenzale *et al.*, 1991; Brown and Smith, 1990). For example, in the analysis of surface drifter data in the Kuroshio extension region in the Pacific Ocean, Osborne *et al.* (1989) detected an anomalous dispersion regime with $\alpha \approx 2/3$. A power-law intermediate dispersion regime, albeit with a different exponent, has also been observed for two-dimensional barotropic turbulence and baroclinic quasi-geostrophic turbulence (Elhmaidi *et al.*, 1993; Bracco *et al.*, 2004). However, a satisfactory explanation of these anomalous regimes is lacking.

4.8 Parameterization of particle dispersion

Lagrangian stochastic models (LSMs) are employed to reproduce the main statistical properties of particle trajectories in turbulent flows, without resolving the full Eulerian dynamics. Individual trajectories computed by a LSM usually do not have the same characteristics of the particles advected by a realistic flow. The similarity is recovered – if ever! – only statistically, after averaging over particle ensembles and over different realizations of the turbulent flow. Thus, one should not expect an individual stochastic trajectory to resemble an individual float trajectory. One simple class of stochastic models describes the process of single-particle dispersion. In this case, the spatial correlations of the advecting flow are discarded insofar as they do not translate into temporal correlations of the Lagrangian velocities (see also Rupolo *et al.*, 1996 for a discussion of how Eulerian spatial correlations are related to Lagrangian time correlations). A more complex approach deals with particle separation processes, i.e., relative dispersion. In this case, the stochastic model describes the time evolution of the separation of a particle pair, and spatial correlations of the turbulent flow become an essential ingredient of the picture. In the following, we shall consider only single-particle dispersion and the related stochastic descriptions. An exhaustive discussion of the atmospheric applications of Lagrangian stochastic models can be found in Rodean (1996); for oceanographic applications see Griffa (1996) and Brickman and Smith (2002).

The simplest stochastic model for single-particle dispersion is the random walk (or Markoff-0 model). In this approach, the particle displacements are randomly extracted from a Gaussian distribution, and there is no temporal correlation between subsequent displacements. If we assume that there is no mean flow advecting the particles and that the turbulent flow is statistically isotropic, we can write a Lagrangian stochastic differential equation for the random walk as:

$$d\mathbf{X}_i = \sqrt{K} d\mathcal{W}_i(t), \tag{4.8}$$

where **X** is the position of the i-th particle and the diffusivity, K, is not allowed to vary in space and time. The incremental Wiener random vector, $dt\mathcal{W}_i$, has zero mean and it is δ-correlated in space and time, $\langle d\mathcal{W}_i(t) \cdot d\mathcal{W}_j(t') \rangle = \delta_{ij}\,\delta(t - t')dt$.

The single-particle stochastic description illustrated above can be framed in terms of a deterministic partial differential equation for the time evolution of the probability density function of particle positions, $P(\mathbf{X}|\mathbf{X}(0), t)$. Defining the particle concentration at $\mathbf{x}$ as $C(\mathbf{x}, t) = \int P(\mathbf{X} = \mathbf{x}|\mathbf{X}(0), t)\, d\mathbf{X}(0)$, the Fokker–Planck equation for the evolution of P gives the well-known diffusion equation:

$$\frac{\partial C(\mathbf{x}, t)}{\partial t} = \frac{1}{2} K \nabla^2 [C(\mathbf{x}, t)]. \tag{4.9}$$

The assumption of uncorrelated displacements is equivalent to the assumption that the Eulerian fluid velocities decorrelate instantaneously, i.e., that the turbulent structure of the flow has no correlations. In general, this assumption is not appropriate for ocean mesoscale flows, where the temporal correlations of the advecting velocity field cannot be discarded. The simplest way of accounting for a memory in Lagrangian velocities is to consider a Markoff-1 model. In this approach, the time evolution of the Lagrangian velocity of the i-th particle, $\mathbf{U}_i$, is described by an Ornstein–Uhlenbeck process:

$$d\mathbf{U}_i = -\frac{\mathbf{U}_i}{T} dt + \sqrt{\frac{2\sigma^2}{T}} d\mathcal{W}_i, \tag{4.10}$$

where T is the Lagrangian correlation time and σ^2 is the variance of the Lagrangian velocities. The first term on the r.h.s. is the (deterministic) fading-memory term, and the second term is the "stochastic kick," or random component, of the velocity fluctuation. For this process, the velocity distribution is a Gaussian with zero mean and variance σ^2, and the velocity autocorrelation is an exponential, $R(\tau) = \exp(-\tau/T)$. The (time-dependent) diffusion coefficient can be computed analytically,

$$K(\tau) = \sigma^2 T \left[1 - \frac{T(1 - e^{-\tau/T})}{\tau} \right], \tag{4.11}$$

see Griffa (1996) for a discussion of this type of stochastic model in the context of oceanographic applications.

In a study of particle dispersion in two-dimensional turbulence, Pasquero *et al.* (2001) showed that the linear Ornstein–Uhlenbeck model provides a good representation of absolute dispersion at short and large times (respectively in the ballistic and Brownian regimes), while at intermediate times it provides estimates of the dispersion coefficient which differ by at most 25% from the values obtained by direct integration of particle dynamics in the turbulent flow. If this discrepancy is acceptable, due for example to uncertain or poorly resolved data, then the use of the Ornstein–Uhlenbeck model is sufficient. To obtain a more precise estimate of the dispersion coefficient, however, a stochastic model that more closely represents the processes of particle dispersion in vortex-dominated mesoscale turbulence is warranted.

Major differences between the Ornstein–Uhlenbeck process and particle dispersion in mesoscale turbulence are related to the facts that the velocity distribution is non-Gaussian (Bracco *et al.*, 2000a), the velocity autocorrelation is non-exponential (Pasquero *et al.*, 2001), and particles get trapped in vortices for long times (Elhmaidi *et al.*, 1993; Babiano *et al.*, 1994). Given these differences, it is indeed surprising that just a 25% discrepancy between the turbulent and the modelled dispersion coefficient is detected.

In an attempt to improve stochastic parameterizations of particle dispersion in mesoscale ocean turbulence, various extensions of the Ornstein–Uhlenbeck model have been proposed. The indications that Lagrangian accelerations in the ocean are correlated in time (Rupolo *et al.*, 1996) have stimulated the development of Markoff-2 models where an Ornstein–Uhlenbeck formulation is written for the acceleration $\mathbf{a}$, with $d\mathbf{U} = \mathbf{a}\, dt$ (e.g., Griffa, 1996). Higher order models have also been proposed, with the aim of better reproducing other statistical properties of Lagrangian motions such as the sub- or superdiffusive behavior at intermediate times (Berloff and McWilliams, 2002). Superdiffusion has also been obtained by Reynolds (2002), using a variation of a Markoff-2 model that includes spin.

Models that include spin have been designed to explicitly describe particle motion in and around coherent structures. In the presence of coherent vortices, particle motion has a rotational component, as evident in the looping trajectories of floats deployed inside mesoscale eddies. The rotational component of the velocity vector along a Lagrangian trajectory is characterized by an acceleration orthogonal to the trajectory. Simple geometrical arguments show that the introduction of the spin in Markoff-1 models corresponds to adding a new term in the stochastic equation for the velocity increment, proportional to the orthogonal velocity component (Sawford, 1999; Reynolds, 2002). The individual trajectories produced by these models display spiralling motion, although the ensemble averaged velocity autocorrelation function is not necessarily oscillatory (Reynolds, 2002). This model has recently been used to reproduce some statistical properties of Northwest Atlantic float trajectories (Veneziani *et al.*, 2004), and to interpret characteristics of trajectories released in a high-resolution ocean model (Veneziani *et al.*, 2005).

On the other hand, it is not clear whether particle spinning inside vortices has any effect on space and time scales larger than those of the vortices themselves. In general, rotational motion inside vortices does not contribute to the large-scale spreading of particles; it is only the motion of the vortex itself that is responsible for particle displacements at large scales. In turn, vortices move because they are advected by other vortices and there is no self-induction of the vortices themselves (Weiss *et al.*, 1998). As a result, the large-time dispersion properties of Lagrangian particles inside or outside the vortices of two-dimensional turbulence are the same.[9] Thus, for the purpose of understanding particle dispersion at scales larger than the size of the individual vortices, the parameterization of particle motion inside a vortex can probably be neglected.

In a study of single-particle dispersion in two-dimensional turbulence, Pasquero *et al.* (2001) proposed a parameterization of dispersion in two-dimensional turbulence at scales larger than those of the individual vortices. In doing so, no a-priori difference between particles inside and outside vortices is drawn. The main point of the approach followed by Pasquero *et al.* (2001) is the observation that the Eulerian velocity at any point is determined by the combined effect of the far field of the vortices and the contribution of the local vorticity field in the background (Bracco *et al.*, 2000a). Thus, even outside vortices, the velocity field induced by the coherent vortices cannot be discarded: on average, 80% of the kinetic energy in the background turbulence outside vortices is due to the velocity field induced by the vortex population. In addition, the non-Gaussian velocities measured in the background turbulence outside vortices are entirely due to the action of the surrounding vortices, which extend their influence far away from their inner cores. This is a signature of the non-locality of the velocity field: a particle moving in a vortex-dominated flow is heavily affected by the vortex dynamics even if it is not located inside them.

[9] The situation is different on the β-plane, where vortices move differently with respect to fluid particles in the background turbulence. Here, significant differences between long-time dispersion properties of particles inside and outside vortices can be detected (Mockett, 1998).

In this approach, the stochastic Lagrangian velocity of a particle at the position $\mathbf{X}(t)$ is produced by the sum of two components,

$$\mathbf{U}(\mathbf{X}) = \mathbf{U}_\mathrm{B}(\mathbf{X}) + \mathbf{U}_\mathrm{V}(\mathbf{X}), \tag{4.12}$$

where $\mathbf{U}_\mathrm{B}(\mathbf{X})$ is the velocity induced by the background turbulence and $\mathbf{U}_\mathrm{V}(\mathbf{X})$ is that induced by the vortices. The background-induced velocity is characterized by small energy and slow dynamics (i.e., long temporal correlations), while the vortex-induced component has large energy and it undergoes fast dynamics (whose temporal scale is of the order of the eddy turnover time). In addition, the vortex-induced component is characterized by a non-Gaussian velocity PDF.

A different stochastic equation has then to be used for each of the two components. Since the background-induced velocity component, $\mathbf{U}_\mathrm{B}(\mathbf{X})$, has a Gaussian distribution, a standard stochastic OU process can be used to describe it. As for the non-Gaussian, vortex-induced component $\mathbf{U}_\mathrm{V}(\mathbf{X})$, a proper description is easily obtained by considering a nonlinear Markoff-1 model (Pasquero *et al.*, 2001). In this case, one needs to consider a generalized Langevin equation

$$\mathrm{d}\mathbf{U}_V = \mathbf{a}_1(\mathbf{U}_\mathrm{V})\mathrm{d}t + \mathbf{a}_2(\mathbf{U}_\mathrm{V})\mathrm{d}\mathcal{W} \tag{4.13}$$

where the functions $\mathbf{a}_1$ and $\mathbf{a}_2$ are functions of the velocity $\mathbf{U}_\mathrm{V}$. The choice of the function $\mathbf{a}(\mathbf{U}_\mathrm{V})$ is (not uniquely) determined by the corresponding Fokker–Planck equation, with the use of the well-mixed condition (Thomson 1987, Pasquero *et al.*, 2001). In the end, the model proposed by Pasquero *et al.* (2001) becomes (we omit the particle index i for simplicity of notation):

$$\begin{aligned}
\mathrm{d}\mathbf{X} &= (\mathbf{U}_\mathrm{B} + \mathbf{U}_\mathrm{V})\,\mathrm{d}t \\
\mathrm{d}\mathbf{U}_\mathrm{B} &= -\frac{\mathbf{U}_\mathrm{B}}{T_\mathrm{B}}\mathrm{d}t + \sqrt{\frac{2\sigma_\mathrm{B}^2}{T_\mathrm{B}}}\mathrm{d}\mathcal{W}_\mathrm{B} \\
\mathrm{d}\mathbf{U}_V &= -\frac{2 + |\mathbf{U}_\mathrm{V}|/\sigma_\mathrm{V}}{(1 + |\mathbf{U}_\mathrm{V}|/\sigma_\mathrm{V})^2}\frac{\mathbf{U}_\mathrm{V}}{T_\mathrm{V}}\,\mathrm{d}t + \sqrt{\frac{2\sigma_\mathrm{V}^2}{T_\mathrm{V}}}\mathrm{d}\mathcal{W}_\mathrm{V}
\end{aligned} \tag{4.14}$$

where $T_\mathrm{B} > T_\mathrm{V}$, $\sigma_\mathrm{V}^2 \gg \sigma_\mathrm{B}^2$, and $\mathcal{W}_\mathrm{B}$ and $\mathcal{W}_\mathrm{V}$ are two independent Wiener processes.

Interestingly, the parameters of the stochastic model depicted above can be obtained from fits to an ensemble of Lagrangian trajectories (i.e., assuming no knowledge of the advecting velocity field). Comparison with particle advection in two-dimensional turbulence shows that this model captures single-particle dispersion with an error of less than 5%, and it does also capture statistical quantities measuring higher-order moments of the dispersion statistics (e.g., the distribution of first-exit times). Note that both the nonlinear nature of the vortex-induced velocity and the presence of a low-energy background-induced velocity are essential ingredients of the model. At shorter times, the vortex-induced velocity dominates and it entirely determines statistical properties such as the non-Gaussian velocity distribution. At longer times, the vortex-induced velocity becomes rapidly uncorrelated and the lower-energy background-induced velocity gives a significant contribution to particle dispersion.

One advantage of the model illustrated above is that it has been built from a detailed knowledge of the dynamics of vortex-dominated flows. That is, it is not obtained by ignoring the structure of the flow, but from an attempt to reproduce, in a stochastic framework, some of the essential ingredients of mesoscale turbulence. In particular, this model fully exploits the two-component nature of mesoscale turbulence.

4.9 Mesoscale vortices and the marine ecosystem

Mesoscale and sub-mesoscale vortices play a potentially important role in the dynamics of the marine ecosystem. In particular, the presence of coherent vortices can have a significant impact on primary productivity in the open ocean (and, consequently, on the carbon cycle). Vortices induce secondary currents that can lead to upwelling and downwelling in and around the vortex. The eddy pumping mechanism (Falkowski *et al.*, 1991; McGillicuddy and Robinson, 1997; McGillicuddy *et al.*, 1998; Siegel *et al.*, 1999) is based on the fact that isopycnals are lifted upwards, towards the surface, in cyclonic eddies. This mechanism can thus bring up nutrients from the deeper waters. In this view, cyclonic eddies act as nutrient pumps for the marine ecosystem. This view has recently been questioned by various authors (e.g., Smith *et al.*, 1996; Lévy, 2003; Williams and Follows, 2003), who showed that nutrient fluxes associated with horizontal secondary circulations should be considered as well. Finally, other oceanic structures such as fronts have been suggested to be more relevant to the functioning of the marine ecosystem (Mahadevan and Archer, 2000). The detailed effect of individual mesoscale eddies on the vertical fluxes of nutrients is thus currently under debate.

In any case, the work mentioned above indicates that mesoscale and sub-mesoscale structures are responsible, one way or another, for significant nutrient fluxes in the marine ecosystem. As a result, nutrient fluxes can thus be highly inhomogeneous both spatially and temporally. The fact that the nutrient supply is concentrated in small individual regions with eddy sizes rather than in a large uniform region has been shown to significantly affect numerical estimates of primary productivity in the ocean (Martin *et al.*, 2002; Pasquero *et al.*, 2005).

When a fluid flow advects nutrients and plankton, the equations describing the dynamics of the biological system are advection–reaction–diffusion partial differential equations. These can be integrated in an Eulerian approach, or by a semi-Lagrangian method which copes with the dynamics of individual fluid particles. In a semi-Lagrangian numerical approach, one integrates the biological reactions in a large ensemble of Lagrangian particles which are advected by a velocity field produced by the Eulerian integration of the momentum equations (see, e.g., Abraham, 1998). Each Lagrangian parcel represents a given water volume (usually taken to have a size comparable with the Eulerian grid spacing), and it is assumed to have homogeneous properties.

In principle, the biological reactions occur within each parcel, and do not depend on the behavior of neighboring particles. This allows for the formation of sharp gradients, when the system is prone to this behavior. When a concentration field is required, the distribution of Lagrangian particles is interpolated onto a regular grid and a concentration field is obtained. Diffusion of the biological components can be accounted for by introducing mixing among nearby water parcels (Pasquero *et al.*, 2004).

The plankton ecosystem model used here includes three components, which represent nutrient, N, phytoplankton, P, and zooplankton, Z. The dynamics of the ecosystem is described by the NPZ equations

$$\begin{aligned}
\frac{\mathrm{d}N}{\mathrm{d}t} &= \Phi_N - \beta\frac{N}{k_N+N}P \\
&\quad + \mu_N\left((1-\gamma)\frac{a\epsilon P^2}{a+\epsilon P^2}Z + \mu_P P + \mu_Z Z^2\right) \\
\frac{\mathrm{d}P}{\mathrm{d}t} &= \beta\frac{N}{k_N+N}P - \frac{a\epsilon P^2}{a+\epsilon P^2}Z - \mu_P P \\
\frac{\mathrm{d}Z}{\mathrm{d}t} &= \gamma\frac{a\epsilon P^2}{a+\epsilon P^2}Z - \mu_Z Z^2 \, .
\end{aligned} \tag{4.15}$$

The terms on the right hand side of the equation for the nutrient represent respectively vertical nitrate supply from deep water, conversion to organic matter through phytoplankton activity, and regeneration of the dead organic

Table 4.1 *List of parameters used in the NPZ ecosystem model adopted here.*

$\beta = 0.66\ \text{day}^{-1}$	$k_N = 0.5\,\text{mmol N m}^{-3}$
$\varepsilon = 1.0\ (\text{mmol N m}^{-3})^{-2}\ \text{day}^{-1}$	$\mu_N = 0.2$
$\gamma = 0.75$	$\mu_P = 0.03\ \text{day}^{-1}$
$a = 2.0\ \text{day}^{-1}$	$\mu_Z = 0.2\ (\text{mmol N m}^{-3})^{-1}\ \text{day}^{-1}$
$s = s_\text{p} = 0.00648\ \text{day}^{-1}$ in nutrient-poor regions	$N_0 = 8.0\,\text{mmol N m}^{-3}$
$s = s_\text{a} = 0.648\ \text{day}^{-1}$ in nutrient-rich regions	

matter into nutrients. The phytoplankton dynamics is regulated by production, depending on available nutrients through a Holling type-II functional response, by a Holling type-III grazing by zooplankton, and by linear mortality. Finally, zooplankton grows when phytoplankton is present (γ is the assimilation efficiency of the zooplankton), and has a quadratic mortality term used to close the system and parameterize the effects of higher trophic levels. The specific form of the terms used in this model is quite standard in marine ecosystem modelling (Oschlies and Garcon, 1999). The term μ_N is smaller than one and it represents the fact that not all biological substance is immediately available as nutrient: the fraction $(1 - \mu_N)$ is lost by sinking to deeper waters. Note, also, that nutrient enters this model by affecting the growth rate of phytoplankton. Since the formulation adopted here is two-dimensional in the horizontal and no vertical structure of the fields is allowed, vertical upwelling has to be parameterized. The parameter values used in the model are listed in Table 4.1.

The nutrient flux is expressed as a relaxation term, $\Phi_N = -s(x, y)(N - N_0)$, where N_0 is the (constant) nutrient content in deep waters and s is the (spatially varying) relaxation rate of the nutrient, which is large in regions of strong vertical mixing and small in regions of weak vertical mixing. This form of the nutrient flux term reflects the fact that nutrient is brought up to the surface from deep water via (isopycnal and diapycnal) turbulent mixing and upwelling. Such a form is the standard formulation used for chemostat models when the reservoir has infinite capacity (Kot, 2001).

The model ecosystem equations are solved, using the semi-Lagrangian method, for each water parcel moving in forced and dissipated, statistically stationary two-dimensional turbulence. The turbulent field used here is forced at wavenumber $k = 40$ and has a resolution of 512^2 grid points (Pasquero *et al.*, 2001, 2004). Assuming that the forcing scale corresponds to a typical eddy size, say about 25 km, then in dimensional units the domain size becomes 1000 km and the resolution is about 2 km. The turbulent velocity field has mean eddy turnover time $T_E = 2.8$ days.

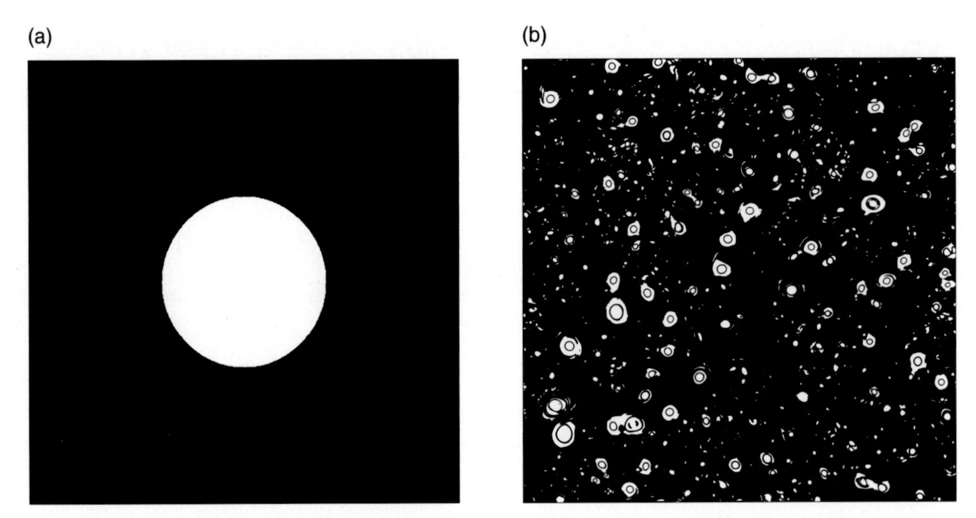

Figure 4.3 White areas mark the regions where intense nutrient flux takes place. Panel (a) refers to the case with a single region of strong upwelling (CP), panel (b) shows a snapshot of the case where upwelling is located in and around eddies (EP). The total area of the regions with strong upwelling is approximately the same in the two cases. [From Pasquero *et al.*, 2004].

In the following, we illustrate the effect of coherent eddies on primary productivity by considering two situations which differ by the spatial distribution of the regions where the nutrient is supplied. While the total area of the "active" region with strong upwelling is kept constant at 12% of the domain area, in the first simulation the intense upwelling is confined to a circular patch at the center of the domain (case CP, Central Patch), while in the second simulation the nutrient is supplied due to upwelling in a number of small patches correlated with eddy structures (Case EP, Eddy Patches). The two types of active regions are shown in Figure 4.3. For the EP case, the figure is a snapshot at a particular time, and the active regions are dynamically moving and changing with the flow.

In the EP case, the intense nutrient flux is assumed to take place in correspondence of the coherent vortices. The active patches with significant upwelling are defined using the value of the Okubo–Weiss parameter, $Q = s^2 - \omega^2$ where s^2 is squared strain and ω^2 is squared vorticity (Okubo, 1970; Weiss, 1991). The nutrient flux is defined to take place in regions for which $|Q| > Q_0$, where Q_0 is a threshold fixed by the requirement that the total area covered by the active regions is $12\% \pm 0.5\%$ at any time. This definition of active patches includes both the vortex cores and small annular regions around the vortices.

In the absence of an advecting flow and diffusion, the system reduces to an ensemble of independent, point-like (homogeneous) ecosystem models described by the system of ordinary differential equations (4.15). Each of these systems is labeled by the (fixed) spatial position of the corresponding fluid parcel and it is characterized by a specific value of the nutrient flux. For the parameter values adopted here, each of these systems tends to a steady state, (N^*, P^*, Z^*), determined by the value of the nutrient input. Note that primary production in the steady state is larger than the nutrient flux, as a consequence of the fact that part of the organic nitrogen content is regenerated into nutrients (such as ammonium).

To study the system behavior in the presence of an advecting flow, we initialize the nutrient, the phytoplankton and the zooplankton in each fluid particle to the appropriate steady solution (which depends on the local value of the nutrient flux). Turbulent advection is then turned on and the evolution of the system is followed for 300 model days.

In Figure 4.4 we show the mean primary production, defined as $PP = \langle \beta NP/(k_n + N) \rangle$, where the angular brackets indicate average over the whole domain, and the ratio between primary production and nutrient upwelling flux, PP/Φ_N, as an indicator of the efficiency of the biological model to convert inorganic into organic matter.

With the form of nutrient flux adopted here, the enhanced stirring increases the mean flux from deep waters, as seen in Figure 4.4a. The enhanced flux originates at active locations when a parcel of water that has low nutrient content is advected over them. To see how this happens, consider two nearby parcels: one is in a region with small nutrient upwelling and characterized by a steady-state nutrient concentration N_{p}^*; the other is in an active region and has a steady-state nutrient concentration N_{a}^*. In this configuration, the total nitrate flux associated with these two parcels of water is $\left(s_{\mathrm{p}}(N_0 - N_{\mathrm{p}}^*) + s_{\mathrm{a}}(N_0 - N_{\mathrm{a}}^*)\right)$, where s_{p} and s_{a} are the relaxation constants for the nutrient-poor and nutrient-rich regions, respectively. Suppose now that, due to advection, the two parcels switch their position: the parcel with small nutrient content gets in a strong upwelling region and vice-versa. In this configuration, the vertical flux is $\left(s_{\mathrm{p}}(N_0 - N_{\mathrm{a}}^*) + s_{\mathrm{a}}(N_0 - N_{\mathrm{p}}^*)\right)$. The net variation of the nutrient flux between the two configurations is $(s_{\mathrm{a}} - s_{\mathrm{p}})(N_{\mathrm{a}}^* - N_{\mathrm{p}}^*)$, which is proportional to $(s_{\mathrm{a}} - s_{\mathrm{p}})$. This term is positive as larger relaxation rates are found in active regions with strong vertical mixing. The enhanced nutrient flux is thus due to the asymmetry in the relaxation times between the active and inactive regions. Note, also, that the exchange rate of water parcels between the two types of region directly affects the increased nutrient flux to the surface, determining larger values of the nutrient flux in case EP than in case CP (Figure 4.4a).

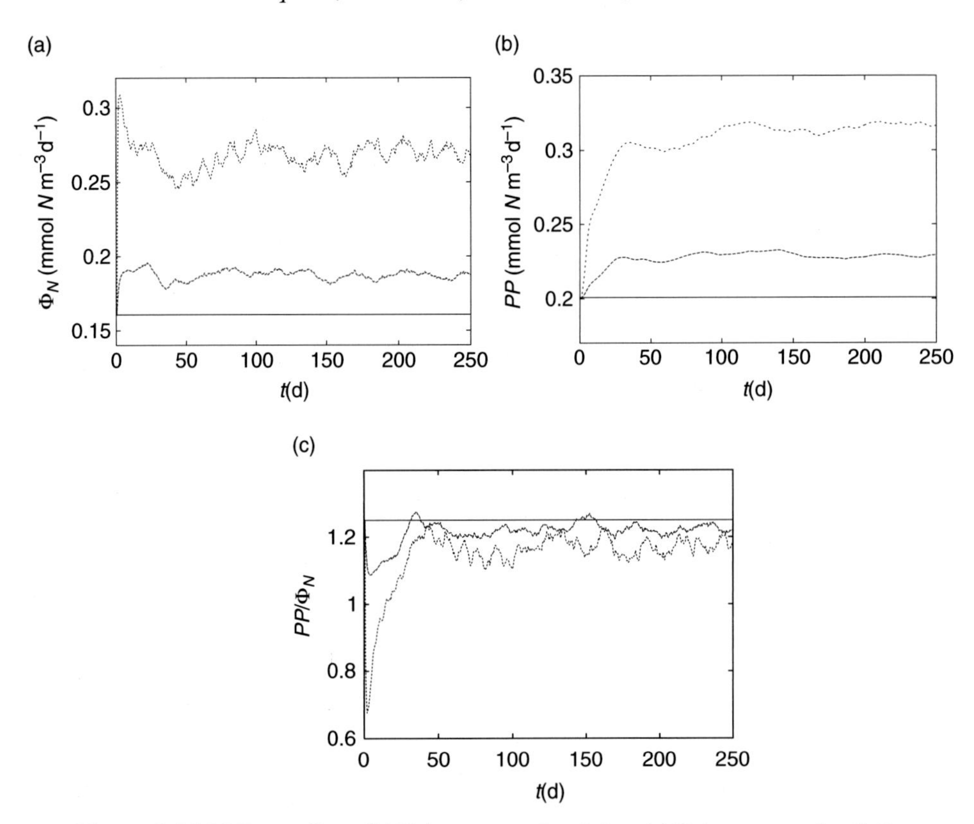

Figure 4.4 (a) Nitrate flux; (b) Primary productivity; (c) Primary productivity per unit nitrate flux. Solid lines show the reference case with no advection; dashed lines refer to the case with horizontal turbulent advection and nutrient flux concentrated in a single circular region (CP); dotted lines refer to the case with horizontal turbulent advection and nutrient flux distributed in small patches associated with the eddies of the turbulent flow (EP). Time is in days. [Adapted from Pasquero *et al.*, 2004.]

Primary productivity follows the increased nutrient flux, although with some delay due to the time taken by the phytoplankton to grow (Figure 4.4b). In case EP, a much larger primary productivity is observed than in case CP or in the absence of horizontal advection, as already observed by Martin *et al.* (2002). The enhanced primary productivity is due to the larger nutrient flux stimulated by the presence of horizontal advection. In fact, after the initial transient the efficiency of the biological system is about the same as in the no-advection case (Figure 4.4c), confirming that the primary production has increased mainly as a response to the larger available nitrate.

If we had considered other forms for the nutrient flux, such as a fixed-flux condition, the results would have been somehow different, see Pasquero *et al.*

(2004, 2005) for a comparison between different forms of the nutrient flux term. In general, however, what emerges is that the ecosystem response, for example in terms of primary productivity, is affected by the spatial and temporal fragmentation of the upwelling regions. Thus, ocean models that do not properly resolve the size of the upwelling regions, be they associated with mesoscale vortices, fronts or submesoscale structures, can provide erroneous estimates of the primary productivity of marine ecosystems.

4.10 Perspectives

In the last twenty years, we have gained some understanding of the role that mesoscale vortices play in ocean dynamics and in ocean transport processes. In particular, we have evolved from a view that considered vortices as interesting but probably irrelevant curiosities to a view of the ocean as "a sea of eddies." Nevertheless, many issues are still open. In the following we list some of the problems (or obsessions) that capture our attention.

First, we should develop better census algorithms capable of extracting information from a few Eulerian and Lagrangian time series and of coping with inhomogeneous conditions. Up to now, vortex census algorithms either require the knowledge of the full vorticity field (Benzi *et al.*, 1987; Farge and Rabreau, 1988; McWilliams, 1990; Siegel and Weiss, 1997; McWilliams *et al.*, 1999; Farge *et al.*, 1999) or work only for homogeneous turbulence (Pasquero *et al.*, 2002).

Second, vortices are not the only structures present in ocean mesoscale turbulence. Rossby waves can play an important role in particle dispersion processes (e.g., Mockett, 1998), and fronts are of overwhelming importance. These structures should be studied as carefully as vortices have been.

An important issue, then, concerns the reason to study the effects of vortices, and of the other mesoscale structures. In our opinion, one central reason is to learn how to parameterize them. Although numerical models of the ocean circulation are attaining increasingly larger resolution, there are limits imposed by both numerical possibilities and the availability of a fine enough network of measured data for model initialization and assimilation. In such conditions, we think it is wise to devote some effort to the development of stochastic parameterizations of the effects that vortices, waves and fronts induce. The works discussed in Section 4.8 are an example of this approach.

A much harder, but even more intriguing issue, is the stochastic parameterization of the effects that mesoscale structures have on the marine ecosystem; i.e., the parameterization of the reaction rates and of the dynamics of an ensemble of (biologically) reacting tracers. In Section 4.9 we saw that the presence of mesoscale vortices has a direct impact on nutrient fluxes and on the primary productivity of the marine ecosystem, and an even larger impact is presumably associated with the presence of fronts. Understanding and parameterizing the biological effects of these mesoscale structures will be a major challenge in the coming years.

Acknowledgments

The authors thank the editors of this volume and two anonymous reviewers who have carefully read the manuscript and suggested how to improve it. This work was supported in part by the European Community's Human Potential Programme under contract HPRN-CT-2002-00300, Stirring and Mixing.

References

Abraham, E. R., 1998. The generation of plankton patchiness by turbulent stirring. *Nature*, **391**, 577–80.

Arhan M., H. Mercier, and J. R. E. Lutjeharms, 1999. The disparate evolution of three Agulhas rings in the South Atlantic Ocean. *J. Geophys. Res. Oceans*, **104**, 20987–1005.

Artale, V., G. Boffetta, A. Celani, M. Cencini, and A. Vulpiani, 1997. Dispersion of passive tracers in closed basins: Beyond the diffusion coefficient. *Phys. Fluids*, **9**, 3162–71.

Babiano, A., G. Boffetta, A. Provenzale, and A. Vulpiani, 1994. Chaotic advection in point vortex models and two-dimensional turbulence. *Phys. Fluids*, **6**, 2465–74.

Babiano, A., J. H. E. Cartwright, O. Piro, and A. Provenzale, 2000. Dynamics of small neutrally buoyant sphere in a fluid and targeting in Hamiltonian systems. *Phys. Rev. Lett.*, **84**, 5764–7.

Bauer, S., M. S. Swenson, and A. Griffa, 2002. Eddy-mean flow decomposition and eddy-diffusivity estimates in the tropical Pacific Ocean. 2. Results. *J. Geophys. Res. Ocean*, **107**, 3154–71.

Benzi, R., S. Patarnello, and P. Santangelo, 1987. On the statistical properties of two-dimensional decaying turbulence. *Europhys. Lett.*, **3**, 811–18.

Berloff, P. S. and J. C. McWilliams, 2002. Material transport in oceanic gyres. Part II: Hierarchy of stochastic models. *J. Phys. Oceanogr.*, **32**, 797–830.

Berloff, P. S., J. C. McWilliams, and A. Bracco, 2002. Material transport in oceanic gyres. Part I: Phenomenology. *J. Phys. Oceanogr.*, **32**, 764–96.

Beron-Vera, F. J., M. J. Olascoaga, and M. G. Brown, 2004. Passive tracer patchiness and particle trajectory stability in incompressible two-dimensional flows. *Nonlinear Proc. Geoph.*, **11**, 67–74.

Bower, A. S., L. Armi, and I. Ambar, 1997. Lagrangian observations of Meddy formation during a mediterranean undercurrent seeding experiment. *J. Phys. Oceanogr.*, **27**, 2545–75.

Bracco, A., J. LaCasce, C. Pasquero, and A. Provenzale, 2000a. The velocity distribution of barotropic turbulence. *Phys. Fluids*, **12**, 2478–88.

Bracco, A., J. H. LaCasce, and A. Provenzale, 2000b. Velocity probability density functions for oceanic floats. *J. Phys. Oceanogr.*, **30**, 461–74.
Bracco, A., A. Provenzale, and I. Scheuring, 2000c. Mesoscale vortices and the paradox of the plankton. *P. Roy. Soc. Lond. B*, **267**, 1795–1800.
Bracco, A., E. P. Chassignet, Z. D. Garraffo, and A. Provenzale, 2003. Lagrangian velocity distributions in a high-resolution numerical simulation of the North-Atlantic. *J. Atmos. Ocean. Tech.*, **20**, 1212–20.
Bracco, A., J. von Hardenberg, A. Provenzale, J. B. Weiss, and J. C. McWilliams, 2004. Dispersion and mixing in quasigeostrophic turbulence. *Phys. Rev. Lett.*, **92**, 084501.
Brickman D. and P. C. Smith, 2002. Lagrangian stochastic modeling in coastal oceanography. *J. Atmos. Ocean. Tech.*, **19**, 83–99.
Brown, M. G. and K. B. Smith, 1990. Are SOFAR float trajectories chaotic? *J. Phys. Oceanog.*, **20**, 139–49.
Colin de Verdiere, A., 1983. Lagrangian eddy statistics from surface drifters in the eastern North-Atlantic. *J. Marine Res.*, **41**, 375–98.
de Steur, L., P. J. van Leeuwen, and S. S. Drijfhout, 2004. Tracer leakage from modeled Agulhas rings. *J. Phys. Oceanog.*, **34**, 1387–99.
Elhmaidi, D., A. Provenzale, and A. Babiano, 1993. Elementary topology of two-dimensional turbulence from a Lagrangian viewpoint and single-particle dispersion. *J. Fluid Mech.*, **242**, 655–700.
Falco, P., A. Griffa, P. M. Poulain, and E. Zambianchi, 2000. Transport properties in the Adriatic Sea as deduced from drifter data. *J. Phys. Oceanogr.*, **30**, 2055–71.
Falkowski, P. G., D. Ziemann, Z. Kolber, and P. K. Bienfang, 1991. Role of eddy pumping in enhancing primary production in the Ocean. *Nature*, **352**, 55–8.
Farge, M. and G. Rabreau, 1988. Wavelet transform to detect and analyze coherent structures in two-dimensional turbulent flows. *C. Acad. Sci. Paris Sér. II*, **307**, 1479–86.
Farge, M., K. Schneider, and N. Kevlahan, 1999. Non-Gaussianity and coherent vortex simulation for two-dimensional turbulence using an adaptive orthogonal wavelet basis. *Phys. Fluids*, **11**, 2187–201.
Flierl, G. R., 1987. Isolated eddy models in geophysics. *Ann. Rev. Fluid Mechanics*, **19**, 493–530.
Fratantoni, D. M., W. E. Johns, and T. L. Townsend, 1995. Rings of the North Brazil Current: their structure and behavior inferred from observations and a numerical simulation. *J. Geophys. Res.*, **100**, 10633–54.
Garraffo, Z. D., A. J. Mariano, A. Griffa, C. Veneziani, and E. P. Chassignet, 2001. Lagrangian data in a high-resolution numerical simulation of the North Atlantic I. Comparison with in situ drifter data. *J. Marine Syst.*, **29**, 157–76.
Garzoli S. L., P. L. Richardson, C. M. D. Rae, D. M. Fratantoni, G. J. Goni, and A. J. Roubicek, 1999. Three Agulhas rings observed during the Benguela Current experiment. *J. Geophys. Res. Oceans*, **104**, 20971–85.
Goni G. J., S. L. Garzoli, A. J. Roubicek, D. B. Olson, and O. B. Brown, 1997. Agulhas ring dynamics from TOPEX/POSEIDON satellite altimeter data. *J. Marine Res.*, **55**, 861–83.
Griffa, A., 1996. Applications of stochastic particle models to oceanographic problems. In *Stochastic Modelling in Physical Oceanography*, ed. R. J. Adler, P. Müller, and R. B. Rozovskii. Cambridge, MA: Birkhäuser Boston, 114–40.

Griffa, A., K. Owens, L. Piterbarg, and B. Rozovskii, 1995. Estimates of turbulence parameters from Lagrangian data using a stochastic particle model. *J. Marine Res.*, **53**, 371–401.

Hogg N. G. and W. B. Owens, 1999. Direct measurement of the deep circulation within the Brazil Basin. *Deep-sea Res. Part II*, **46**, 335–53.

Hooker S. B. and J. W. Brown, 1995. Warm-core ring dynamics derived from satellite imagery. *J. Geophys. Res. Oceans*, **99**(C12), 25181–94.

Isern-Fontanet, J., E. Garcia-Ladona, and J. Font, 2003. Identification of marine eddies from altimetric maps. *J. Atmos. Ocean. Tech.*, **20**, 772–8.

Jiménez, J., 1996. Algebraic probability density tails in decaying isotropic two-dimensional turbulence. *J. Fluid Mech.*, **313**, 223–40.

Kot, M., 2001. *Elements of Mathematical Ecology*. Cambridge: Cambridge University Press.

Lévy, M., 2003. Mesoscale variability of phytoplankton and of new production: Impact of the large-scale nutrient distribution. *J. Geophys. Res. Oceans*, **108**(C11), 3358.

Lévy, M., P. Klein, and A. M. Tréguier, 2001. Impact of sub-mesoscale physics on production and subduction of phytoplankton in an oligotrophic regime. *J. Marine Res.*, **59**, 535–65.

Mahadevan, A. and D. Archer, 2000. Modeling the impact of fronts and mesoscale circulation on the nutrient supply and biogeochemistry of the upper ocean. *J. Geophys. Res. Oceans*, **105**, 1209–25.

Mahadevan, A. and J. W. Campbell, 2002. Biogeochemical patchiness at the sea surface. *Geophys. Res. Lett.*, **29**, 1926.

Mann, K. H. and J. R. N. Lazier, 1996. *Dynamics of Marine Ecosystems: Biological-Physical Interactions in the Oceans*, 2nd edn. Cambridge, MA: Blackwell Science.

Martin, A. P., 2003. Phytoplankton patchiness: the role of lateral stirring and mixing. *Progress in Oceanography*, **57**, 125.

Martin, A. P. and K. J. Richards, 2001. Mechanisms for vertical nutrient transport within a North Atlantic mesoscale eddy. *Deep-sea Res. Part II*, **48**, 757–73.

Martin, A. P., K. J. Richards, A. Bracco, and A. Provenzale, 2002. Patchy productivity in the open ocean. *Global Biogeochem. Cycles*, **16**, 1025.

Maurizi, A., A. Griffa, P. M. Poulain, and F. Tampieri, 2004. Lagrangian turbulence in the Adriatic Sea as computed from drifter data: Effects of inhomogeneity and nonstationarity. *J. Geophys. Res. Oceans*, **109**, C04010.

McGillicuddy, D. J. and A. R. Robinson, 1997. Eddy-induced nutrient supply and new production in the Sargasso Sea. *Deep-sea Res. Part I*, **44**, 1427–50.

McGillicuddy, D. J., A. R. Robinson, D. A. Siegel, H. W. Jannasch, R. Johnson, T. Dickeys, J. McNeil, A. F. Michaels, and A. H. Knap, 1998. Influence of mesoscale eddies on new production in the Sargasso Sea. *Nature*, **394**, 263–6.

McIntyre, M. E., 1989. On the Antarctic ozone hole. *J. Atmos. Terr. Phys*, **51**, 29–43.

McWilliams, J. C., 1985. Submesoscale, coherent vortices in the ocean. *Rev. Geophys.*, **23**, 165–82.

McWilliams, J. C., 1990. The vortices of two-dimensional turbulence. *J. Fluid Mech.*, **219**, 361–85.

McWilliams, J. C., J. B. Weiss, and I. Yavneh, 1999. The vortices of homogeneous geostrophic turbulence. *J. Fluid Mech.*, **401**, 1–26.

Min, I. A., I. Mezic, A. Leonard, 1996. Levy stable distributions for velocity difference in systems of vortex elements. *Phys. Fluids*, **8**, 1169–80.

Mockett, C. R., 1998. Dispersion and reconstruction. In *Astrophysical and Geophysical Flows as Dynamical System*, WHOI Tech. Rep. WHOI-98-00.
Okubo, A., 1970. Horizontal dispersion of floatable particles in the vicinity of velocity singularities such as convergences. *Deep-sea Res.*, **17**, 445–54.
Olson, D. B., 1991. Rings in the ocean. *Annu. Rev. Earth. Planet. Sci.*, **19**, 133–83.
Olson, D. B. and R. H. Evans, 1986. Rings of the Agulhas Current. *Deep-sea Res. Part A*, **33**, 27–42.
Osborne, A. R., A. D. Kirwan, A. Provenzale, and L. Bergamasco, 1986. A search for chaotic behavior in large and mesoscale motions in the Pacific Ocean. *Physica D*, **23**. 75–83.
Osborne, A. R., A. D. Kirwan, A. Provenzale, and L. Bergamasco, 1989. Fractal drifter trajectories in the Kuroshio extension. *Tellus*, **41A**, 416–35.
Oschlies, A. and V. Garcon, 1999. An eddy-permitting coupled physical–biological model of the North Atlantic – 1. Sensitivity to advection numerics and mixed layer physics. *Global Biogeochem. Cy.*, **13**, 135–60.
Pasquero, C., 2005. Differential eddy diffusion of biogeochemical tracers. *Geophys. Res. Lett.*, **32**, L17603, doi:10.1029/2005GL023662.
Pasquero, C., A. Bracco, and A. Provenzale, 2004. Coherent vortices, Lagrangian particles and the marine ecosystem. In *Shallow Flows*, ed. G. H. Jirka and W. S. J. Uijttewaal. Leiden, NL: Balkema Publishers, 399–412.
Pasquero, C., A. Bracco, and A. Provenzale, 2005. Impact of the spatio-temporal variability of the nutrient flux on primary productivity in the ocean. *J. Geophys. Res. – Oceans*, **110**, C07005, doi: 10.129/2004JC002738.
Pasquero, C., A. Provenzale, and A. Babiano, 2001. Parameterization of dispersion in two-dimensional turbulence. *J. Fluid Mech.*, **439**, 279–303.
Pasquero, C., A. Provenzale, and J. B. Weiss, 2002. Vortex statistics from Eulerian and Lagrangian time series. *Phys. Rev. Lett.*, **89**, 284501.
Pedlosky, J., 1987. *Geophysical Fluid Dynamics*, 2nd edn. New York: Springer.
Pickart, R. S., W. M. Smethie, J. R. N. Lazier, E. P. Jones, and W. J. Jenkins, 1996. Eddies of newly formed upper Labrador Sea water. *J. Geophys. Res. Oceans*, **101**(C9), 20711–26.
Poulain, P. M., 2001. Adriatic Sea surface circulation as derived from drifter data between 1990 and 1999. *J. Marine Syst.*, **29**, 3–32.
Provenzale, A., 1999. Transport by coherent barotropic vortices. *Annu. Rev. Fluid Mech.*, **31**, 55–93.
Provenzale, A., A. R. Osborne, A. D. Kirwan, and L. Bergamasco, 1991. The study of fluid parcel trajectories in large-scale ocean flows. In *Nonlinear Topics in Ocean Physics*, ed. A. R. Osborne. Amsterdam: Elsevier, 367–401.
Reynolds, A. M., 2002. On Lagrangian stochastic modelling of material transport in oceanic gyres. *Physica D*, **172**, 124–38.
Richardson, P. L., 1993. A census of eddies observed in North-Atlantic SOFAR float data. *Progress in Oceanography*, **31**, 1–50.
Richardson, P. L., A. S. Bower, and W. Zenk, 2000. A census of Meddies tracked by floats. *Progress in Oceanography*, **45**, 209–50.
Richardson, P. L. and D. M. Fratantoni, 1999. Float trajectories in the deep western boundary current and deep equatorial jets of the tropical Atlantic. *Deep-sea Res. Part II*, **46**, 305–33.
Richardson, P. L., G. E. Hufford, R. Limeburner, and W. S. Brown, 1994. North Brazil current retroflection eddies. *J. Geophys. Res. Oceans*, **99**, 5081–93.

Rodean, H. C., 1996. Stochastic Lagrangian models of turbulent diffusion. *Meteor. Monographs*, **26**(48).

Rupolo, V., B. L. Hua, A. Provenzale, and V. Artale, 1996. Lagrangian velocity spectra at 700 m in the western North Atlantic. *J. Phys. Oceanogr.*, **26**, 1591–1607.

Salmon, R., 1998. *Lectures on Geophysical Fluid Dynamics.* Oxford: Oxford University Press.

Sanderson, B. G. and D. A. Booth, 1991. The fractal dimension of drifter trajectories and estimates for horizontal eddy-diffusivity. *Tellus*, **43A**, 334–49.

Sawford, B. L., 1999. Rotation of trajectories in lagrangian stochastic models of turbulent dispersion. *Bound.-Lay. Meteorol.*, **93**, 411–24.

Shapiro, G. I. and S. L. Meschanov, 1991. Distribution and spreading of Red-Sea water and salt lens formation in the Northwest Indian-Ocean. *Deep-sea Res. Part A*, **38**, 21–34.

Siegel, A. and J. B. Weiss, 1997. A wavelet-packet census algorithm for calculating vortex statistics. *Phys. Fluids*, **9**, 1988–99.

Siegel, D., D. J. McGillicuddy, and E. A. Fields, 1999. Mesoscale eddies, satellite altimetry, and new production in the Sargasso Sea. *J. Geophys. Res. Oceans*, **104**(C6), 13359–79.

Smith, C. L., K. J. Richards, and M. J. R. Fasham, 1996. The impact of mesoscale eddies on plankton dynamics in the upper ocean. *Deep-sea Res. II*, 1807–32.

Stammer, D., 1997. Global characteristics of ocean variability estimated from regional TOPEX/POSEIDON altimeter measurements. *J. Phys. Oceanogr.*, **27**, 1743–69.

Testor, P. and J. C. Gascard, 2003. Large-scale spreading of deep waters in the Western Mediterranean sea by submesoscale coherent eddies. *J. Phys. Oceanogr.*, **33**, 75–87.

Thomson, D. J., 1987. Criteria for the selection of stochastic models of particle trajectories in turbulent flows. *J. Fluid Mech.*, **180**, 529–56.

Veneziani, M., A. Griffa, A. M. Reynolds, and A. J. Mariano, 2004. Oceanic turbulence and stochastic models from subsurface Lagrangian data for the northwest Atlantic Ocean. *J. Phys. Oceanog.*, **34**, 1884–906.

Veneziani, M., A. Griffa, Z. D. Garraffo, and E. P. Chassignet, 2005. Lagrangian spin parameter and coherent structures from trajectories released in a high-resolution ocean model. *J. Marine Res.*, **63**, 753–88.

Weatherly, G., M. Arhan, H. Mercier, and W. Smethie, 2002. Evidence of abyssal eddies in the Brazil Basin. *J. Geophys. Res. Oceans*, **107**(C4), 3027.

Weiss, J. B., 1991. The dynamics of enstrophy transfer in two-dimensional hydrodynamics. *Physica D*, **48**, 273–94.

Weiss, J. B., A. Provenzale, and J. C. McWilliams, 1998. Lagrangian dynamics in high-dimensional point-vortex systems. *Phys. Fluids*, **10**, 1929–41.

Williams, R. G. and M. J. Follows, 2003. Physical transport of nutrients and the maintenance of biological production. In *Ocean biogeochemistry: The role of the ocean carbon cycle in global change*, ed. M. Fasham. Berlin: Springer-Verlag, 19–51.

Zhang, H. M., M. D. Prater, and T. Rossby, 2001. Isopycnal Lagrangian statistics from the North Atlantic current RAFOS float observations. *J. Geophys. Res. Oceans*, **106**, 13817–36.

5

Inertial particle dynamics on the rotating Earth

NATHAN PALDOR
Department of Atmospheric Sciences, The Hebrew University of Jerusalem, Jerusalem, Israel

5.1 Introduction

The study of Newton's second law of motion is a natural basis for all fluid dynamical problems and the Eulerian form of this law is the basis (in addition to the conservation of mass) of Euler (or Navier–Stokes) equations. Despite its primordial importance in Geophysical Fluid Dynamics (GFD, hereafter) the application of Newton's second law of motion to the rotating spherical earth is commonly done only briefly as an addendum to the fluid dynamical problems. A detailed analysis of these equations as applied to the motion of particles on the rotating spherical earth is the subject of the present paper, which summarizes the advances made in the subject in recent years. In particular a comparison between the dynamics on the β-plane and on the sphere will be carried out in order to highlight the ramifications of the inconsistent approximations made in transforming the spherical geometry to a planar one on the β-plane.

The complexity of the spherical geometry is the culprit behind the development of GFD in Cartesian coordinates. Several semi-analytical studies in spherical coordinates were published in the 1960s and 1970s (a review of these works can be found in Moura, 1976) but more recent studies on a sphere are mostly numerical. Recent discussions of the balance between acceleration, the Coriolis force and pressure gradient forces on the elliptical Earth, as well as the subtleties of the Coriolis force itself there, are given in Durran (1993) and Persson (1998). The Inertial dynamics, where no pressure gradient forces are present can be viewed as the lowest order time-dependent motion in any geometry and my intent in this paper is to extend our grasp of the dynamics from the classical (and simple) f-plane into the β-plane and contrast the results with the dynamics on a sphere. Despite the artificial nature of the Inertial motion (which is best regarded as the dynamics of a bead moving without friction on the surface of a rotating sphere) where all fluid dynamical effects

Lagrangian Analysis and Prediction of Coastal and Ocean Dynamics, ed. A. Griffa, D. Kirwan, A. Mariano, T. Özgökmen, and T. Rossby. Published by Cambridge University Press.

(including the generation of pressure gradients by the divergence of the horizontal velocity) are neglected, it provides a clear and concise setup for crystalizing the subtle effects associated with time-dependent motion on the rotating Earth.

Planar simplifications of Earth's curved surface are extensively employed on both the f-plane and the β-plane and in both cases the transformation to Cartesian coordinates filters out the inherent periodicity of the spherical geometry and introduces, instead, an infinite domain where periodicity has to be imposed as an additional simplifying assumption that is not evident a priori. At the same time, the meridional change in the metric terms of the spherical coordinates, where the east–west distance over a unit longitude span shrinks with the increase in latitude, is eliminated by the transformation to Cartesian coordinates. The last simplifying planar assumption is that on the f-plane the Coriolis parameter is constant while on the β-plane it varies linearly with the northward displacement from a central latitude. The ramifications of these simplifications and complications associated with the transformation from spherical to cartesian coordinates have to be assessed before the results of the planar theory can be safely applied to the ocean on the spherical Earth.

Since the β-plane analysis of particle dynamics is commonly derived as a heuristic modification of the analytical f-plane dynamics, a brief review of the main results of particle dynamics on the f-plane is in order. This simplified theory begins with the Inertial motion where the only force acting on a particle moving on the f-plane is the Coriolis force (while gravity and centrifugal acceleration associated with Earth's curved shape and rotation act perpendicular to the plane). An exhaustive review of the resulting Inertial dynamics and the physics of Coriolis force in planar geometry is given in the classic book by Stommel and Moore (1989). The linearity of the governing equations with respect to the velocity yields explicit expressions for the temporal evolution and, in addition, implies that the presence of a prescribed pressure field on the f-plane only adds a steady solution to the time-dependent Inertial motion.

Specifically, the starting point of the inertial dynamics on the f-plane is Newton's second law of motion in the east (coordinate x; velocity component u) and north (coordinate y; velocity component v) directions. The application of Newton's law of motion to a unit volume of fluid yields the well-known equations (which imply that in the absence of rotation the acceleration vanishes):

$$\frac{\mathrm{d}u}{\mathrm{d}t} = f_0 v, \tag{5.1}$$

$$\frac{\mathrm{d}v}{\mathrm{d}t} = -f_0 u, \tag{5.2}$$

where $f_0 = 2\Omega\sin(\phi_0)$ is the Coriolis parameter calculated at some latitude ϕ_0 (and where Ω is Earth's rotation frequency). In addition, the temporal change in the particle coordinates is related to the, time-dependent, velocity components via:

$$\frac{\mathrm{d}x}{\mathrm{d}t} = u, \tag{5.3}$$

$$\frac{\mathrm{d}y}{\mathrm{d}t} = v. \tag{5.4}$$

The solutions of these equations that originate from the initial velocity $u(t=0)=u_0$; $v(t=0) = v_0$ and initial coordinates $x(t=0) = x_0$; $y(t=0) = y_0$ are called Inertial circles and are the counterparts of straight lines in a non-rotating plane. The presence of Coriolis terms turns the Inertial trajectories from straight lines on the former into circles on the (rotating) f-plane. The explicit solution for the trajectory $(x(t), y(t))$ is given by:

$$x(t) = x_0 + \frac{v_0}{f_0} + \frac{1}{f_0}(u_0 \sin(f_0 t) - v_0 \cos(f_0 t)), \tag{5.5}$$

$$y(t) = y_0 - \frac{u_0}{f_0} + \frac{1}{f_0}(u_0 \cos(f_0 t) + v_0 \sin(f_0 t)). \tag{5.6}$$

It is easy to verify that the (x(t), y(t)) trajectory satisfies:

$$\left(x(t) - x_0 - \frac{v_0}{f_0}\right)^2 + \left(y(t) - y_0 + \frac{u_0}{f_0}\right)^2 = \frac{u_0^2 + v_0^2}{f_0^2}, \tag{5.7}$$

which describes a circle of radius $\sqrt{u_0^2 + v_0^2}/f_0$ (i.e. the initial speed divided by f_0) centered on $(x_0 + v_0/f_0, y_0 - u_0/f_0)$.

In order to extend these results to the β-plane a heuristic argument that is often applied is to add the variation of the Coriolis force with latitude to this f-plane result, i.e. to predict the trajectory on the β-plane. The argument is based on the $1/f_0$ dependence of the radius of curvature in Eq. (5.7). According to this well-known argument substituting $f = f_0 + \beta y$ for f_0 implies that the radius of curvature of the trajectory (a circle on the f-plane) is smaller in the northern half of the circle – where f is larger – than in the southern half – where f is smaller. As the simple argument goes, the combined trajectory composed of a half-circle with a smaller radius of curvature in the northern half (where the particle moves primarily eastward) and a half-circle with a larger radius of curvature in the southern half (where the particle moves primarily westward) has a net westward directed drift. However, a stationary, egg-like, trajectory that does not drift at all is also a possible combination of two halves of inertial circles with different radii of curvature so this heuristic argument, which predicts the direction of the zonal drift correctly, is at best incomplete.

Regardless of the details of the arguments invoked to explain the origin of the drift, classical theory predicts that inertial motion on the β-plane, i.e. when the variation of f with y is included to first order in y consists of the f-plane circles and a slow westward directed drift (see for example: von Arx, 1977; Haltiner and Martin, 1957). However, no calculation of the rate of westward drift is given anywhere and, more importantly, counter heuristic arguments can be easily invoked which suggest that the drift should be directed eastward instead of westward! One such argument is based on the fact that the zonal velocity components, u, in the northern and southern tips of the Inertial circle on the f-plane are equal in magnitude and opposite in sign. Taking into account the convergence of longitudes and the relation between u and the change in longitude, $\frac{d\lambda}{dt}u/(acos(\phi))$ (where a is Earth's radius and λ is the longitude), one can infer that at the northern tip, where u is directed eastward, more longitude lines will be travelled in a given time interval compared to the southern tip, where u is directed westward. Thus, one can, erroneously, conclude the net drift should be directed eastward instead of westward. From a casual inspection of Eqs. (5.1)–(5.4), which lead to the explicit expression of the Inertial circle on the f-plane, Eq. (5.7), it becomes obvious that no such explicit solutions can be obtained on the β-plane due to the nonlinearity introduced into these equations when $f_0 + \beta y$ is substituted for f_0.

Since one of the goals of this study is to compare the dynamics on a sphere to that on the β-plane I will ignore the interesting (and difficult) case encountered on the equator $f_0 = 0$ and limit the discussion throughout the remainder of this work to the mid-latitudes, where f_0 does not vanish. The equatorial β-plane geometry is as complex as the entire sphere and no explicit solutions can be derived there. The interested reader is referred to Dvorkin and Paldor (1999), where the cross-equatorial dynamics on a sphere is analyzed using similar methodology. The goal of the present work is to provide a physical, quantifiable, description of the westward drift of Inertial circles on a rotating sphere and to contrast this theory with its β-plane counterpart. As we shall see, contrary to intuition and despite the complex mathematical structure of the equations on a sphere, the quantification of westward drift, and the analysis of the structure of the particle dynamics, is simpler on the sphere than on the β-plane since physical insight and intuition is available only in the former.

5.2 The nondimensional equations of Inertial dynamics on a sphere

The dimensional form of the momentum equations in spherical coordinates when rotation is included is (see e.g. Holton, 1992; Gill, 1982):

$$\frac{\mathrm{d}u}{\mathrm{d}t} = v\sin(\phi)\left(2\Omega + \frac{u}{a\cos(\phi)}\right), \tag{5.8}$$

$$\frac{\mathrm{d}v}{\mathrm{d}t} = -u\sin(\phi)\left(2\Omega + \frac{u}{a\cos(\phi)}\right), \tag{5.9}$$

where all variables and parameters were already defined earlier. The temporal change in the particle's coordinates is given by:

$$\frac{\mathrm{d}\lambda}{\mathrm{d}t} = \frac{u}{a\cos(\phi)}, \tag{5.10}$$

$$\frac{\mathrm{d}\phi}{\mathrm{d}t} = \frac{v}{a}. \tag{5.11}$$

The natural choice of the scales to be used in nondimensionalizing these equations are those appearing explicitly in the problem (i.e. either the equations or the initial/boundary conditions). Accordingly I choose the radius of Earth, $a = 6.4 \times 10^6$ m, as the length scale and $(2\Omega)^{-1} = (6/\pi)$ hours $\sim$1.9 hours as the time scale. The resulting velocity scale is $2\Omega a = 930$ m/s so a nondimensional velocity (e.g. in the initial conditions) of O(0.001) is sufficient to ensure the applicability of the results to ocean dynamics where velocities are O(1 m/s). Reverting to nondimensional variables (to simplify the notation the same symbols used earlier for designating dimensional variables will be used hereafter to designate the corresponding nondimensional variables) one gets:

$$\frac{\mathrm{d}u}{\mathrm{d}t} = v\sin(\phi)\left(1 + \frac{u}{\cos(\phi)}\right), \tag{5.12}$$

$$\frac{\mathrm{d}v}{\mathrm{d}t} = -u\sin(\phi)\left(1 + \frac{u}{\cos(\phi)}\right), \tag{5.13}$$

$$\frac{\mathrm{d}\lambda}{\mathrm{d}t} = \frac{u}{\cos(\phi)}, \tag{5.14}$$

$$\frac{\mathrm{d}\phi}{\mathrm{d}t} = v. \tag{5.15}$$

System (5.12)–(5.15) is a nonlinear dynamical system that can be easily integrated numerically to any time starting from a set of initial conditions (u_0, v_0, λ_0, ϕ_0) prescribed at $t=0$. However, the system's underlying dynamical structure can be best deciphered by transforming it to a Hamiltonian form. Before doing so I note that the system has two integrals of motion, i.e. quantities that remain fixed at their initial values. These quantities are: E, the (kinetic) energy, and the angular momentum, D. The energy conservation is easily derived by multiplying Eq. (5.12) by u and Eq. (5.13) by v and then adding the two equations. The resulting integral of motion is:

$$E = \frac{1}{2}(u^2 + v^2), \tag{5.16}$$

which satisfies $\mathrm{d}E/\mathrm{d}t = 0$.

The second integral of motion is obtained by dividing Eq. (5.12) through by Eq. (5.15) to get a relationship between $u(t)$ and $\cos(\phi(t))$ along the solution curve in the 4-dimensional, (u, v, λ, ϕ), space. The angular momentum, D, that results from this relation (and satisfies $\mathrm{d}D/\mathrm{d}t = 0$) is given by:

$$D = \cos(\phi)\left(\frac{\cos(\phi)}{2} + u\right). \tag{5.17}$$

It can be easily verified that D is, in fact, the component of the angular momentum vector parallel to the axis of rotation in nondimensional form. The existence of two integrals of motion, E and D, is of paramount significance in deciphering the structure of inertial dynamics on a sphere and in transforming it to a Hamiltonian form, which is the subject of the next section.

5.3 Hamiltonian form of the Inertial dynamics on a sphere: westward drift

The natural choice for a Hamiltonian of a dynamics system is the energy, which for the Inertial dynamics on a sphere is given by Eq. (5.16). However, in order for the Hamiltonian to provide an efficient description of the dynamics the coordinates and momenta have to be arranged in pairs, each of which satisfies equations known as canonical equations that determine their temporal evolution. The coordinates (ϕ, λ) and velocity components (u, v) can not be arranged in such a way. On the other hand, the inversion of the relationship between D and u in Eq. (5.17) yields:

$$u = \frac{D}{\cos(\phi)} - \frac{\cos(\phi)}{2}, \tag{5.18}$$

which can be employed to substitute D for u in the dynamical system Eqs. (5.12)–(5.15) as well as in the expression for the energy Eq. (5.16). The latter leads to the definition of the Hamiltonian function $H(\phi, v; \lambda, D)$ as:

$$H = \frac{1}{2}v^2 + \frac{1}{2}\left(\frac{D}{\cos(\phi)} - \frac{\cos(\phi)}{2}\right)^2. \tag{5.19}$$

The temporal evolution of the two pairs of (conjugate) variables (ϕ, v) and (λ, D) now follows the canonical equations:

$$\frac{d\phi}{dt} = v = \frac{dH}{dv}, \tag{5.20}$$

$$\frac{dv}{dt} = -\frac{1}{2}\sin(2\phi)\left(\frac{1}{4} - \frac{D^2}{\cos^4(\phi)}\right) = -\frac{dH}{d\phi}, \tag{5.21}$$

$$\frac{d\lambda}{dt} = \frac{D}{\cos^2(\phi)} - \frac{1}{2} = \frac{dH}{dD}, \tag{5.22}$$

$$\frac{dD}{dt} = 0 = -\frac{dH}{d\lambda}. \tag{5.23}$$

Since D is constant (i.e. λ, its conjugate coordinate, does appear in the Hamiltonian function Eq. 5.19) one can ignore the temporal changes in λ in the study of the (ϕ, v) dynamics: Eqs. (5.20), (5.21) in which D appears as a parameter. The study of the (ϕ, v) dynamics (called 1-Degree-Of-Freedom, or 1DOF, since only 1 pair of conjugate variables is present) begins with its steady states. Setting $d/dt = 0$ in Eqs. (5.20), (5.21) one immediately finds out that $v = 0$ in all steady states and that for $\phi \neq 0$ the steady state has to satisfy $D = \cos^2(\phi)/2$ (the negative root, $D = -\cos^2(\phi)/2$, is associated with unphysically large speed since $u = -\cos(\phi)$ according to Eq. 5.18). Therefore, for any value of $0 < D < 1/2$ there exists a latitude (in each hemisphere) $\phi_{ell} = \arccos(\sqrt{2D})$ where the system has a steady state. Linearizing the equations in the vicinity of these steady states one finds that the $(\phi(t), v(t))$ curve merely oscillates there with frequency $\sin(\phi_{ell})$. These oscillations are the Inertial oscillations on the f-plane solved analytically in the Introduction with the only change that the dimensional oscillation frequency, $f_0 = 2\Omega \sin(\phi_0)$, is the counterpart of $\sin(\phi_{ell})$ in the linearized nondimensional spherical system. The elliptic designation of ϕ_{ell} follows from the realization that only pure oscillations are possible near the steady state $(\phi = \phi_{ell}, v = 0)$.

The dynamical system approach to the inertial motion on a sphere goes beyond an alternate, perhaps more general, formulation of the problem. It also offers a way of quantifying the westward drift of the inertial circles. In contrast to the planar approach, an account of the zonal drift on the sphere does not require an expansion of f_0 in Eqs. (5.3)–(5.4) to first order in $(\phi - \phi_0)$. Instead, one needs to examine closely the behavior of $\lambda(t)$ that results from Eq. (5.22) for $D = \cos^2(\phi_{\mathrm{ell}})/2$. Since $\phi(t)$ is oscillatory one can expect that these (inertial) oscillations will also dominate the $\lambda(t)$ evolution. However, as in other, primarily oscillatory, dynamical systems such as a charged particle in a variable magnetic field these oscillations can be compounded by a slow drift that results from nonlinear dependence of $\mathrm{d}\lambda/\mathrm{d}t$ on $\phi(t)$. This sought drift in $\lambda(t)$ is obtained as the non-oscillatory remnant that remains as the net motion after its oscillatory part has been filtered out. Inspecting Eq. (5.22) one notices that this filtering out of the oscillatory part from $\lambda(t)$ can only be achieved by eliminating the inertial oscillations in $\phi(t)$.

A standard way of eliminating the oscillatory part of the $(\phi(t), v(t))$ trajectory is to transform the (ϕ, v) variables to another set of conjugate variables called action-angle variables (see e.g. Tabor, 1989). These two new conjugate variables are proportional to the area (I) and angle (θ) encompassed in (ϕ, v) phase space by the solution trajectory. For the inertial oscillations (or any other purely oscillatory solution in (ϕ, v)) the phase space curve is a circle whose radius is constant (recall that the circumference of the circle is the conserved energy, H). Thus, the action variable, I, is constant whose value is determined by this radius while the angle θ (the conjugate variable to I) increases monotonically with time as the solution goes around the circle. The importance of the action-angle variables is due to the fact that the energy (i.e. the value of the Hamiltonian) is independent of θ which implies that these canonical variables satisfy:

$$\frac{\mathrm{d}I}{\mathrm{d}t} = \frac{\mathrm{d}H}{\mathrm{d}\theta}; \qquad \frac{\mathrm{d}\theta}{\mathrm{d}t} = -\frac{\mathrm{d}H}{\mathrm{d}I}, \tag{5.24}$$

where $H(I, \theta; D)$ is the transformed (ϕ, v) Hamiltonian in which D is a parameter. The details of this transformation are given in Paldor (2001) and the basic steps of both the transformation and its application to the Inertial motion are briefly summarized here. The first step is to assign a numerical value to the (conserved) Hamiltonian, $EH(\phi, v; D)$, (note that λ does not appear in the expression for H) and to invert the expression for H in Eq. (5.19) to get an expression for $v(E, \phi, D)$. In the second step the

ϕ-dependence of this expression is expanded to second order in $\phi - \phi_{\text{ell}}$. In the third step the action, I, is calculated from the expression:

$$I(E, D) = \frac{1}{2\pi} \oint v(E, \phi - \phi_{\text{ell}}, D) \mathrm{d}\phi, \tag{5.25}$$

where the closed integral is carried out over an entire inertial circle. The second-order expansion in $(\phi - \phi_{\text{ell}})$ guarantees that the integral can be simply evaluated as the area of a circle (a higher-order expansion in $(\phi - \phi_{\text{ell}})$ requires the evaluation of elliptic integrals). The fourth step in the calculation is a simple inversion of the $I(E, D)$ expression in Eq. (5.25) to obtain $E(I, D)$, which is the transformed Hamiltonian.

Having completed the transformation of (ϕ, v) to the action-angle variables, (I, θ), we note that Eq. (5.24) implies that the θ-evolution, and with it the dynamics of this (ϕ, v) degree of freedom, is completely eliminated if I is kept constant so $\mathrm{d}H/\mathrm{d}I = 0$. Since the evolution equation $\lambda_t = H_D$ (Eq. 5.22) only requires that D is varied, the way to filter out the (inertial) oscillation from $\lambda(t)$ is to keep I constant when taking the derivative $\frac{\mathrm{d}E(I, D)}{\mathrm{d}D}$ i.e. calculate $\left.\frac{\partial E(I, D)}{\partial D}\right|_{I=\text{Const.}}$. The expression for the zonal drift, $\mathrm{d}\lambda_{\text{drift}}/\mathrm{d}t$, that results from this calculation is (see Eq. 5.8 in Paldor, 2001):

$$\frac{\mathrm{d}\lambda_{\text{drift}}}{\mathrm{d}t} = \left.\frac{\partial E(I, D)}{\partial D}\right|_{I=\text{Const.}} = -\frac{E}{1 - 2D} = -\frac{E}{\sin^2(\phi_{\text{ell}})}. \tag{5.26}$$

How well does this analytic expression approximate the zonal drift that obtains from numerical integration of system (5.12)–(5.15)? The results shown in Fig. 5.1a demonstrate that the analytic expression Eq. (5.26) fits the drift of the numerically calculated $\lambda(t)$ time-series excellently, to the point where the difference between the two curves cannot be detected by the naked eye. The underlying reason for the excellent fit is that the expansion of the Hamiltonian, Eq. (5.19), to second order in $(\phi - \phi_{\text{ell}})$ provides an accurate approximation to the actual curve traced out by the solution in (ϕ, v) phase space. The elliptical form of this curve is demonstrated in Fig. 5.1b.

The elliptical form of the phase space curve in Fig. 5.1b is of paramount importance in assuring that the analytical estimate of the zonal drift, which is derived from a second-order expansion of the Hamiltonian in $(\phi - \phi_0)$ is accurate. This procedure should also apply to the dynamics on the β-plane but as a physical entity the angular momentum (that provides the expression for the second conserved quantity on the sphere) does not apply to the β-plane and a counterpart has to be concocted there.

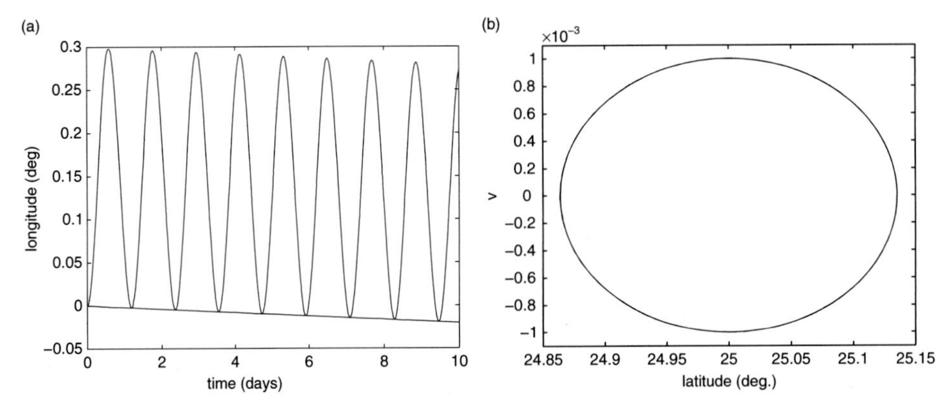

Figure 5.1 (a) The numerically calculated $\lambda(t)$ time-series (oscillatory curve) and the analytic approximation to its drift, Eq. (5.26) (straight line of constant slope). (b) The (ϕ, v) phase space portrait of the solution curve, which guarantees the accuracy of the second-order expansion of the Hamiltonian, whose value determines the ellipse's circumference, in $(\phi - \phi_{ell})$. Initial conditions: $\phi_0 = 25°$; $v_0 = 0.001$ (dimensional velocity of 1 m/s); $\lambda_0 = 0$; $D_0 = \frac{1}{2}\cos^2(\phi_0) = 0.4107$ (so $\phi_{ell} \equiv \arccos(\sqrt{2D}) = \phi_0 = 25°$).

5.4 Hamiltonian formulation on the β-plane and the zonal drift there

In principle, the mathematical procedure outlined in the previous sections for calculating the zonal drift on a sphere can also be applied to the β-plane. It is easy to repeat the above procedure with the planar equations, including the nondimensionalization and conservation properties. Using the exact same time, length and velocity scales as in Section 5.2, omitting the spherical geometry terms $u/(a\cos(\phi))$ from the momentum equations and keeping the $1/\cos(\phi)$ metric coefficients independent of ϕ (i.e. keeping dx equal to $d\lambda\cos(\phi_0)$ so it is independent of y or $(\phi - \phi_0)$) one finds that the nondimensional system on the β-plane is:

$$\frac{du}{dt} = v(\sin(\phi_0) + y\cos(\phi_0)), \tag{5.27}$$

$$\frac{dv}{dt} = -u(\sin(\phi_0) + y\cos(\phi_0)), \tag{5.28}$$

$$\frac{dx}{dt} = u, \tag{5.29}$$

$$\frac{dy}{dt} = v, \tag{5.30}$$

where y and x are the northward and eastward displacement, respectively, scaled on a.

The energy conservation follows directly from Eqs. (5.27), (5.28) in the same way it was derived on a sphere. It is easy to show that $\mathrm{d}E/\mathrm{d}t = 0$ where E has the same expression as on the sphere, Eq. (5.16):

$$E = \frac{1}{2}(u^2 + v^2). \tag{5.31}$$

The second conserved quantity, which on the β-plane has no physical meaning, can be derived by following the mathematical procedure employed on the sphere for deriving the angular momentum conservation. Accordingly, we eliminate v between Eqs. (5.27) and (5.30) to get a relation between u and y

$$D = u - y\left(\sin(\phi_0) + \frac{y\cos(\phi_0)}{2}\right), \tag{5.32}$$

where it can be easily verified that $\mathrm{d}D/\mathrm{d}t = 0$. Although the value of D is conserved, its expression in Eq. (5.32) does not provide any physical meaning to D as was the case on the sphere, where D is the angular momentum.

Inverting this relationship to express u as a function of D and substitution in the expression for the kinetic energy one gets the Hamiltonian:

$$H = \frac{1}{2}v^2 + \frac{1}{2}\left(D + y\left(\sin(\phi_0) + \frac{y\cos(\phi_0)}{2}\right)\right)^2. \tag{5.33}$$

It is straightforward to show that (y, v) and (x, D) are indeed two pairs of conjugate variables that satisfy the canonical equations associated with this Hamiltonian:

$$\frac{\mathrm{d}y}{\mathrm{d}t} = v = \frac{\mathrm{d}H}{\mathrm{d}v}, \tag{5.34}$$

$$\frac{\mathrm{d}v}{\mathrm{d}t} = -\left(D + y\left(\sin(\phi_0) + \frac{y\cos(\phi_0)}{2}\right)\right)(\sin(\phi_0) + y\cos(\phi_0)) = -\frac{\mathrm{d}H}{\mathrm{d}y}, \tag{5.35}$$

$$\frac{\mathrm{d}x}{\mathrm{d}t} = D + y\left(\sin(\phi_0) + \frac{y\cos(\phi_0)}{2}\right) = \frac{\mathrm{d}H}{\mathrm{d}D}, \tag{5.36}$$

$$\frac{\mathrm{d}D}{\mathrm{d}t} = 0 = -\frac{\mathrm{d}H}{\mathrm{d}x}. \tag{5.37}$$

In this system all $\cos(\phi_0)$ terms originate from the β term in the momentum equation and are not present on the f-plane. Since the dynamics on the β-plane has a 2DOF Hamiltonian form as on the sphere, the zonal drift can be calculated there in the same way it was done in Section 5.3. This analysis is new and I will present it in more details than the spherical analysis.

The steady states of system (5.34)–(5.37) are obtained by setting $\mathrm{d}/\mathrm{d}t=0$ in all four equations. Accordingly, one immediately finds out that all steady states satisfy $v=0$ (from Eq. 5.34) and $D=\text{Const.}$ (from Eq. 5.37). The remaining two nonlinear equations, Eqs. (5.35) and (5.36), are satisfied simultaneously when the condition $u=0$ is translated into a relationship between y_{ell} and D via Eq. (5.32). The two roots of this quadratic equation are:

$$y_{\mathrm{ell}} = -\tan(\phi_0) \pm \sqrt{\tan^2(\phi_0) - \frac{2D}{\cos(\phi_0)}}. \tag{5.38}$$

The value of D is restricted only by the condition that the determinant remains nonnegative, $D \geq \frac{\sin^2(\phi_0)}{2\cos(\phi_0)}$, and every such D value is an acceptable steady state provided that y_{ell} is given by Eq. (5.38). (Recall that on the sphere the physically acceptable values of the angular momentum were bounded between 0 and 1/2.) However, in contrast to the sphere where one of the two steady states is associated with unphysically large speeds, on the β-plane both steady states have $u=0$. Having found the steady states, the 1DOF system (5.34)–(5.35) can now be linearized about these steady states to find the system's evolution **near** them. The eigenvalues of the linearized system, μ, satisfy: $\mu^2 = -(\sin(\phi_0)+y_{\mathrm{ell}}\cos(\phi_0))^2$, where y_{ell} is one of the two steady states in Eq. (5.38) and substitution of the latter equation in this expression for μ^2 yields:

$$\mu^2 = -\sin^2(\phi_0)(1 - 2D\cot(\phi_0)). \tag{5.39}$$

Thus, $D=0$ (i.e. $u_0=0=y_0$) yields oscillations at the (nondimensional) frequency $\sin(\phi_0)$ as in the inertial oscillations on both the f-plane and on the sphere.

Expanding y near y_{ell} by letting $y=y_{\mathrm{ell}}+\epsilon$ in the Hamiltonian (5.33) and evaluating the action $I \equiv \frac{1}{2\pi}\oint v\mathrm{d}y$ (where the integration is taken around an inertial circle) yields:

$$I = \frac{1}{\pi}\int_{-\max(\epsilon)}^{\max(\epsilon)} \sqrt{2E - \left[D + (y_{\mathrm{ell}}+\epsilon)(\sin(\phi_0) + \frac{\cos(\phi_0)(y_{\mathrm{ell}}+\epsilon)}{2}\right]^2}\,\mathrm{d}\epsilon. \tag{5.40}$$

Expanding the integrand to $O(\epsilon^2)$ and imposing the steady-state condition $D + y_{\mathrm{ell}}\left(\sin(\phi_0) + \frac{y_{\mathrm{ell}}\cos(\phi_0)}{2}\right) = 0$, one gets:

$$I = \frac{1}{\pi}\int_{-\max(\epsilon)}^{\max(\epsilon)} \sqrt{2E - (\sin^2(\phi_0) - 2D\cos(\phi_0))\epsilon^2 \mathrm{d}\epsilon}. \tag{5.41}$$

This last integral can be simply calculated as half the area of a circle whose radius-squared is $\frac{2E}{\sin^2(\phi_0)-2D\cos(\phi_0)}$, which yields the expression for the action:

$$I = \frac{E}{\sqrt{\sin^2(\phi_0) - 2D\cos(\phi_0)}}, \tag{5.42}$$

which can be easily inverted to yield: $E = I\sqrt{\sin^2(\phi_0) - 2D\cos(\phi_0)}$.

Calculating the rate of drift in x, $\mathrm{d}x_{\mathrm{drift}}/\mathrm{d}t$, that follows from Eq. (5.36) with I held fixed to eliminate the (y, v) inertial oscillations (i.e. evaluating $\frac{\partial E}{\partial D}|_{I=\mathrm{Const}}$) one gets:

$$\frac{\mathrm{d}x_{\mathrm{drift}}}{\mathrm{d}t} = \left.\frac{\partial E}{\partial D}\right|_{I=\mathrm{Const.}} = -\cos(\phi_0)\frac{E}{\sin^2(\phi_0) - 2D\cos(\phi_0)}. \tag{5.43}$$

For the initial conditions stated above $u_0 = 0 = y_0$ one gets $D = 0$ in this expression in which case this expression for the drift has the same form as the expression for the drift in λ – Eq. (5.26):

$$\frac{\mathrm{d}x_{\mathrm{drift}}}{\mathrm{d}t} = -\cos(\phi_0)\frac{E}{\sin^2(\phi_0)}. \tag{5.44}$$

It is obvious that the $\cos(\phi_0)$ coefficient in the β-plane drift, that originates from the β term of the Coriolis acceleration, merely transforms the drift in λ – Eq. (5.26) – to a drift in x. This transformation is only required because of the trivial geometric correspondence $\mathrm{d}x = \mathrm{d}\lambda \cos(\phi_0)$. One can conclude that the sole contribution of the, so-called, β-effect is to transform the λ-coordinate of the spherical geometry to the x-coordinate of the planar geometry.

The results shown in Fig. 5.2 demonstrate that the analytic estimate for the drift on the β-plane provides an accurate approximation for the mean (over several inertial oscillations) of the numerically calculated $x(t)$ time-series. As in the case of the sphere, the reason for the excellent fit is that the phase space (y, v) solution curve (Fig. 5.2b) is accurately approximated by the second-order expansion of the Hamiltonian in $(y - y_{\mathrm{ell}})$.

5. Concluding remarks

Analytical estimates for the zonal drift of Inertial motion on the sphere and on the β-plane can be obtained by employing a similar procedure in the two geometries based on Hamiltonian formulation of the momentum equations and a transformation to action-angle variables. This transformation is made possible by the existence of a second (in addition to the Hamiltonian itself) conserved quantity, which guarantees the system's integrability. The resulting

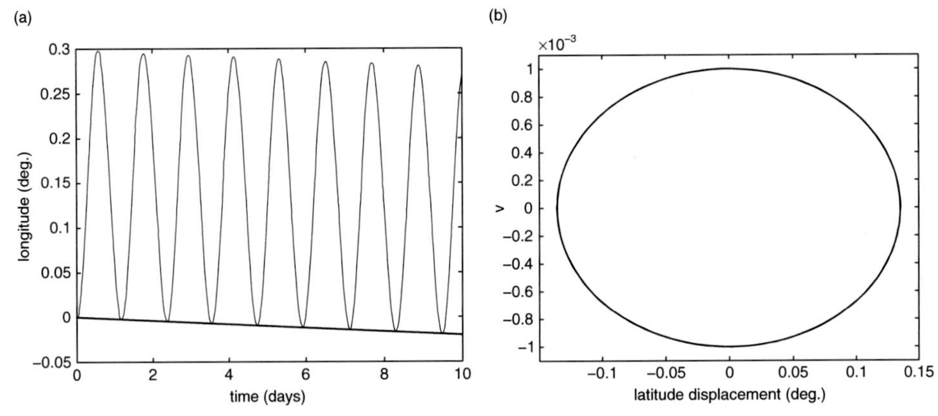

Figure 5.2 (a) The numerically calculated $x(t)$ time-series and the analytic approximation to its drift, Eq. (5.44), both translated to degrees of longitude by dividing the nondimensional x and x_{drift} values by $\cos(\phi_0)$. (b) The (y, v) phase space portrait of the solution curve, which guarantees the accuracy of the second-order expansion of the Hamiltonian (whose value determines the ellipse's circumference). Since $y_{\mathrm{ell}}=0$ the nondimensional $y(t)$ time series is the latitude displacement of the particle from $\phi_0=25°$. Initial conditions: $y_0=0$; $u_0=0$ (i.e. $D=0$); $v_0=0.001$ (dimensional velocity of 1 m/s) and $x_0=0$.

estimates for the drift are amazingly similar in the two geometries and are extremely accurate as is evident upon comparing them to direct numerical integration of the longitudinal coordinates: $\lambda(t)$ on the sphere and $x(t)$ on the β-plane. The accurate agreement between the sphere and the β-plane is due to the smallness of the terms on the right-hand side of the governing equations that are neglected or approximated on the latter. The neglected term is $u/\cos(\phi_0)$ and the inconsistently approximated terms are the $\cos(\phi)$ metric coefficients that are evaluated at $\phi=\phi_0$ while the $\sin(\phi)$, Coriolis, term is expanded to first order in $(\phi-\phi_0)$. Since for oceanic values of the velocity the neglect, or inconsistent approximation, of these terms yields only tiny differences on the right-hand side of the governing equations in the two geometries the (inertial) oscillation frequency and the westward drift estimate are identical.

There is, however, a fundamental difference between the sphere and the β-plane. While on the sphere the second conserved quantity is the angular momentum the conserved quantity on the β-plane does not have a physical meaning and it is only a mathematical concoction that happens to be conserved by the dynamics. Moreover, the conserved quantity on the β-plane does not accurately approximate the expression of the (physically meaningful) angular momentum as is the case on the sphere.

To demonstrate this difference I show in Fig. 5.3 the time series of the conserved quantity on the β-plane Eq. (5.32) and the physical angular

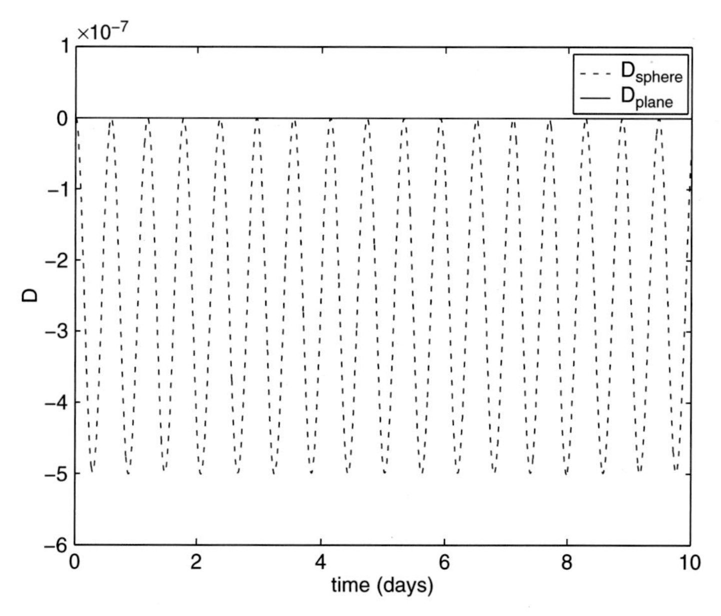

Figure 5.3 The β-plane dynamics: Evolution of the angular momentum D_{sphere} and the unphysical conserved quantity, D_{plane}, there. Clearly, the β-plane dynamics does not conserve the angular momentum $D_{\text{sphere}} = \cos(\phi_0 + y)\left(\frac{1}{2}\cos(\phi_0 + y) + u\right)$ whereas it does conserve the concocted quantity $D_{\text{plane}} = u - y\left(\sin(\phi_0) + \frac{y\cos(\phi_o)}{2}\right)$ (Eq. (5.32)). The maximal error in the physical D_{sphere} on the β-plane is of order $\frac{1}{2}v_0^2$. Initial conditions and ϕ_0 as in Fig. 5.2.

momentum Eq. (5.17) as computed from the β-plane dynamics, Eqs. (5.27)–(5.30). Clearly, the β-plane accurately conserves the wrong physical quantity. The error in the angular momentum conservation on the β-plane can be estimated analytically by expanding the expression for u in Eq. (5.18) to second order in $(\phi - \phi_0)$ (with $D = \frac{1}{2}\cos^2(\phi_0)$) and comparing it to its planar counterpart Eq. (5.32). The resulting two expression for u_{sphere} and u_{plane} are:

$$u_{\text{sphere}} = \sin(\phi_0)(\phi - \phi_0) + \frac{1}{2\cos(\phi_0)}(\phi - \phi_0)^2, \tag{5.45}$$

$$u_{\text{plane}} = \sin(\phi_0)y + \frac{\cos(\phi_0)}{2}y^2. \tag{5.46}$$

Since the nondimensional y (on the β-plane) equals $(\phi - \phi_0)$ (on the sphere) one can immediately conclude from this comparison that on the β-plane the $u(y)$ relationship is in second-order error.

A more detailed (and revealing) comparison between the $u(y)$ relationship on the β-plane and the $u(\phi - \phi_0)$ relationship on the sphere is shown in Fig. 5.4.

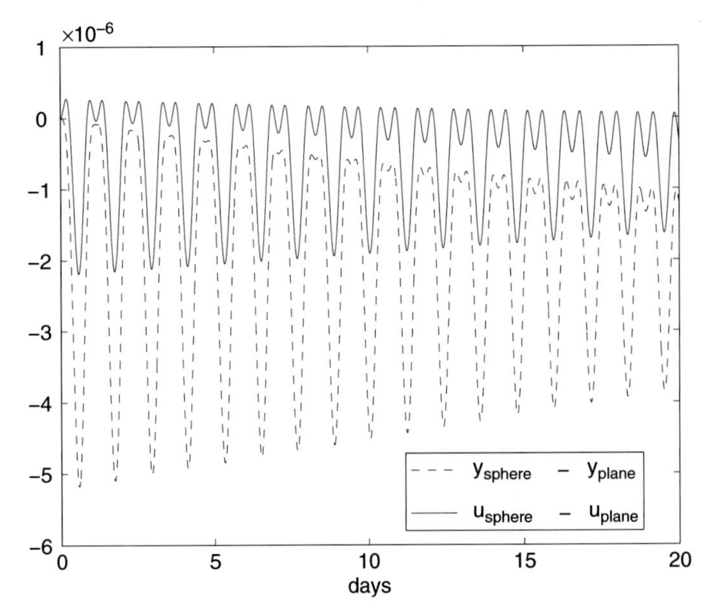

Figure 5.4 The zonal velocity and meridional displacement differences between the sphere and the β-plane. Recall that the nondimensional value of u is 0.001 and that the meridional displacement from ϕ_0 (or y on the plane) is order 0.1° i.e. 0.002 Rad. so the absolute difference of order 10^{-6} in this figure represents a relative difference between the sphere and the (β-)plane of order 10^{-3} only. Initial conditions: $v_0 = 0.001$ (dimensional velocity of 1 m/s); $u_0 = 0$; $y_0 = 0$; $\phi_0 = 25°$; and $x_0 = 0$. These initial conditions guarantee that the conserved quantities satisfy: $D_{\text{plane}} = 0$ and $D_{\text{sphere}} - \frac{1}{2}\cos^2(\phi_0) = 0$.

The curves of $(u_{\text{sphere}} - u_{\text{plane}})$ and $(y_{\text{sphere}} - y_{\text{plane}})$ demonstrate both the smallness of errors in u and y, and the high correlation between the errors in the two variables, on the β-plane. The second frequency in the y-error develops only after about ten days in response to its presence in u.

The results developed in this work can easily be extended to the case when a meridional pressure gradient is present. Given the small velocities involved in oceanic applications of the theory the sole contribution of this pressure gradient is to add a steady zonal velocity to the westward drift. The reader is referred to Paldor (2001) for a detailed demonstration of how the present calculations can be accurately extended to include a geostrophic zonal velocity.

Acknowledgments

Financial support for this work was granted by the Office of Naval Research under grants N00014-99-1-0049 and N00014-03-1-0284 to University of Miami.

References

Durran, D. R., 1993. Is the Coriolis force really responsible for the inertial oscillations? *Bull. Am. Met. Soc.*, **74(11)**, 2179–84.
Dvorkin, Y. and N. Paldor, 1999. Analytical considerations of Lagrangian cross-equatorial flow. *J. Atmos. Sci.*, **56(9)**, 1229–37.
Gill, A. E., 1982. *Atmosphere-Ocean Dynamics*. San Diego: Academic Press.
Haltiner, G. J. and F. L. Martin, 1957. *Dynamical and physical meteorology*. New York: McGraw-Hill.
Holton, G., 1992. *An introduction to Dynamic Meteorology*. Second Edition. New York: Academic Press.
Moura, A. D., 1976. The eigensolutions of the linearized balance equations over a sphere. *J. Atmos. Sci.*, **33(6)**, 877–907.
Paldor, N., 2001. The zonal drift associated with time-dependent particle motion on the earth. *Q. J. Roy. Meteor. Soc.*, **127(577A)**, 2435–50.
Persson, A., 1998. How do we understand the Coriolis force? *Bull. Am. Met. Soc.*, **79(7)**, 1373–85.
Stommel, H. M. and D. W. Moore 1989. *An Introduction to the Coriolis Force*. New York: Columbia University Press.
Tabor, M., 1989. *Chaos and Integrability in Nonlinear Dynamics: An Introduction*. New York: J. Wiley and Sons Inc.
von Arx, W. S., 1977. *An Introduction to Physical Oceanography*. Reading MA: Addison-Wesley.

6

Predictability of Lagrangian motion in the upper ocean

LEONID I. PITERBARG
Department of Mathematics, University of Southern California, Los Angeles, California, USA
TAMAY M. ÖZGÖKMEN
Rosenstiel School of Marine and Atmospheric Science, University of Miami, Miami, Florida, USA
ANNALISA GRIFFA
Rosenstiel School of Marine and Atmospheric Science, University of Miami, Miami, Florida, USA
ISMAR/CNR, La Spezia, Italy
AND
ARTHUR J. MARIANO
Rosenstiel School of Marine and Atmospheric Science, University of Miami, Miami, Florida, USA

6.1 Introduction

The prediction particle trajectories in the ocean is of practical importance for problems such as searching for objects lost at sea, tracking floating mines, designing oceanic observing systems, and studying ecological issues such as the spreading of pollutants and fish larvae (Mariano *et al.*, 2002). In a given year, for example, the US Coast Guard (USCG) performs over 5000 search and rescue missions (Schneider, 1998). Even though the USCG and its predecessor, the Lifesaving Service, have been performing search and rescue operations for over 200 years, it has only been in the last 30 years that Computer Assisted Search Planning has been used by the USCG. The two primary components are determining the drift caused by ocean currents and the movement caused by wind. The results presented in this review are motivated by the drift estimation problem.

A number of authors (e.g., Aref, 1984; Samelson, 1996) have shown that prediction of particle motion is an intrinsically difficult problem because Lagrangian motion often exhibits chaotic behavior, even in regular and simple Eulerian flows. In the ocean, the combined effects of complex time-dependence (Samelson, 1992; Meyers, 1994; Duan and Wiggins, 1996) and three-dimensional structure (Yang and Liu, 1996) are likely to induce chaotic transport. Chaos implies strong dependence on the initial conditions, which are usually not known with great accuracy, so that the task of predicting particle motion is often extremely difficult. Also, the temporal limit for predicting particle positions is reduced as velocity errors accumulate in time due to the nonrandom aspect of Lagrangian errors.

Lagrangian Analysis and Prediction of Coastal and Ocean Dynamics, ed. A. Griffa, D. Kirwan, A. Mariano, T. Özgökmen, and T. Rossby. Published by Cambridge University Press.

One of the approaches for representing chaotic ocean dynamics is to use Lagrangian stochastic models (LSM). Under this approach, the Lagrangian velocity of an infinitely small fluid parcel, a particle, is decomposed into a large-scale deterministic component and a velocity fluctuation described by a stochastic differential equation. Single particle LSMs were first developed in atmospheric research (Thomson, 1986, 1987), and then applied to ocean studies for estimating transport and mixing parameters (Griffa *et al.*, 1995; Griffa, 1996). These models, often indicated as "random flight" models, provide information about the probability of finding particles in a specific area, based on the known statistics of the velocity field, such as the turbulent velocity variance and Lagrangian decorrelation time scale. Reasonable success has been achieved in reproducing the distribution of real drifters based on such models (e.g., Falco *et al.*, 2000). LSMs with spin or of higher order are needed to describe Lagrangian motion in more energetic regions characterized by coherent structures such as rings and vortices (e.g., Berloff and McWiliams, 2002; Berloff *et al.*, 2002; Berloff and McWilliams, 2003; Veneziani *et al.*, 2004). As for Lagrangian predictability, single particle LSMs have a serious limitation since they do not account for possible correlations between different particles. Such correlations are present, for example, when the particles are within a Rossby radius of deformation of each other.

The first effort to account for correlations in particle motion was made by Özgökmen *et al.* (2000) in the framework of a process study using synthetic drifter trajectories from a double-gyre Miami Isopycnic Ocean Model (MICOM) simulation. In that study, drifter data in the vicinity of predicted trajectories were used to improve the accuracy of prediction. Further testing has been conducted (Castellari *et al.*, 2001) with real drifter data collected in the Adriatic Sea (Poulain, 1999), a semi-enclosed sub-basin of the Mediterranean Sea, to show the validity of this method. In particular, using other particles (predictors) in the vicinity of the predicted trajectory (predictand) is motivated by search operations in the sea. An important part of such operations is to properly narrow the search area based on the best possible prediction of the position of a lost object given its approximate initial position (Schneider, 1998). Knowledge of only the mean current leads to inaccurate prediction of trajectories because of strong velocity fluctuations advecting the object away from the path indicated by the mean velocity field. One can expect significant help from other floating objects in the same area, like debris or buoys/drifters that can be tracked by planes or satellites.

Results from Özgökmen *et al.* (2000) stimulated development of multi-particle LSMs intended for the predictability studies. A multi-particle LSM describes the joint motion of two or more particles in a flow containing a stochastic component. A first order multi-particle LSM, extending that of Thomson (1987) and Griffa (1996), has been put forth and investigated by Piterbarg (2001a). Piterbarg (2001b) applied the multi-particle LSM to the predictability problem. Some earlier efforts in constructing multi-particle LSMs were focused on fitting observations and ensuring the well-mixed condition (Thomson, 1990; Sawford, 1993). Those models are hardly appropriate for prediction studies in the ocean.

In view of the suggested multi-particle LSM, the prediction algorithm used in Özgökmen *et al.* (2000) turned out to be a modification of the well-known Kalman filter. In addition, the theoretical analysis in the multi-particle framework has allowed a slightly improved algorithm. This new procedure has been applied to real drifter data in the Tropical Pacific Ocean (Özgökmen *et al.*, 2001). The results of this work, as well as others by the authors, are reviewed as detailed below.

The paper is organized in accordance with our research philosophy: to develop a rigorous approach to a problem, to test theoretical conclusions on synthetic data generated by LSMs, to proceed to synthetic data from simplified ocean circulation models, and finally, to apply to real *in-situ* data. Section 6.2 addresses a mathematical framework for the Lagrangian prediction problem. Multi-particle LSMs used in the theoretical predictability study are presented in Section 6.3. Specific prediction algorithms are introduced in Section 6.4. In Section 6.5, results of testing these algorithms on synthetic data are presented. Examples of real data analysis are given in Section 6.6. Optimal predictor sampling is discussed in Section 6.7, where new theoretical findings are presented. Finally, we discuss future research directions in Section 6.8.

6.2 Problem statement

There can be different formulations of the prediction problem depending on the known information and real-life application. Paldor *et al.* (2004) used a surface current climatology and a slab model, forced by surface wind fields to forecast particle position. In recent studies (Piterbarg, 2004a; Griffa *et al.*, 2004), the Lagrangian predictability problem is considered as an uncertainty in a particle position due to uncertainty of the Eulerian velocity field. The related problem of estimating and reconstructing Eulerian velocity fields from Lagrangian data is attacked by Toner *et al.* (2001a,b).

Here, we concentrate on a specific formulation for the application of search and rescue operations, namely that predictor data natural or drifters/buoys are available. The methodology for the evaluation of different Lagrangian trajectory prediction algorithms is as follows. Several drifters are released simultaneously at different, known positions in a turbulent flow. One of the floats, called the predictand, is unobservable, while the remaining floats, called the predictors, are observed. The problem is to predict the position of the unobservable float at any time given its initial position and the predictor observations. Apart from practical needs, this problem is of great importance from the theoretical viewpoint since it addresses the predictability issue for the Lagrangian motion in turbulent flows. Here, we assume that the turbulent velocity field is described by stochastic differential equations. Thus, a least-squares kinematic approach is employed: given a set of flow statistics and observations of nearby drifter positions, minimize the mean square error of a prediction algorithm.

Let $\mathbf{u}(t, \mathbf{r})$ be a time-varying, two-dimensional random velocity field. Consider $M > 1$ Lagrangian particles starting at the time $t = 0$ from different positions $\mathbf{r}_1^0, \mathbf{r}_2^0, \ldots, \mathbf{r}_M^0$. Their motion is governed by the following system of $2M$ equations:

$$\frac{d\mathbf{r}_j}{dt} = \mathbf{u}(t, \mathbf{r}_j), \qquad \mathbf{r}_j(0) = \mathbf{r}_j^0, \tag{6.1}$$

$j = 1, \ldots, M$. Assume that trajectories of the first $p = M - 1$ particles $\mathbf{r}_1(t)$, $\mathbf{r}_2(t), \ldots, \mathbf{r}_p(t)$ are completely observed during time interval $(0,T)$ while the trajectory of the last one, $\mathbf{r}_M(t)$, is not observed. The problem is to find a reasonable prediction of the position of the unobserved particle given the above predictor observations and the initial predictand position. The optimal prediction in the mean square sense

$$E|\hat{\mathbf{r}}_M(T) - \mathbf{r}_M(T)|^2 \to \min$$

is given by the conditional expectation (e.g., Liptser and Shiryaev, 2000)

$$\hat{\mathbf{r}}_M(T) = E\big(\mathbf{r}_M(T) \mid \mathbf{r}_M(0), \mathbf{r}_1(t), \mathbf{r}_2(t), \ldots, \mathbf{r}_p(t), \quad 0 \leq t \leq T\big),$$

based on all the observations. Theoretically speaking, this conditional expectation is expressed in terms of the joint probability distribution of all M trajectories, which is a distribution in a functional space of $2M$-vector functions. If the Lagrangian displacement is a Markovian process, which is the case for a white noise velocity fluctuation, then the expectations depend on $\mathbf{r}_1(T)$, $\mathbf{r}_2(T), \ldots, \mathbf{r}_p(T)$ only. Under some conditions a stochastic differential equation can be derived for the optimal prediction in this case (e.g. Liptser and Shiryaev, 2000). Finally, if the joint distribution of $\mathbf{r}_1(T)$, $\mathbf{r}_2(T), \ldots, \mathbf{r}_p(T)$ is Gaussian, then the dependence is linear and the corresponding weighting coefficients are determined by the mean flow and statistics of the second order, such as the Lagrangian velocity covariance tensor. Unfortunately, Markovian and

Gaussian approximations for the particle displacement are not satisfactory for real oceanic flows. As for the former, there is a strong indication (e.g., Griffa *et al.*, 1995, Bauer *et al.*, 2002, Veneziani *et al.*, 2004) that the Lagrangian *velocity* has a significant correlation time of 1–10 days and its time evolution can be approximated well by a stationary Markovian process. Thus, the *displacement should be viewed as an integral of a Markovian process.* Regarding the Gaussian approximation, note that even if the Eulerian velocity fluctuation is a Gaussian white noise in time, the *joint distribution of a particle pair is not Gaussian.* Therefore, one should rule out any effort to get an exact expression for the optimal prediction. However, if the joint process of Lagrangian velocities $(\mathbf{v}_1(t), \mathbf{v}_2(t),\ldots, \mathbf{v}_M(t))$ is diffusive, then so is the joint process of velocities and positions, $(\mathbf{v}_1(t), \mathbf{r}_1(t), \mathbf{v}_2(t), \mathbf{r}_2(t), \ldots, \mathbf{v}_M(t), \mathbf{r}_M(t))$, and one can try to apply an Extended Kalman filter (EKF). An EKF for multi-particle Lagrangian stochastic model is now formulated.

6.3 Multi-particle LSM

Let $\mathbf{w}(t, \mathbf{r})$ be a continuous family of two-dimensional, Brownian motions parameterized by the space coordinate $\mathbf{r}$. In other words, for any two points in the space, $\mathbf{r}_1$ and $\mathbf{r}_2$, the stochastic processes $\mathbf{w}_1(t) = \mathbf{w}(t, \mathbf{r}_1)$ and $\mathbf{w}_2(t) = \mathbf{w}(t, \mathbf{r}_2)$ are correlated Brownian motions and their covariance is given by a fixed tensor $\mathbf{B}(\mathbf{r}_1, \mathbf{r}_2)$ such that

$$E\{\mathrm{d}\mathbf{w}_1\mathrm{d}\mathbf{w}_2^{\mathrm{T}}\} = \mathbf{B}(\mathbf{r}_1, \mathbf{r}_2)\mathrm{d}t,$$

where the superscript stands for transposition and all vectors are column vectors. The following stochastic differential equations determine a flow in the position/velocity phase space

$$\mathrm{d}\mathbf{r} = (\mathbf{U}(t, \mathbf{r}) + \mathbf{v})\mathrm{d}t, \quad \mathrm{d}\mathbf{v} = -(\mathbf{v}/\tau)\mathrm{d}t + \mathrm{d}\mathbf{w}(t, \mathbf{r}), \tag{6.2}$$

where $\mathbf{U}(t, \mathbf{r})$ is a deterministic large-scale velocity field (the mean flow) and τ is the Lagrangian correlation time. For any collection of particles (finite, or countable, or even continuous), with given positions and velocities at moment $t = 0$, equations (6.2) allow us to find their positions and velocities at any prescribed moment $t > 0$.

It can be shown that the Lagrangian formulation (6.2) is equivalent to a non-linear partial differential equation for the underlying Eulerian velocity field, $\mathbf{u}(t, \mathbf{r})$ driven by a white noise forcing distributed in space and characterized by space covariance $\mathbf{B}(\mathbf{r})$ (Piterbarg, 2001a). This equation differs from the 2D Navier–Stokes equation in two aspects: (1) the viscous term ($\nu\Delta\mathbf{u}$) is replaced by the low-frequency friction ($-\tau^{-1}\mathbf{u}$); (2) the pressure gradient is assumed to be known and included in the forcing rather than being determined by using the continuity equation. Regarding the latter, note that the continuity equation cannot be added to (6.2) since that would make this system over-determined. For this reason, the degree of divergence of (6.2) is measured by the Lyapunov exponent that is introduced and discussed below. A deterministic model incorporating effects of wind forcing, friction and Coriolis terms on particle motion has been used in Paldor *et al.* (2004). Note also that the Coriolis term, or in general spin, in the Eulerian equation can be readily accounted for in (6.2) (see the cited paper and Veneziani *et al.*, 2004), however for our present purposes we consider the model without spin. Overall, LSM is essentially a simple parameterization of the general dynamics for particle motion, in which the effect of the classical forcing terms can be included also in the statistically determined model parameters.

It follows from (6.2) that the positions and velocities of any M particles are covered by the following finite system of stochastic differential equations (Piterbarg, 2001a, b)

$$\mathrm{d}\mathbf{r}_m = (\mathbf{U}(t, \mathbf{r}_m) + \mathbf{v}_m)\mathrm{d}t, \quad \mathrm{d}\mathbf{v}_m = -(\mathbf{v}_m/\tau)\mathrm{d}t + \sigma_{mj}\mathrm{d}\mathbf{w}_j, \quad m = 1, \ldots, M, \tag{6.3}$$

where $\mathbf{w}_j$, $j = 1, \ldots, M$ are mutually independent two-dimensional Brownian motions, the summation over j is meant and for 2×2 matrices σ's satisfy

$$\sigma_{kj}\sigma_{lj} = \mathbf{B}(\mathbf{r}_k - \mathbf{r}_l).$$

Thus the motion of each single particle is described by the classical Langevin equation for its velocity, however the noise terms for different particles are correlated, thereby the particle positions and velocities in (6.3) are correlated as well. The only exception is the case of diagonal (σ_{ij}), which corresponds to the absence of spatial correlation. In other words, for any M particles, their positions and velocities form a classical multi-diffusion process in $4M$ dimensions with drift explicitly expressed in terms of the mean flow and τ and a diffusion matrix completely determined by the covariance tensor $\mathbf{B}$. Finally, note that the M velocity equations in (6.3) can be written as a single Langevin equation for the "aggregate" velocity vector including the velocities of all particles.

In the case of zero mean flow and isotropic forcing $\mathbf{w}(t, \mathbf{r})$, the model is completely defined by τ and longitudinal and normal covariance functions corresponding to tensor $\mathbf{B}(\mathbf{r})$. If one fixes the shape of these functions (for example, a Gaussian bell), then the multi-particle model is determined by only three parameters: Lagrangian correlation time τ, variance of one of the velocity components σ_u^2, and the forcing correlation space scale R, which can also be viewed as the correlation scale of the Eulerian velocity field.

Let us sum up the main features of the model (6.3) important from the application viewpoint. First, the one-particle motion is described by a "random flight" model that is very common and well founded in oceanography and meteorology (Thomson, 1986; Griffa, 1996; Rodean, 1996; Reynolds, 1998). Second, the model can mimic all the relative dispersion regimes known from the classical hydrodynamics approach, such as Richardson t^3 (Piterbarg, 2004b). Third, the model can be deduced from a hydrodynamic type stochastic equation for the corresponding Eulerian velocity field. Hence, it has a solid physical background. Next, in the isotropic case, it depends on a few parameters that can be directly estimated from ocean data (e.g. Griffa *et al.*, 1995; Bauer, 1998). Finally, it yields a mathematically consistent description of the Lagrangian motion of any number of particles.

Note that even the two-particle motion modeling is a non-trivial problem. Some physically reasonable models have been formulated for this purpose (e.g., Thomson, 1990; Sawford, 1993; Borgas and Sawford, 1994). These models are three-dimensional and are also based on diffusion equations like (6.3) with a general drift and diagonal covariance matrix of the stochastic forcing, i.e., the forcing is assumed to be a white noise in space as well as in time. The above papers address finding a drift such that the two-point Eulerian velocity distribution is Gaussian. Note, first, that in the case of general space covariance, there is no such drift and, second, that the hypothesis of joint Gaussianity is hardly acceptable for the velocity in the upper ocean or for fluid mechanics in general. Another approach to the two-particle motion is addressed in Pedrizzetti and Novikov (1994). In the same Markovian framework, the drift is found from physical considerations and then the corresponding two-point Eulerian velocity distribution is computed from the corresponding Fokker–Planck equation. The obtained solution is significantly different from Gaussian distribution, even though the initial condition was Gaussian distributed. Similar results have been presented by Pope (1987).

As it was noticed by Thomson (1987), each reasonable LSM should satisfy the so-called, well-mixed condition that states the initial uniform distribution of tracer remains uniform under the action of a flow described by the LSM. It is important to point out that this condition can be treated not as an independent assumption, but rather as a consequence of the Markovian property in the case of the divergence free Eulerian velocity field (Pedrizzetti and Novikov, 1994). In other words, no additional constraints on the drift and diffusion matrix are needed to ensure the well-mixed condition in the incompressible case.

It can be shown that the model ((6.2), (6.3)) satisfies the well-mixed condition at least for the divergence free forcing $\mathbf{w}(t, \mathbf{r})$ even though the flow itself is not incompressible. We believe that this condition holds true in any case when the top Lyapunov exponent of the flow (see definition in (6.5)) is positive. Moreover, in the framework ((6.2), (6.3)), the joint distribution of the positions and velocities of several particles could be fruitfully investigated as well as multi-point statistics of the corresponding Eulerian velocity field. Here we do not discuss these physically important issues because the present intention of multi-particle LSM ((6.2), (6.3)) is quite different. Namely, it is used as a mathematically consistent and effective tool for solving the applied Lagrangian prediction problem formulated above.

A key role in this problem belongs to the separation of two particles and its mean square value defined, respectively, as

$$\Delta\mathbf{r}(t,\mathbf{a}) = \mathbf{r}(t,\mathbf{a}) - \mathbf{r}(t,\mathbf{0}), \quad \rho(t,\mathbf{a}) = E\{\Delta\mathbf{r}(t,\mathbf{a})^2\}, \tag{6.4}$$

where $\mathbf{r}(t, \mathbf{a})$ is a trajectory started at $\mathbf{a}$ and $E\{\}$ is the expectation or ensemble average. For small initial separations $|\mathbf{a}| = d_0 \ll R$, both quantities behave exponentially at the initial stage

$$|\Delta\mathbf{r}(t,\mathbf{a})| \sim |\mathbf{a}| \exp(\lambda t), \quad \rho(t,\mathbf{a})^{1/2} \sim |\mathbf{a}| \exp(\Lambda t) \tag{6.5}$$

where λ and Λ are so called the top Lyapunov exponent and the second Lyapunov moment, respectively. Note that λ is not random, despite the fact that the separation, $\Delta\mathbf{r}$ is random. Thus, the rate of divergence (convergence) of trajectories is the same for any realization. As for signs, Lyapunov second moment Λ is always positive for isotropic flows while λ can be either positive or negative. The corresponding conditions, as well as exact expressions for λ and Λ are given in Piterbarg (2001a) and Falkovich and Piterbarg (2004). Note only that if the forcing is divergence free, then $\lambda > 0$, which is assumed to be true for the rest of the paper. The dependence of λ^{-1} and Λ^{-1} on τ and the turnover time $\tau_0 = R/\sigma_u$ is shown in Fig. 6.1a,b.

Both Lyapunov metrics exhibit a similar behavior with a strong dependence on τ_0 and a weak dependence on τ. However, values of λ^{-1} are in general 2.5 times greater than Λ^{-1}. Hence, individual Lagrangian trajectories diverge at a much slower rate than that indicated by the Lyapunov second moment. The quantity Λ^{-1} is more appropriate for estimating the predictability horizon when the predictability skill is measured by the mean square error. Asymptotics (6.5) hold until

$$t < \lambda^{-1} \ln(R/d_0), \quad t < \Lambda^{-1} \ln(R/d_0)$$

respectively. For $t \gg \Lambda^{-1} \ln(R/d_0)$, the usual inertial asymptotic

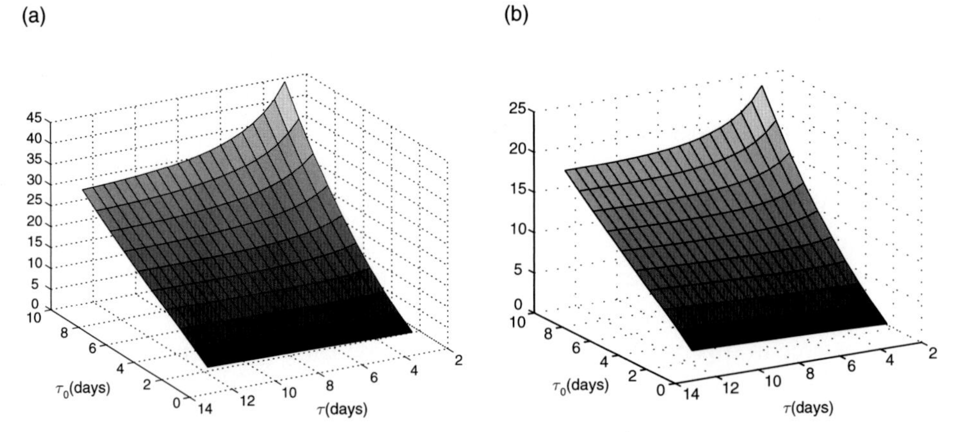

Figure 6.1 (a) Dependence of the reciprocal λ^{-1} of the top Lyapunov exponent on Lagrangian correlation time τ and turnover time τ_0. (b) Dependence of the reciprocal Λ^{-1} of the Lyapunov second moment on Lagrangian correlation time τ and turnover time τ_0.

$$\rho(t, \mathbf{a})^{1/2} \sim 2\sqrt{2}\sigma_u(\tau t)^{1/2}$$

holds true. Note that the dispersion in the inertial regime (e.g., Zambianchi and Griffa, 1994) is

$$D(t) = E\{\mathbf{r}(t, \mathbf{a})^2\}^{1/2} \sim 2\sigma_u(\tau t)^{1/2} \,. \tag{6.6}$$

Thus, λ^{-1} can be viewed as a predictability limit for an individual particle when a nearby particle is observed and the prediction error is defined as the distance between the real and predicted positions, while Λ^{-1} has the same meaning when the prediction error is measured by the mean-square deviation.

6.4 Prediction algorithms

6.4.1 Kalman filter

Given a system of stochastic differential equations (6.3) describing the behavior of the particle cluster and characterized by the mean flow $\mathbf{U}(\mathbf{r})$ and the Lagrangian time scale τ, and the observations, the Extended Kalman Filter (EKF) can be applied for the prediction of an unobserved component (e.g. Jazwinski, 1970). This cannot be done exactly, because the corresponding diffusion matrix depends on the state variable. In other words, the velocity covariance matrix in (6.3) depends on the particle positions at any time. This fact makes the direct application of EKF impossible since it assumes that the covariance matrix of the stochastic forcing does not depend on the state vector. However, the main idea of optimally combining the dynamics and observations can be realized by using the predicted and observed positions from the previous time step for evaluating the stochastic forcing covariance. The result is (see details in Piterbarg, 2001b)

$$\begin{aligned} \mathbf{v}_M^{\mathrm{a}}(n) &= \mathbf{v}_M^{\mathrm{f}}(n) + \sum_{j=1}^{p} \mathbf{K}_j(\mathbf{v}_j(n) - \mathbf{v}_j^{\mathrm{f}}(n)), \\ \mathbf{r}_M^{\mathrm{a}}(n) &= \mathbf{r}_M^{\mathrm{f}}(n) + \sum_{j=1}^{p} \mathbf{K}_j(\mathbf{r}_j(n) - \mathbf{r}_j^{\mathrm{f}}(n)). \end{aligned} \tag{6.7}$$

The superscripts a and f stand for "analyzed" and "forecasted," respectively, in accordance with the common Kalman filter notation. Note that for the observed particles the analyzed variables coincide with observational ones $\mathbf{v}_i^{\mathrm{a}}(n) = \mathbf{v}_i(n)$, $\mathbf{r}_i^{\mathrm{a}}(n) = \mathbf{r}_i(n)$, $\quad i = 1, \ldots, p$. The forecasted values are computed from

$$\begin{aligned} &\mathbf{v}_k^{\mathrm{f}}(n) = (1-\Delta t/\tau)\mathbf{v}_k^{\mathrm{a}}(n-1), \\ &\mathbf{r}_k^{\mathrm{f}}(n) = \mathbf{r}_k^{\mathrm{a}}(n-1) + (\mathbf{U}(\mathbf{r}_k^{\mathrm{a}}(n-1)) + \mathbf{v}_k^{\mathrm{a}}(n-1))\Delta t, \quad k = 1, \ldots, M \end{aligned} \tag{6.8}$$

and

$$\mathbf{K}_j = \sum_{k=1}^{p} \mathbf{B}(\mathbf{r}_M^{\mathrm{a}}(n-1), \mathbf{r}_k(n-1))\mathbf{B}_{kj}^{-1}(n-1), \qquad j = 1, \ldots, p, \tag{6.9}$$

where $\mathbf{B}_{kj}^{-1}(n)$ are entries of $(\mathbf{B}(\mathbf{r}_k(n), \mathbf{r}_j(n))^{-1}$ and variable n points to moment $n\Delta t$, where Δt is the observation interval. Note that first prediction formula (6.7) was suggested in Özgökmen *et al.* (2000) with no second term in (6.7) for the analyzed velocity and no reference to multi-particle LSMs at all. In contrast, this paper became an impulse for developing multi-particle LSMs and as a result, algorithm ((6.7)–(6.9)) got a solid theoretical boost. Moreover, the additional term for the velocity correction was found to give a slight improvement in the prediction skill in some cases, as shown in our other papers (Piterbarg, 2001b and Özgökmen *et al.*, 2001). Finally, note that this scheme can be applied to both homogeneous and inhomogeneous environments like jets and gyres. However, in practical applications we assumed $\mathbf{B}(\mathbf{r}_1, \mathbf{r}_2) = \mathbf{B}(\mathbf{r}_1 - \mathbf{r}_2)$ depending on the difference only, therefore assuming local homogeneity of fluctuations. In most cases the Gauss function $\exp(-|\mathbf{r}_1 - \mathbf{r}_2|^2/R^2)$ was assumed for the diagonal entries, while the off-diagonal entries were set to zero.

Of course, this approach does not yield the optimal estimator because of the nonlinearity of the deterministic dynamics and because of the covariance dependence of the particle positions. Moreover, it is very difficult to find analytically how close the suboptimal prediction is to the optimal one. In this situation there is nothing better than using stochastic simulations to evaluate the prediction skill. The dependence of the prediction error on R, τ, initial cluster radius d_0, and the number of predictors was analyzed by using the Monte Carlo approach (Piterbarg, 2001b). The results are briefly reviewed in the next section, Section 6.5.

6.4.2 Regression

The drawback of the previous algorithm is that it requires knowledge of the mean flow as well as the statistics of the velocity fluctuations that are often not available. For this reason another prediction procedure, which is model independent, was suggested (Piterbarg and Özgökmen, 2002). Assume the following classical regression scheme for the motion of M particles

$$\mathbf{r}_i(t) = \mathbf{A}(t)\mathbf{r}_i(0) + \mathbf{b}(t) + \mathbf{y}_i(t), \quad i = 1, \ldots, M \tag{6.10}$$

where $\mathbf{A}(t)$ and $\mathbf{b}(t)$ are unknown, random in general, 2×2-matrix and 2-vector respectively, $\mathbf{y}_i(t)$ are stochastic processes with zero mean uncorrelated for any fixed t. Note that this model does not follow from motion equations (6.1) in the general case. Moreover, it even contradicts (6.1) for a nonlinear velocity field. On the other hand, the system (6.1) can be linearized on short times so it can be expected that a prediction algorithm based on (6.10) might be useful for short-term predictability. Our strategy then is to construct a prediction algorithm based on (6.10) and then to investigate its performance for some important particular cases of (6.1).

The least-square estimators of $\mathbf{A}(t)$ and $\mathbf{b}(t)$ based on the observed particles at the moment t are given by

$$\hat{\mathbf{A}}(t) = \mathbf{S}(t)\mathbf{S}(0)^{-1}, \quad \hat{\mathbf{b}}(t) = \mathbf{r}_c(t) - \hat{\mathbf{A}}(t)\mathbf{r}_c(0),$$

where

$$\mathbf{r}_c(t) = \frac{1}{p}\sum_{i=1}^{p} \mathbf{r}_i(t)$$

is the center of mass of the predictor cluster and

$$\mathbf{S}(t) = \sum_{i=1}^{p} (\mathbf{r}_i(t) - \mathbf{r}_c(t))(\mathbf{r}_i(0) - \mathbf{r}_c(0))^{\mathrm{T}}.$$

It is assumed that $p > 2$ to have a non-degenerate matrix $\mathbf{S}(0)$. The obtained estimators then are used to predict the unobservable particle

$$\hat{\mathbf{r}}_M(t) = \mathbf{r}_c(t) + \mathbf{S}(t)\mathbf{S}(0)^{-1}(\mathbf{r}_M(0) - \mathbf{r}_c(0)). \tag{6.11}$$

This prediction formula, referred to as RA (Regression Algorithm), is optimal in the framework of the model (6.10) if $\mathbf{A}$ and $\mathbf{b}$ are deterministic. In particular, if the predictand is initially located at the center of the mass (CM) of the cluster, then the predicted position coincides with the CM of the predictors for all time. In other words, the RA algorithm can be viewed as a CM algorithm with adjustment for the initial position of the unobservable particle. It was shown in Piterbarg and Özgökmen (2002) that in the framework of isotropic LSM (6.3), for the initial stage

$$s^2(t) = E|\hat{\mathbf{r}}_M(t) - \mathbf{r}_M(t)|^2 \sim C\rho(t) \tag{6.12}$$

where the constant C depends on the initial positions only and $\rho(t) = \exp(2\Lambda t)$. As for large t, the same formula (6.12) is valid with $\rho(t) = t$ and a different C, also depending on the initial configuration. In this case the asymptotic is the same as for the CM prediction. Finally, note that when the top Lyapunov exponent is negative, the prediction error is still increasing, but slower as $t/\ln t$.

6.5 Monte Carlo experiments with LSM

6.5.1 EKF

Extensive experiments on the EKF performance have been performed based on LSM simulated trajectories (Piterbarg, 2001b). The experimental conditions are as follows. Initially, the observed particles (predictors) are located in the vertices of a right polygon with radius d_0, while the unobserved particle is released at the center of the polygon at the same time. The initial velocities for all the particles are Gaussian independent variables. The isotropic case with zero mean flow is the focus in the first series of EKF experiments. The following parameters are fixed during all the experiments: simulation step, $\mathrm{d}t = 1$ hour, assimilation step, $\Delta t = 12$ hours, root-mean-square velocity, $\sigma_u = 20\,\mathrm{km\,day}^{-1}$, and the maximum prediction time $T_m = N\mathrm{d}t = 15$ days. The remaining parameters typically vary in the following range: Lagrangian correlation time, $\tau = 1$ to 10 days, velocity space scale, $R = 10$ to 500 km, initial cluster radius, $d_0 = 10$ to 100 km, number of predictors $p = M - 1 = 1$ to 6, and prediction time $T = 1$ to 15 days.

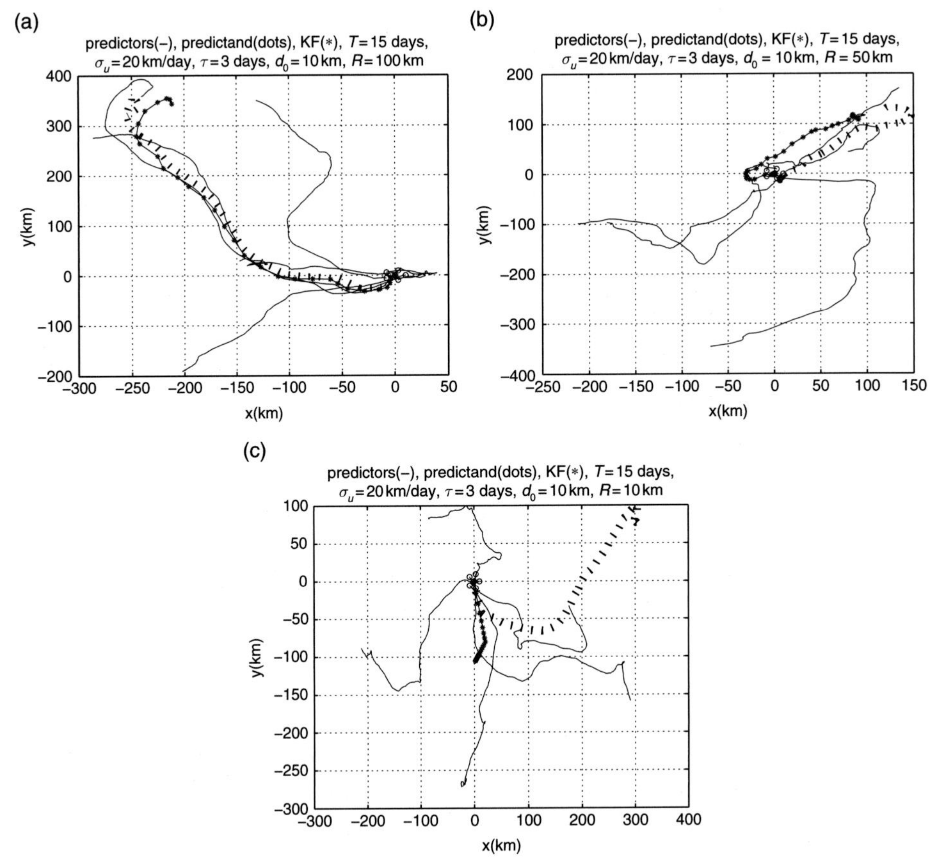

Figure 6.2 Trajectories of Lagrangian particles in stochastic isotropic flow. Predictor trajectories (solid lines), predictand trajectory (dots), and Kalman filter prediction (asterisks). Predictors start from the vertices of a right pentagon (circles) and predictand starts from its center. The observation time $T = 15$ days, Lagrangian correlation time $\tau = 3$ days, mean-root-square of velocity fluctuations $\sigma_u = 20\,\text{km day}^{-1}$, number of predictors $p = 5$, pentagon radius $d_0 = 10$ km, and (a) velocity space correlation radius $R = 100$ km, (b) $R = 50$ km, (c) $R = 10$ km.

Figures 6.2a,b,c demonstrate trajectories of the observed particles (solid unmarked lines), the predictand (dots) and the EKF prediction (asterisks) for small, medium and large ratio R/d_0 respectively, in the case of five predictors. In the first case, $R = 100$ km, while $d_0 = 10$ km is fixed for all three experiments. The prediction in this case is very good due to the fact that three out of five predictors closely follow the predictand because of high correlation between their velocities. In fact, for all the prediction times, the error of prediction is less than 15 to 20 km.

When $R = 50$ km the prediction is still very good. In fact, it is not significantly worse than the previous case, even though only two out of five predictors stay close to the predictand during the 15 days, while three particles leave the initial cluster. These experiments have shown that the EKF algorithm works in a way that a good prediction is ensured by having a single, nearby predictor. It would be naive to suppose that the unobservable particle should follow the majority of the observable ones, which is the implicit assumption for the center of mass method. In fact, the prediction skill is completely determined by whether one or more predictors are staying close to the predictand during the observation time. We call these predictors significant. The algorithm automatically picks up significant predictors by adjusting weights $\mathbf{K}_{ij}$ in (6.9) that decay fast with the distance between predictand and the corresponding predictor.

Finally, in the case of $R = 10$ km, the prediction completely fails. The particles become uncorrelated from the very beginning. All the predictors are quickly separating from each other as well as from the predictand. The result is unfortunate: the error is about 180 km in 15 days, i.e., close to the dispersion. Thus, the ratio R/d_0 is crucial for the prediction skill. Now we look at this dependence in more detail.

The dependence of the prediction root-mean-square error s on the prediction time for fixed $d_0 = 10$ km and different $R = 10, 110, 210, 310$ km is shown in Fig. 6.3a. The diamond line shows the dispersion D. In agreement with Taylor's classical theory, the dispersion is linear at the initial stage and proportional to $\sqrt{t}$ for large t. It is about 250 km at $T = 15$ days. Note that the displacement can be theoretically estimated as $2\sigma_u\sqrt{(T-\tau)\tau} \approx 240$ km. The curves line up with the correlation radius: the larger is R, the less is the error $s(t)$ for any time t. The explanation is simple: in the absence of the mean flow, for the large correlation scale, the probability is high that one of the predictors is close enough to the predictand ensuring good prediction conditions. The error of the best prediction corresponding to the largest $R = 310$ km, is just about 25 km, i.e., 10% of the dispersion for the prediction time of 15 days. The dependence of the prediction error on the prediction time for fixed $R =$ 150 km and different $d_0 = 10, 30, 50, 70$ is shown in Fig. 6.3b. First, note that the dispersion in this study is exactly 240 km in agreement with the theory and the previous experiment. The latter indicates that the sample size of 100 runs used in experiments is adequate for reasonable statistical inference. Recall that the statistics of an individual particle is determined by σ_u and τ only, while the two-particle statistics essentially depend on R and d_0. The result in Fig. 6.3b is similar to that of Fig. 6.3a in the sense that the error lines up with the value of d_0: the less is d_0, the better is the prediction.

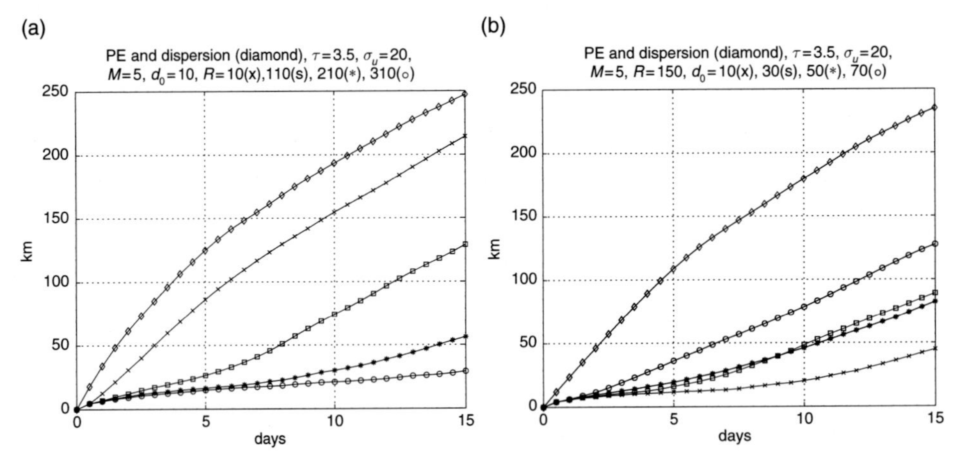

Figure 6.3 (a) Dependence of the prediction error on the prediction time and the velocity space correlation R and its comparison with dispersion. Dispersion (diamonds), error for $R = 10$ km (x), $R = 110$ km (squares), $R = 210$ km (∗), $R = 310$ km (circles), $\tau = 3$ days, $d_0 = 10$ km. (b) Dependence of the relative prediction error on the prediction time and the initial cluster radius R_0 and its comparison with dispersion. Dispersion (diamonds), error for $d_0 = 10$ km (x), $d_0 = 30$ km (squares), $d_0 = 50$ km (∗), $d_0 = 70$ km (circles), $\tau = 3$ days, $R = 150$ km.

However, there is an essential difference. Changes in d_0 in the wide range from 10 to 50 km do not cause drastic changes in the prediction skill; the error grows from 45 km to 80 km. There is almost no difference in the error for initial values of the cluster radius of 30 and 50 km. The simulation results show that the error is not determined by the ratio R/d_0, but rather by a complex function of R and d_0.

Introduce the relative error $s_r = s/D$. Fig. 6.4 demonstrates the dependence of the 15-day relative prediction error on τ and R. The prediction skill is much more sensitive to variations in R rather than variations in τ. Note the agreement between Fig. 6.4 and Fig. 6.1b since R is directly proportional to τ_0, and the prediction error decreases as Λ^{-1} increases. Also, both s_r and Λ^{-1} are weakly dependent on τ.

Finally, the prediction error dependence of the number of predictors, $p = M - 1$, was studied. Experiments with $R = 100$ km demonstrated that the prediction error decreases from 0.42 to only 0.36 as p increases from 7 to 20. Prediction errors for the cases with $R = 200$ km (not shown) even increased from 0.14 to 0.28. Such a trend is typical for numerous experiments not shown in the figures, which covered a wider range of R. The same conclusion follows from our consideration of initial predictor configurations different from the right polygon configuration. First, the predictors were located along the straight line distanced

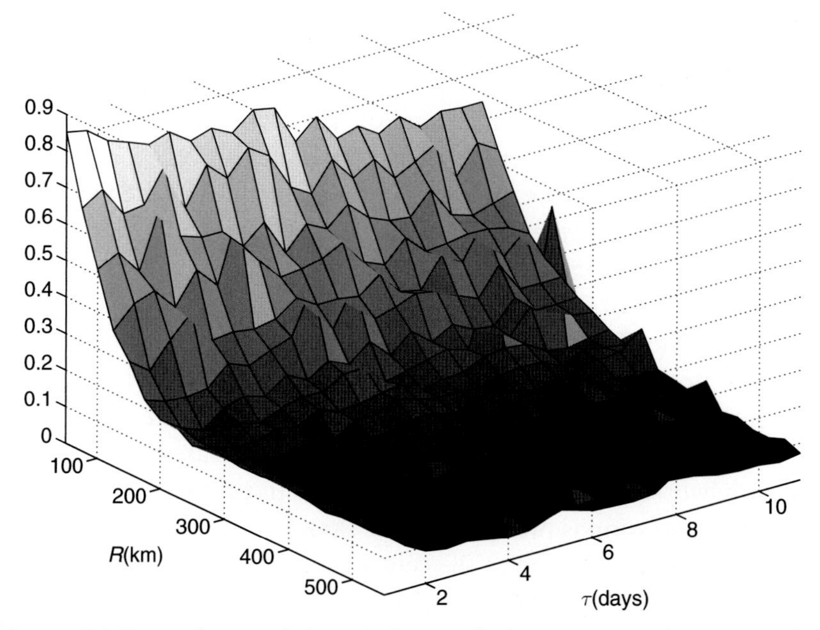

Figure 6.4 Dependence of the relative prediction error on the Lagrangian correlation time and the velocity space correlation radius. $d_0 = 50$ km, $M = 5$, $\sigma_u = 20$ km day^{-1}.

from the predictand. Then, we tried a random distribution of the predictors around the predictand. In both cases the dependence of the error on the number of predictors remains basically the same. Approximately, for $p > 7$ the error is in the range of 0.2 to 0.3 for $R = 200$ km, and 0.35 to 0.45 for $R = 100$ km. The initial right polygon configuration gives slightly better results than the linear and random ones. There is no indication that increasing the number of predictors beyond four to seven significantly improves the prediction skill.

In summary, we can conclude that the prediction skill improves with increasing ratio R/d_0, where, in general, d_0 is a characteristic radius of the cloud of predictors. The least prediction error is due to small Lagrangian correlation time $\tau = 1$ to 3 days for the prediction time $T = 15$ days. The error slowly decreases as the number of predictors grows for ratio R/d_0 around 1 to 2, while the prediction skill slightly worsens or does not change for p larger than four to seven in the case $R/d_0 > 4$. The increase in the prediction error for a large number of predictors is a computational issue. As we can see from (6.9) the EKF prediction formula involves the inverse of the predictor covariance matrix, which can become ill-posed when two of the predictors come close one to another.

6.5.2 Comparison of RA with EKF using LSM

The goal of the next set of experiments, discussed in this section, is to compare the performance of RA (Section 6.4.2) and EKF (Section 6.4.1) for two cases, zero mean flow and linear shear mean flow (Piterbarg and Özgökmen, 2002). In the zero mean flow experiments we fixed $\tau = 3$ days, $R = 250$ km, $d_0 = 50$ km, and $M = 7$. As before, initially the predictors are located at the vertices of the right hexagon and the velocities are proportional to the position vectors.

If the predictand is placed some distance from the center ($r_0 = 25$ km), then the mean-square error of the regression algorithm is slightly lower than that of the EKF for the first seven days while after that EKF slightly prevails (Fig. 6.5a). Both algorithms are doing quite well comparatively with the dispersion (diamonds). This is because the initial cluster radius is five times less than the spatial correlation radius. The center of mass prediction (crosses) gives clearly worse prediction caused by a bias. In summary the performance of EKF and regression is pretty much the same.

If the predictand is placed at the center under the same experiment conditions, then the EKF prediction turns out to be visibly better (Fig. 6.5b). However, if we introduce a mean shear flow, not very strong, the picture changes drastically. Namely, set

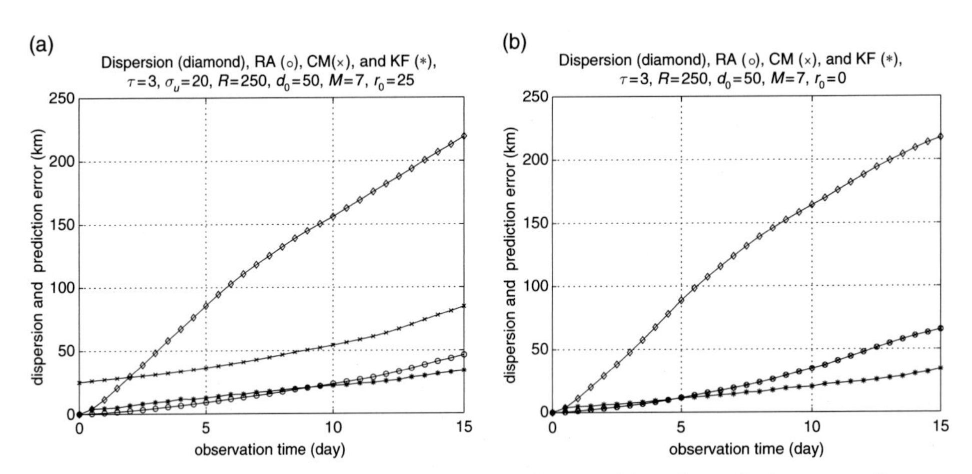

Figure 6.5 Comparison of the dispersion (diamonds) and prediction error for the regression algorithm (circles), the center mass (×), and the sub-optimal Kalman filter method (∗) for the maximum observation time $T = 15$ days and zero mean flow. The number of predictors $p = 6$. Lagrangian correlation time $\tau = 3$ days, Eulerian velocity correlation radius $R = 250$ km, initial hexagon radius $d_0 = 50$ km, distance of the predictand from the hexagon center $r_0 = 25$ km. (a) The case when the predictors are located in vertices of a right hexagon, and (b) with the predictor initially located at the center ($r_0 = 0$).

$$\mathbf{U}(\mathbf{r}) = \mathbf{U} + \mathbf{G}\mathbf{r}, \quad \mathbf{G} = \begin{pmatrix} \gamma & \omega \\ -\omega & -\gamma \end{pmatrix},$$

where $\mathbf{U}$ is a constant flow and the shear is characterized by stretching and rotation parameters γ, ω. For a gyre given by $\omega = 0.1$ and $\gamma = 0$, the performance of EKF is very poor (Fig. 6.6a). The error reaches 160 km for 15 day forecast, almost 80% of the dispersion (220 km), while the error of the regression algorithm is acceptable (60 km). Consider a different shear with no rotation: $\omega = 0$ and $\gamma = 0.05$. The conclusion is the same: the performance of the regression is clearly better (Fig. 6.6b). The error of EKF is about 140 km, while for the regression it is some 60 km. Finally, combining both cases ($\omega = 0.1$ and $\gamma = 0.05$), we observe that the EKF error is almost twice as much as the RA

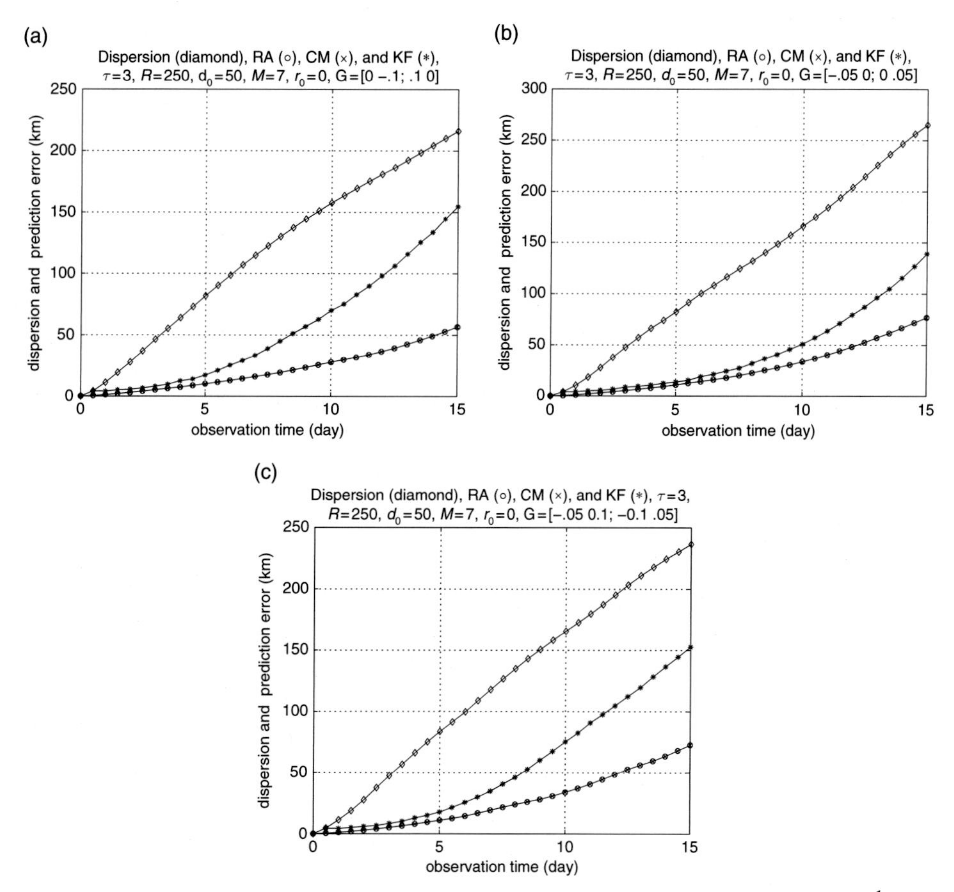

Figure 6.6 (a) Same as in Fig. 6.5b with added linear shear: $\gamma = 0.05$ day^{-1}, $\omega = 0$. (b) Same as in Fig. 6.6a for $\gamma = 0$, $\omega = 0.1$ day^{-1}. (c) Same as in Fig. 6.6a for $\gamma = 0.05$, $\omega = 0.1$ day^{-1}.

error (Fig. 6.6c). Thus, with no doubt the regression algorithm performs better in the presence of a deterministic linear shear flow. This is because it is based on the assumption of linear dependence of the particle current position on the initial position. In fact, for purely linear flow the RA gives the exact prediction. Therefore, the presence of a linear shear mean flow implies the better performance of the regression algorithm.

6.6 Ocean circulation model simulations and *in-situ* data

6.6.1 *Simulations with Miami isopycnic coordinate ocean model*

A simplified version of the EKF algorithm ((6.7)–(6.9)) was used to investigate the Lagrangian predictability in the context of a primitive-equation, idealized, double-gyre ocean model (Özgökmen *et al.*, 2000). Synthetic drifters were released in the double-gyre flow and the problem of predicting one trajectory using information from other nearby trajectories was considered.

The performance of the prediction scheme is quantified as a function of a number of factors: (i) dynamically different flow regimes: interior gyre, western boundary current and mid-latitude jet regions (Fig. 6.7); (ii) density of drifter data used in the EKF; and (iii) uncertainties in the knowledge of the

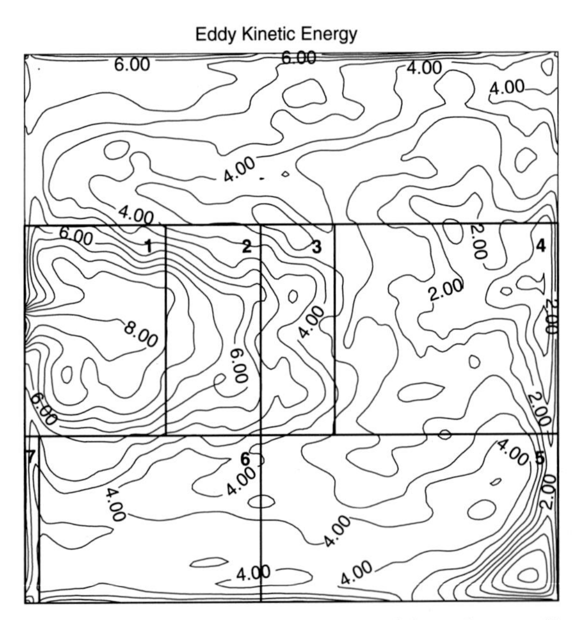

Figure 6.7 Quasi-homogeneous regions 1–7 partitioned according to the eddy kinetic energy distribution [EKE $=(u'^2+v'^2)/2$] in the numerical ocean model. Plotted is ln(EKE), where EKE is in cm^2 s^{-2}.

mean flow field and the initial conditions. The data density is quantified by the number of drifter data per degrees of freedom N_R, defined as the number of drifters within the typical Eulerian space scale R from the predictand particle.

An example for the gyre interior is given in Fig. 6.8 where the dependence of the prediction error on time and on N_R is demonstrated. These and other simulations in different regions of the open ocean indicate that the actual WOCE sampling (1 particle/[5° × 5°] or $N_R \ll 1$) does not improve particle prediction, while predictions improve significantly when $N_R > 1$. For instance, a coverage of 1 particle/[1° × 1°] or $N_R \sim O(1)$ is already able to reduce the errors by about a third to one-half. If the sampling resolution is increased to 1 particle/[0.5° × 0.5°] or 1 particle/[0.25° × 0.25°] or $N_R \gg 1$, reasonably accurate predictions (rms errors of less than 50 km) can be obtained for periods ranging from one week (western boundary current and mid-latitude jet regions) to three months (interior gyre region). Even when the mean flow field and initial turbulent velocities are not known accurately, the information derived from the surrounding drifter data is shown to compensate when $N_R > 1$.

6.6.2 In-situ *drifter data: applications to Adriatic Sea and Pacific Ocean clusters*

The Lagrangian prediction schemes summarized in Section 6.4 have been tested using clusters of *in-situ* drifter data. Two geographical regions with very different characteristics in terms of circulation have been considered: the Adriatic Sea, which is a mid-latitude semi-enclosed basin characterized by steep topography and coastal currents, and the Pacific Ocean in an open equatorial area. The Adriatic Sea analysis (Castellari *et al.*, 2001) directly applies the methodology of Özgökmen *et al.* (2000), while a more extensive analysis including different tests and different methodologies was performed using the drifter data from the Pacific Ocean (Özgökmen *et al.*, 2001).

Adriatic Sea clusters

Three clusters each consisting of five to seven drifters (Poulain, 1999) have been considered, each characterized by a different value of the initial data density N_R. The trajectory of each particle within a cluster is predicted using the other ones, and the prediction error is estimated as the root-mean-square prediction error for all floats in the cluster. Prediction periods of 1–2 weeks are considered, during which the initial cluster distance tends to increase and N_R to decrease.

The results indicate that the prediction is effective as long as $N_R \geq 1$, in good quantitative agreement with the results of Özgökmen *et al.* (2000), while for $N_R < 1$ the method does not lead to an improvement since the particles are too

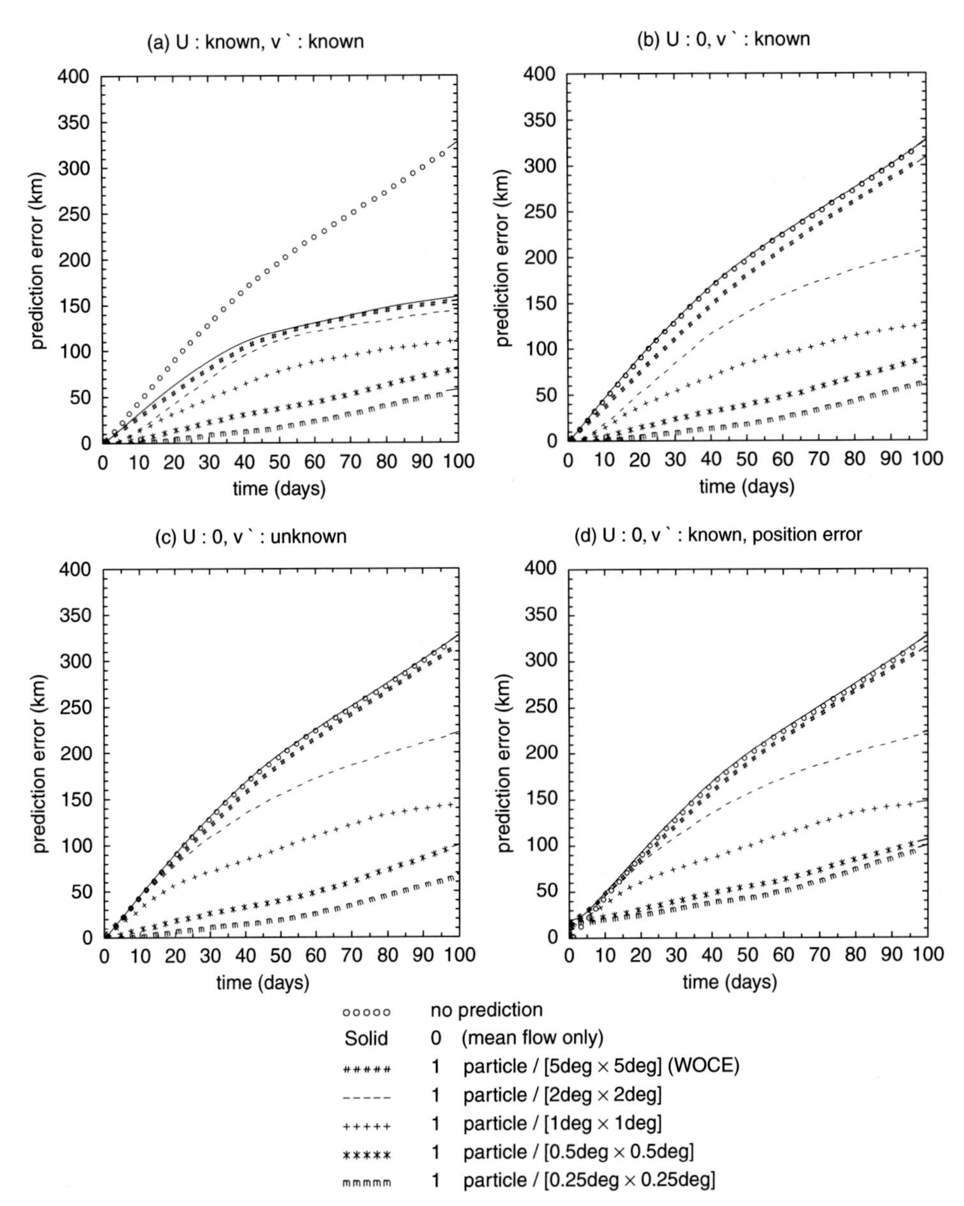

Figure 6.8 Prediction (rms) error versus time as a function of data density in the gyre interior (Region 4) for (a) the "ideal case" with accurate mean flow and accurate initial conditions for the particle, i.e. known initial position and turbulent velocity, (b) the case with zero mean flow and accurate initial conditions for the particle, (c) the case with zero mean flow and unknown initial turbulent velocities (estimated from the surrounding data), and (d) the case with zero mean flow, unknown initial turbulent velocities and an initial position error of 20 km.

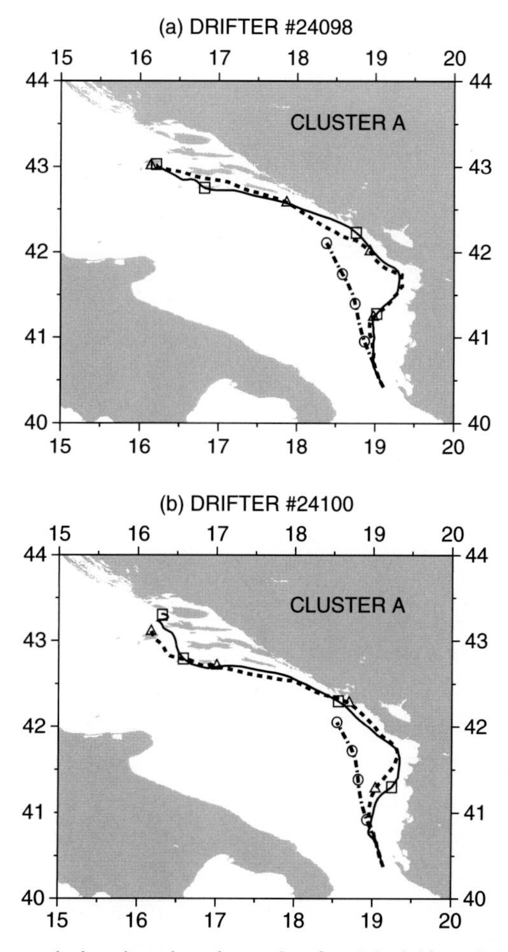

Figure 6.9 True and simulated trajectories for (a) drifter # 24098 and (b) # 24100 of cluster A. Solid lines indicate the true trajectories (square symbols at five-day-intervals), predicted trajectories using only the mean flow information are shown with dashed-dotted lines (circle symbols at five-day-intervals), and dashed lines depict predicted trajectories using data assimilation (triangle symbols at five-day-intervals).

far apart to exhibit correlated motion. An example of prediction for a cluster of particles launched in the eastern boundary current of the basin is shown in Fig. 6.9a,b. The results are found to be robust with respect to uncertainties in the knowledge of the mean flow and the initial release position.

Pacific Ocean clusters

Three clusters, each consisting of 5–10 drifters, are selected from the 1988–96 WOCE data obtained from the NOAA Atlantic Oceanographic and Meteorological Laboratory, Global Drifter Center (Fig. 6.10).

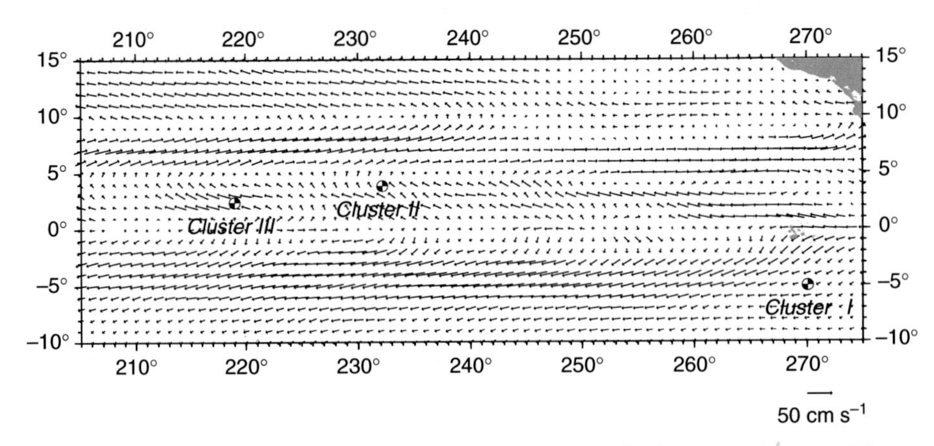

Figure 6.10 The mean flow field **U** used in the prediction experiments. The circles mark the initial positions of clusters I, II and III.

The drifter design has been calibrated to an accuracy of 1–2 $\mathrm{cm\,s^{-1}}$ for conditions that prevail in the tropical Pacific (Niiler *et al.*, 1995). The raw drifter data then undergo standard quality control procedures (Hansen and Poulain, 1996) and are archived with positions and velocities given at six-hour intervals. We have searched the entire 1988–96 data set in the tropical Pacific by plotting annual maps of release positions. The clusters are defined by a group of at least five drifters released within a time interval less than or equal to decorrelation time scale (three days) and/or within the Eulerian decorrelation space scale, or approximately the radius of deformation ($\approx$250 km in the equatorial Pacific, Cushman-Roisin, 1994). We found that this criterion leads to only about half a dozen clusters. This is because drifter clusters cannot be formed from a fixed or random deployment pattern from ships of opportunity, but require specific preparation. In the specific cases that the drifters have been deployed on tight grids, a real-time analysis of a variety of data sets including current meter profilers and satellite images have been necessary for a detailed dynamical analysis (e.g., Flament *et al.*, 1996) due to strong currents in this region.

The climatological mean field is calculated from the 1988–96 WOCE drifter data set by using a least-squares bicubic spline interpolation technique developed by Bauer *et al.* (1998). When compared to the traditional binning technique, the spline interpolation technique minimizes the energy in the fluctuation field and maintains a degree of smoothness for the mean flow even in regions of strong currents and strong shears, such as the equatorial Pacific. The well-known major currents are clearly visible in the mean field: the westward North Equatorial Current (NEC) north of about 10° N, the eastward North Equatorial Countercurrent (NECC) between approximately

4° N and 9° N, and the westward South Equatorial Current (SEC), which extends across the equator to 10° S. The reader is referred to Reverdin *et al.* (1994) for the temporal variability of the large-scale velocity field derived from this drifter data. Drifters in the first cluster (Cluster I) were released in the SEC, whereas the others in clusters II and III were launched just south of the NECC (Fig. 6.10). The other model parameters are chosen as follows. The Lagrangian time scale τ is taken as three days for all clusters, in agreement with the values computed by Bauer *et al.* (1998). As explained above, since the mean flow field is calculated as climatology, the fluctuating velocity field is primarily representative of mesoscale turbulence, and therefore the Eulerian space scale R corresponds approximately to the radius of deformation, and R is taken as 250 km in this study, the deformation radius in the equatorial Pacific. The results presented in the following are not sensitive to the specific value of τ and R in the range of ±25% of the above cited numbers. When working with clusters, a more reasonable measure of the data density, N_d, is defined as the number of drifters over an area scaled by the mean diameter of the cluster. The results for Cluster II are presented in Fig. 6.11.

These results as well as results for other clusters indicate that three general regimes can be characterized using N_d. In the first regime, which corresponds to the period after the release of drifters in a tight cluster when $N_d \gg 1$ drifter/degree^2, the center of mass technique performs nearly as well as the EKF, and both methods yield very accurate predictions with rms errors $\leq$15 km over a time scale of a few days: about 7 days for clusters I and II with $\sigma_v =$ 24–31 cm s^{-1}, and about 3 days for Cluster III with $\sigma_v = 54$ cm s^{-1}. When the drifters start to disperse, i.e., in the regime where $N_d \geq 1$ drifter/degree^2, the EKF is the only method that gives accurate results. This result shows the utility of the EKF technique and implies that this technique is "cost effective," i.e., it can give accurate results without the need of releasing a very high number of drifters in practical applications. Finally, when $N_d \ll 1$ drifter/degree^2, no method investigated in this study is effective. Probably more complete dynamical models including the effect of rotation and *in-situ* wind forcing can be useful. We also find in general that advection by the mean flow field is not a good indicator of drifter motion. Given that this area has relatively large amounts of drifter data to calculate the mean flow, data assimilation methods and dynamical data will be necessary for most of the world's oceans.

As for uncertainties in the knowledge of the mean flow field and initial release positions on the predictions, it is found that the effect of the mean flow on the EKF is negligible; assimilation of high density data compensates for the lack of knowledge of the mean flow, as concluded also in the MICOM experiments (Özgökmen *et al.*, 2000) and in the Adriatic Sea data analysis (Castellari

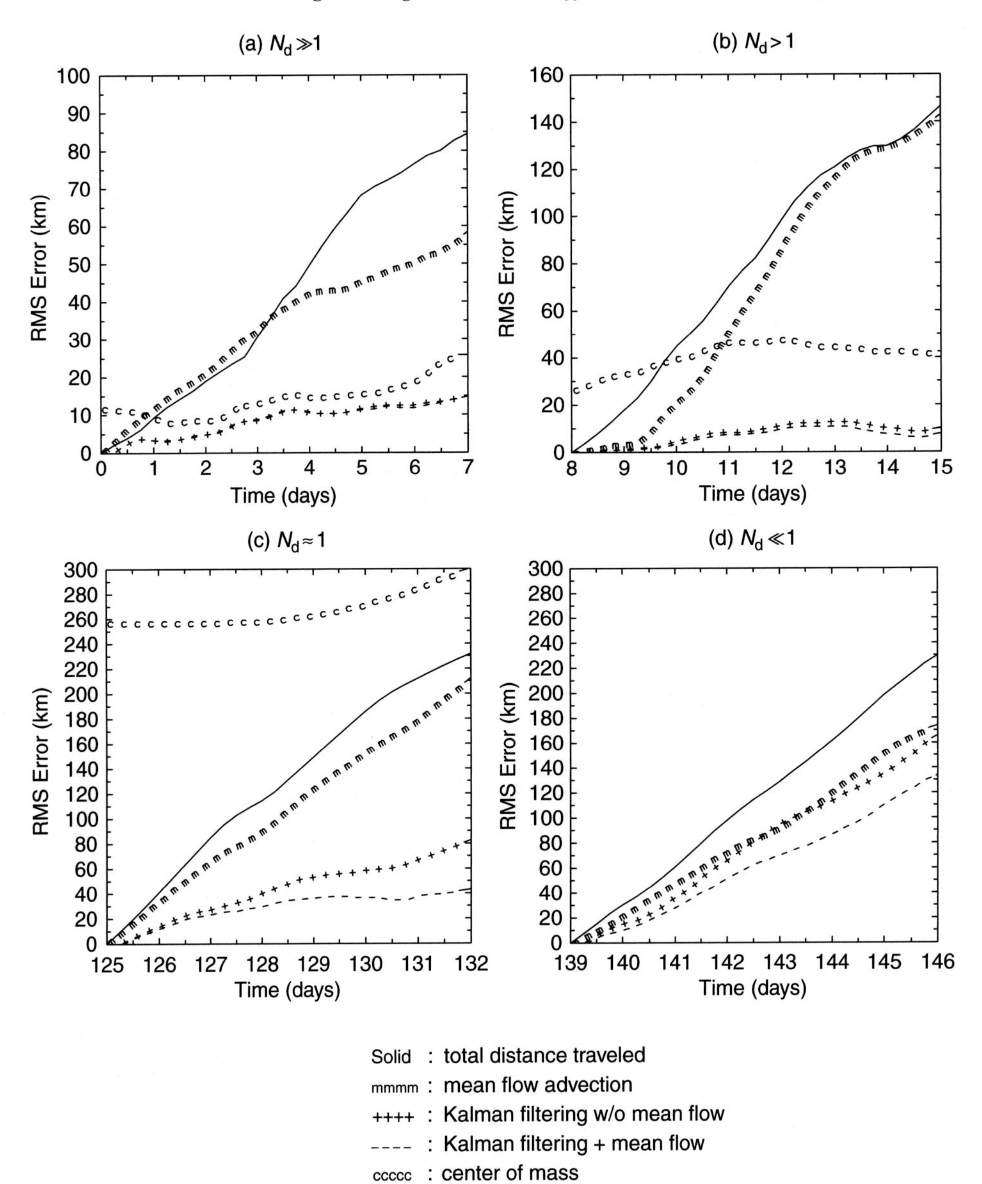

Figure 6.11 Total distance traveled s, prediction error with mean flow advection s_m, prediction error based on the center of mass s_c, prediction error using data assimilation s_a with and without mean flow $\mathbf{U}$ for Cluster II during (a) $0 \le t \le 7$ days ($N_d \gg 1$ drifter/degree^2), (b) $8 \le t \le 15$ days ($N_d > 1$ drifter/degree^2), (c) $125 \le t \le 132$ days ($N_d \approx 1$ drifter/degree^2) and (d) $139 \le t \le 146$ days ($N_d \ll 1$ drifter/degree^2).

et al., 2001). It appears in general that the initial position errors are not recovered, even though the error tends to be confined given a high enough data density.

We note that EKF does not show a significant difference between the regimes $N_{\mathrm{d}} \gg 1$ and $N_{\mathrm{d}} > 1$ drifter/degree^2. While this can be interpreted as being due to the fact that the model accuracy is reaching a degree of saturation, a series of experiments conducted with assimilation periods of 6, 12, 24 and 48 hours indicate that the data-availability or assimilation frequency is likely to be an important factor for improving prediction accuracy in regions of strong currents. It is shown that assimilating surrounding drifter data every 6 hours gives significantly better predictions (i.e., reducing the rms error by half or more) when compared to that from carrying out the assimilation procedure every 12 hours (Fig. 6.12). Consequently, three-day sampling periods used by a number of new drifters will probably lead to unsatisfactory Lagrangian prediction results using the EKF type techniques described in this study.

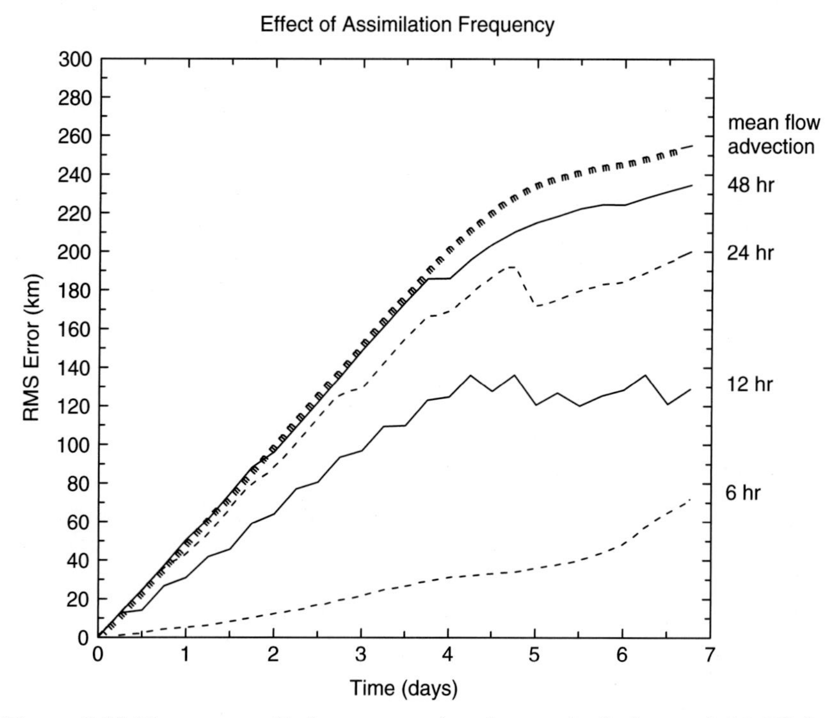

Figure 6.12 The rms prediction error using data assimilation s_{a} (with $\mathbf{U}$) for different data assimilation frequencies: every 6 hours (dashed line), 12 hours (solid line), 24 hours (dashed line) and 48 hours (solid line) during $0 \leq t \leq 7$ days for Cluster III. The line marked with *mmm* shows the prediction error using mean flow advection (i.e., when data are not assimilated).

As final remarks, we point out that the deterministic part of the present model consists only of advection by the climatological flow field. There is not sufficient information for accurate prediction capability in the absence of surrounding drifter information; i.e., once the drifters in the cluster have dispersed, or between the assimilation moments. Therefore, improvement of deterministic model dynamics and employing other data (e.g., *in-situ* winds) are expected to help improve prediction results and they can be considered as next steps.

6.6.3 Comparison of EKF and RA using Pacific Ocean clusters

Finally, we apply RA to the same clusters. It was found in Özgökmen *et al.* (2001) that during the period in which drifters remain close to one another as a tight cluster (quantified by the number of drifters within the velocity space correlation scale R), the center of mass method is a simple yet effective means of predicting the drifter location. It was investigated in Piterbarg and Özgökmen (2002) how the prediction accuracy of RA compares to that of center of mass and that of more complicated technique EKF for oceanic drifters.

Given that $d_0 \ll R$ and the low velocity variance, one can anticipate very good performance by RA based on the results from theory and stochastic simulations. Dispersion $D(t)$ and prediction errors $s(t)$ from regression algorithm RA, center of mass technique CM and EKF are calculated by sequentially selecting each drifter as predictand and the remaining others in the cluster as predictors, and correspond to the root mean square of that of all cluster particles. The results are shown in Fig. 6.13a,b for Cluster I for an observation time of seven days. This figure shows that prediction errors of both EKF and RA are less than that of CM during the observation period, and that RA is as accurate as EKF. More quantitatively, dispersion reaches approximately 51 km at $T = 7$ days, error from CM is 8 km ($s_r = 0.16$), and error from EKF and RA are about 5 km ($s_r = 0.1$).

Cluster II consists of seven drifters that are also released with a mean diameter of approximately 10 km, but disperses much faster than Cluster I due to a higher velocity variance of $\sigma_u^2 = 720\,\text{km}^2\,\text{day}^{-2}$, and the mean cluster diameter reaches 25 km and 50 km after 7 and 14 days of observation time, respectively (Fig. 6.14a). Dispersion and prediction errors for Cluster II over an observation period of 14 days are shown in Fig. 6.14b, and the conclusion remains the same as for Cluster I; prediction errors of both EKF and RA are less than that of CM during the observation period, and RA is as accurate as EKF. Dispersion reaches approximately 136 km at $T = 14$ days, error from CM is 44 km ($s_r = 0.32$) and errors from EKF and RA are about 26 km ($s_r = 0.19$). The sensitivity of the prediction accuracy of RA to the number of

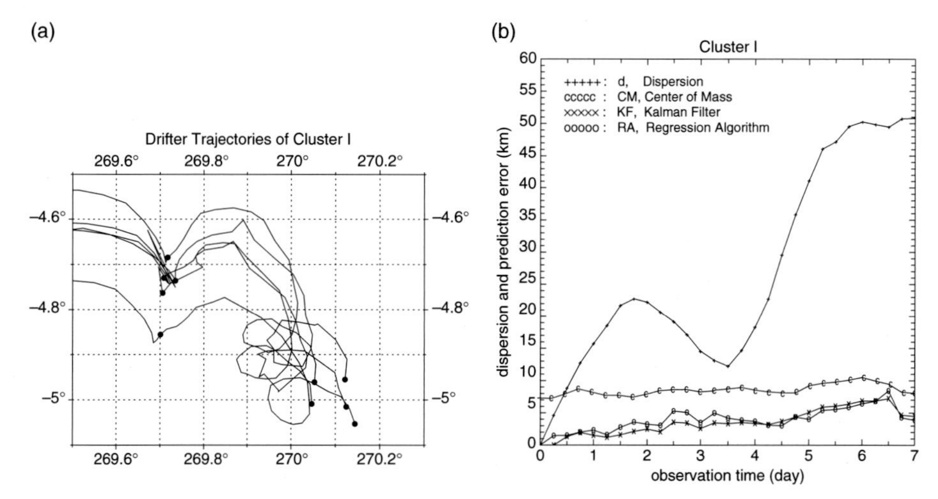

Figure 6.13 (a) Drifter trajectories in Cluster I. The circles mark seven-day intervals. (b) Comparison of the dispersion, $d(t)$, and prediction errors, $s(t)$, of the regression algorithm RA, the center of mass technique CM and Kalman filter type method KM for an observation time of seven days for Cluster I.

predictors p is investigated by randomly eliminating drifters from Cluster II. Fig. 6.14c shows the dispersion curve based on the entire cluster and prediction errors calculated for $p=6$ (same as in fig. 6.12b), $p=5$, $p=4$ and $p=3$. When $p=3$, a drastic reduction of prediction accuracy takes place, which is found to be independent of the combination of chosen predictors in this cluster. Otherwise, the prediction accuracy gradually decreases as the number of predictors is decreased from 6 to 4, but the accuracy of the method using 4 to 6 predictors remains essentially constant for $T \leq \tau$, or for $T \leq 3$ days.

The motion of Cluster III, consisting of ten drifters, is investigated for 21 days during which velocity variance is $\sigma_u^2 = 2240\,\text{km}^2\,\text{day}^{-2}$, or the highest of the three clusters. These drifters were released over an area with an approximate diameter of 30 km, but this scale increases to approximately 100, 180, and 250 km after 7, 14 and 21 days, respectively (Fig. 6.15a). Dispersion and prediction errors for Cluster III are shown in Fig. 6.15b. During the first 10 days, prediction errors of both EKF and RA are approximately the same and less than that of CM. But during the second half of the observation period, the error of EKF increases faster than that of RA. This increase appears to be related to the inability of the EKF algorithm to follow the bifurcation of 1–2 drifters from a group of 8 as effectively as the RA technique. Dispersion is 426 km (543 km), error from CM is 99 km (235 km) or $s_r=0.23$ ($s_r=0.43$), error from EKF is 55 km (176 km) or $s_r=0.13$ ($s_r=0.32$), and error from RA is 54 km (80 km) or $s_r=0.13$ ($s_r=0.15$) at $T=7$ days ($T=21$ days).

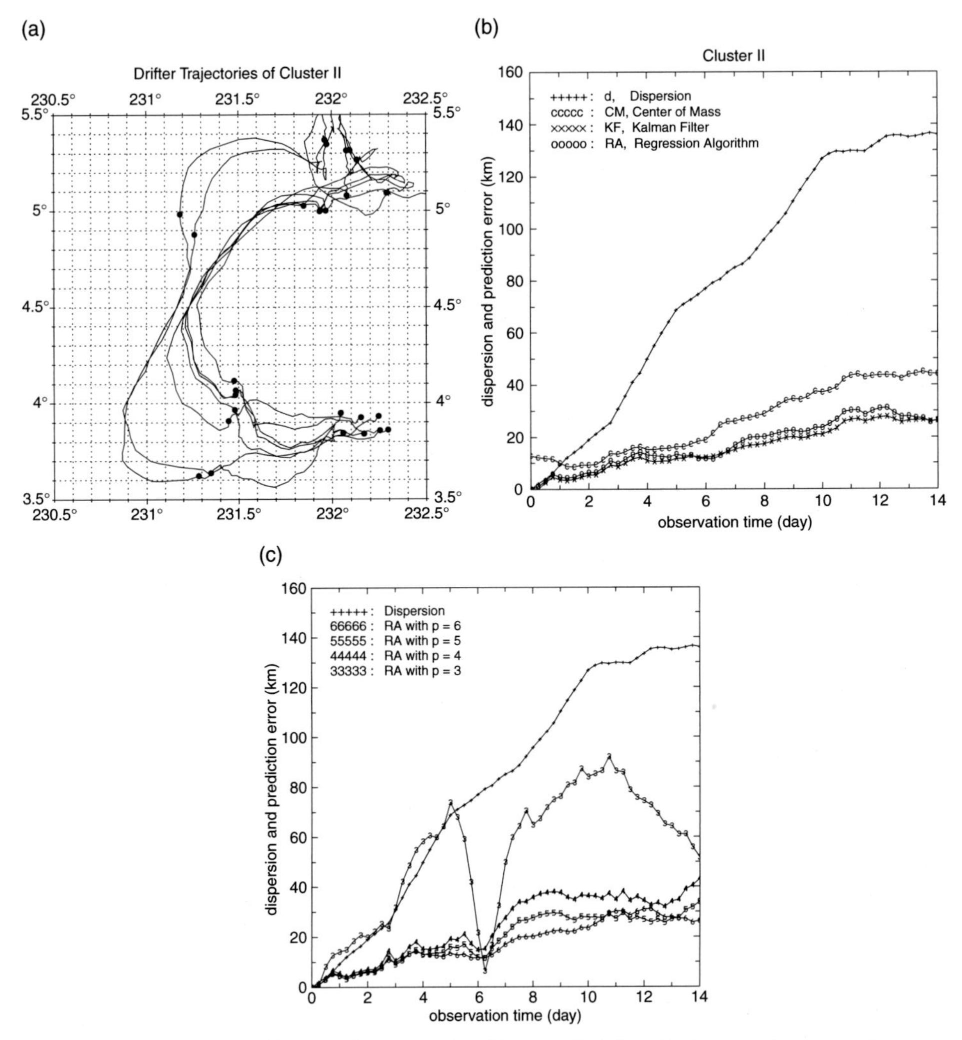

Figure 6.14 (a) Drifter trajectories in Cluster II. The circles mark seven-day intervals. (b) Comparison of the dispersion, $d(t)$, and prediction errors, $s(t)$, of the regression algorithm RA, the center of mass technique CM and Kalman filter type method KM for an observation time of 14 days for Cluster II. (c) Sensitivity of the prediction error of the regression algorithm RA to the number of predictors in Cluster II.

All in all, the real data comparison of different prediction algorithms are in good qualitative agreement with the simulation results. Even the prediction error values are of the same order as our simple error theory concludes. Deviations are related to simplifications in the considered stochastic model such as linearity of the shear flow and isotropy of the fluctuations.

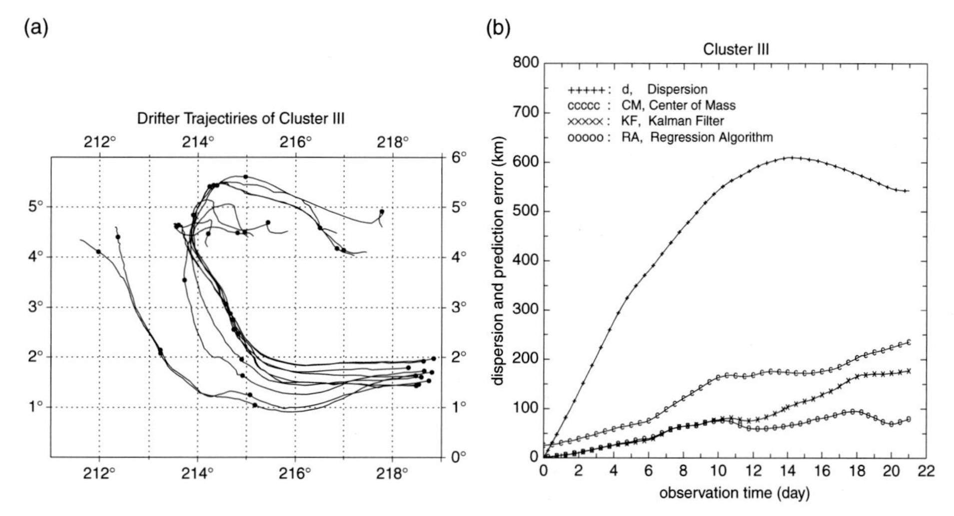

Figure 6.15 (a) Drifter trajectories in Cluster III. The circles mark seven-day intervals. (b) Comparison of the dispersion, $d(t)$, and prediction errors, $s(t)$, of the regression algorithm RA, the center of mass technique CM and Kalman filter type method KM for an observation time of 21 days for Cluster III.

In summary, RA has several important simplifications with respect to the EKF: (i) RA does not require any parameters, such as the Lagrangian parameters describing the characteristics of the underlying flow; the velocity correlation space scale R; and the Lagrangian correlation time scale τ. (ii) RA does not utilize the mean flow field, the calculation of which requires large data sets and the associated sub-grid scale interpolation introduces further errors. Instead it absorbs information on the currents from the predictors. (iii) RA does not need to be initialized with turbulent velocity fluctuations at the launch location. (iv) RA is not based on the integration of velocity field to estimate the particle position, which necessarily leads to accumulation of velocity errors as errors of drifter location, and (v) consequently RA is computationally far simpler than EKF. Despite these simplifications, it is found on the basis of several oceanic clusters that RA outperforms CM, and that RA is as accurate as EKF. Also, predictions from RA appear to remain applicable over a time scale of $T \gg \tau$, or much longer than one would anticipate.

6.7 Optimal sampling

In the theoretical development of Section 6.4, we have assumed that the initial position of the predictand is exactly known. It is explicitly used in both EKF and regression procedures. This assumption was partially relaxed in the MICOM simulations (Özgökmen *et al.*, 2000 and in the *in-situ* data analysis (Castellari *et al.*, 2001; Özgökmen *et al.*, 2001), where the impact of uncertainty on the initial release was explored numerically. Here we introduce the (realistic) assumption that we have only some idea about the place where an object has been lost and we formalize it as a probabilistic distribution of the predictand initial position. Next, pretend that the initial predictor configuration can be controlled and optimized, in order to minimize the prediction error. That leads to the optimal sampling problem formulation as follows. Given statistics of the predictand initial position, find initial predictor positions minimizing the prediction error. An answer depends on the prediction algorithm used. It is natural to start with the simplest one, namely the center of mass (CM) method first systematically tested in Castellari *et al.* (2001) and given by

$$\hat{\mathbf{r}}_M(t) = \frac{1}{p}\sum_{i=1}^{p} \mathbf{r}_i(t), \tag{6.13}$$

i.e., the prediction is taken as the center of mass of the predictors for any given time. At first glance the procedure looks naive, however it gives reasonable results in cases when the initial cluster is tight enough and the flow is significantly correlated in space.

Assume that the probabilistic distribution of the initial predictand position is centered at the point $\mathbf{r}_0$ and its scattering around $\mathbf{r}_0$ is characterized by covariance matrix $\mathbf{\Sigma}$. We found theoretically that in the case of isotropic flow and tight initial cluster ($d_0 \ll R$), an optimal initial configuration $\mathbf{r}_1(0), \ldots, \mathbf{r}_p(0)$ should satisfy the following conditions

$$\frac{1}{p}\sum_{i=1}^{p} \mathbf{r}_i(0) = \mathbf{r}_0, \quad \frac{1}{p^2}\sum_{i,j=1}^{p} (\mathbf{r}_i(0) - \mathbf{r}_0)(\mathbf{r}_j(0) - \mathbf{r}_0)^{\mathrm{T}} = \mathbf{\Sigma}. \tag{6.14}$$

In other words, any optimal configuration should have the center of mass at $\mathbf{r}_0$ and its scattering ellipse should be similar to that of the predictand initial position. In particular, if the uncertainty distribution is isotropic with

$$\mathbf{\Sigma} = \begin{pmatrix} \sigma^2 & 0 \\ 0 & \sigma^2 \end{pmatrix},$$

then one of the possible solutions is given by a right p-polygon centered at $\mathbf{r}_0$ with radius $d_0 = c(p)\sigma$, where $c(2) = \sqrt{4/3}$ and $c(p) = \sqrt{2}$ for $p > 2$. This approximate solution is examined by the LSM.

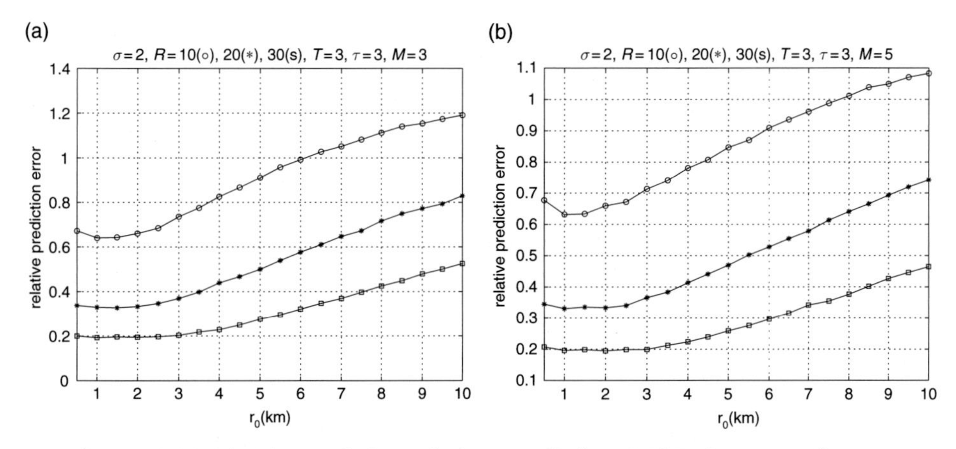

Figure 6.16 (a) Plots of the relative prediction MSE δ versus distance of predictors from the center d_0. First order Markov flow with the same parameters $\sigma=2$, $p=2$, $R=10$ (circles), 20 (asterisks), 30 (squares). (b) Results with $p=4$.

In Fig. 6.16a,b the outcome of such experiments is given for two and four predictors respectively with $\sigma=2$, $\tau=3$, observation time $T=3$ days and different $R=10$, 20, 30 km. Two predictors start from points $(-d_0, 0)$ and $(d_0, 0)$ in the first case (Fig. 6.14a) and from the vertices of a square with diagonal $2d_0$ centered at the origin (Fig. 6.14b) while the predictand starts at a random point with coordinates normally distributed with zero mean and standard deviation σ. The initial velocities for all particles are normally distributed variables with zero means and standard deviation 5 km day^{-1}. Both cases demonstrate that the crucial parameter for the prediction skill is the ratio R/σ and the optimal value d_0^* of the initial cluster radius is between 0 and 2σ.

In summary theoretical solutions (not given here) and our Monte Carlo experiments support that a minimum point d_0^* for MSE $s^2 = s^2(d_0)$ exists, however, the difference between $s^2(d_0^*)$ and $s^2(d_0)$ in the range $0 < d_0 < 1.5\sigma$ is very subtle. The only useful practical conclusion is that one should keep d_0 not greater than 2σ.

6.8 Summary and discussion

In this chapter we discussed the problem of predicting Lagrangian trajectories based on the knowledge of other nearby particle observations. This can be considered as a particular case of the general Lagrangian prediction problem. There are several other cases and formulations where we intend to apply our experience and the theoretical methods developed.

First, consider the setup, where "prediction" means computing Lagrangian trajectories using an ocean general circulation model. Obviously, such prediction would not be perfect because the model is not so, in particular because its space resolution is limited. For applications, quantitative estimates of uncertainty are needed to evaluate the deviation of modeled trajectories from real ones. This issue has been investigated by considering a classical twin experiments methodology by Griffa *et al.* (2004). In such experiments "model" trajectories are generated by the model velocity field which is a smoothed version of the "real" velocity field generated by the same model with higher resolution.

The next step is to consider predicting continuously distributed tracer that can be viewed as a continuum of Lagrangian particles. This problem is of great practical importance because of application to pollutant spreading. First, the problem of forecasting the CM of a pollutant patch and its area is considered given knowledge of the large scale currents and statistics of velocity fluctuations at the given region. Under some assumptions one can write down a Fokker–Planck equation for the mean tracer distribution. That's not what we mean by "forecasting." Our goal is to predict an individual realization of the tracer rather than its statistics. For example, in the simplest homogeneous situation, the Lyapunov exponent, λ, can be found based on the velocity fluctuation statistics. In turn, it can be used for predicting the length of the patch boundary since it grows exactly as $\exp(\lambda t)$ (e.g., Baxendale and Harris, 1986).

Another problem is close to the one discussed in the previous paragraph. Assume that the patch evolution is predicted via a numerical circulation model. How accurate is this prediction given a limited space resolution of the model? Also, we can incorporate *in-situ* drifter observations for predicting a continuously distributed tracer.

The use of other data sources was only briefly mentioned in this review. Paldor *et al.* (2004) demonstrated for the Pacific clusters (Figs. 6.10, 6.13, 6.14, 6.15), that the inclusion of a simplified wind-forced slab model leads to better drifter trajectory predictions than just climatological currents alone. They also showed what was expected and what we've seen (e.g. Fig. 6.12), that more frequent data, wind data in Paldor's case, leads to better prediction. High temporal resolution surface velocity data will be available from High Frequency (HF) radar-based observing systems. Given that O(100) such sites are planned for the American coastline over the next few years, how to best use HF radar-derived velocity data for Lagrangian prediction will be an important research issue. Planned analysis of velocity fields from numerical simulations should lay a solid foundation for application to real data.

Satellite-based estimates of velocity from altimeters is another data source where because of its coarse space-time resolution, the numerical results of how to correct the effects of smoothing (Griffa *et al.*, 2004) will be relevant. Besides the different resolution of the different data sets, other outstanding research issues arise when only partial information is available from data. For example, Feature Displacement Velocities, hereafter FDVs, (Yang *et al.*, 2001) from a sequence of satellite images, e.g. AVHRR or ocean color, contain only a component of the full velocity field. Given the size of this potential data set for applied Lagrangian prediction problems, algorithms to incorporate FDVs need to be formulated.

Finally, even though the procedures developed are comprehensively tested by synthetic data, there certainly is a lack of validation using real data. We hope to have various representative cluster observations in the near future and analyze it using our approaches. Synergistic research between observationalists, modelers, and theoreticians, centered on real-time experiments, will be needed to significantly improve applied Lagrangian prediction.

Acknowledgments

The authors greatly appreciate the support of the Office of Naval Research under grants N00014-97-1-0620 and N00014-99-1-0049 (T. M. Özgökmen, A. Griffa and A. Mariano), and N00014-99-1-0042 (L. I. Piterbarg). We also thank the two anonymous reviewers for their constructive comments that helped greatly improve the manuscript. We thank Linda Steel for editing the manuscript.

References

Aref, H., 1984. Stirring by chaotic advection. *J. Fluid Mech.*, **143**, 1–21.

Bauer, S., M. S. Swenson, A. Griffa, A. J. Mariano, and K. Owens, 1998. Eddy-mean flow decomposition and eddy-diffusivity estimates in the tropical Pacific Ocean. *J. Geophys. Res.*, **103**, 30855–71.

Bauer, S., M. S. Swenson, and A. Griffa, 2002. Eddy-mean flow and eddy diffusivity estimates in the tropical Pacific Ocean. 2: Results. *J. Geophys. Res.*, **107**, (c10), 3154–71.

Baxendale, P. and T. Harris, 1986. Isotropic stochastic flows, *Ann. Prob.*, **14**, No. 4, 1155–79.

Berloff, P. and J. C. McWilliams, 2002. Material transport in oceanic gyres. Part II: Hierarchy of stochastic models. *J. Phys. Oceanogr.*, **32**, 797–830.

Berloff, P., J. C. McWilliams, and A. Bracco, 2002. Material transport in oceanic gyres. Part I: Phenomenology. *J. Phys. Oceanogr.*, **32**, 764–96.

Berloff, P. and J. C. McWilliams, 2003. Material transport in oceanic gyres. Part III: Randomized stochastic models. *J. Phys. Oceanogr.*, **33**, 1416–45.

Borgas, M. S. and B. L. Sawford, 1994. Stochastic equations with multifractal random increments for modeling turbulent dispersion. *Phys. Fluids*, **6**, 618.

Castellari, S., A. Griffa, T. M. Özgökmen, and P.-M. Poulain, 2001. Prediction of particle trajectories in the Adriatic Sea using Lagrangian data assimilation. *J. Mar. Sys.*, **29**, 33–50.

Cushman-Roisin, B., 1994. *Introduction to Geophysical Fluid Dynamics*. Englewood Cliffs, NJ: Prentice-Hall.

Duan, J. and S. Wiggins, 1996. Fluid exchange across a meandering jet with quasiperiodic variability. *J. Phys. Oceanogr.*, **26**, 1176–88.

Falco, P., A. Griffa, P.-M. Poulain, and E. Zambianchi, 2000. Transport properties in the Adriatic Sea deduced from drifter data. *J. Phys. Oceanogr.*, **30**, 2055–71.

Falkovich, G. and L. I. Piterbarg, 2004. The Lyapunov exponent for inertial particles and an explosive ergodic diffusion. Submitted to *Comm. Math. Phys.*

Flament, P. J., S. C. Kennan, R. A. Knox, P. P. Niiler, and R. L. Bernstein, 1996. The three-dimensional structure of an upper ocean vortex in the tropical Pacific Ocean. *Nature*, **383**, 610–13.

Griffa, A., 1996. Applications of stochastic particle models to oceanographic problems. In *Stochastic Modelling in Physical Oceanography*, ed. R. Adler, P. Muller, and B. Rozovskii. Cambridge, MA: Birkhauser Boston, 113–28.

Griffa, A., K. Owens, L. Piterbarg, and B. Rozovskii, 1995. Estimates of turbulence parameters from Lagrangian data using a stochastic particle model. *J. Mar. Res.*, **53**, 212–34.

Griffa, A., L. Piterbarg, and T. M. Özgökmen, 2004. Predictability of Lagrangian particles: effects of smoothing of the underlying Eulerian flow. *J. Mar. Res.*, **62**/1, 1–35.

Hansen, D. V. and P.-M. Poulain, 1996. Quality control and interpolations of WOCE/TOGA drifter data. *J. Atmos. Oceanic Tec.*, **13**, 900–9.

Jazwinski, A. H., 1970. *Stochastic Processes and Filtering Theory*. New York: Academic Press.

Liptser, R. S. and A. N. Shiryaev, 2000. *Statistics of Random Processes.*, second edition. Berlin: Springer-Verlag.

Mariano, A. J., A. Griffa, T. M. Özgökmen, and E. Zambianchi, 2002. Lagrangian analysis and predictability of coastal and ocean dynamics 2000. *J. Atmos. Ocean. Tech.*, **19**, 1114–26.

Monin, A. S. and A. M. Yaglom, 1975. *Statistical Fluid Mechanics: Mechanics of Turbulence*. Cambridge, MA: MIT Press.

Meyers, S., 1994. Cross-frontal mixing in a meandering jet. *J. Phys. Oceanogr.*, **24**, 1641–6.

Niiler, P. P., A. S. Sybrandy, K. Bi, P.-M. Poulain, and D. S. Bittermam, 1995. Measurements of the water-following capability of holey-sock and TRISTART drifters. *Deep Sea Res.*, **42**, 1951–64.

Özgökmen, T. M., A. Griffa, L. I. Piterbarg, and A. J. Mariano, 2000. On the predictability of the Lagrangian trajectories in the ocean. *J. Atmos. Ocean. Tech.*, **17**/3, 366–83.

Özgökmen, T. M., L. I. Piterbarg, A. J. Mariano, and E. H. Ryan, 2001. Predictability of drifter trajectories in the tropical Pacific Ocean. *J. Phys. Oceanogr.*, **31**, 2691–720.

Paldor, N., Y. Dvorkin, A. J. Mariano, T. M. Özgökmen, and E. Ryan, 2004. A practical hybrid model for predicting the trajectories of near-surface drifters in the Pacific Ocean. *J. Ocean. Atmos. Sci.*, **21**, 1246–58.

Pedrizzetti, G. and E. A. Novikov, 1994. On Markov modelling of turbulence, *J. Fluid Mech.*, **280**, 69–93.

Piterbarg, L. I., 2001a. The top Lyapunov exponent for a stochastic flow modeling the upper ocean turbulence. *SIAM J. Appl. Math.*, **62**, 777–800.

Piterbarg, L. I., 2001b. Short term prediction of Lagrangian trajectories. *J. Atmos. Ocean. Tech.*, **18**, 1398–410.
Piterbarg, L. I. and T. M. Özgökmen, 2002. A simple prediction algorithm for the Lagrangian motion in two-dimensional turbulent flows. *SIAM J. Appl. Math.*, **63**, 116–48.
Piterbarg, L. I., 2004a. On predictability of particle clusters in a stochastic flow, to appear in *Stochastics and Dynamics*.
Piterbarg, L. I., 2005. Relative dispersion in 2D stochastic flows. *J. of Turbulence*, **6**(4), doi:10.1080/14685240500103168.
Pope, S. B., 1987. Consistency conditions for random walk models of turbulent dispersion. *Phys. Fluids*, **30**, 2374–9.
Poulain, P.-M., 1999. Drifter observations of surface circulation in the Adriatic Sea between December 1994 and March 1996. *J. Mar. Sys.*, **20**, 231–53.
Reverdin, G., C. Frankignoul, and E. Kestenare, 1994. Seasonal variability in the surface currents of the Equatorial Pacific. *J. Geophys. Res.*, **99**(10), 20323–44.
Reynolds, A. M., 1998. On the formulation of Lagrangian stochastic models of scalar dispersion within plant canopies. *Boundary-Layer Meteor.*, **86**, 333–44.
Rodean, H. C., 1996. *Stochastic Lagrangian models of turbulent diffusion*, Meteorological Monographs, 26, n.48. Boston: AMS.
Samelson, R. M., 1992. Fluid exchange across a meandering jet. *J. Phys. Oceanogr.*, **22**, 431–40.
Samelson, R. M., 1996. Chaotic transport by mesoscale motions. In *Stochastic Modelling in Physical Oceanography*, ed. R. J. Adler, P. Müller, and B. L. Rozovoskii. Cambridge, MA: Birkhäuser Boston, 423–38.
Sawford, B. L., 1993. Recent developments in the Lagrangian stochastic theory of turbulent dispersion. *Boundary-Layer Meteor.*, **62**, 197–215.
Schneider, T., 1998. Lagrangian drifter models as search and rescue tools. M. S. Thesis, Dept. of Meteorology and Physical Oceanography, University of Miami.
Thomson, D. J., 1986. A random walk model of dispersion in turbulent flows and its application to dispersion in a valley. *Quat. J. R. Met. Soc.*, **112**, 511–29.
Thomson, D. J., 1987. Criteria for the selection of stochastic models of particle trajectories in turbulent flows. *J. Fluid Mech.*, **180**, 529–56.
Thomson, D. J., 1990. A stochastic model for the motion of particle pairs in isotropic high-Reynolds-number turbulence, and its application to the problem of concentration variance. *J. Fluid Mech.*, **210**, 113–53.
Toner, M., A. C. Poje, A. D. Kirwan, C. K. R. T. Jones, B. L. Lipphardt, and C. E. Grosch, 2001a. Reconstructing basin-scale Eulerian velocity fields from simulated drifter data. *J. Phys. Oceanogr.*, **31**, 1361–76.
Toner, M., A. D. Kirwan, L. H. Kantha, and J. K. Choi, 2001b. Can general circulation models be assessed and their output enhanced with drifter data? *J. Geophys. Res. Oceans*, **106**, 19563–79.
Veneziani, M., A. Griffa, A. M. Reynolds, and A. J. Mariano, 2004. Oceanic turbulence and stochastic models from subsurface Lagrangian data for the North-West Atlantic Ocean, *J. Phys. Oceanogr.*, **34**, 1884–1906.
Yang, H. and Z. Liu, 1996. The three-dimensional chaotic transport and the great ocean barrier. *J. Phys. Oceanogr.*, **27**, 1258–73.
Yang, Q., B. Parvin, and A. J. Mariano, 2001. Detection of vortices and saddle points in SST data. *Geophys. Res. Lett.*, **28**, 331–4.
Zambianchi, E. and A. Griffa, 1994. Effects of finite scales of turbulence on dispersion estimates. *J. Mar. Res.*, **52**, 129–48.

7

Lagrangian data assimilation in ocean general circulation models

ANNE MOLCARD
LSEET, University of Toulon, France
TAMAY M. ÖZGÖKMEN
Rosenstiel School of Marine and Atmospheric Science, University of Miami, Miami, Florida, USA
ANNALISA GRIFFA
Rosenstiel School of Marine and Atmospheric Science, University of Miami, Miami, Florida, USA
ISMAR/CNR, La Spezia, Italy
LEONID I. PITERBARG
Department of Mathematics, University of Southern California, Los Angeles, California, USA
AND
TOSHIO M. CHIN
Rosenstiel School of Marine and Atmospheric Science, University of Miami, Miami, Florida, USA
Jet Propulsion Laboratory, California Institute of Technology, Pasadena, California, USA

7.1 Introduction

In the last 20 years, the deployment of surface and subsurface buoys has increased significantly, and the scientific community is now focusing on the development of new techniques to maximize the use of these data. As shown by Davis (1983, 1991), oceanic observations of quasi-Lagrangian floats provide a useful and direct description of lateral advection and eddy dispersal. Data from surface drifters and subsurface floats have been intensively used to describe the main statistics of the general circulation in most of the world ocean, in terms of mean flow structure, second-order statistics and transport properties (e.g. Owens, 1991; Richardson, 1993; Fratantoni, 2001; Zhang *et al.*, 2001; Bauer *et al.*, 2002; Niiler *et al.*, 2003; Reverdin *et al.*, 2003). Translation, swirl speed and evolution of surface temperature in warm-core rings, which are ubiquitous in the oceans, have also been studied with floats by releasing them inside of the eddies (Hansen and Maul, 1991). Trajectories of freely drifting buoys allow estimation of horizontal divergence and vertical velocity in the mixed layer (Poulain, 1993). Also, data from drifters allows investigation of properties and statistics of near-inertial waves, which provide much of the shear responsible for mixing in the upper thermocline and entrainment at the base of the mixed layer (Poulain *et al.*, 1992). Drifters have proved to be robust autonomous platforms with which to observe ocean circulation and return data from a variety of

Lagrangian Analysis and Prediction of Coastal and Ocean Dynamics, ed. A. Griffa, D. Kirwan, A. Mariano, T. Özgökmen, and T. Rossby. Published by Cambridge University Press.

sensors. They have proved to be useful in oil spill or floating debris tracking, discharge dispersement calculations, and similar studies.

In this paper, we focus on Lagrangian instruments that provide real-time information via satellite, because they are directly relevant to the nowcasting/hindcasting/forecasting problem. These include near-surface drifters and profiling floats (e.g. Davis 1996, 1998) moving in the subsurface at a certain level and resurfacing every Δt to communicate. In addition to the position, they provide other types of information, such as temperature and salinity profiles. All these data can be used for assimilation in numerical models. In this study, we focus on the assimilation of float positions into ocean general circulation models (OGCMs) for improving nowcasts of the Eulerian velocity field.

Traditional ocean measurement techniques, such as current meter arrays, offer data at fixed-points, but show limitations via small coverage or sparse sampling. Promising emerging new techniques, such as Doppler radar, can provide accurate high-resolution velocity data, but are restricted to the ocean surface and to coastal areas. Lagrangian data, and in particular surface drifters, presumably contain records of turbulent motion at all scales down to several meters, in particular due to steadily increasing sampling rate with GPS technology. Floats also offer wide coverage, both horizontally and vertically, and availability of significant data sets collected under operational programs (e.g., WOCE, GOOS, CLIVAR, MFSTEP, ACCE, ARGO). These characteristics make Lagrangian float data desirable to be incorporated in OGCMs, which have been becoming increasingly more realistic (Smith *et al.*, 2000; Garraffo *et al.*, 2001; McClean *et al.*, 2002).

However, Lagrangian data are difficult to interpret, owing to the well-known fact that Lagrangian motion often exhibits chaotic behavior (e.g., Aref, 1984; Samelson, 1996), in particular because of the complex time dependence in the ocean (e.g., Samelson, 1992). Many different factors such as sub-mesoscale processes unresolved by OGCMs and details of surface wind forcing are likely to pose further challenges for processing Lagrangian data in the future. Consequently, it is difficult to infer the corresponding Eulerian field responsible for the Lagrangian transport. This is why special techniques need to be developed for each particular use of Lagrangian data.

Here, we summarize recent results obtained by the authors (Molcard *et al.*, 2003, 2005; Özgökmen *et al.*, 2003) in the development of a methodology for assimilating Lagrangian drifters position data in OGCMs with a direct impact on ocean state forecasting problem. Other on-going efforts are summarized in Ide and Ghil (1997a,b), Ide *et al.* (2002), and Kuznetsov *et al.* (2002). Three main issues regarding assimilation of Lagrangian information to OGCMs are covered in this chapter:

(1) One of the main problems is posed by the nonlinear nature of the relationship between the path followed by the drifters leading to position information $\mathbf{r}$ at selected sampling intervals Δt, and the corresponding Eulerian flow field $\mathbf{u}$, which is a prognostic variable set in OGCMs. A simple and common solution to this problem is to approximate the Eulerian field by $\Delta\mathbf{r}/\Delta t$, and to assimilate this velocity estimate into OGCMs (the so-called "pseudo-Lagrangian" assimilation, e.g., Hernandez *et al.*, 1995; Ishikawa *et al.*, 1996). This method works well provided that the sampling period Δt is much shorter than the Lagrangian correlation time scale (estimated as the *e*-folding time scale of the Lagrangian autocorrelation function of turbulent velocities) T_{L}, or $\Delta t \ll T_{\mathrm{L}}$. At the Global Drifter Center (NOAA Atlantic Oceanographic and Meteorological Laboratory, Miami), surface drifter positions data are posted at the interval of 6 h (hours), after data gaps of up to 3 d (days) are removed by interpolation (Hansen and Poulain, 1996). For subsurface profiling floats, the sampling period can vary from a few days to 1–2 weeks depending on applications. In comparison, T_{L} has a nominal range of 1 to 3 d for ocean surface and 7 to 14 d for subsurface flows (Griffa, 1996; Veneziani *et al.*, 2004). Given that in practice Δt is not significantly smaller than T_{L} for surface drifters and it can actually be of the same order as T_{L} for subsurface data, an alternative technique for assimilation of drifter data in OGCMs was developed (Molcard *et al.*, 2003) and is discussed here.

(2) Using float positions to correct solely the OGCM's velocity field is generally not adequate. Once the model velocity is modified, the other mass variables such as the layer thickness or the density field must also be modified accordingly. Otherwise, the imprint of the velocity corrections will not last and the fields will quickly be restored back under the strong influence of pressure gradients associated with the layer thickness. A simple technique to accomplish this in midlatitude large-scale circulation models is discussed (Özgökmen *et al.*, 2003).

(3) Another important problem is how to use the information provided by Lagrangian instruments in a given layer (which can be either at the surface or in the ocean interior) by projecting them onto other layers. This point has also been addressed and tested (Molcard *et al.*, 2005).

The goal of the Lagrangian data assimilation is to use the position information given by drifters or profilers to correct the Eulerian velocity field forecasted by the OGCM. The effectiveness of the developed Lagrangian assimilation techniques is tested in the framework of the classical twin-experiment approach considering a hierarchy of models, from a reduced gravity quasi-geostrophic model (QG) to a multi-layer Miami Isopycnic Ocean Model (MICOM). Finally, we stress that an important consideration for method development is computational efficiency and portability. Complex and computationally expensive techniques are not likely to be cost-effective for Lagrangian assimilation, given that float data are usually not the main data source in conventional observational programs.

7.2 General formulation for Lagrangian data assimilation

The data assimilation problem can be approached in different ways (Bennett, 1992; Ghil and Malanotte-Rizzoli, 1991). In particular, three main approaches can be recognized on the basis of the relative importance which an investigator assigns to a model and observations. If it is required that the assimilated field satisfies the model equations, exactly or approximately, one readily arrives at the control theory setup, which in turn leads to multiple forward and backward integrations of the original and adjoint equations. The observations are used to "correct" the model parameters, initial and boundary conditions, etc. This approach is hard to apply to nonlinear systems and is computationally expensive. However, when these difficulties are overcome, the method is very efficient (e.g. the review and references therein by Evensen *et al.* 1998). An example of application of the adjoint method to Lagrangian data assimilation is provided by Kamachi and O'Brien (1995), for a simplified model of the equatorial Pacific Ocean.

The second approach is the Kalman filter method, in which the assimilated field is not required to be the global solution of the model equations anymore, but instead the equations are used as a physically reasonable interpolator between sequential observations. There are also some computational difficulties, which have been effectively attacked lately (Cane *et al.*, 1996; Chin *et al.* 1999). As for the Lagrangian data assimilation, the Kalman filter was used by Carter (1989) for assimilating "RAFOS" floats in a simple shallow water model of the Gulf Stream.

Finally, if the model and observations are treated as equally important (maybe with different degrees of confidence), then the problem is to find a best (linear) combination of them as the true field representation. Analytically and computationally this is the simplest assimilation method, so we have chosen it for the initial stage of our efforts. The Lagrangian data assimilation problem is a difficult one and it is quite reasonable to start with a simple method to establish a benchmark for implementing more sophisticated approaches.

A theoretical basis for an optimal interpolation (OI) of model output and observations is the following relationship based on general Bayesian theory (e.g., Lorenc, 2000):

$$\mathbf{u}^{\mathrm{a}} = \mathbf{u}^{\mathrm{b}} + \mathbf{R}^{\mathrm{b}}\mathbf{G}^{\mathrm{T}}(\mathbf{G}\mathbf{R}^{\mathrm{b}}\mathbf{G}^{\mathrm{T}} + \mathbf{R}^{\mathrm{o}})^{-1}(\mathbf{y} - \mathbf{H}(\mathbf{u}^{\mathrm{b}})), \tag{7.1}$$

where $\mathbf{u}^a$ is the model velocity vector after assimilation, $\mathbf{u}^b$ is the model velocity vector before assimilation, $\mathbf{y}$ is the vector of observations, $\mathbf{H}(\mathbf{u}^b)$ is the operator that relates model state variables to the observations, $\mathbf{R}^o$ is the observation error covariance matrix, $\mathbf{R}^b$ is the covariance matrix of the model uncertainty, superscript T stands for transposition, and, finally

$$\mathbf{G} = \frac{\delta \mathbf{H}(\mathbf{u}^b)}{\delta \mathbf{u}^b} \tag{7.2}$$

is the derivative of the model-to-observation operator (sensitivity matrix). We interpret "vectors" $\mathbf{u}^a$ and $\mathbf{u}^b$ as assimilated and model Eulerian vector fields on a given grid during run time T_{ob}, respectively, and $\mathbf{y}$ as observed position increments computed from a set of particle trajectories observed during the same time T_{ob}.

Formula (7.1) is optimal in the maximum likelihood sense under the following conditions:

(i) The prior distribution of the true velocity vector $\mathbf{u}$ is Gaussian with mean $\mathbf{u}^b$ and covariance $\mathbf{R}^b$. Thus, we suppose that a model gives an unbiased estimate of the real velocity field with Gaussian error characterized by $\mathbf{R}^b$.

(ii) The distribution of the observation vector $\mathbf{y}$ is Gaussian with mean $\mathbf{H}(\mathbf{u})$ and covariance $\mathbf{R}^o$. Hence, it is supposed that the observations are also unbiased and their error is characterized by $\mathbf{R}^o$. This error is determined by instrument accuracy and resolution.

(iii) The operator $\mathbf{H}(\mathbf{u})$ is linear, which implies that $\mathbf{G}$ does not depend on $\mathbf{u}$. Even if the Gaussianity conditions (i–ii) do not hold true, estimator (7.1) is still optimal under the linearity condition in the mean square sense. However, in the discussed problem the conjecture (iii) is not true in general, even though it might approximately hold locally. Thus, in the considered case, formula (7.1) is not optimal. Nevertheless, we choose to pursue (7.1), because in nonlinear stochastic problems the optimal solution is almost never available and the hope is that the solution optimal in the linear case would perform in an acceptable way also in the non linear case. There is a full analogy between (7.1) and formulas for the extended Kalman filter (EKF) (e.g. Carter, 1989). Although the EKF is not the optimal (in the mean square sense) nonlinear filtering method, it is widely used in assimilation applications.

Consider M Lagrangian particles released at the same time $t = 0$ from different positions $\mathbf{r}_1^0, \mathbf{r}_2^0, \ldots, \mathbf{r}_M^0$ on the plane or isopycnic surface. Their motion is covered by the following system of $2M$ equations:

$$\frac{\mathrm{d}\mathbf{r}_m}{\mathrm{d}t} = \mathbf{u}(t, \mathbf{r}_m), \quad \mathbf{r}_m(0) = \mathbf{r}_m^o, \quad m = 1, \ldots, M,$$

where $\mathbf{u}(t, \mathbf{r})$ is a Eulerian velocity field, and $\mathbf{v}_m(t) = \mathrm{d}\mathbf{r}_m/\mathrm{d}t$ is the horizontal Lagrangian velocity of the m-th particle. Assume that the trajectories are observed with some errors in discrete moments $n\Delta t$, $n = 1, 2, \ldots, N$ and denote observations by $\mathbf{r}_m^{\mathrm{o}}(n)$, while the corresponding quantities obtained from the model for the same initial conditions are denoted by $\mathbf{r}_m^{\mathrm{b}}(n)$.

Introduce finite difference Lagrangian velocity obtained from the position increments for observations and model respectively

$$\mathbf{v}_m^{\mathrm{o}}(n) = \frac{\Delta \mathbf{r}_m^{\mathrm{o}}}{\Delta t} = \frac{\mathbf{r}_m^{\mathrm{o}}(n) - \mathbf{r}_m^{\mathrm{o}}(n-1)}{\Delta t}, \quad \mathbf{v}_m^{\mathrm{b}}(n) = \frac{\Delta \mathbf{r}_m^{\mathrm{b}}}{\Delta t} = \frac{\mathbf{r}_m^{\mathrm{b}}(n) - \mathbf{r}_m^{\mathrm{b}}(n-1)}{\Delta t}$$

and let $\mathbf{u}_{ij}(n) = \mathbf{u}(n\Delta t, ih, jh)$ be the Eulerian velocity values on a grid with step h at moment n. Let ϵ be a small dimensionless parameter characterizing spatial velocity gradients comparatively to the observation frequency. In a rigorous approach it can be defined as

$$\epsilon = \Delta t \left\| \frac{\mathbf{Du}}{\mathbf{Dr}} \right\|, \tag{7.3}$$

where $\mathbf{Du}/\mathbf{Dr}$ is the matrix of velocity spatial gradients and $|| \cdot ||$ is a matrix norm. Assume

$$\epsilon \ll 1. \tag{7.4}$$

Condition (7.4) is true if the frequency of measurements is high enough to resolve the spatial gradients of the current. In this study, we mostly consider and evaluate the simplest assimilation formula. This formula can be obtained from the general relations ((7.1), (7.2)) accounting for only zeroth-order terms in ϵ.

The zeroth-order assimilation formula is

$$\mathbf{u}_{ij}^{\mathrm{a}}(n) = \mathbf{u}_{ij}^{\mathrm{b}}(n) + \alpha^{-1} \sum_{m=1}^{M} \gamma_{ijm} (\mathbf{v}_m^{\mathrm{o}}(n) - \mathbf{v}_m^{\mathrm{b}}(n)) \,. \tag{7.5}$$

Here

$$\gamma_{ijm} = E_h(x_m^{\mathrm{b}}(n) - ih, y_m^{\mathrm{b}}(n) - jh), \quad E_h(x, y) \equiv \exp(-\frac{x^2}{2h^2} - \frac{y^2}{2h^2})$$

and

$$\alpha = 1 + \sigma_{\mathrm{o}}^2 / \sigma_{\mathrm{b}}^2,$$

where σ_{b}^2 is the modeling velocity mean square error and σ_{o}^2 is the error for the Lagrangian velocity which is related to the error of independent positions, say σ_{r}^2, by $\sigma_{\mathrm{o}}^2 = \sigma_{\mathrm{r}}^2 / \Delta t^2$. Our derivation of (7.5) also implies certain restrictions on the correlation scales. We assume, first, that the errors of both model and observed variables are uncorrelated in time on scales longer than Δt and, second, the space correlation radius of the model errors is of order h while the observations are uncorrelated in space. The assumption of short correlations in time allows us to find explicitly the inverse matrix in (7.1) thereby avoiding time consuming numerical operations. Coefficients γ in (7.5) come from parameterization of the space correlation function of the velocity by the Gauss function.

Notice that the simplified expression (7.5) takes into account only one assimilation time step Δt, or equivalently only two successive data points. Conceptually, then, it does not fully introduce the information on particle paths. Rather, it converts the position information into Lagrangian velocity information $\mathbf{v}$, i.e. velocity averaged along particle trajectories during the time Δt. The velocity $\mathbf{v}^b$ is computed in the model by generating trajectories during Δt and using the two end points. The simplified algorithm (7.5) can be considered as a first step in the implementation of the general formulation ((7.1), (7.2)). Possible generalizations to multi-step algorithms, including more extended path informations, are discussed in Molcard *et al.* (2003).

7.3 Methodology

Surface and subsurface Lagrangian data provide successive position informations at intervals Δt, at a particular depth, giving highly scattered data. The assimilation procedure developed here, is concerned with three main issues: (i) the nonlinear relationship existing between the observed variable, the positions r, and the model variables to be corrected; (ii) the dynamical compatibility between multiple prognostic model variables; and (iii) the vertical projection of the corrections in a multi-layer model. We will discuss these three fundamental issues separately, using different numerical tools according to the problem. The various tools will then be used together in the applications in Section 7.4.

7.3.1 Correction of Eulerian velocity field from float position data

Implementation of the Lagrangian assimilation scheme (7.1) in its zeroth-order version (7.5) has been done first with a quasi-geostrophic reduced gravity model in a double gyre configuration (Molcard *et al.*, 2003). In the quasi-geostrophic reduced gravity model the only prognostic variable is the streamfunction, which is linearly related to the velocity $\mathbf{u}$.

The model is configured in a square domain of 2000 km $\times$ 2000 km and centered at 30°N. The equilibrium layer thickness is taken as 1000 m, and the

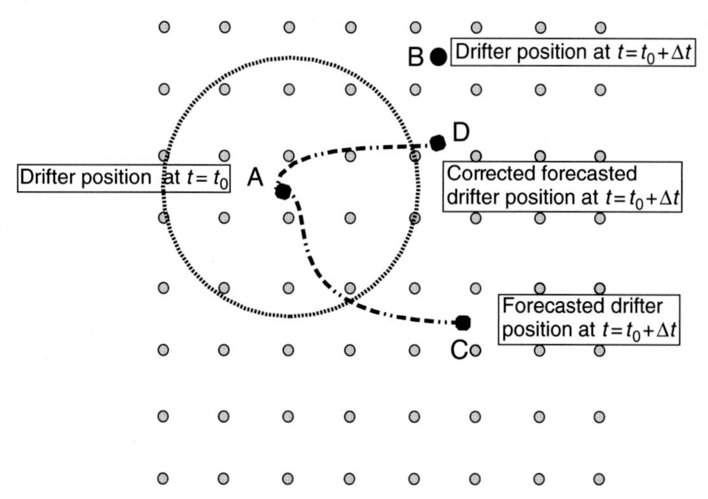

Figure 7.1 Schematic illustration of the assimilation algorithm. Given the drifter positions at $t=t_0$ (A) and $t=t_0+\Delta t$ (B), the model forecast at $t=t_0+\Delta t$ is improved from (C) to (D) by modifying the model Eulerian circulation field at $t=t_0$ within a circle of influence (model grid layout is shown in the background) using algorithm (7.5), which acts to minimize the distance between positions (B) and (C). The drifter position data are given at discrete time intervals Δt, whereas the model simulated drifters can follow paths as shown between forecasted and corrected positions, (A)–(C) and (A)–(D), respectively.

stratification is such that the Rossby radius of deformation is approximately 42 km, typical of midlatitude circulation. An idealized double-gyre configuration is adopted for the experiments, since this is probably the best-known setup in ocean modeling since Holland (1978) and its dynamics, i.e. Sverdrup gyres, western boundary currents, midlatitude jet and mesoscale eddies, are familiar to oceanographers. The model is purely wind-driven by a steady wind stress that varies sinusoidally with the latitude, which drives subtropical and subpolar gyres.

From the practical point of view, the assimilation (7.5) is implemented in the following way (Fig. 7.1). Let's assume that a drifter position $\mathbf{r}^o(t_0)$ is observed at time t_0 (corresponding to point A in Fig. 7.1). The model is forwarded in time from t_0 to $t_0+\Delta t$, providing a forecast of the drifter path starting from $\mathbf{r}^o(t_0)$. The forecasted drifter will reach a certain position $\mathbf{r}^b(t_0+\Delta t)$, (C in Fig. 7.1), and it will be characterized by a position increment $\Delta\mathbf{r}^b$ and a Lagrangian velocity $\mathbf{v}^b=\Delta\mathbf{r}^b/\Delta t$. The forecasted $\mathbf{v}^b$ is then compared with the observed Lagrangian velocity $\mathbf{v}^o$ at $t_0+\Delta t$, obtained from the observed position $\mathbf{r}^o(t_0+\Delta t)$ (B in Fig. 7.1). The model Eulerian velocity $\mathbf{u}^b(t_0)$ is modified in the vicinity of $\mathbf{r}^o(t_0)$, as a function of the difference between forecast (C) and observation (B) using (7.5). In the case of the quasi-geostrophic model, from the modified $\mathbf{u}^a(t_0)$ field, first the modified relative vorticity, and then the streamfunction is computed inverting the finite-difference matrix representation of the potential vorticity equation. The model is then advanced forward again, starting from t_0 to $t_0 + \Delta t$, using the modified velocity. The procedure can be iterated to optimize the results.

The parameter α defined above has been chosen on the basis of the ratio σ_r^2/σ_b^2, which is varied between 10^5–10^9 s^2. This corresponds to a regime in which the Lagrangian position error σ_r is on the order of 10–100 m and modeling velocity error σ_b is on the order of 1–10 cm/s. Those values give for α defined as $\alpha = 1 + \frac{1}{\Delta t^2}\frac{\sigma_r^2}{\sigma_b^2}$ a range of values going from 1.01 for small sampling period to 1.001 for higher Δt.

7.3.2 *Dynamical compatibility between corrected velocity and layer thickness*

In the QG application discussed in Section 7.3.1, the only dynamical variable to be corrected is the streamfunction, which is linearly related to the velocity. For more complex primitive equation models, on the other hand, other prognostic variables, e.g. mass variables, have to be considered and simultaneously corrected. The dynamical compatibility of multiple model variables is analyzed with a reduced gravity version of the MICOM primitive equation, layered ocean model (Bleck *et al.*, 1992). The model integrates the momentum and layer-thickness conservation equations. The model domain is chosen identical to the QG model (see Molcard *et al.*, 2003 or Özgökmen *et al.*, 2003 for numerical details).

Given the velocity corrections (Δu, Δv) derived from the drifter data, we wish to find the layer thickness correction Δh, that satisfies the momentum equations approximately. Assuming that the dynamical relation between model variables is satisfied also by the respective corrections, it is possible to deduce the layer thickness correction directly from the velocity corrections. In order to reduce the complexity of the problem, geostrophic balance and mass conservation assumptions have been used.

The geostrophic equations are combined and the following relation is obtained:

$$\nabla^2(\Delta h_g) = \frac{f}{g'}\left[\frac{\partial(\Delta v)}{\partial x} - \frac{\partial(\Delta u)}{\partial y}\right], \tag{7.6}$$

where $g' = g\frac{\Delta\rho}{\rho}$ is the reduced gravity, g the gravitational acceleration, ρ the layer density, $\Delta\rho$ the density difference between the active and motionless layers, and $f = f_0 + \beta_0\, y$ the Coriolis frequency with beta-plane approximation. Eq (7.6) is solved for Δh_g. In order to ensure mass conservation, that could be disturbed by equation (7.6) by an artifact of sampling, we applied a formulation of constraint imposed by the continuity equation of the model to obtain the final correction Δh.

7.3.3 Vertical projection of corrections in multi-layer models

Finally, we have to treat the vertical projection of the velocity corrections obtained from the Lagrangian data, in the general case of a multi-layer OGCM. A general approach is developed and tested with a 3-layer primitive equation MICOM (Molcard *et al.*, 2005). The model setup is similar to the two previous models: a box domain 3000 km × 3000 km with a flat bottom and straight sidewalls. Three isopycnal layers simulate the vertical density stratification and the only forcing is a steady zonal wind stress that drives a classical double-gyre circulation. The upper layer L1 has an initial depth of 400 m, L2 goes from 400 to 1100 m, and the bottom layer is 4000 m deep. In a multi-layer system, the corrections obtained in a single layer need to be projected to the other layers. Since there is no simple analytical relationship between layer velocities, we choose to correct the velocity in layers which do not contain Lagrangian floats using a simple statistical linear regression, with parameters determined empirically. A similar approach has been successfully used by Oschlies and Willebrand (1996), projecting observed corrections of surface geostrophic velocities onto deep velocity corrections.

For each individual column of model grid points, and each velocity component (u, v), we define the vertical regression coefficients as

$$S_{li} = \frac{< \delta u_l \delta u_i >}{< (\delta u_l)^2 >}, \tag{7.7}$$

where l is the layer in which float positions are measured, $i = 1, \ldots, N$ is the layer index, $< >$ indicates an ensemble average approximated as a time average, and δu denotes the deviation of the velocity from its mean, which is provided by the model.

The regression coefficient S_{li} represents the rate of change of one variable as a function of changes in the other. Assuming that the model corrections on average have the same vertical structure as the model fluctuation velocity, one can estimate the velocity correction in all layers using (7.7).

In the layer which contains observations, velocities are corrected via (7.5), while in the layers which contain no data, the velocity is projected using:

$$\Delta u_i = S_{li}\, \Delta u_l. \tag{7.8}$$

In practice, S_{li} are computed considering time averages of model outputs, typically over a one year period or more (e.g., Chin *et al.*, 2002). Also, the S_{li} can be further averaged in space assuming homogeneity, with the advantage of filtering out gravity waves and noise effects. The results of Chin *et al.* (2002) suggest that spatial averages as wide as the whole basin can provide satisfactory results, while simplifying the computational problem.

7.3.4 Twin experiment approach and error analysis

In order to exactly quantify the performance of the data assimilation scheme, the identical twin experiment approach is used (Fig. 7.2). The main characteristic of this approach is that synthetic floats are used instead of real data, and its great advantage is that the time evolution of the "real" ocean is assumed fully known, so that a direct comparison with the forecast results can be exactly carried out. For this reason the approach has been widely used to test methodologies, even though it should be noticed that the use of synthetic data can be misleading and the approach has the tendency of giving optimistic results because of the simplistic model setup.

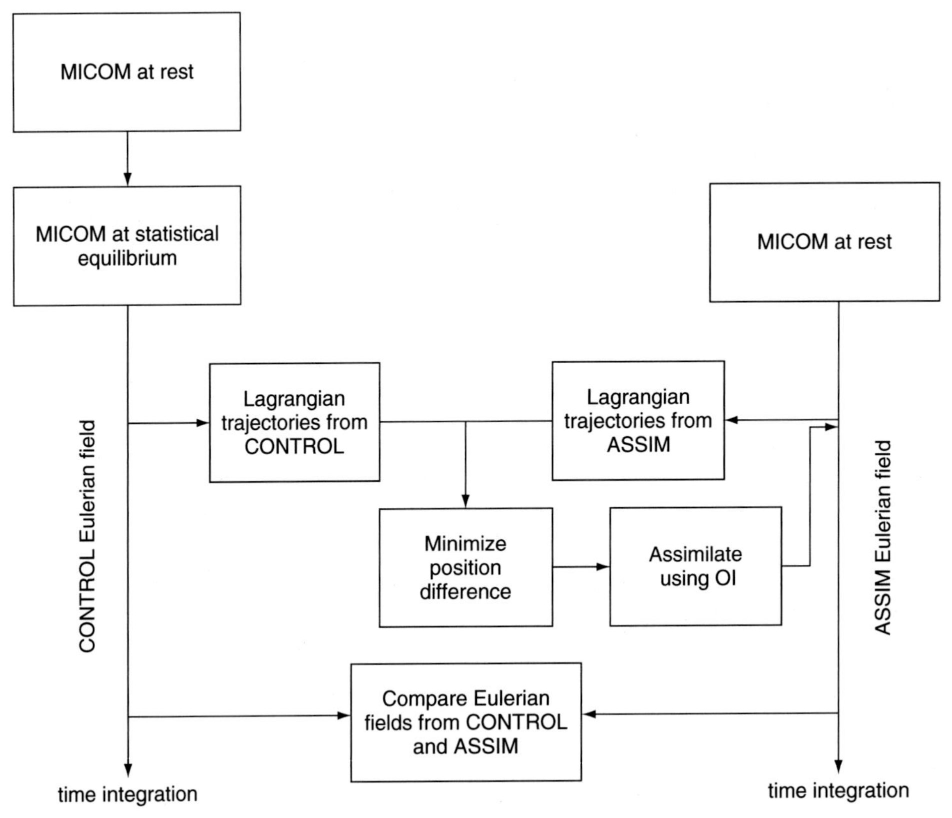

Figure 7.2 Schematic illustration of identical twin experiments using Lagrangian assimilation.

First, the model is integrated for a period of time long enough to reach a statistically steady state, which requires approximately ten years. Then, a successive simulation period is considered and regarded as the "true ocean." In this run, called "control," a set of synthetic floats are launched and advected with a fourth-order Runga–Kutta scheme. They generate the "observed data," i.e. the successive float positions.

The float data are then assimilated into another model run, the so-called "assimilation" run, which has the same parameters as the control case, but starts from a different set of initial conditions during the statistically steady state. This simulates the effect of not knowing the exact state of the ocean in reality. The third experiment, the so-called "no-assimilation" run, depicts the state of the model evolution without assimilation of drifter data.

The success of the assimilation in the experiments is evaluated both qualitatively and quantitatively. Qualitatively, the streamfunction, layer thickness patterns or velocity field of the control, assimilation and no-assimilation runs are visually compared at different times. Quantitatively, we use a performance criterion based on the (normalized) difference between the control and the assimilation, either in terms of root mean square velocity,

$$\mathrm{Eru}(t) = \frac{\sqrt{\sum_{i,j}^{K}\left[(u_{ij}^{\mathrm{C}}(t) - u_{ij}^{\mathrm{A}}(t))^2 + (v_{ij}^{\mathrm{C}}(t) - v_{ij}^{\mathrm{A}}(t))^2\right]}}{\sqrt{\sum_{i,j}^{K}(u_{ij}^{\mathrm{C}}(t) + v_{ij}^{\mathrm{C}}(t))^2}},$$

or in terms of layer thickness

$$h_{\mathrm{error}}(t) = \frac{\sqrt{\sum_{i,j}^{K}(h_{ij}^{\mathrm{C}}(t) - h_{ij}^{\mathrm{A}}(t))^2}}{\sqrt{\sum_{i,j}^{K}(h_{ij}^{\mathrm{C}}(t) - H)^2}},$$

where K is the number of model grid points, subscripts "C" and "A" stand for CONTROL and ASSIM, respectively, and $H = 1000\,\mathrm{m}$ is the background layer thickness. In the examples discussed in Section 7.4, h_{error} is used to characterize the reduced gravity MICOM experiments, while in all the other cases the velocity error Eru is used.

7.4 Results

Results of the assimilation method are shown in a hierarchy of models, which allow to illustrate and test the various issues and techniques discussed in Section 7.3. Results from a reduced gravity QG model are shown first, followed by a reduced gravity primitive equation MICOM model and finally by a 3-layer MICOM.

7.4.1 Impact of velocity field correction in single-layer QG

An example of results from the assimilation of surface drifter positions in a reduced-gravity QG model are plotted in Fig. 7.3. A cluster of 25 drifters located in the most energetic western-central region is released and advected

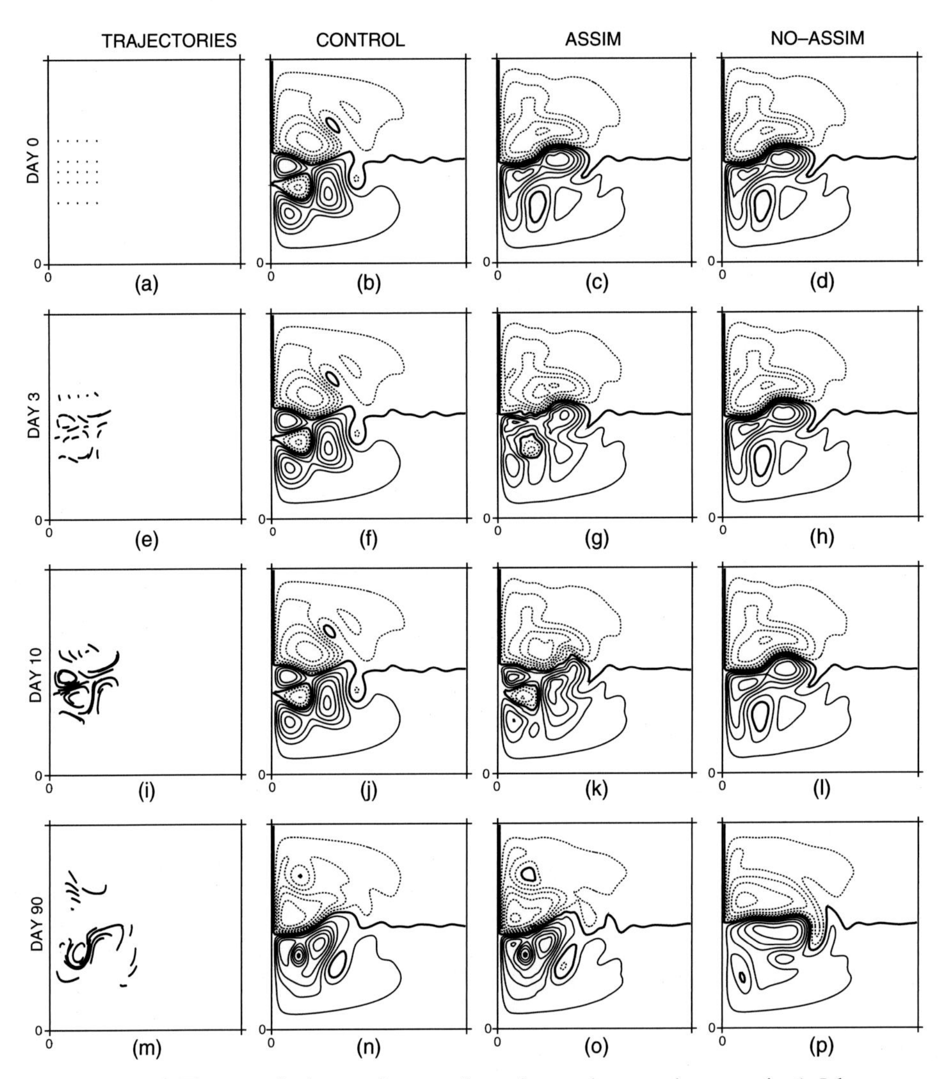

Figure 7.3 Time evolutions of control, assim and no-assim runs in 1.5-layer QG experiments. Drifter trajectories (a, e, i, m), transport streamfunction (contour interval: 10 Sv) for the control (b, f, j, n), assimilation (c, g, k, o) and no-assimilation (d, h, l, p) oceans at selected times ($t=0$, $t=3$, $t=10$, $t=90$ days). Spaghetti diagram m is shown from $t=70$ days to $t=90$ days. Note the rapid transition of the assimilation ocean circulation from the no-assimilation case to the control case.

for 90 days in the CONTROL run, and the resulting data set will be considered as our "real" data (Fig. 7.3 a, e, i, m). These synthetic drifters are characterized by a Lagrangian decorrelation time scale $T_{\mathrm{L}} \approx 10\,\mathrm{d}$, which are longer by a factor 2–3 compared to real ocean data, due to the simplified dynamics and the steady forcing (Garraffo *et al.*, 2001; Berloff and McWilliams, 2002). By taking into account this off-set between simulated and real drifters, we can use model results on the sampling period Δt dependence, to infer real data information using an appropriate scaling.

The initial conditions for the no-assim and assim runs are identical (Fig. 7.3c and 7.3d), but different from that for the control run (Fig. 7.3b). After 3 days of simulation, the assim run is modified by the velocity corrections, and the assimilation has the effect of starting to reconstruct some of the control features (Fig. 7.3g). The assimilation is highly effective after 10 days, since most of the main structures of the control (Fig. 7.3j) are reproduced (Fig. 7.3k), and the patterns of control and assim runs are nearly identical at 90 days (Fig. 7.3 n, o), while the no-assim run appears completely different (Fig. 7.3p). The convergence of the assimilation toward the control is very rapid in the first 10 days, and then followed by a slower but continuous slope. This evidence will be confirmed in the next section.

7.4.2 Sensitivity experiments

Since the sampling period of Lagrangian data, and the launching strategy are important issues for experimental oceanographers, we investigated the sensitivity of the assimilation to the sampling period and to the number and initial distribution of drifters.

A series of experiments have been performed keeping the same configuration as for the experiment shown in Fig. 7.3, but changing the value of the sampling period Δt, from 2 days to 10 days. The convergence characteristics are shown in Fig. 7.4 in which the velocity error Eru is plotted for 90 days. For comparison, the time evolution of the no-assim experiment error is also displayed which shows always high values (over 95%). In contrast, in all assimilation runs the error decreases very quickly in the first 10 days and then it keeps a slower decreasing rate. The best result is obtained with $\Delta t = 2$ days with a final velocity rms error of $\approx$20%. Also for significantly smaller values of Δt (not shown), of the order of the model time step, the results are not found to improve further. This shows the existence of an optimal range, $\Delta t \approx T_{\mathrm{L}}/5$. For the ocean surface, using appropriate

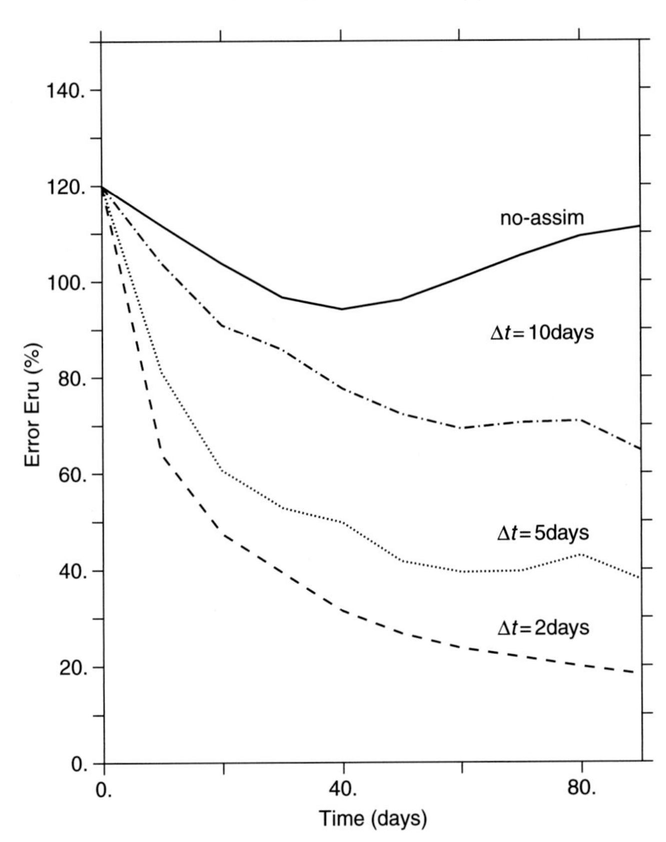

Figure 7.4 Convergence error as a function of observation time using different sampling periods $\Delta t = 2$, 5, and 10 days in assimilation runs (dashed lines) and that of the no-assimilation runs (solid line).

scaling by T_L, this corresponds to $\Delta t \approx 5$–15 h, while for the subsurface $\Delta t \approx 1.5$–2 days.

In order to analyze the sensitivity of the assimilation procedure to the number of drifters, a set of seven experiments have been performed with 9, 16, 36, 49, 100, 144, and 196 drifters that were homogeneously released in the energetic western midlatitude jet region. The velocity error at the end of the simulation (90 days) is plotted as a function of the number of drifters in Fig. 7.5, which shows the asymptotic behavior of error as the number of drifters is increased. The assimilation performance is improved from 55% to 25% from 9 to 16 drifters, followed by a linear decrease of the error up to 16% for 49 drifters, and finally a slow decay up to 11% for the maximum number of drifters tested.

To explore the impact of the launching strategy on the assimilation error, 6 further experiments are conducted, in which 25 drifters distributed over the

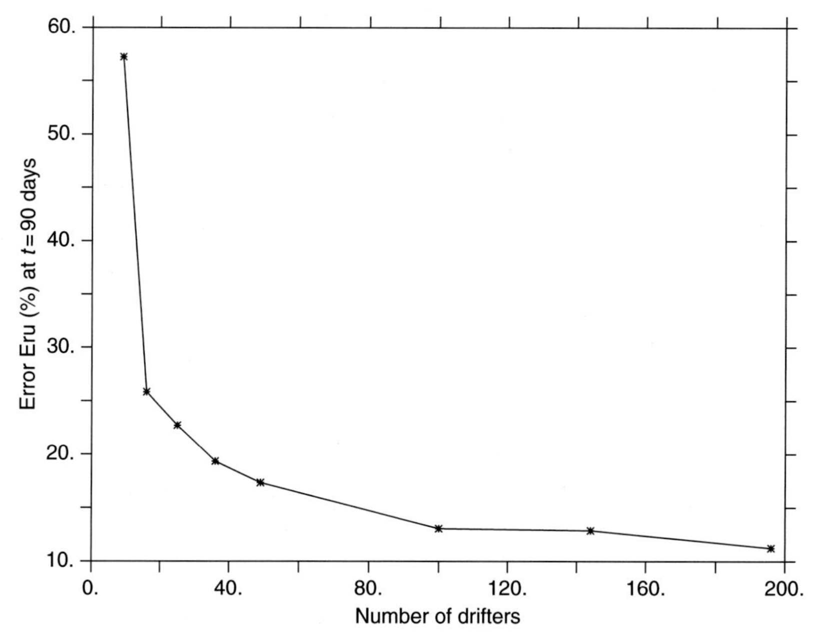

Figure 7.5 Convergence error Eru (%) at $t = 90$ days as a function of the number of assimilated drifters.

same area as in the first experiment (see Section 7.4.1 and Fig. 7.3) are moved around the domain (regions a, b, c, d, e, f, g in Fig. 7.6) according to the kinetic energy distribution. In addition an experiment with 25 drifters launched homogeneously over the entire domain is performed (region h in Fig. 7.6). The error Eru at $t = 90$ days is then plotted as a function of the average kinetic energy of these subdomains normalized by that of the entire domain in Fig. 7.6 (lower panel). This figure shows the general trend of better assimilation performance when the drifters are released in energetic regions. The importance of effectively sampling the energetic regions in assimilation problems has been pointed out in a number of previous studies (e.g. Malanotte-Rizzoli and Holland, 1988). Notice that, despite the clear general trend, a considerable scatter can be seen in the results of Fig. 7.6 (lower panel). This appears to be indicative of the sensitivity of Lagrangian motion to the detailed characteristics of the flow field, suggesting that other factors play an important role, such as for instance dispersion and data voids occurring during the drifter history. The comprehensive investigation of optimal sampling for Lagrangian data assimilation described in this study is a difficult task beyond the scope of the present study. A recent paper by Poje *et al.* (2002) provides detailed considerations for drifter launch strategies to reconstruct Eulerian model flow fields using a least squares minimization of the difference between model and drifter

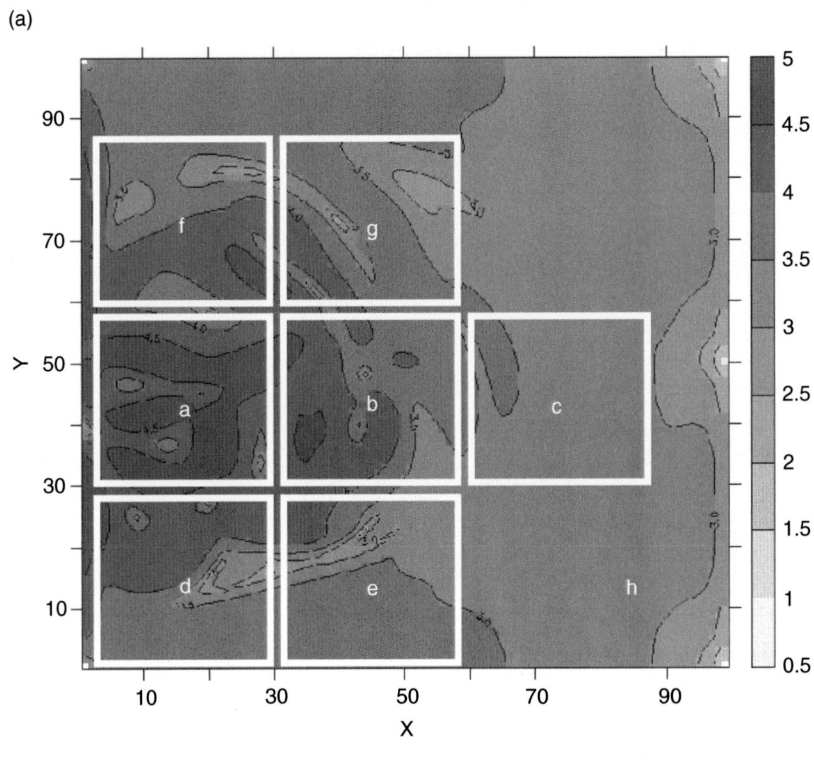
(a)
f
g
a
b
c
d
e
h
Y
X
10
30
50
70
90
5
4.5
4
3.5
3
2.5
2
1.5
1
0.5

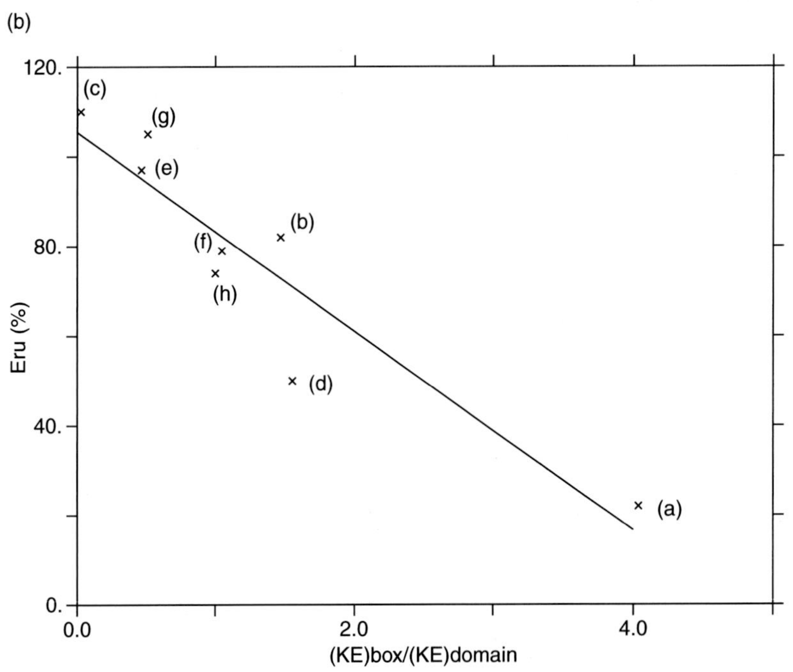
(b)
120.
80.
40.
0.
Eru (%)
0.0
2.0
4.0
(KE)box/(KE)domain
(c)
(g)
(e)
(b)
(f)
(h)
(d)
(a)

velocities, as described by Toner *et al.* (2001). Furthermore a collaboration has started to study the optimal sampling deployment of Lagrangian drifters in terms of assimilation efficiency, based on the computation of the manifolds: the main results of this study is that by seeding the outflowing manifold with the initial location of the drifters, the convergence of the assimilation is consistently improved.

For applications in the real ocean, the results generally indicate the importance of the initial sampling strategy and suggest that a sampling targeted at high-energy regions is likely to be more efficient than a homogeneous sampling. Also, the results suggest that the assimilation can be highly effective even with a relatively low data density, provided that the drifters are concentrated in the energetic regions. It should be pointed out, however, that the residence time of the drifters in the energetic structures might be significantly different in the model and in the real ocean. In the model, drifters released in the western recirculation tend to be trapped for a relatively long time (on the order of months), whereas in the real ocean (especially at the surface) drifters tend to be advected away from the energetic regions much more quickly. As a consequence, the sampling problem in the ocean is certainly more complex, and probably implies repeated samplings in order to maintain a certain data density in the regions of interest.

7.4.3 Impact of layer thickness correction in single-layer MICOM

The experiments presented in previous sections illustrate the effectiveness of the assimilation scheme. In this section we present results of experiments performed with a primitive equation model MICOM in a reduced gravity configuration. The methodology presented in Section 7.3.2 is used to correct both the velocity and the layer thickness, and the impact of this second step is analyzed. If the impact is found to be small, then the conclusion would be that this step is not very crucial, and velocity/position information alone are adequate. On the other hand, if the impact is large, then this indicates the importance of simultaneously correcting h. This has implications for many

---

Figure 7.6 (Upper panel) Initial locations of the 25 drifters released in a box of 600 km $\times$ 600 km moved around the domain (a–g). In experiment h 25 drifters are homogeneously released in the 2000 km $\times$ 2000 km basin. The contours in the background denote log (KE [cm^2/s^2]) at $t=0$. (Lower panel) Convergence error Eru (%) at $t=90$ days from these experiments as a function of the spatial average of the kinetic energy in the launch box normalized by the the basin-averaged kinetic energy. The solid line represents a least-squares fit to data.

applications, including cases when geostrophy doesn't hold (e.g., at low latitudes). In order to address this question, a number of experiments are conducted, in which drifter data are assimilated using only the velocity correction, and without simultaneously correcting the layer thickness.

Here, the results from the assimilation with 121 drifters homogeneously released in the whole domain are shown, as this case is the least-likely to show degradation of results due to sampling issues and inaccuracies of the method, and the change in accuracy results primarily from the missing h-correction. The results in terms of time evolution of h_{error} for a 1-year simulation, are illustrated for both $\Delta t = 6$ h and $\Delta t = 3$ d cases in Fig. 7.7. In both cases, the lack of h-correction has a significant impact leading to drastically slower response of the model to the assimilation. The greatest gain from h-correction is during the first 50 d, and h-correction has a bigger effect when $\Delta t = 3$ d than when $\Delta t = 6$ h. The results are qualitatively similar for other drifter densities as well (not shown). This result not only supports the validity of the formulation

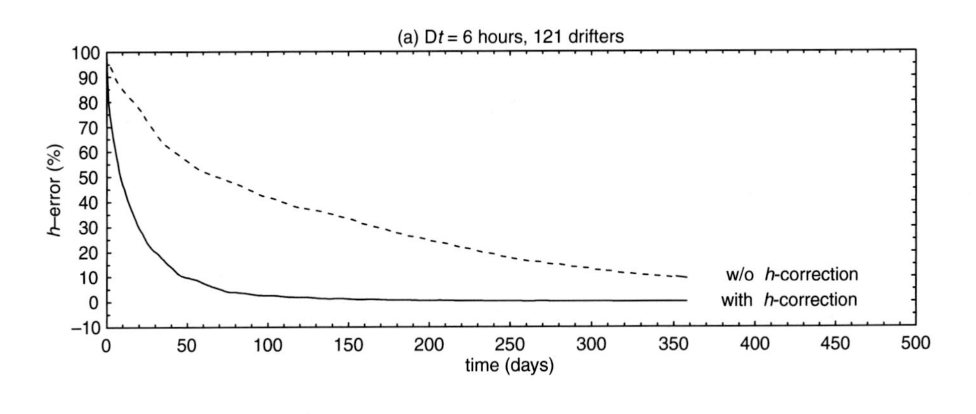

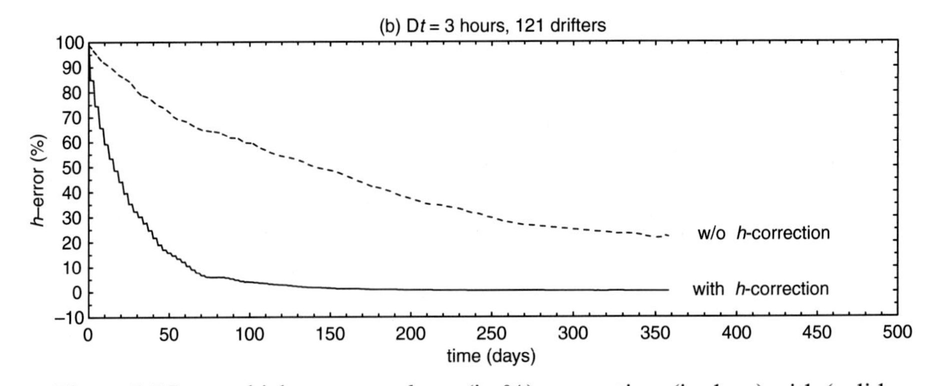

Figure 7.7 Layer thickness error h_{error} (in %) versus time (in days) with (solid curve) and without (dashed curve) the correction of layer thickness h in the case of 121 drifters and (a) $\Delta t = 6$ h and (b) $\Delta t = 3$ days.

(7.6) for the specific case considered in this study (midlatitude circulation, 1.5 layer model), but it also indicates that the correction of the model velocity field using drifters must be accompanied with an appropriate correction of the layer thickness field (or mass field, pressure, etc., depending on the model formulation) for such assimilation to be effective.

7.4.4 Comparison with pseudo-Lagrangian assimilation methods

The same framework of reduced gravity MICOM has been used to compare the fully Lagrangian formulation scheme presented in Section 7.1, which is indicated as `OI-Lag`, with more conventional pseudo-Lagrangian techniques (e.g. Ishikawa *et al.*, 1996). In pseudo-Lagrangian assimilation, synthetic model trajectories and Lagrangian velocities are not computed, but rather the model Eulerian velocity is directly used to compute the velocity correction. This is equivalent to replacing in equation (7.5) the model Lagrangian velocity $\mathbf{v}_m^b$ by the Eulerian $\mathbf{u}_m^b$.

We consider two algorithms for implementation of pseudo-Lagrangian assimilation. The first one uses the same basic OI formulation, `OI-PsLag`, and the second one is based on Kalman filter, `KF-PsLag`. Due to the number of variables, the Kalman filter algorithm must be approximated (usually by parameterization of the large covariance matrix) to be practical for data assimilation purposes in present-day computers. The particular realization used here, called *reduced-order information filter* (ROIF), has been demonstrated effective in reconstruction of mesoscale features in MICOM, as detailed by Chin *et al.* (1999, 2002).

In the limit of small Δt, Lagrangian and Eulerian velocities indeed coincide, and therefore the pseudo-Lagrangian assimilation is expected to provide similar results with respect to the full Lagrangian assimilation. For finite Δt, though, the differences between the methods are expected to be relevant.

A set of experiments has been performed, with different number of drifters and different sampling period, and the performances of the three assimilation techniques have been compared. Since the time evolution of the layer thickness error h_{error} (not shown) for all techniques appears to exhibit a nearly-exponential convergence, $\ln(h_{\mathrm{error}})$ is plotted linearly in time in the particular case of 64 drifters and $\Delta t = 3$ days (others show a similar behavior). Fig. 7.8 indicates that the convergence of ASSIM to CONTROL can be characterized in three stages, which exhibit piecewise and nearly-constant slopes (or exponential convergence). In the initial stage ($0 \leq t < 60$ d), all methods show an exponential decay in error, and the rate of decay reduces somewhat in the intermediate stage ($60 \leq t \leq 200$ d). Near the end of the integration ($t > 200$ d),

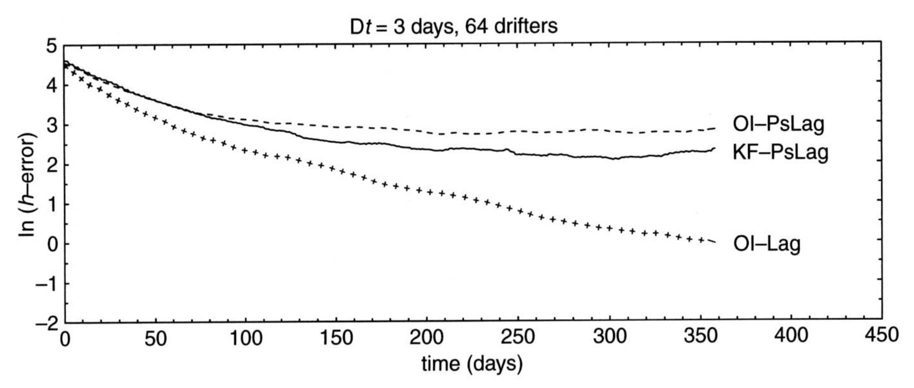

Figure 7.8 $\ln(h_{\text{error}})$ vs time in the case of $\Delta t = 3$ days and 64 drifters for OI-PsLag, (KF-PsLag, and OI-Lag).

pseudo-Lagrangian methods asymptote to a constant value, whereas OI-Lag continues to reduce the error at the same exponential rate. As the major reduction of error takes place in the first few months of assimilation for all techniques, all cases are simply approximated by

$$h_{\text{error}} \sim \exp\left(-\frac{t}{t_0}\right) + h^*_{\text{error}} ,$$

where t_0 is the e-folding time scale of error, calculated by estimating the slope of the lines (Fig. 7.8) in the interval of $0 \leq t \leq 60$ d using a least-square fit, and h^*_{error} is the residual convergence error calculated in the interval $300\text{ d} \leq t \leq 360$ d. Therefore, t_0 and h^*_{error} serve as simple gauges to illustrate the differences between the three techniques. The results are shown in Fig. 7.9 for all experiments with $\Delta t = 3$ d. The comparison of the e-folding time scales (Fig. 7.9a) indicates that t_0 decreases as the number of drifters increases, at approximately the same rate for OI-Lag and OI-PsLag, and at a higher rate for KF-PsLag. The most notable aspect in Fig. 7.9a is that OI-Lag using only 36 drifters yields equivalent performance (and much better with 121 drifters) as for OI-PsLag using 121 drifters. This demonstrates the improvement in performance due to Lagrangian technique. The impact of differences in t_0 is then clearly reflected in h^*_{error} (Fig. 7.9b). The residual errors for both OI methods reduce with increasing number of drifters, but they are in the range of 16–23% for OI-PsLag, whereas OI-Lag converges to a very low 0.5–4%. Residual errors for KF-PsLag are nearly constant at 10% and are relatively independent of the number of drifters. Overall, the implementation of the Lagrangian assimilation technique in an otherwise conventional OI method appears to have improved significantly the convergence characteristics and accuracy of the

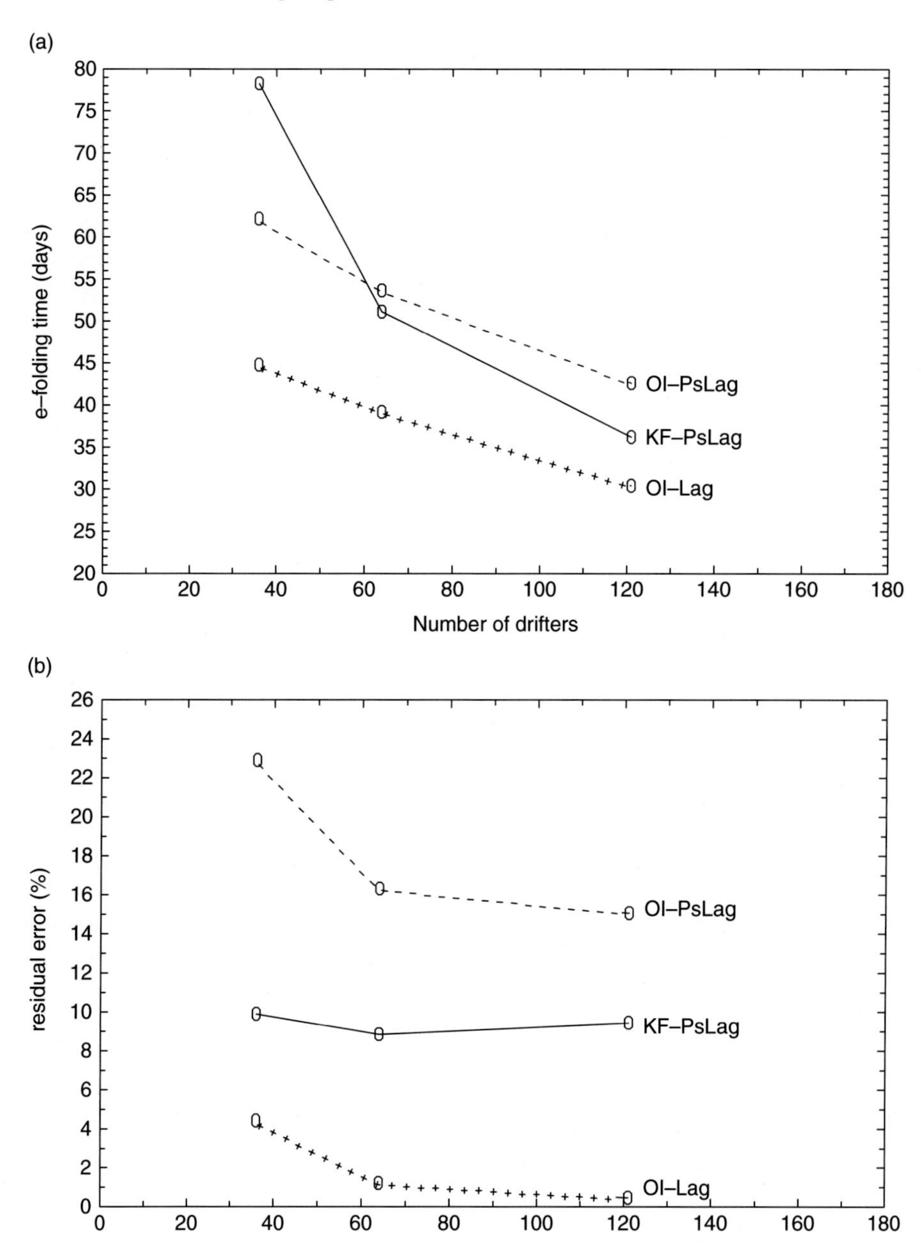

Figure 7.9 (a) The e-folding time scales t_0, and (b) residual errors h^*_{error} as a function of the number of drifters and assimilation methods in the case of $\Delta t = 3$ days.

method in the range where assimilation period is significant when compared to the minimum Lagrangian correlation time in the system.

The conceptual reason for this result is discussed in Molcard *et al.* (2003), where simplified error estimates for the Lagrangian and pseudo-Lagrangian formulations are derived. It is shown that the error for the Lagrangian assimilation can in principle converge to zero for finite Δt, provided that the model velocity gradient is sufficiently similar to the truth gradient. The pseudo-Lagrangian error, on the other hand, is always finite for finite Δt, suggesting the existence of a bias that prevents the full convergence of the assimilation to the control. The Lagrangian assimilation, then, converges more efficiently once the model reaches a condition that is relatively similar to the control and in these cases it is clearly superior to the pseudo-Lagrangian method. For $\Delta T \approx T_L$, on the other hand, the model solution maintains quite distant from the control and this effect plays a less significant role. It might be expected that the effect is less evident also for the case of real floats, characterized by significant errors. This aspect will have to be tested in the future using real data.

7.4.5 Impact of vertical projection in multi-layer MICOM

In this application, we consider a 3-layer MICOM model. In this case, all the three issues discussed in Section 7.3 are considered and the method can be summarized in terms of the following successive steps:

(a) The Eulerian velocity is corrected in the same layer where Lagrangian data are measured, minimizing the difference between observed and simulated Lagrangian velocity during the sampling period Δt using (7.5). The simulated Lagrangian velocity is computed by advecting synthetic floats in the model velocity field.
(b) The velocity correction in the other layers is computed using statistical regression coefficients between layers (7.7). These coefficients are computed from model outputs as time and spatial averages.
(c) The layer thickness is corrected in each layer using a dynamical balancing technique based on geostrophy and mass conservation, similarly to (7.6).

The model described in Section 7.3.3 is used for this analysis. Fig. 7.10 depicts snapshots of layer thickness and velocity field in the three layers at the initial time and after 120 days of simulation for the control run in the central-western energetic region. Only this sub-region is plotted as the float positions are measured essentially in this area. The first layer is the most energetic one and it is characterized by the fastest time scale. In the upper layer (L1) the western boundary current is very intense while it loses strength with depth (L2 and L3).

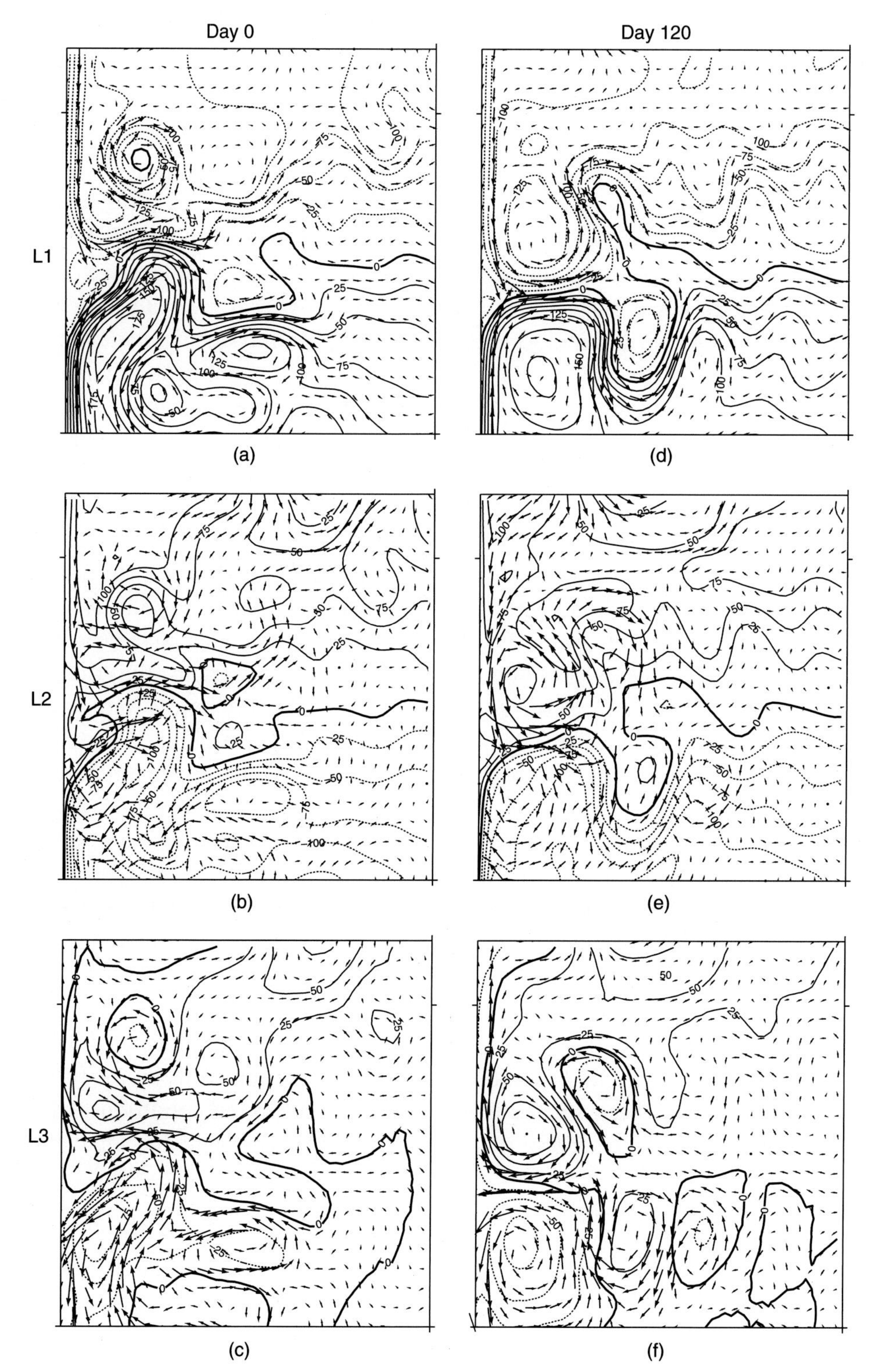

Figure 7.10 Control run in the 3-layer MICOM. Layer thickness displacement ($h - h_{rest}$) superimposed to the velocity field in each layer (L1, L2, L3), at initial time (Day 0) and at the end (Day 120) of the simulation. Contour intervals are 25 m and the maximum velocity field is approximately 50 cm/s in L1, 10 cm/s in L2, 5 cm/s in L3.

Most of the circulation features are present in all three layers: the western boundary current and the meandering jet in L1 and L2, the gyres' centers in L1 and L3. While the current is clearly geostrophic in the first layer, the direct balance between pressure gradient and velocity in the bottom layer is more ambiguous. This is mainly due to the rigid-lid approximation in MICOM, in which the barotropic mass flux divergence is removed from all layer velocities (Bleck and Boudra, 1986; Bleck *et al.*, 1989). This term gives an important contribution in the third and second layer and can lead to significant deviations from geostrophy. Notice also that L1 and L3 appear significantly anti-correlated, having the main gyres with opposite layer thickness anomaly sign and corresponding opposite velocity field (Fig. 7.10a, c and Fig. 7.10d, f), while L2 appears less significantly correlated and at the same time less clearly in geostrophic balance. While estimating the vertical regression coefficients (7.7) from this simulation, the mean flow is computed over a one-year period, and removed from the velocity field to obtain the deviation δu for each velocity component. S_{li} are first computed at all grid points and then spatially averaged over the domain (Chin *et al.*, 2002). Preliminary experiments have been done considering different scales of homogeneity in space, by averaging the grid point regression coefficients over sub-regions, and no major differences were found.

The correlation between L1 and L2 is quite weak, reaching a maximum value of 0.5, while the correlation between L1 and L3 is 0.75. The efficiency of the vertical projection of the velocity using equation (7.8) is very much dependent on the correlation between vertical layers, since the velocity corrections are estimated from it. One can anticipate good results from assimilating experiments when floats data are launched in either L1 or L3, since the correlation factor between these two layers is high, whereas launching floats in L2 could lead to a less efficient improvement in the forecast.

A set of experiments has been performed, and the assimilation has been tested for launches in the various layers and for different values of the sampling period, $\Delta t = \mathrm{d}t$, 3d, 6d. In order to increase the generality of the results, the assim run has been started from three different initial conditions, which are characterized by a six-month lag, and the error is presented as an average over them.

For these experiments we adopted a relative error definition,

$$\epsilon_{\mathrm{rel}}(l, t) = \frac{\mathrm{Eru(C - A)}}{\mathrm{Eru(C - NA)}},$$

where l indicates the layer and Eru(C – A) (Eru(C – NA)) represents the rms of the velocity difference between control and assim (noassim) defined as in section 7.3.4. With this normalized value all the experiments can be directly compared, or averaged, since the initial condition dependence is removed. If the assimilation is successful, $\epsilon_{rel} < 1$, and as long as it is effective, the slope of the error curve $\partial\epsilon_{rel}(t)/\partial t < 0$. The spatial average of ϵ_{rel} is done over the energetic sub-region surrounding the area where the cluster of Lagrangian data are almost confined during the simulation period.

We are investigating the performance of the assimilation algorithm when launching Lagrangian data in either the upper layer (LaunchL1), the central layer (LaunchL2) or in the bottom one (LaunchL3).

Figure 7.11 depicts the time evolution of ϵ_{rel} for the various experiments. Each column represents a specific experiment characterized by the layer where Lagrangian floats are launched. Each row indicates the layer in which the error value is measured. In each panel the three lines represent

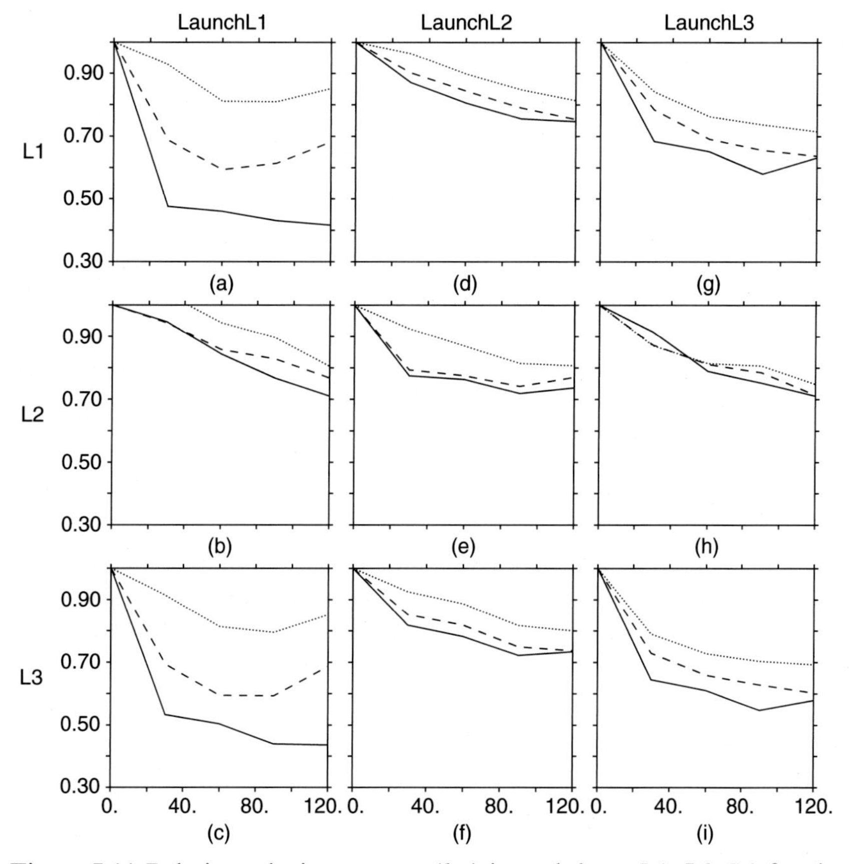

Figure 7.11 Relative velocity error $\epsilon_{rel}(l, t)$ in each layer L1, L2, L3 for the three launchings, LaunchL1, LaunchL2, LaunchL3. The three lines in each figure show results from different sampling periods Δt.

results from the different sampling periods. Since the final relative errors for all experiments are smaller than 1, the velocity assimilation is effective. Furthermore, the general trend is a negative slope, meaning a continuous improvement of the forecast during the assimilation experiment. As in Fig. 7.4 and Section 7.4.2, the convergence is fast in the initial period, and the curves have exponential trends. In the experiment LaunchL1, the dependence on the sampling period Δt is more accentuated. The Lagrangian time scale of the floats in the upper layer is about 5 days, so the deterioration of the assimilation occurs when the sampling period is close to or higher than this value. Notice the high performance of the assimilation in the bottom layer, justified by the high correlation existing between L1 and L3. Since the upper layer is the most energetic, surface floats are capturing a higher number of structures of the real ocean, and the spatial and temporal coverage is higher. This leads to the absolute minimum error of 40%, obtained when the launching is done in this upper layer, and when the sampling period is as small as the model time step.

Experiment LaunchL2 is the least successful, even though the assimilation is improving the forecast. The ϵ_{rel} reaches $\approx$70–80% in all layers, and is almost independent of the sampling period, since the Lagrangian time scale characterizing the data floating in the middle layer is around 10 days.

All together, the launching in L3 gives the best averaged result, as the improvement of the forecast is consistent in all layers even at high sampling period and the difference between the crude forecast and the true ocean has been reduced almost by a factor of 2, at least in the third layer.

As in Section 7.4.1, in order to interpret these results for real ocean applications, the results should be renormalized in terms of ocean values for T_L.

7.5 Conclusions

The increasing number of Lagrangian data released in the ocean, and the recent advances in the OGCMs provide the main motivation for development of new methods of using drifter data in these models. Assimilation of real data to improve forecasted ocean circulation is an important issue, and the particular use of Lagrangian data is a new and challenging problem. Here we presented a new assimilation algorithm, developed in order to take into account the Lagrangian nature of the observations. Furthermore, the algorithm has been reduced to its simplest form, in order to have high portability of the assimilation module to any OGCM. The general method can be considered as composed of three steps which have been developed and tested separately: assimilation of Lagrangian position in data-containing layer, vertical projection of this information to other layers using statistical correlations, and establishment of dynamical compatibility between model prognostic variables.

The methodology has been tested using the twin experiment approach. The assimilated data are synthetic floats advected in a particular setup of the numerical model, regarded as the true ocean and indicated as "control." The model setup is then modified choosing a different initial condition, and this configuration is used for the forecast. If the assimilation of the observations is successful, the corrected forecast should converge toward the initial simulation. This approach has the great advantage that the full time evolution of the true Eulerian fields is known, and the performance of the assimilation is easily measured, either visually and quantitatively.

We have shown that the assimilation algorithm is improving the Eulerian velocity field in ocean models, provided that the sampling period is smaller than the Lagrangian time scale characterizing the observations. The assimilation is particularly efficient for an optimal range of the sampling period, $\Delta t \approx 20$–50% of the Lagrangian integral timescale T_L of the flow field.

By including the dynamical balances of the corrected variables, the response of the ocean is found to improve, which indicates that the correction of model velocity field should be coupled with an appropriate correction of the other prognostic variables of the OGCM for such assimilation to be effective.

In the case of a multi-layer model, we have shown that the vertical projection of the corrected variables done by correlating the vertical layers can improve the forecast in the whole water column, provided of course that the correlation between layers is significant.

The assimilation algorithm has been compared with two pseudo-Lagrangian methods, which compare directly the Lagrangian velocity computed from the observations, with the Eulerian model velocity. The pure Lagrangian scheme shows a better performance than either the simplest pseudo-Lagrangian method, and the more sophisticated Kalman filter based method. This result confirms the advantage of using this method, which takes into account the Lagrangian nature of the observations.

Varying the number of floats and their initial sampling shows that the assimilation is effective even with a small number of floats, provided that they are concentrated in the energetic regions. When the floats are released in high-energy regions (meandering jet, recirculation regions), and their residence time in those regions is high, the success of the assimilation is guaranteed. In the real ocean, the sampling strategy might be more complex, including repeated samplings, given that the residence time in the energetic structures is likely to be smaller than in the model, so that other factors may be important like dispersion and data voids.

This extensive study has shown the effectiveness of the use of a simple and portable algorithm which takes into account the Lagrangian nature of the observations. The positive results are encouraging, and the present work has pointed out a number of interesting issues that could be investigated in the future. Some approximations of the present algorithm could be improved to gain in efficiency and realism. For instance, the impact of data errors is assumed to be constant and based on empirical values from observations, while more sophisticated error covariance matrices, time and space dependent, could improve the assimilation performance. Furthermore, for profiling floats, the displacement due to vertical shear during the diving and emerging processes as well as the surface drift between successive satellite fixes, should be considered and corrected to improve the results.

Also, the present algorithm is implemented in its simplest form, local both in time and space. Only observations made at the same instant and situated in the surrounding area are considered. The general assimilation formula developed in Molcard *et al.* (2003) could be easily rewritten for multi-step algorithms, in order to allow a more complete use of path information. The improvement extent of this higher-order algorithm compared to the complexity of the coding and the computational cost, is an interesting issue for this problem.

Acknowledgments

The authors greatly appreciate the support of the Office of Naval Research under grants N00014-97-1-0620 and N00014-99-1-0049 (A. Molcard, T. M. Özgökmen and A. Griffa), N00014-99-1-0042 (L. I. Piterbarg), and the support of the EU-MFSTEP project (A. Molcard and A. Griffa). The authors thank N. Pinardi for helpful discussions.

References

Aref, H., 1984. Mixing by chaotic advection. *J. Fluid Mech.*, **143**, 1–24.

Bauer, S., M. S. Swenson, and A. Griffa, 2002. Eddy-mean flow decomposition and eddy-diffusivity estimates in the tropical Pacific Ocean. 2. Results. *J. Geophys. Res.*, **107**, 3154–71.

Bennett, A. F., 1992. *Inverse Methods in Physical Oceanography*. New York: Cambridge University Press.

Berloff, P. and J. C. McWilliams, 2002. Material transport in oceanic gyres. Part II: Hierarchy of stochastic models. *J. Phys. Oceanogr.*, **32**, 797–830.

Bleck, R. and D. B. Boudra, 1986. Wind-driven spin-up eddy-resolving ocean models formulated in isopycnic and isobaric coordinates. *J. Geophys. Res.*, **91/C**, 7611–21.

Bleck, R., H. P. Hanson, D. Hu, and E. B. Kraus, 1989. Mixed layer-thermocline interaction in a three-dimensional isopycnic coordinate model. *J. Phys. Oceanogr.*, **19C**, 1417–39.

Bleck, R., C. Rooth, D. Hu, and L. T. Smith, 1992. Ventilation and mode water formation in a wind- and thermohaline-driven isopycnic coordinate model of the North Atlantic. *J. Phys. Oceanogr.*, **22**, 1486–1505.

Cane, M. A., A. Kaplan, R. N. Miller, B. Tang, E. C. Hackert, and A. J. Busalacchi, 1996. Mapping tropical Pacific sea level: Data assimilation via a reduced state space Kalman filter. *J. Phys. Oceanogr.*, **101**, 22599–617.

Carter, E. F., 1989. Assimilation of Lagrangian data into a numerical model. *Dyn. Atmos. Oceans*, **13**, 335–48.

Chin, T. M., A. J. Mariano, and E. P. Chassignet, 1999. Spatial regression with Markov Random Fields for Kalman filter approximation in least-squares solution of oceanic data assimilation problems. *J. Geophys. Res.*, **104**, 1233–57.

Chin, T. M., A. C. Haza, and A. J. Mariano, 2002. A reduced-order information filter for multi-layer shallow water models: profiling and assimilation of sea surface height. *J. Atmos. Ocean. Tech.*, **19**(4), 517–33.

Davis, R. E., 1983. Oceanic property transport, Lagrangian particle statistics, and their prediction. *J. Mar. Res.*, **41**, 163–94.

Davis, R. E., 1991. Observing the general circulation with floats. *Deep-Sea Res.*, **38**, 5531–71.

Davis, R. E., 1996. Comparison of Autonomous Lagrangian Circulation Explorer and fine resolution Antarctic model results in the South Atlantic. *J. Geophys. Res.*, **101**, C1, 855–84.

Davis, R., 1998. Preliminary results from directly measuring mid-depth circulation in the Tropical and South Pacific. *J. Geophys. Res.*, **103**, 24619–39.

Evensen, G., D. P. Dee, and J. Schroter, 1998. Parameter estimation in dynamical models. In *Ocean Modeling and Parameterization*, ed. E. P. Chassignet and J. Veron. Dordrecht: Kluwer Academic Publishers, 373–98.

Fratantoni, D. M., 2001. North Atlantic surface circulation during the 1990s observed with satellite-tracked drifters. *J. Geophys. Res.*, **106**, 22,067–93.

Garraffo, Z. D., A. J. Mariano, A. Griffa, C. Veneziani, and E. P. Chassignet, 2001. Lagrangian data in a high-resolution numerical simulation of the North Atlantic. 1. Comparison with in situ drifter data. *J. Mar. Sys.*, **29**, 157–76.

Ghil, M. and P. Malanotte-Rizzoli, 1991. Data assimilation in meteorology and oceanography. *Adv. Geophy.*, **33**, 141–266.

Griffa, A., 1996. Applications of stochastic particle models to oceanographic problems. In *Stochastic Modeling in Physical Oceanography*, ed. R. Adler, P. Muller, B. Rozovskiim. Cambridge, MA: Birkhauser Boston, 113–28.

Hansen, D. V. and G. A. Maul, 1991. Anticyclonic current rings in the eastern tropical Pacific Ocean. *J. Geophys. Res.*, **96**, 6965–79.

Hansen, D. V. and P.-M. Poulain, 1996. Quality control and interpolations of WOCE/TOGA drifter data. *J. Atmos. Oceanic Tec.*, **13**, 900–9.

Hernandez, F., P. Y. Le Traon, and N. H. Barth, 1995. Optimizing a drifter cast strategy with a genetic algorithm. *J. Atmos. Ocean Tech.*, **12**, 330–45.

Holland, W. R., 1978. The role of mesoscale eddies in the general circulation of the ocean. *J. Phys. Oceanogr.*, **22**, 1033–46.

Ide, K. and M. Ghil, 1997a. Extended Kalman filtering for vortex systems. Part I: Methodology and point vortices. *Dyn. Atm. Oceans*, **27**, 301–32.

Ide, K. and M. Ghil, 1997b. Extended Kalman filtering for vortex systems. Part II: Rankine vortices and observing system design. *Dyn. Atm. Oceans*, **27**, 333–50.
Ide, K., L. Kuznetsov, and C. K. R. T. Jones, 2002. Lagrangian data assimilation for point-vortex system. *J. Turbulence*, **3**, 053.
Ishikawa, Y. I., T. Awaji, and K. Akimoto, 1996. Successive correction of the mean sea surface height by the simultaneous assimilation of drifting buoy and altimetric data. *J. Phys. Oceanogr.*, **26**, 2381–97.
Kamachi, M. and J. J. O'Brien, 1995. Continuous assimilation of drifting buoy trajectories into an equatorial Pacific Ocean model. *J. Mar. Sys.*, **6**, 159–78.
Kuznetsov, L., M. Toner, A. D. Kirwan, C. K. R. T. Jones, L. H. Kantha, and J.Choi, 2002. The Loop Current and adjacent rings delineated by Lagrangian analysis of near-surface flow. *J. Mar. Res.*, **60**, 405–29.
Lorenc, A. C., 2000. A Bayesian approach to observation quality control in variational and statistical assimilation. *Proceedings of Aha Huliko Hawaiian Winter Workshop*, 249–63.
Malanotte-Rizzoli, P. and W. R. Holland, 1988. Data constraint applied to models of the ocean general circulation. Part 2. The transient, eddy resolving case. *J. Phys. Oceanogr.*, **18**, 1093–107.
McClean, J. L., P.-M. Poulain, and J. W. Pelton, 2002. Eulerian and Lagrangian statistics from surface drifters and a high-resolution POP simulation in the North Atlantic. *J. Phys. Oceanogr.*, **32**, 2472–91.
Molcard, A., A. Griffa, and T. M. Özgökmen, 2005. Lagrangian data assimilation in multi-layer primitive equation ocean models. *J. Atmos. Ocean. Tech.*, **22**, 1, 70–83.
Molcard, A., L. I. Piterbarg, A. Griffa, T. M. Özgökmen, and A. J. Mariano, 2003. Assimilation of drifter positions for the reconstruction of the Eulerian circulation field. *J. Geophys. Res.*, **108**, C3, 3056.
Niiler, P. P., N. A. Maximenko, G. G. Panteleev, T. Yamagata, and D. B. Olson, 2003. Near surface dynamical structure of the Kuroshio extension. *J. Geophys. Res.*, **108**, C6.
Oschlies, A. and J. Willebrand, 1996. Assimilation of Geosat altimeter data into an eddy-resolving primitive equation model of the North Atlantic Ocean. *J. Geophys. Res.*, **101/C6**, 14,175–90.
Owens, W. B., 1991. A statistical description of the mean circulation and eddy variability in the northwestern Atlantic using SOFAR floats. *Prog. Oceanogr.*, **28**, 257–303.
Özgökmen, T. M., A. Molcard, T. M. Chin, L. I. Piterbarg, and A. Griffa, 2003. Assimilation of drifter positions in primitive equation models of midlatitude ocean circulation. *J. Geophys. Res.*, **108**, C7, 3238.
Poje, A. C., M. Toner, A. D. Kirwan Jr., and C. K. R. T. Jones, 2002. Drifter launch strategies based on Lagrangian templates. *J. Phys. Oceanogr.*, **32**, 1855–69.
Poulain, P. M., 1993. Estimates of horizontal divergence and vertical velocity in the equatorial Pacific. *J. Phys. Oceanogr.*, **23**/4, 601–7.
Poulain, P. M., D. S. Luther, and W. C. Patzert, 1992. Deriving inertial wave characteristics from surface drifter velocities – frequency variability in the Tropical Pacific. *J. Geophys. Res.*, **97**(C11), 17947–59.
Richardson, P. L., 1993. A census of eddies observed in North Atlantic SOFAR float data. *Prog. Oceanogr.*, **31**, 1–50.
Reverdin, G., P. P. Niiler, and H. Valdimarsson, 2003. North Atlantic Ocean surface currents. *J. Geophys. Res.*, **108**, C1, 3002.

Samelson, R. M., 1992. Fluid exchange across a meandering jet. *J. Phys. Oceanogr.*, **22**, 431–40.
Samelson, R. M., 1996. Chaotic transport by mesoscale motions. In *Stochastic Modeling in Physical Oceanography*, ed. R. J. Adler, P. Müuller, and B. Rozovskii. Cambridge, MA: Birkhäuser Boston, 423–38.
Smith, R. D., M. E. Maltrud, F. O. Bryan, and M. W. Hecht, 2000. Numerical simulation of the North Atlantic ocean at 1/10°. *J. Phys. Oceanogr.*, **30**, 1532–61.
Toner, M., A. C. Poje, A. D. Kirwan, C. K. R. T. Jones, B. L. Lipphardt, and C. E. Grosch, 2001. Reconstructing basin-scale Eulerian velocity fields from simulated drifter data. *J. Phys. Oceanogr.*, **31**, 1361–76.
Veneziani, M., A. Griffa, A. M. Reynolds, and A. J. Mariano, 2004. Oceanic turbulence and stochastic models from subsurface Lagrangian data for the North-West Atlantic Ocean. *J. Phys. Oceanogr.*, **34**(8), 1884–906.
Zhang, H.-M., M. D. Prater, and T. Rossby, 2001. Isopycnal Lagrangian statistics from the North Atlantic Current RAFOS floats observations. *J. Geophys. Res.*, **106**, 13, 817–36.

8

Dynamic consistency and Lagrangian data in oceanography: mapping, assimilation, and optimization schemes

TOSHIO M. CHIN

Rosenstiel School of Marine and Atmospheric Science, University of Miami, Miami, Florida, USA

KAYO IDE

University of California at Los Angeles, Los Angeles, California, USA

CHRISTOPHER K. R. T. JONES

University of North Carolina at Chapel Hill, Chapel Hill, North Carolina, USA

LEONID KUZNETSOV

University of North Carolina at Chapel Hill, Chapel Hill, North Carolina, USA

AND

ARTHUR J. MARIANO

Rosenstiel School of Marine and Atmospheric Science, University of Miami, Miami, Florida, USA

8.1 Introduction

As illustrated throughout this book, Lagrangian data can provide us with a unique perspective on the study of geophysical fluid dynamics, particle dispersion, and general circulation. Drifting buoys, floats, and even a crate-full of rubber ducks or athletic shoes lost in mid-ocean (Christopherson, 2000) may be used to gain insights into ocean circulation. All Lagrangian instruments will be referred to as "drifters" hereafter for simplicity. Because movement of a drifter tends to follow that of a water parcel, the primary attributes of Lagrangian measurements are (i) horizontal coverage due to dispersion in time, (ii) that many of the observed variables obey conservation laws approximately over some lengths of time, and (iii) their ability to trace circulation features such as meanders and vortices at a wide range of spatial scales. Due mainly to inherently irregular spatial distributions, the Lagrangian measurements must first be interpolated for most applications. As we will see, the design of interpolation and mapping schemes that can preserve the Lagrangian attributes is often non-trivial.

To observe finer dynamical details of oceanic and coastal phenomena and to forecast drifter trajectories more accurately (for search-and-rescue operation, spill containment, and so on), Lagrangian data afford a particularly informative and novel perspective if they are combined with a dynamical model, rather than mapped by a standard synoptic-scale interpolation procedure which

Lagrangian Analysis and Prediction of Coastal and Ocean Dynamics, ed. A. Griffa, D. Kirwan, A. Mariano, T. Özgökmen, and T. Rossby. Published by Cambridge University Press.

can smear some details at smaller and faster scales. Data assimilation can be viewed as a methodology for imposing dynamical consistency upon observed data for the purpose of space-time interpolation. Assimilation techniques are developed as numerical optimization, which is usually performed on an Eulerian grid. The Lagrangian attributes such as mass conservation can be lost when the data are simplistically projected onto the Eulerian coordinates required by the traditional interpolation and assimilation techniques.

Data assimilation techniques are based on an Eulerian coordinate system because numerical circulation models are usually Eulerian based. For Lagrangian data, the equations that relate the data with model variables (often called the *observation operators*) are almost always coupled, nonlinear, and difficult to deal with in general. A fundamental formulation that allows effective exchange of information between the two coordinate systems is not straightforward and is still an open research topic. Techniques designed to preserve and exploit the Lagrangian information have been proposed by a number of researchers. In this chapter, we examine a variety of these techniques for data mapping and assimilation. Our main aim here is to document some of the progress made in numerically projecting the Lagrangian information onto the standard Eulerian grid. We offer an insight into *optimization* techniques that are formulated to capture some aspects of the Lagrangian attributes in the data. We also expose a technique in which the system is enlarged by appending the Lagrangian drifter coordinates to the Eulerian flow field.

Section 8.2 is a short survey of some early works that serves also as a motivation for the discussion later in the chapter. In Section 8.3 we describe an analysis technique based on the material-conservation attribute. We then describe a contour-based optimization technique to estimate a velocity field from drifter position data in Section 8.4. The approach in which Lagrangian data is assimilated directly using a Kalman filter algorithm is then discussed in Sections 8.5 through 8.7. In particular, in Section 8.7, we describe a new method in which the dynamic equations for the drifter are appended to the Eulerian dynamic model, so that this augmented dynamic model can become the basis for assimilation of both Lagrangian and Eulerian data. As we will see, the augmented model can fit straightforwardly into the standard Kalman filter formulation. Section 8.8 contains some concluding remarks.

8.2 Background and history

In practice, the traditional first step in application of drifter data is a local interpolation of the observed quantities to the nearest Eulerian grids. This can result in artificially induced diffusion and hence compromise the material tracking attribute of the Lagrangian data. However, a full exploitation of the Lagrangian information was not always considered necessary in early attempts to map the velocity field. For example, during the MODE and POLYMODE experiments, the trajectories of the float instruments were used to determine a reference horizontal velocity field (at each depth of 700, 1300, 1500 m) which was then summed with the relative velocity field derived from hydrographic data to estimate the absolute velocity values. The horizontal coverage from an array of floats was adequate for simple Eulerian-based analysis and assimilation schemes to map regionally. Typically, velocity estimates from the floats are obtained by differencing the position measurements. In most float tracking algorithms, there are substantial levels of systematic (bias) errors as well as random errors. When differencing the positions for velocity estimates, the bias errors tend to cancel out while the random errors are amplified.

Use of Lagrangian measurements such as drifter position data, and the corresponding velocity estimates, in the simulation of ocean circulation has been under consideration for over thirty years. An early research focus had been on the initialization problem in which drifter data distributed irregularly over space are to be combined with each other, as well as with other types of data, to produce a regularly gridded field for the initial and boundary conditions for regional numerical models (Bretherton *et al.*, 1976; Freeland and Gould, 1976; Carter and Robinson, 1987). Freeland and Gould (1976) have interpolated the 1500-meter SOFAR float data from the MODE survey and produced a time sequence of the stream function maps. They assume that the velocity field has no horizontal divergence so that a stream function ψ can be introduced. The two components of horizontal velocity data (u,v), along with a regularization scheme, are then used to estimate the scalar ψ field. Hua *et al.* (1986) have used cross-correlation functions, based on quasi-geostrophic dynamics, to combine hydrographic data, float data, and current meter data from the POLYMODE experiment to produce a time sequence of gridded fields of u, v, and ψ for various depths. Also, Pinardi and Robinson (1987) used a simpler multi-variate objective analysis scheme to produce similar fields for initialization and boundary conditions of a regional, quasi-geostrophic circulation model.

In such early works, the Lagrangian data are simply interpolated onto Eulerian grids with little regards for dynamical consistency with the numerical model into which the data are being incorporated. Initialization fields that are not dynamically consistent with the model are known to produce inaccurate forecasts due to excitation of spurious dynamic modes (of both physical and numerical origins). Such a problem is noted especially in operational meteorological forecasting, and in the early 1980s this has created the research topic now known as "data assimilation."

8.2.1 Assimilation of "pseudo-Lagrangian" velocity

An example of the early development of Lagrangian data assimilation is the work by Carter (1989), who used a circulation model based on the shallow water equations and the RAFOS float data from the main thermocline of the Gulf Stream. The circulation model has a one-dimensional and cyclic spatial domain over a constant latitude around the globe. The RAFOS floats are tracked acoustically and designed to measure the horizontal velocity (u,v), temperature (T), and pressure (P) isopycnally (along a constant density surface). The measured data are then transmitted via the Argos satellite system. The (u,v) velocity components are estimated by differencing the float positions, which are sampled every ten days. Through the use of an average cross-stream hydrographic transect determined from Pegasus data (Halkin and Rossby, 1985), the (u,v,T,P) values are converted to the state variables of the shallow-water model (u,v,ϕ), where ϕ is the geopotential. Results of the one-dimensional simulations show that it takes a couple of months for the model to acquire the correct dominant positive west-to-east flow characteristic of the Gulf Stream.

Using a similar assimilation technique that first converts the Lagrangian location sequence into velocity data, Ishikawa *et al.* (1996) have shown that assimilation of velocity data derived from drifting-buoy trajectories can improve the accuracy in simulated mean sea surface height. The drifter-based velocity data are shown to be more effective than mooring-based (fixed) velocities, due to horizontal coverage of the former. The error in mean sea surface height is reduced by nearly 40% when both the drifter-based velocity and altimetry data are assimilated, while assimilation of altimetry data alone does not significantly reduce the model forecast error.

In the above attempts of Lagrangian data assimilation, velocity values estimated from the drifter position data are mapped onto an Eulerian grid before assimilation. The drifters are hence used effectively as moving current-meters. Specifically, the velocity information is first obtained by numerical differencing of the drifter trajectory, and the velocity estimate is then interpolated to the nearest grid location(s), while this grid location changes with time as the drifter travels. The advantage of this approach is primarily for numerical simplicity, as analysis and assimilation methods developed for Eulerian data can be used with very little modification. The disadvantage is that the property-tracking attribute of Lagrangian data cannot be fully exploited because of the interpolation procedure. Also, velocity estimation by a simple differencing can introduce aliasing of high frequency motions (many Argo float positions are sampled at a ten-day interval; see also Section 8.4).

A more desirable approach would be to assimilate the position information directly into the model.

8.2.2 Assimilation of drifter positions

To assimilate the drifter position data, a common strategy is to make incremental corrections to the state variables to steer the model-simulated drifter trajectory toward the observed trajectory. For example, Kamachi and O'Brien (1995) assimilated drifter position data by iteratively adjusting the initialization fields (initial conditions for the model equations) to minimize the total distance between the observed and simulated trajectories. Some recent assimilation studies (Ide *et al.*, 2002; Kuznetsov *et al.*, 2003; Molcard *et al.*, 2003; Ozgokmen *et al.*, 2003) have focused more on sequential, rather than iterative, computational approaches. For example, Molcard *et al.* (2003) have used the difference between the simulated and observed drifter position data to adjust the velocity field within a given influential radius to minimize the position differences. The adjustment is made directly on the model state (rather than the initialization fields) whenever a drifter observation becomes available. They show, based on experimentation with model-simulated drifter data, that assimilation of position data can reduce the estimation error by a factor of two over the pseudo-Lagrangian ("moving current meter") approach. Such reduction in errors is shown to be possible when the data sampling is on the order of 20 to 50 percent of the Lagrangian velocity integral time scale.

At present, assimilation methodology for Lagrangian positions is an active research topic. To find a fundamental formulation to integrate Lagrangian and Eulerian information is regarded as an open question. In particular, an aspect that remains to be exploited is the geometrical information provided by the drifter trajectories. Drifter trajectories are highly indicative of the *shapes* of dynamical features such as a vortex, current meander, and front. These geometrical patterns are of importance to societal needs (e.g., fishery, transportation, climate) as well as academic research including validation of model-simulated ocean circulations. Mariano and Chin (1996) have demonstrated that a contour-based assimilation technique can preserve the dynamical features better than a traditional Eulerian interpolation which tends to smear them by smoothing. Another approach is to append equations of the drifter motion to the Eulerian model, so that the augmented model has the capability of simulating the Lagrangian dynamics. With such augmentation, a Lagrangian assimilation scheme can be formulated straightforwardly using standard techniques such as the Kalman filter (Section 8.7). We show that it presents new challenges by uncovering a key cause in cases where it breaks down due to the presence of saddle points in the flow.

8.3 Analysis methods based on conservation laws

Conservation of material or energy in a water parcel is fundamentally a Lagrangian notion. Mathematical formulas for conservation laws can allow an Eulerian expression for some Lagrangian attributes. Specifically, for a scalar quantity $Q(x,y,z,t)$, the Lagrangian derivative (or "material derivative") $\frac{\mathrm{D}}{\mathrm{D}t}$ can be expanded as

$$\frac{\mathrm{D}Q}{\mathrm{D}t} = \frac{\partial Q}{\partial t} + u\frac{\partial Q}{\partial x} + v\frac{\partial Q}{\partial y} + w\frac{\partial Q}{\partial z} = \frac{\partial Q}{\partial t} + \mathbf{u} \cdot \nabla Q \tag{8.1}$$

in terms of the local velocity $\mathbf{u} = (u,v,w)$. It is evident from (8.1) that changes measured by an instrument in a Lagrangian framework will be equal to the Eulerian time derivative, $\partial/\partial t$, if the advection velocity $\mathbf{u}$ is zero. In practice, a small enough advection velocity can allow treating the Lagrangian data as moving Eulerian data. The advective term, however, cannot be ignored when the flow is strong or when the data sampling is coarse, and the "moving Eulerian data" approach is found to be error-prone for Lagrangian data assimilation in controlled experiments (Molcard *et al.*, 2003). We examine here numerical techniques that conserve some scalar properties.

8.3.1 Conservation of temperature

Conservation laws are fundamental to oceanographic analysis whenever they are numerically realizable. They are especially applicable to Lagrangian measurements as the drifters are designed to track the same water parcel. A generic conservation law for a scalar quantity Q would be

$$\frac{\mathrm{D}Q}{\mathrm{D}t} = \text{Sources} - \text{Sinks}, \tag{8.2}$$

where the source–sink term usually includes some expressions for dissipation, since invariance of Q is seldom perfect in fluid dynamics. In other words, unlike solid mass, "water parcel" is mostly a conceptual object due to practical approximation. By applying a conservation law to the temperature T, Rossby *et al.* (1986) and Mariano and Rossby (1989) analyzed subsurface SOFAR floats in the northwest Atlantic at 700 and 1300 m and estimated the vertical velocity w from the drifter trajectories and some hydrographic data. Since the SOFAR drifters are isobaric they measure both the local temporal rate of change and the horizontal advection of temperature, $\partial T/\partial t + u\partial T/\partial x + v\partial T/\partial y$. Temperature invariance, $DT/Dt = 0$, is assumed for the subsurface thermocline waters over eddy time scales. The vertical derivative $\partial T/\partial z$ is evaluated from the hydrographic data. Then from (8.1),

$$w = -\left(\frac{\partial T}{\partial t} + u\frac{\partial T}{\partial x} + v\frac{\partial T}{\partial y}\right) \Big/ \left(\frac{\partial T}{\partial z}\right).$$

Estimation techniques like this are considered to be practical since the vertical velocity w is typically much smaller in magnitude than its horizontal counterpart (u, v) and is hence difficult to measure directly.

8.3.2 Analysis based on vorticity conservation

Consider the problem of interpolating irregularly sampled data to map a scalar field. The data to be analyzed contain both random and systematic errors. Although optimal interpolation (OI) methods are effective for removing the random error, the systematic error ("bias") remains. A time series of analyses produced at consecutive epochs would not be consistent with known dynamical constraints such as the Navier–Stokes equations. Such inconsistency can also result due to time-dependent sampling patterns for the data in data-scarce regions at the time of analysis. Thompson (1969) proposed a method for maintaining a dynamical consistency in two consecutive analyses by evaluating corrections necessary to satisfy a conservation equation for absolute vorticity. The assumed dynamic constraint is $D(\zeta + f)/Dt = 0$ which is expanded as

$$\frac{\partial \zeta}{\partial t} + \mathbf{u}\cdot\nabla\zeta + \mathbf{u}\cdot\nabla f = 0, \tag{8.3}$$

where $\mathbf{u} \equiv (u, v)$ is the horizontal velocity, $\zeta(x, y, t) \equiv \partial v/\partial x - \partial u/\partial y$ is the relative vorticity, and $f(x, y)$ is the time-invariant Coriolis parameter. Since $\mathbf{u}$ is an integrated quantity with respect to ζ, it is considered to be affected less by the observational and analysis error than ζ itself. The horizontal velocity $\mathbf{u}$ is hence assumed to be error free and constant over two consecutive analyses, and it is denoted as $\bar{\mathbf{u}}$ hereafter. Let the vorticity function $\zeta(t)$ be expanded as $\zeta(n\Delta t) = \bar{\zeta}_n + \zeta'_n$, where n is the time index, Δt is the sampling interval, $\bar{\zeta}_n$ is the original analysis, and ζ'_n is the adjustment necessary for dynamical consistency. The problem is to evaluate ζ'_{n-1} and ζ'_{n+1} for two consecutive analyses. The dynamical constraint (8.3) is discretized as

$$\frac{\bar{\zeta}_{n+1} + \zeta'_{n+1} - \bar{\zeta}_{n-1} - \zeta'_{n-1}}{2\Delta t} + \bar{\mathbf{u}} \cdot \nabla\left(\frac{\bar{\zeta}_{n+1} + \zeta'_{n+1} + \bar{\zeta}_{n-1} + \zeta'_{n-1}}{2}\right) + \bar{\mathbf{u}} \cdot \nabla f = 0, \tag{8.4}$$

which can be written as

$$D = -(E + \Delta t \bar{\mathbf{u}} \cdot \nabla S), \tag{8.5}$$

by defining $D \equiv \zeta'_{n+1} - \zeta'_{n-1}$, $S \equiv \zeta'_{n+1} + \zeta'_{n-1}$, and

$$E \equiv 2\Delta t\left[\frac{\bar{\zeta}_{n+1} - \bar{\zeta}_{n-1}}{2\Delta t} + \bar{\mathbf{u}} \cdot \nabla\left(\frac{\bar{\zeta}_{n+1} + \bar{\zeta}_{n-1}}{2}\right) + \bar{\mathbf{u}} \cdot \nabla f\right].$$

For small Δt, E can be considered as a measure of the degree to which $\bar{\zeta}_{n-1}$ and $\bar{\zeta}_{n+1}$ are dynamically inconsistent. If $E = 0$, the two analyses are consistent with (8.3), and no adjustment is required.

Only a single equation (8.5) is available for the two unknowns, D and S. To obtain an additional constraint, Thompson (1969) argues that, if the initial analyses are not seriously in error, then they should be adjusted only by a small amount. In particular, the two (unknown) adjustment fields are specified to have the smallest possible combined mean-square magnitude, so that the integral

$$I = \int_A (\zeta'_{n-1})^2 + (\zeta'_{n+1})^2 \ \mathrm{d}A, \tag{8.6}$$

is minimized over the analysis area A. Since the integrand can be rewritten as $(S^2 + D^2)/2$, substitution of the main constraint (8.5) into (8.6) yields

$$I = \frac{1}{2}\int_A [S^2 + (E + \Delta t\, \bar{\mathbf{u}} \cdot \nabla S)^2] \ \mathrm{d}A \tag{8.7}$$

which can be minimized (unconstrained) for the single variable S. Minimization can be performed using variational calculus (see Appendix A), which specifies the optimal S as the solution of the Euler–Lagrange equation

$$S - (\Delta t)^2 \bar{\mathbf{u}} \cdot \nabla(\bar{\mathbf{u}} \cdot \nabla S) = \Delta t\, \bar{\mathbf{u}} \cdot \nabla E, \tag{8.8}$$

an elliptic partial differential equation with respect to S, given an appropriate boundary conditions for ζ. After solving for S, D is evaluated from (8.5). The adjustments ζ'_{n-1} and ζ'_{n+1} can then be calculated.

Thompson (1969) shows that, if one of the analyses has no bias error, the mean square error in each analysis would be halved in a linear system due to the adjustment scheme described above. The dynamical constraint is thus an independent source of information that can reduce errors in the two analyses significantly over the accuracy attainable by each individual analysis. Lewis (1982) has also observed halving of the total analysis errors due to inclusion of a dynamical constraint, when Thompson's basic scheme is extended to conservation of potential vorticity in a baroclinic model. To obtain an equation analogous to (8.8) for the baroclinic case, a fixed "steering current" $\bar{\mathbf{u}}$ needs to be evaluated for each baroclinic layer. The steering current is the large-scale velocity field that is primarily responsible for advecting features, and a set of drifter trajectories may be used to determine the steering current (e.g., Bauer *et al.*, 1998). Potential vorticity from the model simulation would then be corrected using the conservation scheme along each drifter trajectory. Mariano and Rossby (1989) have observed that potential vorticity is indeed conserved along the tracks of subsurface thermocline floats to within Rossby number of a few percent. Finally, a similar conservation scheme can also be applied to variables other than vorticity. For example, the temperature, measured by most floats, can be considered invariant for some applications with discretion. Temperature conservation has been used to formulate an algorithm to estimate the surface velocity field from a time-sequence of temperature maps (Kelly, 1989). Conversely, the series of temperature analyses can be improved with respect to dynamical consistency by imposition of conservation along the drifter trajectories using the scheme described in this section.

We emphasize that the Euler–Lagrange equation (8.8) can be naturally solved over a curvilinear coordinate system along the flow, or a Lagrangian coordinate system, rather than the standard two-dimensional Eulerian grids. Specifically, (8.8) can be written as

$$\ell \frac{\partial}{\partial s}\left(\ell \frac{\partial S}{\partial s}\right) - S = -\ell \frac{\partial E}{\partial s}, \tag{8.9}$$

where s is the distance coordinate ("arclength") along the trajectory and $\ell(s) \equiv \Delta t|\bar{\mathbf{u}}|$ (see Thompson, 1969, for more details including a general solution of the differential equation). Note that the dimension of the domain of differentiation has been reduced from two in (8.8) to one in (8.9), implying that the along-flow arclength is a more natural coordinate. The adjustment scheme described here is thus simply a smoothing procedure over a Lagrangian coordinate that tracks the flow trajectories. Algebraically, note that all spatial derivatives in the Eulerian form (8.8) are given as the inner product of the gradient with the velocity vector, or $\bar{\mathbf{u}} \cdot \nabla$. All gradient vectors are thus projected onto ℓ whose direction is along the coincident velocity vector. Because of this fundamentally Lagrangian nature, the adjustment scheme avoids the smearing problem by advecting the fields before averaging

(Mariano and Chin, 1996). If an Eulerian average of the two fields was used to reduce estimation error, the fields would be smeared. By incorporating dynamics into the estimation problem, fields from different times can be combined with less danger of smearing frontal features. An underlying assumption is that the initial analysis is reasonably accurate so that only small adjustments are needed.

8.4 Optimal trajectory between two positions

Estimation of circulation velocity along drifter tracks is an important application of the Lagrangian data. Typically, consecutive samples of drifter locations are converted into estimate of velocities by assuming uniform velocity between the samples. This is often called the "pseudo-Lagrangian" approach in drifter-based velocity estimation. Earlier in this chapter we have noted that a potential problem with such an approach is that the sampling interval may be too large for estimating the instantaneous velocity accurately. The realistic sampling intervals in drifter instruments available today can range from a day to as much as two weeks. Temporal and spatial variability of the oceanic velocity field can be significant during such an interval and over the distance covered by drifters. In addition, a related issue is that the exact location of the trajectory is not known between consecutive samples of the drifter locations. As a result, there is an uncertainty in the placement (on the Eulerian space where analysis or assimilation takes place) of the velocity estimate. This issue of placement to determine cross-referencing of Eulerian and Lagrangian objects is known generically in the image processing literature as the "association (or assignment) problem". Solution usually involves a numerical optimization task, where a measure of physical or geometrical inconsistency among all matching pairs of objects is defined and minimized. Previously (Section 8.3.2), we have seen that the association problem can be addressed *locally* by using an advection formula, and a regularized solution can be obtained by optimization. In this section, we examine general optimization formulations for the drifter location data, with emphasis on addressing the two main issues in velocity estimation, namely, (i) accuracy of drifter-based velocity and (ii) placement of the estimated velocity.

8.4.1 Optimization formulation with standard constraints

Drifter sampling intervals (on the order of days) are substantially larger than the typical time step (minutes) of a circulation model. Finding drifter location and velocity at the time resolution of the model thus tends to be an under-determined problem. For example, the red contour segment in Figure 8.1 is

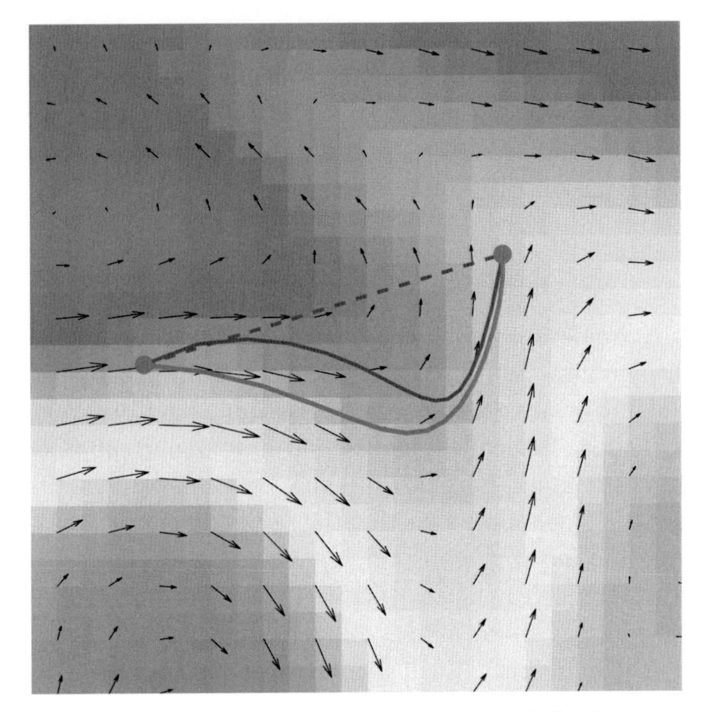

Figure 8.1 A drifter trajectory (red line) from a model-simulated circulation and time-averages of the simulated velocity (arrows) and pressure (color coded) fields. The blue lines are the drifter trajectory estimated using the active contour approach (see text), indicating the initial estimate (dashed line) and the estimate after 20 iterations (solid blue line). See Plate 18 for color version.

a simulated drifter trajectory between two sample points (red circles). The mean circulation pattern averaged over this sampling interval is also shown. A straight line connecting the two position samples (blue dashed line) would be the basis for the velocity estimate and placement under the pseudo-Lagrangian approach. Clearly, the two location samples which are processed too simplistically do not reveal correct information on the underlying circulation pattern in this case. For a general case, let us denote the Lagrangian (drifter) position as $\mathbf{r}(t)$, the underlying Eulerian velocity field as $\mathbf{u}(\mathbf{s}, t)$ where $\mathbf{s}$ is the Eulerian position, and the two sample times for the drifter as t_0 and t_1. We can then write

$$\mathbf{r}(t_1) = \mathbf{r}(t_0) + \int_{t_0}^{t_1} \mathbf{u}(\mathbf{r}(t), t) \, \mathrm{d}t \tag{8.10}$$

whose last term is a line integral. One difficulty is that the path of integration is the unknown! Specifically, the problem to find $\mathbf{r}(t)$ and $\mathbf{u}(\mathbf{r}, t)$ at each model update time t, $t_0 \leq t \leq t_1$, given the drifter location observations $\mathbf{r}(t_0)$ and $\mathbf{r}(t_1)$ is under-determined (i.e., many possible $\mathbf{r}$ and $\mathbf{u}$ can satisfy the integral equation), and extra constraints are required for a unique solution.

Standard regularization schemes (generic mathematical techniques to reformulate an ill-posed problem) can be applied to ensure a unique solution. Approaches based on entropy maximization or energy minimization, however, tend to yield a solution identical to the pseudo-Lagrangian velocity estimation, which we wish to improve upon. Specifically, consider a one-dimensional and discrete version of the track estimation problem of (8.10): Given r_0 and r_N, find r_n for $n = 1, \ldots, (N-1)$ based on the relations

$$r_N = r_0 + \sum_{n=1}^{N-1} u_n \Delta t, \quad u_n \equiv r_n - r_{n-1},$$

where Δt is the time step of the circulation model and n is the time index. Obviously the positions r_n are under-constrained. The *maximum entropy principle* is an information-theoretic argument that, in most cases, favors an uniformly distributed set of unknowns as the most likely solution. A realization of this principle is to constrain the velocities to be constant, as $u_n = c$. Application of a regularization scheme usually leads to an over-constrained problem that requires optimization for solution; however, in our case a direct substitution of the regularization constraint yields the solution r_n which constitutes regularly spaced positions along the line segment between the two end points r_0 and r_N. The maximum entropy approach would thus lead to the undesirable solution of the blue dashed line in Figure 8.1. Similarly, the *minimum energy principle* in regularization argues to favor solutions that have the minimal mean squared magnitude. A realization for our case is to minimize $\sum_{n=1}^{N-1} \|u_n\|^2$, and the solution can be evaluated by a technique for constrained optimization such as application of the Lagrangian multiplier. The minimum energy solution again traces the straight line between the end points. In summary, most standard information-theoretic constraints are consistent with the uniform velocity assumption employed by the pseudo-Lagrangian velocity estimation, and thus do not optimize the Lagrangian information content of the data.

8.4.2 Constraints based on background statistics

Another source of the extra information is the statistical parameters such as the autocorrelation of the velocity field. Such statistics may be evaluated from climatological observations. In principle, a space-time correlation function over the entire Eulerian domain would be required to cover all possible locations of the drifter. In practice, such a correlation function is parameterized to reduce the computational complexity. Unless the geometry of the streamlines are somehow incorporated into the parameterization, however, the Lagrangian information can be lost due (once again) to smearing (Mariano and Chin, 1996).

Alternatively, a Lagrangian formulation for trajectory estimation could often be able to maintain geometrical integrity of the circulation features while reducing computational burden significantly. We consider here a simple optimization model for the drifter trajectory based on smoothness. Minimizing the term $||\mathrm{d}\mathbf{r}/\mathrm{d}t||^2$ can maintain continuity of the trajectory path, while minimization of $||\mathrm{d}^2\mathbf{r}/\mathrm{d}t^2||^2$ would serve as a model for (local) persistence of the drifter motion. More sophisticated trajectory models can also be considered (e.g., Storvik, 1994). The trajectory must also be consistent with a given Eulerian background velocity field $\bar{\mathbf{u}}$, and a term like $||\mathrm{d}\mathbf{r}/\mathrm{d}t - \bar{\mathbf{u}}(\mathbf{r}, t)||^2$ can impose such a consistency condition. We then consider finding the optimal contour based on the objective functional

$$J(\mathbf{r}) = \frac{1}{2}\int_{t_0}^{t_1} \alpha_0 \left\| \frac{\mathrm{d}\mathbf{r}}{\mathrm{d}t} - \bar{\mathbf{u}}(\mathbf{r}, t) \right\|^2 + \alpha_1 \left\| \frac{\mathrm{d}\mathbf{r}}{\mathrm{d}t} \right\|^2 + \alpha_2 \left\| \frac{\mathrm{d}^2\mathbf{r}}{\mathrm{d}t^2} \right\|^2 \mathrm{d}t \tag{8.11}$$

where α_0, α_1, and α_2 are given constant weights. The corresponding Euler–Lagrange equation from variational calculus (Appendix A) is

$$\alpha_2 \frac{\mathrm{d}^4\mathbf{r}}{\mathrm{d}t^4} - (\alpha_0 + \alpha_1)\frac{\mathrm{d}^2\mathbf{r}}{\mathrm{d}t^2} + \alpha_0 \bar{\mathbf{u}} \frac{\mathrm{d}\bar{\mathbf{u}}}{\mathrm{d}\mathbf{r}} = 0 \tag{8.12}$$

whose solution, with the location data $\mathbf{r}(t_0)$ and $\mathbf{r}(t_1)$ as boundary conditions, would be the estimate for the drifter trajectory. Solution is difficult because of the highly nonlinear last term on the left-hand side. Kass *et al.* (1988) have introduced an iterative solution method that allows evaluation of this term using the trajectory estimate from the previous iteration, or

$$\alpha_2 \frac{\mathrm{d}^4\mathbf{r}_i}{\mathrm{d}t^4} - (\alpha_0 + \alpha_1)\frac{\mathrm{d}^2\mathbf{r}_i}{\mathrm{d}t^2} + \alpha_0 \left(\bar{\mathbf{u}} \frac{\mathrm{d}\bar{\mathbf{u}}}{\mathrm{d}\mathbf{r}} \right)\Bigg|_{\mathbf{r}=\mathbf{r}_{i-1}} = -\kappa(\mathbf{r}_i - \mathbf{r}_{i-1}) \tag{8.13}$$

where i is iteration index and κ is a given constant. This is an example of the *active contour method* (Kass *et al.*, 1988; Storvik, 1994) in which a given initial contour is updated incrementally. The blue contour on Figure 8.1 has been evaluated by solving the ordinary differential equation (8.13) for $i = 20$ iterations, starting with the straight line (dashed blue line) as the initial contour. The parameter values used are $\alpha_0 = 1$, $\alpha_1 = \alpha_2 = 0.01$, and $\kappa = 0.1$.

The background velocity $\bar{\mathbf{u}}$ used (arrows) is an average of 15 velocity fields coincident with the drifter trajectory. Several initial trajectory estimates which are radically different from the dashed blue lines are also considered, and all are observed to converge to the single solid blue contour displayed. This solution given by the active contour method is a considerably better estimate of the true drifter trajectory (red contour) than the one implied by the pseudo-Lagrangian approach. The former, optimized solution is more useful for prediction of the drifter trajectory. For example, consider predicting the future drifter location by regression (e.g., motion persistence). Knowing the more accurate track would allow a better short-term prediction of the drifter location than the linear extension of the straight line (dashed blue line). This demonstrates that the active contour approach offers a practical method to optimally combine Eulerian background information (e.g., $\bar{\mathbf{u}}$) with Lagrangian models (e.g., smoothness of the drifter trajectory). The optimization problem considered in this section is much more difficult to solve than those involving the conservation laws discussed previously (Section 8.3.2), because the unknown $\mathbf{r}$ appears as an argument of $\bar{\mathbf{u}}(\mathbf{r}, t)$ in (8.11) This is a manifestation of the "association problem" mentioned previously.

8.5 Sequential data assimilation using the Kalman filter

The accuracy of the optimal solutions described previously in Sections 8.3.2 and 8.4 are dependent on the quality of the background/steering velocity field $\bar{\mathbf{u}}$. An ideal source of such a velocity field is an ocean circulation model. For example, in (8.11), $\bar{\mathbf{u}}(\mathbf{r}, t)$ can be evaluated as the model *forecast* over $t_0 \leq t \leq t_1$. The Lagrangian trajectory can then be optimized as described above based on the position data $\mathbf{r}(t_0)$ and $\mathbf{r}(t_1)$. Let the resulting optimal contour be $\hat{\mathbf{r}}$, i.e., $J(\hat{\mathbf{r}}) = \min J(\mathbf{r})$. Then the velocity field $\mathbf{u}(\mathbf{r}, t)$ can be *updated* by assimilating the contour data $\hat{\mathbf{r}}$ into the circulation model. Specifically, the contour derivative $\mathrm{d}\hat{\mathbf{r}}/\mathrm{d}t$ is used as the observation of the velocity $\mathbf{u}(\hat{\mathbf{r}}, t)$ along the contour. The Lagrangian velocity estimate $\mathrm{d}\hat{\mathbf{r}}/\mathrm{d}t$ would be more accurate than the pseudo-Lagrangian estimate (along a straight line) due to the finer space-time resolution of the former. Once the data-updated velocity field $\hat{\mathbf{u}}(\mathbf{s}, t)$, $t_0 \leq t \leq t_1$, is obtained, the value of $\hat{\mathbf{u}}(\mathbf{s}, t_1)$ could be used to initialize the circulation model to forecast the background velocity field $\bar{\mathbf{u}}(\mathbf{s}, t)$, $t_1 \leq t \leq t_2$, until the next observation time, t_2, of the drifter position. This new background field can then be used by (8.11) to find the next segment of the optimal drifter trajectory given the new Lagrangian data $\mathbf{r}(t_2)$, and the optimization/assimilation cycle can continue recursively.

8.5.1 Extended Kalman filter algorithm

Such an alternating sequence of model-forecast and data-update is the main feature of a *sequential data assimilation* scheme. For a general ocean data assimilation problem, the *Kalman filter* is the theoretical framework on which most sequential assimilation schemes are based (Ghil and Malanotte-Rizzoli, 1991). The Kalman filter (KF) approach to data assimilation can be considered as a dynamically extended version of optimal interpolation (OI), a standard mapping technique based on statistical estimation theory. The OI technique usually interpolates the data based on a prescribed, spatially homogeneous covariance function. The KF method, on the other hand, maintains its own covariance function in such a way that the spatial and heterogeneous structures of the covariance are dynamically consistent with the circulation model at hand. Also, while the OI technique is instantaneous in time, the KF analysis evolves in time using the dynamics of the circulation model.

The KF algorithm prescribes an equation for time-evolution of the covariance statistics of the model variables and a formula for statistically optimal corrections to the model state based on the data. Specifically, let $\frac{\mathrm{d}}{\mathrm{d}t}\mathbf{x} = \mathbf{m}(\mathbf{x}, t)$ be the dynamics of the circulation model where the vector $\mathbf{x}(t)$ is the state variable (which include the Eulerian velocity field). To each evaluation of the model state $\mathbf{x}(t)$ we assign a vector of statistical uncertainty (or error) $\Delta\mathbf{x}(t)$ whose covariance matrix is denoted as $\mathbf{P}(t)$. The error is assumed to be unbiased, so that the mean of $\Delta\mathbf{x}$ is zero. Formally, the KF approach assumes that the model equation also contains some uncertainty as

$$\mathrm{d}\mathbf{x} = \mathbf{m}(\mathbf{x}, t)\mathrm{d}t + \mathrm{d}\boldsymbol{\eta} \tag{8.14}$$

where the random vector $\mathrm{d}\boldsymbol{\eta}$ has a mean of zero and covariance of $\mathbf{Q}$. The model uncertainty is crucial for sensitivity to instantaneous observations. Without a finite amount of uncertainty in the model (i.e.,$\mathbf{Q} = \mathbf{0}$), the KF analysis would asymptotically ignore the observations, since the forecasts from the "perfect model" would be perceived to be overwhelmingly more accurate than any noisy observation. We denote the time when the observations become available as t_k and the corresponding vector of the observations as $\mathbf{y}_k$, which is related to the model state as

$$\mathbf{y}_k = \mathbf{h}_k(\mathbf{x}_k) + \boldsymbol{\epsilon}_k, \tag{8.15}$$

according to a given observation operator $\mathbf{h}_k$, where $\mathbf{x}_k \equiv \mathbf{x}(t_k)$ and $\boldsymbol{\epsilon}_k$ is the zero-mean observation error whose covariance is denoted as $\mathbf{R}_k$.

The Kalman filter is a recursive algorithm to find an optimal function (or time series) $\mathbf{x}(t)$ that minimizes the total time-integrated magnitudes of the model and data errors, i.e., $d\boldsymbol{\eta}(t)$ and $\boldsymbol{\epsilon}_k$. Recursion of the following two-step cycle, for $k = 1,2, \ldots$, can perform this optimization:

Step 1 Forecast from t_{k-1} to t_k: Starting from initial conditions $\mathbf{x}^{\mathrm{f}}(t_{k-1}) = \mathbf{x}^{\mathrm{a}}_{k-1}$ and $\mathbf{P}^{\mathrm{f}}(t_{k-1}) = \mathbf{P}^{\mathrm{a}}_{k-1}$, $\mathbf{x}^{\mathrm{f}}(t)$ and $\mathbf{P}^{\mathrm{f}}(t)$ are obtained by integrating

$$\begin{aligned} \frac{\mathrm{d}}{\mathrm{d}t}\mathbf{x}^{\mathrm{f}} &= \mathbf{m}(\mathbf{x}^{\mathrm{f}}, t) \\ \frac{\mathrm{d}}{\mathrm{d}t}\mathbf{P}^{\mathrm{f}} &= \mathbf{M}\mathbf{P}^{\mathrm{f}} + (\mathbf{M}\mathbf{P}^{\mathrm{f}})^{\mathrm{T}} + \mathbf{Q} \end{aligned} \tag{8.16}$$

where the superscripts "a" and "f" denote the analysis and forecast evaluations, respectively (see Ide *et al.*, 1997). Also, $\mathbf{M}(t) \equiv \frac{\partial}{\partial \mathbf{x}}\mathbf{m}(\mathbf{x}, t)|_{\mathbf{x}^{\mathrm{f}}(t)}$ is the Jacobian of the model dynamics and is often referred to as the *tangent linear model* (TLM) of $\mathbf{m}(\mathbf{x}, t)$.

Step 2 Data update at t_k: Using the model forecast $\mathbf{x}^{\mathrm{f}}_k = \mathbf{x}^{\mathrm{f}}(t_k)$ with $\mathbf{P}^{\mathrm{f}}_k = \mathbf{P}^{\mathrm{f}}(t_k)$ and observation $\mathbf{y}_k$ with $\mathbf{R}_k$, the optimal analysis $\mathbf{x}^{\mathrm{a}}_k$ and corresponding error covariance matrix $\mathbf{P}^{\mathrm{a}}_k$ are evaluated as

$$\mathbf{K}_k = \mathbf{P}^{\mathrm{f}}_k\mathbf{H}^{\mathrm{T}}_k(\mathbf{H}_k\mathbf{P}^{\mathrm{f}}_k\mathbf{H}^{\mathrm{T}}_k + \mathbf{R}_k)^{-1} \tag{8.17}$$

$$\mathbf{x}^{\mathrm{a}}_k = \mathbf{x}^{\mathrm{f}}_k + \mathbf{K}(\mathbf{y}_k - \mathbf{h}_k(\mathbf{x}^{\mathrm{f}}_k))$$

$$\mathbf{P}^{\mathrm{a}}_k = (\mathbf{I} - \mathbf{K}_k\mathbf{H}_k)\mathbf{P}^{\mathrm{f}}_k \tag{8.18}$$

where $\mathbf{H}_k = \frac{\partial}{\partial \mathbf{x}}\mathbf{h}_k(\mathbf{x})|_{\mathbf{x}^{\mathrm{f}}_k}$ is the TLM of the observation operator.

At time t_k, $\mathbf{x}^{\mathrm{a}}_k$ is the optimal estimate of $\mathbf{x}_k$. In the next cycle, $\mathbf{x}^{\mathrm{a}}_k$ and $\mathbf{P}^{\mathrm{a}}_k$ become the initial conditions for repeating Step 1.

The algorithm displayed above is known as the *extended Kalman filter* whose distinguishing feature is the TLM approximations of the dynamic and observation operators $\mathbf{m}$ and $\mathbf{h}$, which are assumed to be smooth functions with respect to the model state $\mathbf{x}$. The original Kalman filter (Kalman, 1960) has been developed for the case where $\mathbf{m}$ and $\mathbf{h}$ are both linear in $\mathbf{x}$, and the solution in this case is also optimal in a (non-statistical) least-squares sense (see Sorenson, 1970). Also, the original Kalman filter is developed for a discrete-time dynamic equation, a more convenient form for numerical realization. Volumes of general discussions on variations and intricacies of KF are available and should be consulted for practice (e.g., Gelb, 1974; Lewis, 1986).

For applications in data assimilation, the extended Kalman filter algorithm must be approximated further due to the computational complexity resulted from the large dimension of the state vector. In particular, a typical ocean circulation model has on the order of 10^4 to 10^8 variables, and the dimension of the corresponding covariance matrix would be the square of that number. Forecasting of such a large number of variables through integration of the dynamic equations (8.16) is usually the main computational challenge. Several disciplined as well as heuristic approaches to drastically reduce the computational complexity have been proposed and practiced (see Section 3 of Chin *et al.*, 1999). At present, the approximation approach to solve the dynamic equations (8.16) by a Monte-Carlo integration (known as the *ensemble Kalman filter* approach; Evensen, 1994) is gaining popularity due mainly to simplicity in numerical realization.

8.5.2 Combining KF with the optimal contour

Following up on the discussion at the beginning of this section, we consider obtaining the background velocity field $\bar{\mathbf{u}}$ in (8.11) from the forecast (Step 1) of the KF algorithm. (The accompanying covariance forecasts may also be used to refine the weights for the term $\|\mathrm{d}\mathbf{r}/\mathrm{d}t - \bar{\mathbf{u}}(\mathbf{r}, t)\|^2$.) The active contour method is then used to obtain the optimal contour $\hat{\mathbf{r}}$, which sets up the KF observation equation (cf. 8.15) as

$$\frac{\mathrm{d}\hat{\mathbf{r}}}{\mathrm{d}t} = \mathbf{u}(\hat{\mathbf{r}}, t) + \boldsymbol{\epsilon}_{\mathbf{u}}(t)$$

for the time period $t_0 \leq t \leq t_1$. (The equation may be sampled in time to fit the form of Step 2 above.) Thus, the computation in the forecast step of KF needs to be performed again to incorporate the velocity observation $\mathrm{d}\hat{\mathbf{r}}/\mathrm{d}t$. This re-evaluation of the KF forecast as well as the execution of the active contour method can be considered as the extra computational steps, over the traditional KF implementations, necessary to optimally transfer the Lagrangian information to the Eulerian model-state variables. Molcard *et al.* (2003) and Özgökmen *et al.* (2003) have used a similar computational structure that evaluates the forecast *twice* between adjacent times of drifter observations. They have shown that, although such a double-evaluation of forecast would not be necessary if pseudo-Lagrangian velocity were used as the observations, the extra forecast pass is instrumental in maintaining the accuracy in the state estimates, especially when the observation interval is longer.

8.6 Lagrangian model dynamics and Kalman filter

Innovative re-formulations of the Lagrangian data assimilation problem can lead to a solution method that conforms more naturally to the KF computational framework. In particular, application of KF to Lagrangian data assimilation is considerably more straightforward (at least conceptually) if the equations of ocean dynamics can be expressed in the same Lagrangian coordinate system as that of the observations. It has been shown that an ocean circulation model constructed on a Lagrangian coordinate system does indeed simplify the observation operator for the drifter data (e.g., Bennett and Chua, 1999; Mead and Bennet, 2001). The advantage is, however, offset by the complexity in evaluating such terms as dissipation and pressure gradients of the Navier–Stokes equations in a Lagrangian framework. Still, regional forecasts over several eddy time scales have been successfully performed by Mead and Bennett (2001).

There are practical advantages to Lagrangian dynamic models as well. When the flow can be represented by a few isolated vortices, however, assimilating the state of the fluid in an Eulerian frame may be computationally more expensive as well as less accurate. Localized coherent vortices are common phenomena in oceanography (McWilliams, 1991). For example, the Gulf Stream rings are nearly axisymmetric vortices generated by the breaking of a meander from the Gulf Stream. For such flows, tracking vortices is far more efficient and accurate in describing some macroscopic aspects of flow dynamics. A distinction should be made between macroscopic and microscopic Lagrangian data. In the macroscopic scales, the Lagrangian signatures are identified by coherent structures such as meso-scale eddies and larger-scale structures such as ocean jets and fronts. Boundaries of these Lagrangian structures are generally defined by sharp gradients of ocean properties, such as temperature and velocity. As a result, the Lagrangian structures are readily identifiable in an instantaneous Eulerian field. Individually, they are likely to satisfy the conservation laws over the regions they occupy, as discussed in Section 8.3, and hence are intimately connected with the background ocean field. In the microscopic scales, the Lagrangian signature is the trajectory of an individual water particle traced by a drifter. Unlike macroscopic Lagrangian structures, microscopic Lagrangian trajectories are generally considered to be passive. Specifically, while the circulation velocity field determines the shapes of trajectories, the trajectories have no influence on the evolution of the flow pattern, as implied by (8.10).

The KF algorithm has been applied to a system governed by the macroscopic Lagrangian coherent structures using both macroscopic Lagrangian data and instantaneous Eulerian data, to study the estimation of ocean flows dominated by a limited number of the Lagrangian structures (Ide and Ghil, 1997a). In particular, consider a two-dimensional domain where the position is represented by a complex number z. The flow dynamics is governed by N point-vortices located at $\bar{z}_n$ with given circulation values Γ_n for $n = 1, \ldots, N$. At location z, the flow velocity is given by the Biot–Savart law

$$\frac{\mathrm{d}}{\mathrm{d}t} z = \sum_{n=1}^{N} \frac{\mathrm{i}\Gamma_n}{2\pi(z - \bar{z}_n)^*} \tag{8.19}$$

where the superscript $*$ denotes the complex conjugate and $\mathrm{i} \equiv \sqrt{-1}$. The strong nonlinearity in the velocity leads to rich vortex dynamics, from regular through quasi-regular to chaotic, depending on N and initial conditions, even without noise.

Such point-vortex systems with both regular and chaotic motions can be successfully tracked using the KF formulas described earlier, given Lagrangian vortex-position observations and Eulerian velocity observations at a few fixed stations. Data type, accuracy, and frequency of the observations are the key factors in achieving a sufficiently accurate ocean estimation. It is found that stochastic forcing ($\mathrm{d}\boldsymbol{\eta}$ in 8.14) is necessary for the Eulerian velocity observations to remain effective in the KF cycles. If only Eulerian velocity observations are available, sudden filter divergence can develop due to nonlinearity in the observation operator (8.19) from the following two causes: (i) the observed or forecast speed becomes too small with respect to the observation noise, and (ii) the point-vortices come too close to the observing stations.

The first cause can trigger filter divergence near the instantaneous saddle points of the Eulerian velocity field. This mechanism has a ready geometric explanation. Because this is an instantaneous cause, its detection is straightforward. The prevention method can then be easily implemented by simply enforcing the minimum speed criterion on the Eulerian velocity observations. It is worth mentioning that a related issue also arises in the assimilation of Lagrangian trajectories (microscopic data) as we shall see in Section 8.7. However, the detection and prevention become more complex for the assimilation of the Lagrangian trajectories. They require Lagrangian analysis to extract the hidden time-integrated information.

The second cause can be avoided using a more sophisticated yet still idealized vortex system. The Rankine vortices have finite core in solid-body rotation and hence remove the velocity singularity associated with the point vortices (Ide and Ghil, 1997b). An accurate tracking of Lagrangian structures represented by the Rankine vortices can be achieved using the limited number of Eulerian velocity observations only, comparable to that of the Rankine vortices in the flow. This success naturally leads to an observing system design by predicting in advance a combination of the observing stations that give the most effective result in estimating the position of Lagrangian structures.

8.7 An augmented state approach for Lagrangian positions

A technique that augments the (Eulerian) state vector with (Lagrangian) drifter coordinates gives a straightforward, and easily implemented, way of assimilating Lagrangian data. Data assimilation techniques, such as the Kalman filter, can then be applied to this enlarged system. We show here that, while such a technique offers a powerful framework in which the (augmented) dynamics possess both Eulerian and Lagrangian characteristics, it also presents new technical challenges such as the saddle points which can cause filter divergence.

The "association problem" encountered in Section 8.4, particularly in optimization of $\|d\mathbf{r} / dt - \mathbf{u}(\mathbf{r}, t)\|^2$, is difficult because the Lagrangian position $\mathbf{r}$ is one of the unknowns. An idea to address such difficulty is to include $\mathbf{r}$ into the state vector of the KF formulation. Let the Eulerian ocean dynamics be (cf. 8.14)

$$d\mathbf{x}_F = \mathbf{m}_F(\mathbf{x}_F, t)dt + d\boldsymbol{\eta}_F \tag{8.20}$$

where the subscript F denotes "flow". We then consider augmenting the model dynamics with equations for the Lagrangian trajectories

$$d\mathbf{x}_D = \mathbf{m}_D(\mathbf{x}_F, \mathbf{x}_D, t)dt + d\boldsymbol{\eta}_D \tag{8.21}$$

where the vector $\mathbf{x}_D(t)$ collectively represents all drifter positions and the subscript D denotes "drifter".

The traditional KF that uses only $\mathbf{x}_F$ as its state variable would have difficulty incorporating the Lagrangian trajectory observation whose observation equation would be $\mathbf{y} = \mathbf{x}_D$. A method developed by Ide *et al.* (2002) and Kuznetsov *et al.* (2003) overcomes this difficulty by augmenting $\mathbf{x}_F$ by $\mathbf{x}_D$ and forming a new state vector $\mathbf{x}(t)$ and corresponding error covariance matrix $\mathbf{P}(t)$

$$\mathbf{x} \equiv \begin{bmatrix} \mathbf{x}_F \\ \mathbf{x}_D \end{bmatrix}, \quad \mathbf{P} \equiv \begin{bmatrix} \mathbf{P}_{FF} & \mathbf{P}_{FD} \\ \mathbf{P}_{DF} & \mathbf{P}_{DD} \end{bmatrix}. \tag{8.22}$$

Note that KF with this augmented state can utilize not only Lagrangian trajectories but also the conventional (Eulerian) observations simultaneously.

Step 1 Forecast from t_{k-1} to t_k: Substituting $\mathbf{x}$ and $\mathbf{P}$ defined in (8.22) into (8.16) yields:

$$
\begin{aligned}
\frac{\mathrm{d}}{\mathrm{d}t}\begin{bmatrix} \mathbf{x}_{\mathrm{F}}^{\mathrm{f}} \\ \mathbf{x}_{\mathrm{D}}^{\mathrm{f}} \end{bmatrix} &= \begin{bmatrix} \mathbf{m}_{\mathrm{F}}(\mathbf{x}_{\mathrm{F}}^{\mathrm{f}}, t) \\ \mathbf{m}_{\mathrm{D}}(\mathbf{x}_{\mathrm{F}}^{\mathrm{f}}, \mathbf{x}_{\mathrm{D}}^{\mathrm{f}}, t) \end{bmatrix} \\
\frac{\mathrm{d}}{\mathrm{d}t}\begin{bmatrix} \mathbf{P}_{\mathrm{FF}}^{\mathrm{f}} & \mathbf{P}_{\mathrm{FD}}^{\mathrm{f}} \\ \mathbf{P}_{\mathrm{DF}}^{\mathrm{f}} & \mathbf{P}_{\mathrm{DD}}^{\mathrm{f}} \end{bmatrix} &= \left[\begin{matrix} \mathbf{M}_{\mathrm{FF}}^{\mathrm{f}}\mathbf{P}_{\mathrm{FF}}^{\mathrm{f}} + (\mathbf{M}_{\mathrm{FF}}\mathbf{P}_{\mathrm{FF}}^{\mathrm{f}})^{\mathrm{T}} \\ (\mathbf{M}_{\mathrm{FF}}\mathbf{P}_{\mathrm{FD}}^{\mathrm{f}})^{\mathrm{T}} + \mathbf{M}_{\mathrm{DD}}\mathbf{P}_{\mathrm{DF}}^{\mathrm{f}} + \mathbf{M}_{\mathrm{DF}}\mathbf{P}_{\mathrm{FF}}^{\mathrm{f}} \end{matrix}\right. \\
&\qquad \left.\begin{matrix} \mathbf{M}_{\mathrm{FF}}\mathbf{P}_{\mathrm{FD}}^{\mathrm{f}} + (\mathbf{M}_{\mathrm{DD}}\mathbf{P}_{\mathrm{DF}}^{\mathrm{f}})^{\mathrm{T}} + (\mathbf{M}_{\mathrm{DF}}^{\mathrm{f}}\mathbf{P}_{\mathrm{FF}}^{\mathrm{f}})^{\mathrm{T}} \\ \mathbf{M}_{\mathrm{DD}}\mathbf{P}_{\mathrm{DD}}^{\mathrm{f}} + (\mathbf{M}_{\mathrm{DD}}\mathbf{P}_{\mathrm{DD}}^{\mathrm{f}})^{\mathrm{T}} + \mathbf{M}_{\mathrm{DF}}\mathbf{P}_{\mathrm{FD}}^{\mathrm{f}} + (\mathbf{M}_{\mathrm{DF}}\mathbf{P}_{\mathrm{FD}}^{\mathrm{f}})^{\mathrm{T}} \end{matrix}\right] \\
&\qquad + \begin{bmatrix} \mathbf{Q}_{\mathrm{FF}} & \mathbf{Q}_{\mathrm{FD}} \\ \mathbf{Q}_{\mathrm{DF}} & \mathbf{Q}_{\mathrm{DD}} \end{bmatrix}
\end{aligned}
\tag{8.23}
$$

where $\mathbf{M}_{\mathrm{FF}} = \frac{\partial}{\partial \mathbf{x}_{\mathrm{F}}}\mathbf{m}_{\mathrm{F}}(\mathbf{x}_{\mathrm{F}}, t)|_{\mathbf{x}_{\mathrm{F}}^{\mathrm{f}}(t)}$ is the TLM of the flow dynamics with respect to $\mathbf{x}_{\mathrm{F}}$, $\mathbf{M}_{\mathrm{DF}} = \frac{\partial}{\partial \mathbf{x}_{\mathrm{F}}}\mathbf{m}_{\mathrm{D}}(\mathbf{x}_{\mathrm{F}}, \mathbf{x}_{\mathrm{D}}, t)|_{\mathbf{x}_{\mathrm{F}}^{\mathrm{f}}(t), \mathbf{x}_{\mathrm{D}}^{\mathrm{f}}(t)}$ and $\mathbf{M}_{\mathrm{DD}} = \frac{\partial}{\partial \mathbf{x}_{\mathrm{D}}}\mathbf{m}_{\mathrm{D}}(\mathbf{x}_{\mathrm{F}}, \mathbf{x}_{\mathrm{D}}, t)|_{\mathbf{x}_{\mathrm{F}}^{\mathrm{f}}(t), \mathbf{x}_{\mathrm{D}}^{\mathrm{f}}(t)}$ are the TLM of the Lagrangian trajectories with respect to $\mathbf{x}_{\mathrm{F}}$ and $\mathbf{x}_{\mathrm{D}}$, respectively.

Step 2 Data update at t_k: Dropping time index k for simplicity in notations, (8.17) and (8.18) become:

$$
\mathbf{K} = \begin{bmatrix} \mathbf{K}_{\mathrm{FD}} \\ \mathbf{K}_{\mathrm{DD}} \end{bmatrix} = \begin{bmatrix} \mathbf{P}_{\mathrm{FD}}^{\mathrm{f}}(\mathbf{P}_{\mathrm{DD}}^{\mathrm{f}} + \mathbf{R})^{-1} \\ \mathbf{P}_{\mathrm{DD}}^{\mathrm{f}}(\mathbf{P}_{\mathrm{DD}}^{\mathrm{f}} + \mathbf{R})^{-1} \end{bmatrix} \tag{8.24}
$$

$$
\begin{aligned}
\begin{bmatrix} \mathbf{x}_{\mathrm{D}}^{\mathrm{a}} \\ \mathbf{x}_{\mathrm{F}}^{\mathrm{a}} \end{bmatrix} &= \begin{bmatrix} \mathbf{x}_{\mathrm{D}}^{\mathrm{f}} \\ \mathbf{x}_{\mathrm{F}}^{\mathrm{f}} \end{bmatrix} + \begin{bmatrix} \mathbf{K}_{\mathrm{FD}} \\ \mathbf{K}_{\mathrm{FD}} \end{bmatrix}(\mathbf{y}_k - \mathbf{x}_{\mathrm{F}}k) \\
\begin{bmatrix} \mathbf{P}_{\mathrm{FF}}^{\mathrm{a}} & \mathbf{P}_{\mathrm{FD}}^{\mathrm{a}} \\ \mathbf{P}_{\mathrm{DF}}^{\mathrm{a}} & \mathbf{P}_{\mathrm{DD}}^{\mathrm{a}} \end{bmatrix} &= \begin{bmatrix} \mathbf{P}_{\mathrm{FF}}^{\mathrm{f}} - \mathbf{K}_{\mathrm{FD}}\mathbf{P}_{\mathrm{FD}}^{\mathrm{f}} & \mathbf{P}_{\mathrm{FD}}^{\mathrm{f}} - \mathbf{K}_{\mathrm{FD}}\mathbf{P}_{\mathrm{DD}}^{\mathrm{f}} \\ \mathbf{P}_{\mathrm{DF}}^{\mathrm{f}} - \mathbf{K}_{\mathrm{DD}}\mathbf{P}_{\mathrm{DF}}^{\mathrm{f}} & \mathbf{P}_{\mathrm{DD}}^{\mathrm{f}} - \mathbf{K}_{\mathrm{DD}}\mathbf{P}_{\mathrm{DD}}^{\mathrm{f}} \end{bmatrix}.
\end{aligned}
\tag{8.25}
$$

This KF strategy has been applied to point-vortex flows in Kuznetsov *et al.* (2003). As described earlier, the flows are dominated by a small number of coherent structures. We discuss here the simplest case of having two vortices in the dynamics and a single drifter trajectory in the observation. The equations for the vortex centers can be derived from (8.19) where we take $N=2$. The tracers (drifters) satisfy the same equations. A set of random samples of the stochastic components of these equations is obtained to simulate a reference solution which we take as the "truth." The random perturbation is then exactly the same in the drifter equations as for the vortices themselves and it is this perturbation that is being potentially uncovered through the assimilation process.

Without any data assimilation, the vortices governed by the truth and the forecast will diverge significantly within a period or two of the vortex rotation (Kuznetsov *et al.*, 2003). On space scales less than the Taylor microscale, the forecast error grows as $O(t)$ initially; the growth rate increases to $O(t^{3/2})$ after the effects of the shear have been fully established. The main question is then whether sampling one drifter trajectory from the true field and assimilating that information, with some observational error, using the KF strategy outlined above could reduce the divergence of the forecast from the truth. The results of the assimilation experiments are promising. In the first case considered, the drifter was observed every $\Delta T = 1$ time unit (the vortices rotate approximately every 12 time units). The analyzed vortex centers, which represent the Eulerian flow field, tracked the true centers well for an indefinite period of time. Out of a set of randomly generated experiments (different realizations of the truth) a 0% failure rate was recorded (a failure being defined as one of the centers drifting from the truth by more than the distance between the original centers). When the observation interval was increased to $\Delta T = 1.5$, a non-zero failure rate of 2% was observed. In some of the failed cases, the KF estimates of the vortex centers could not track the truth effectively; these failures are due to the saddle effect which we describe later. Also, the new KF strategy outlined here was compared to a method of assimilating the pseudo-Lagrangian velocity, i.e., by approximating the velocity data as time-differences of the drifter positions and assimilating that information in a purely Eulerian fashion. The time between observations was used as an indicator of efficacy. In comparable experiments, the pseudo-Lagrangian approach could maintain a reasonably low failure rate only up to about $\Delta T = 0.7$, while the KF strategy performed well at $\Delta T = 1$ and $\Delta T = 1.5$ as described above.

We now examine the cause(s) of the failure of the KF approach, namely the *saddle effect*. For the $\Delta T = 1.5$ case discussed above, the failures occur when the drifters have come too close to the saddle point that lies between the two vortex centers and, in some sense, separates the regions of dynamic influence by the two vortices. When the KF method fails completely, the true and forecast trajectories are found on different sides of the saddle. The two trajectories then diverge in an exponential fashion. Unless an observation becomes available in a short enough time after such an event, the error to be corrected by KF would be too enormous for the linearly approximated dynamics (i.e., TLM) to capture correctly. Such a case is illustrated in Figure 8.2 which depicts an experiment intended only to correct bad initial conditions for the vortex centers. Note that there is no time dependent randomness in the truth. Two different initial drifter approximations are taken for

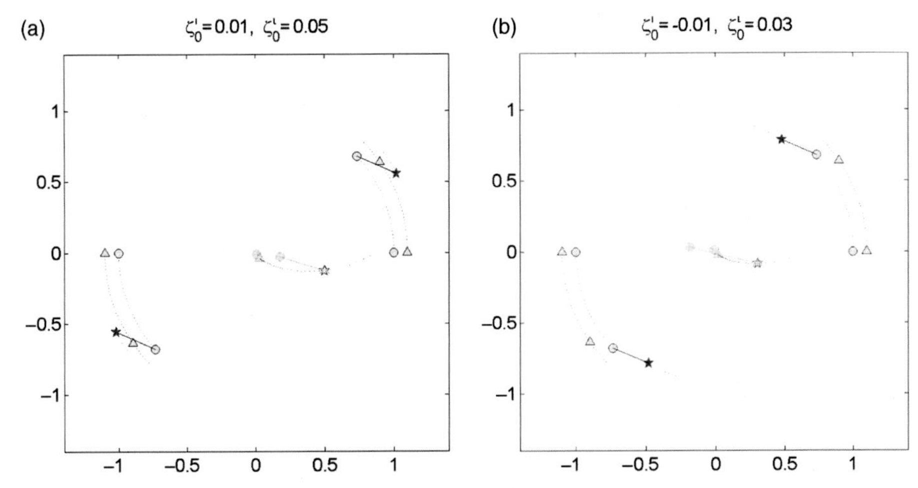

Figure 8.2 Illustration of the saddle effect for the two cases where the forecast and true drifter flow the same way (left panel) and opposite way (right panel). Truth positions are marked with triangles, forecasts are circles, and analyses are stars. The difference in the positions of the dark stars in the two panels demonstrates the divergence that has resulted from having the initial conditions in the opposite sides of the saddle.

the same true initial drifter location and the same (Eulerian) vortex center initial placements. Their difference is almost imperceptible in the figure (the cluster of the symbols at the center of each panel), but they lie on different sides of the saddle point. This fact is revealed in their subsequent evolution. In the left panel (Fig. 8.2), the forecast drifter flows the same way as the truth. The analysis corrections (lines between dark stars and circles) contain some overshoot and are hence not perfect; however, they are pointing the right directions to track the the true positions (triangles). On the other hand, the assimilation of the drifter data from the other side of the saddle leads to corrections on the vortex centers that send them in opposite directions from the truth. This is illustrated in the right panel (Fig. 8.2), where the analysis locations (dark stars) are away from the truth (triangles) with respect to the forecast locations (circles). This "saddle effect" is a manifestation of the known problem that the extended Kalman filter has in regions of high Lyapunov exponents. However, since the saddle effect is usually quite pronounced, identifiable, and localized, there is possibility that it can be anticipated and corrected.

8.8 Concluding remarks

Lagrangian assimilation techniques are still work in progress. Each of the methods presented here have advantages and disadvantages. For example, methods based on conservation laws require that the variables are actually conserved. This becomes more problematic for highly dispersive variables with local sinks and sources like biochemical tracers (e.g., chlorophyll concentration observed through surface color) or when the Lagrangian sampling is sparse in time. Feature-based assimilation, such as the approach involving the active contour method, has the desirable property of preserving the shape and strength of dynamical phenomena. Its computational complexity may be, however, too high for practical applications and especially for assimilation of hundreds of drifters in global numerical ocean circulation models. In general, reliable pattern recognition and matching algorithms would be the key elements of a feature-based scheme. To ensure preservation of the strength of dynamical features, these need to be somehow combined with a methodology that maintains dynamical consistency of the state variables during data updates. Material conservation can impose some dynamical consistency through the advection equation given a background velocity field. Combination of the conservation-based and feature-based approaches might hence merit some consideration. To this end, for example, Thompson's (1969) conservation method can be generalized through reformulation onto a feature-based coordinate system, which would be a more natural numerical environment for Lagrangian data update.

Recent studies on the use of the Kalman filter to assimilate the drifter position data have yielded positive results. Further research is still underway to determine how to best formulate the Lagrangian position in a mathematical form conducive to incorporation into a traditional Eulerian assimilation method. The use of an augmented state (Section 8.7) is an example of such effort. Degradation of the Kalman filter performance near "saddle" regions might be expected due to a high degree of dynamic divergence as well as undersampling by the drifters. Algorithms that can detect such potentially problematic regions by, for example, testing for a large dispersion in drifter positions from an ensemble of simulated drifter trajectories, may be one possible approach to improve the filter performance.

Appendix A: Calculus of variation

The purpose of this section is to derive some general formulas relevant to Sections 8.3 and 8.4 using variational calculus. Consider the problem of finding a function $Q(x)$ that minimizes the definite integral

$$I = \int_{x_0}^{x_1} F(x, Q, Q_x, Q_{xx})\, \mathrm{d}x \tag{8.26}$$

where F is a functional (function of functions) whose arguments are Q and its derivatives up to the second order (higher order derivatives can certainly be considered also). For the minimization problem to be well-posed, F is usually assumed to possess certain general properties such as positivity and reciprocity so that it behaves mathematically like energy. The basic minimization procedure is as follows: If $Q(x)$ is perturbed as $Q(x)+\epsilon\eta(x)$, we expect $\mathrm{d}I/\mathrm{d}\epsilon=0$ at the minimum of I for any choice of the test function $\eta(x)$. To approximate the derivative of I, the perturbed functional is expanded as a Taylor series

$$F(x, Q+\epsilon\eta, Q_x+\epsilon\eta_x, Q_{xx}+\epsilon\eta_{xx}) = F+\epsilon\eta F_Q+\epsilon\eta_x F_{Q_x}+\epsilon\eta_{xx}F_{Q_{xx}}+\cdots,$$

where the argument for each functional on the right-hand side is (x, Q, Q_x, Q_{xx}), and then only the first- and lower-order terms with respect to ϵ are retained. The truncated series is substituted for F in the definite integral (8.26) to obtain

$$\frac{\mathrm{d}I}{\mathrm{d}\epsilon} = \int_{x_0}^{x_1} \eta F_Q+\eta_x F_{Q_x}+\eta_{xx}F_{Q_{xx}}\,\mathrm{d}_x = 0,$$

which is a condition that must be satisfied by the solution $Q(x)$. To eliminate the test function η from the condition, all derivatives of $\eta(x)$ are removed from the integrand of the last equation by application of integration by parts as

$$\left[\eta F_{Q_x}+\eta_x F_{Q_{xx}}-\eta\frac{\partial}{\partial x}F_{Q_{xx}}\right]_{x_0}^{x_1}+\int_{x_0}^{x_1}\eta\left(F_Q-\frac{\partial}{\partial x}F_{Q_x}+\frac{\partial^2}{\partial x^2}F_{Q_{xx}}\right)\mathrm{d}x=0.$$

The term in the square brackets is nulled by assuming further that $\eta(x)$ and its first derivative are zero at the two boundary points. The remaining parts of the equation holds true if and only if

$$F_Q-\frac{\partial}{\partial x}F_{Q_x}+\frac{\partial^2}{\partial x^2}F_{Q_{xx}}=0 \tag{8.27}$$

which is called the *Euler–Lagrange equation* associated with (8.26). With appropriate boundary conditions, (8.27) specifies the optimal $Q(x)$ that minimizes the definite integral.

Similar derivation applies for the multi-dimension function $Q(x, y)$. In particular, for the functional $F(x, y, Q, Q_x, Q_y, Q_{xx}, Q_{yy}, Q_{xy})$ the Euler–Lagrange equation becomes

$$F_Q-\frac{\partial}{\partial x}F_{Q_x}-\frac{\partial}{\partial y}F_{Q_y}+\frac{\partial^2}{\partial x^2}F_{Q_{xx}}+\frac{\partial^2}{\partial y^2}F_{Q_{yy}}+\frac{\partial^2}{\partial x\partial y}F_{Q_{xy}}=0. \tag{8.28}$$

Plate 1 The aluminum sphere SOFAR float with the transducer underneath (mostly concealed by launch frame).

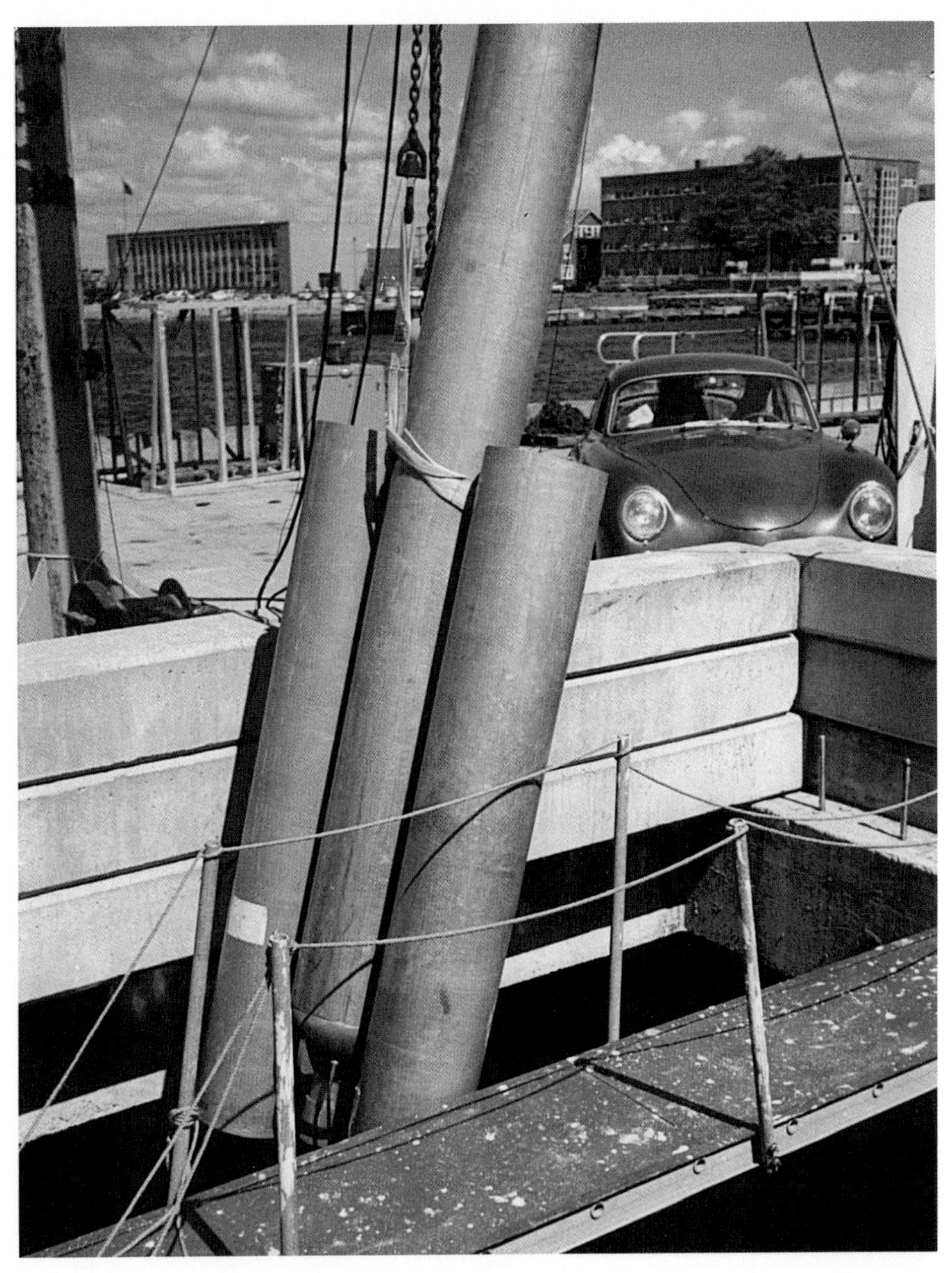

Plate 2 The MODE SOFAR float with two resonator tubes mounted on opposite sides of the flotation tube.

Plate 3 The POLYMODE SOFAR float with its single resonator tube mounted end-to-end.

Plate 4 Phil Richardson (WHOI) holding a RAFOS float. The drop weight (not shown) is attached just before launch.

Plate 5 The COOL float. The float has a compressee underneath so that the float rides on an isopycnal surface. But to the extent that the float does cross isopycnal surfaces, the slanted vanes will force it to rotate indicating how much relative motion there is. The bright spot at the top is the flasher. Photo courtesy Dave Hebert.

Plate 6 Deployment of the bottom-following float. The red 'sails' enhance the float's ability to drift with the waters in the presence of drag forces from the bottom-following line (faintly visible) below the float. Photo courtesy Mark Prater.

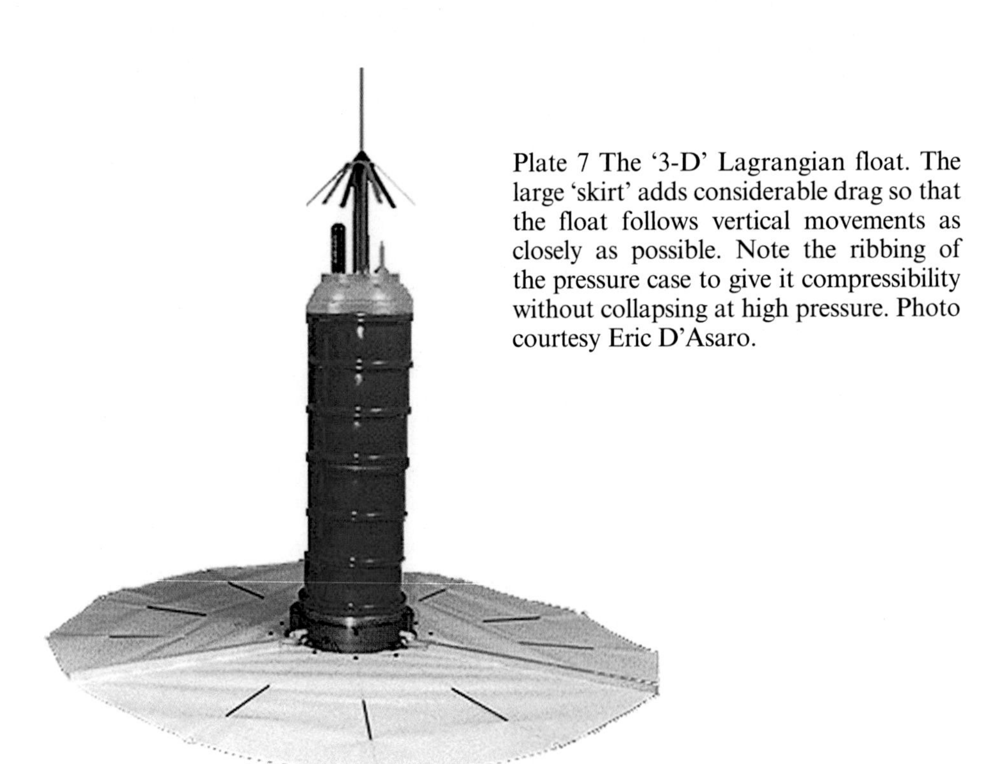

Plate 7 The '3-D' Lagrangian float. The large 'skirt' adds considerable drag so that the float follows vertical movements as closely as possible. Note the ribbing of the pressure case to give it compressibility without collapsing at high pressure. Photo courtesy Eric D'Asaro.

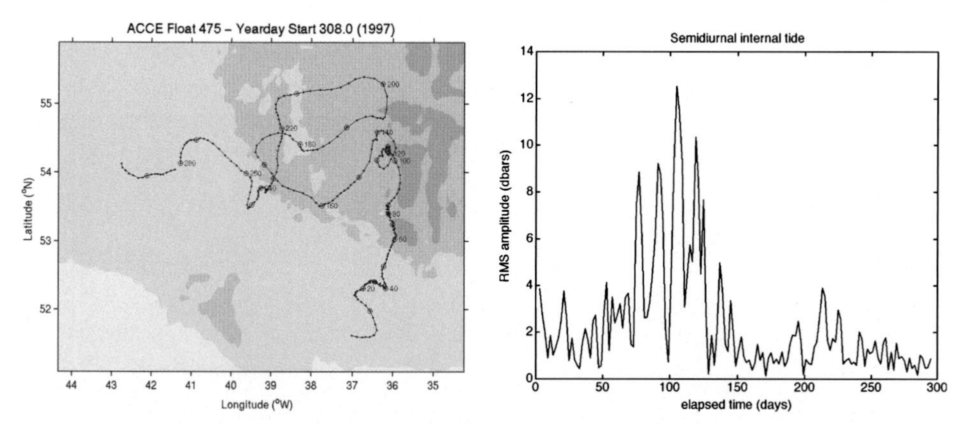

Plate 8 Left panel: trajectory of float 475 just west of the mid-Atlantic Ridge (depth contours at 2, 3, and 4 km). Right panel: the semi-diurnal pressure fluctuations along the track of isopycnal float #475. Note the large amplitude fortnightly variations while the float is over the western slope of the Reykjanes Ridge and only there.

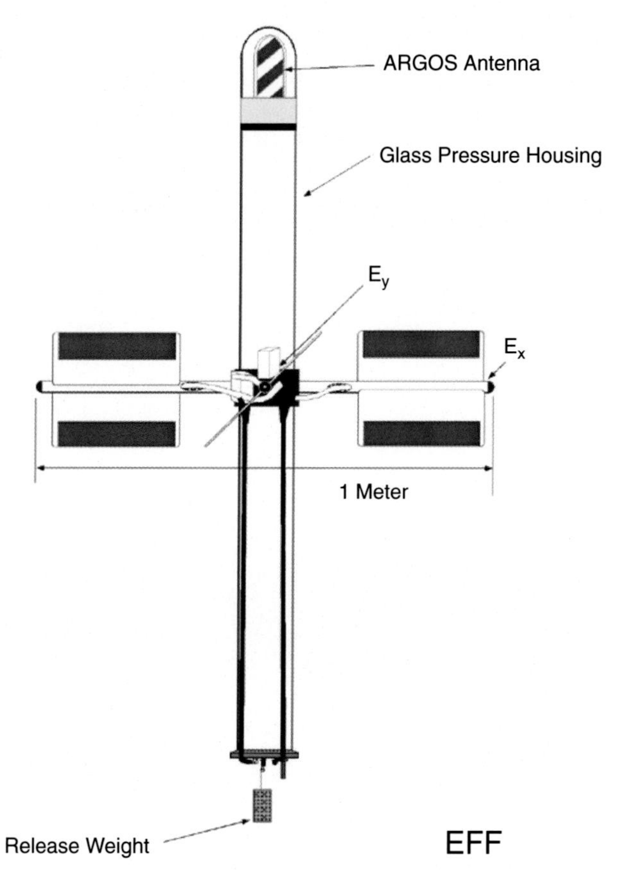

Plate 9 The Electric field float. The electrodes, which sit at the outer end of each arm, measure the in-situ electric field. The slanted vanes cause the float to rotate (induced by the internal wave field) so the electric field measurements can be correlated with direction thereby removing ("chopping" to use an electrical engineering term) errors due to electrochemical potentials. Drawing courtesy Tom Sanford.

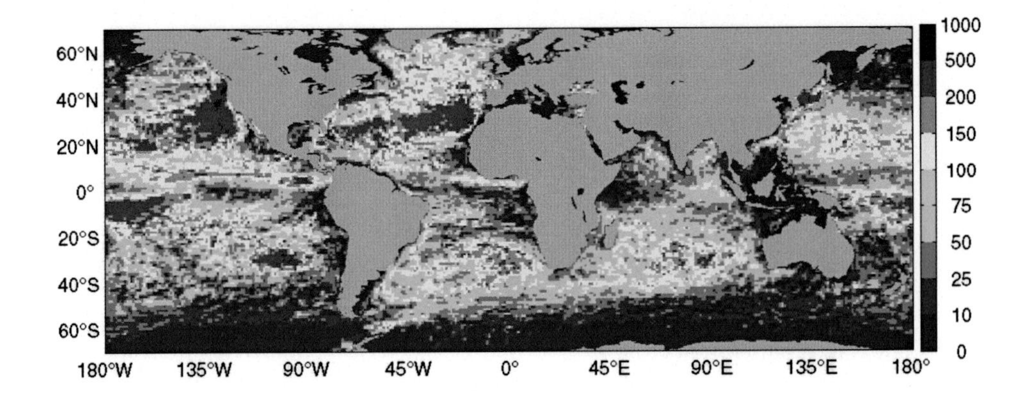

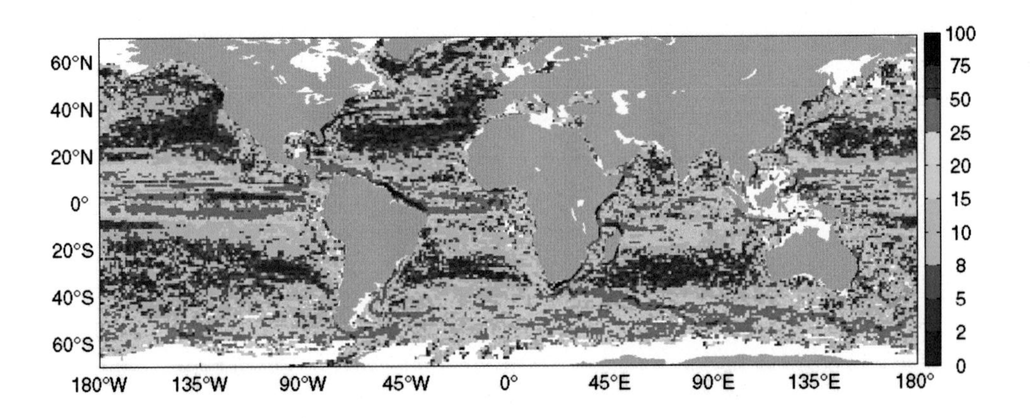

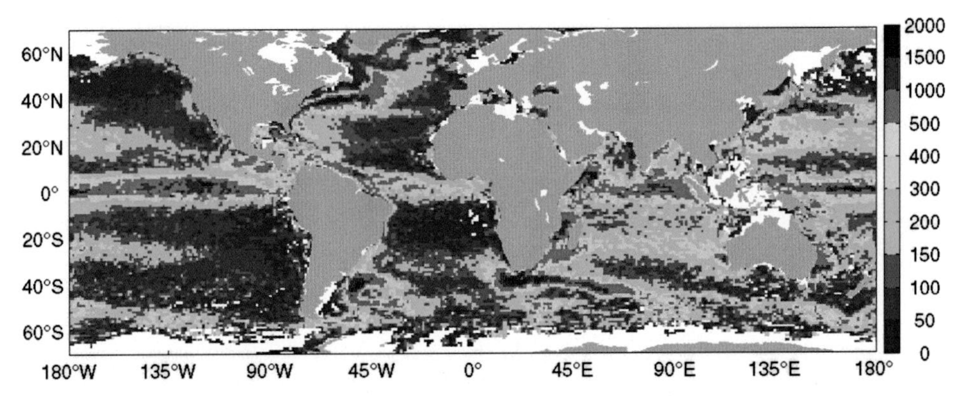

Plate 10 Fields calculated from the most recent decade (to October 31, 2004) of quality-controlled SVP drifter observations. Top: density of observations (drifter days per square degree). Middle: mean current speed (cm/s). Bottom: mean eddy kinetic energy (cm^2/s^2).

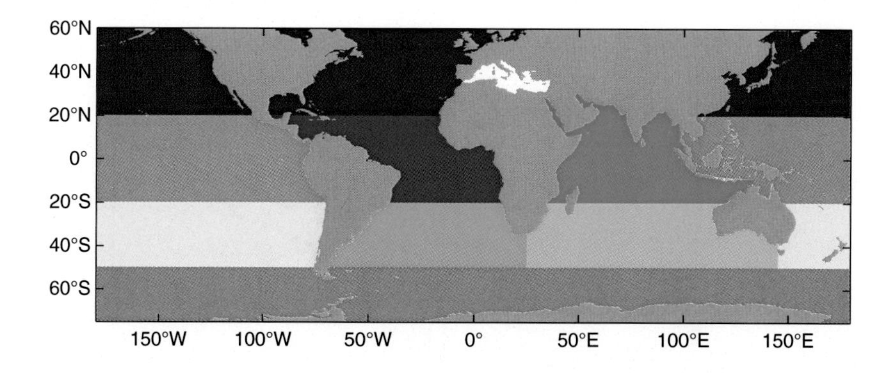

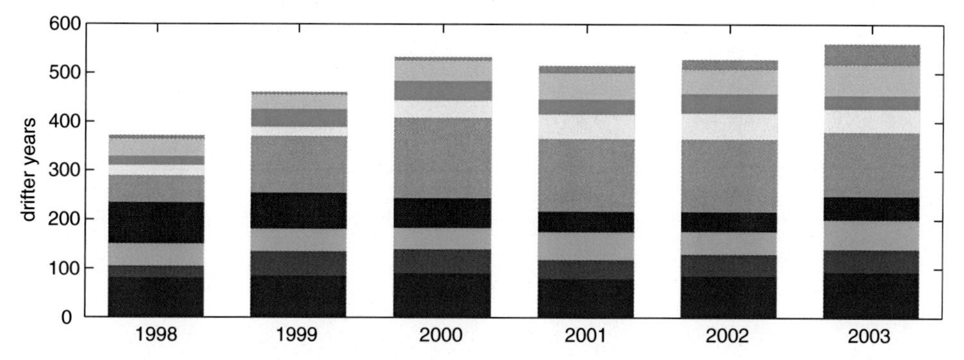

Plate 11 Top: regions (colored) used to subdivide data in bottom panel. Bottom: number of observations (drifter-years) per year in the different regions.

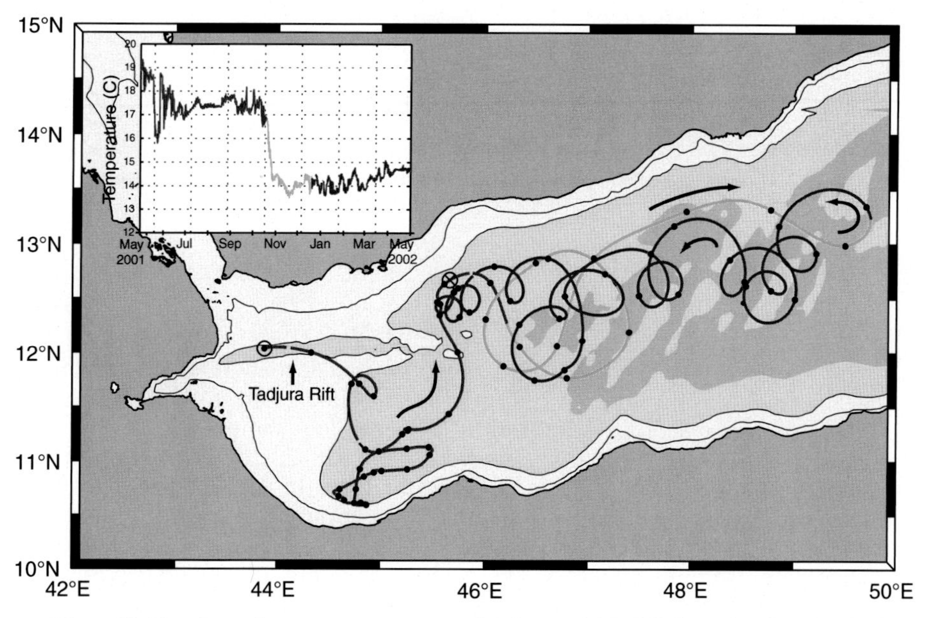

Plate 12 Track and temperature record of one RAFOS float in the Gulf of Aden. The multi-colored track reveals the rotation and translation of one cyclonic eddy from the mouth of the gulf westward. Colors allow the reader to match different segments of the track with the temperature record. Black dots indicate position every 10 days. Isobaths are 200 and every 1000 m.

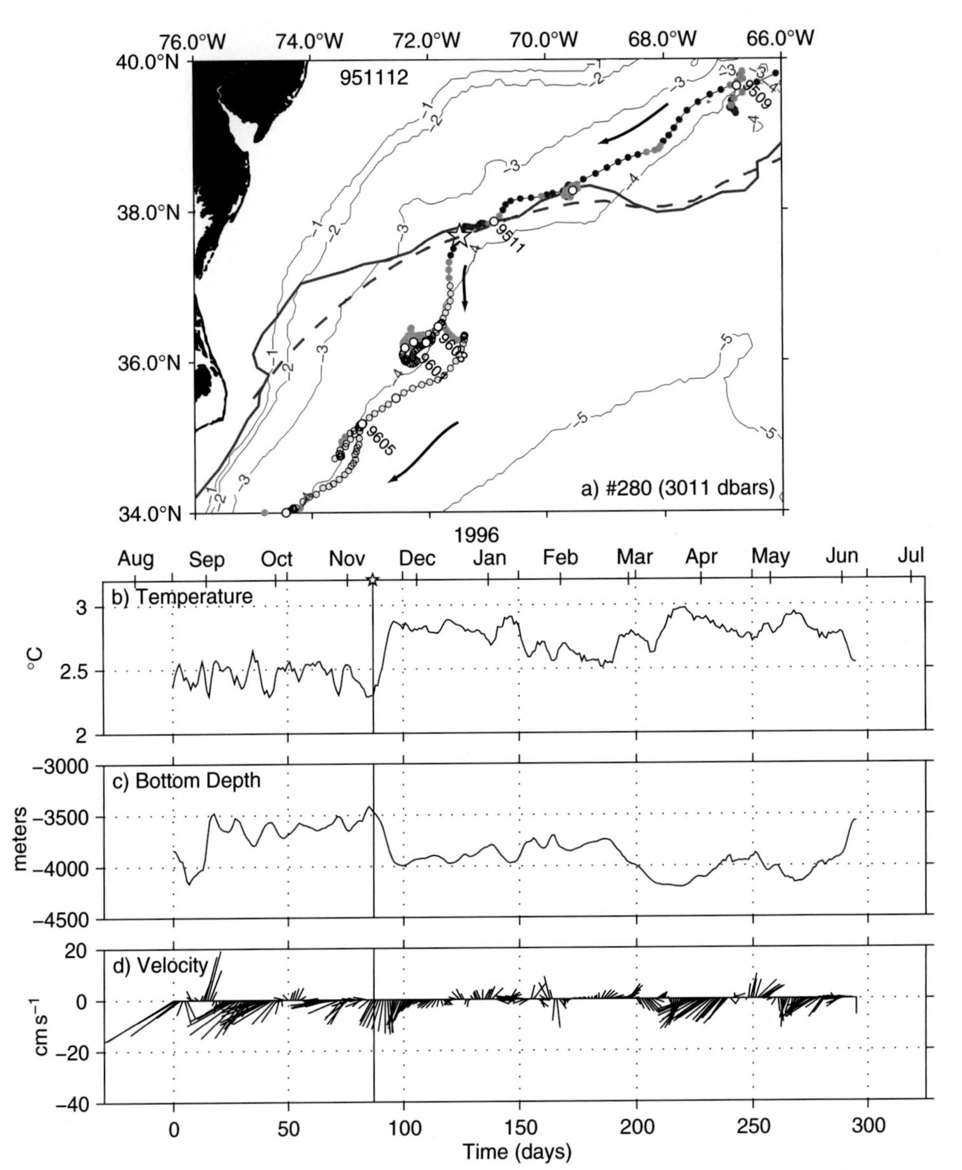

Plate 13 Trajectory segment, temperature, bottom depth and velocity of float #280, which was deployed at 3000 m in June 1995 over the continental slope south of Nova Scotia. Colored dots indicate float temperature in 0.5 °C intervals (e.g., blue $T \leq 2.5$ °C, green $2.5 < T \leq 3.0$ °C, etc). Dashed red line shows long-term mean position of Gulf Stream north wall as observed with satellite AVHRR, the solid red line shows the instantaneous location of the north wall on November 12, 1995, and the star shows the float position on the same date. Isobaths are every 1000 m. Bottom depth at each float position was determined from the digital bathymetric data base ETOPO5.

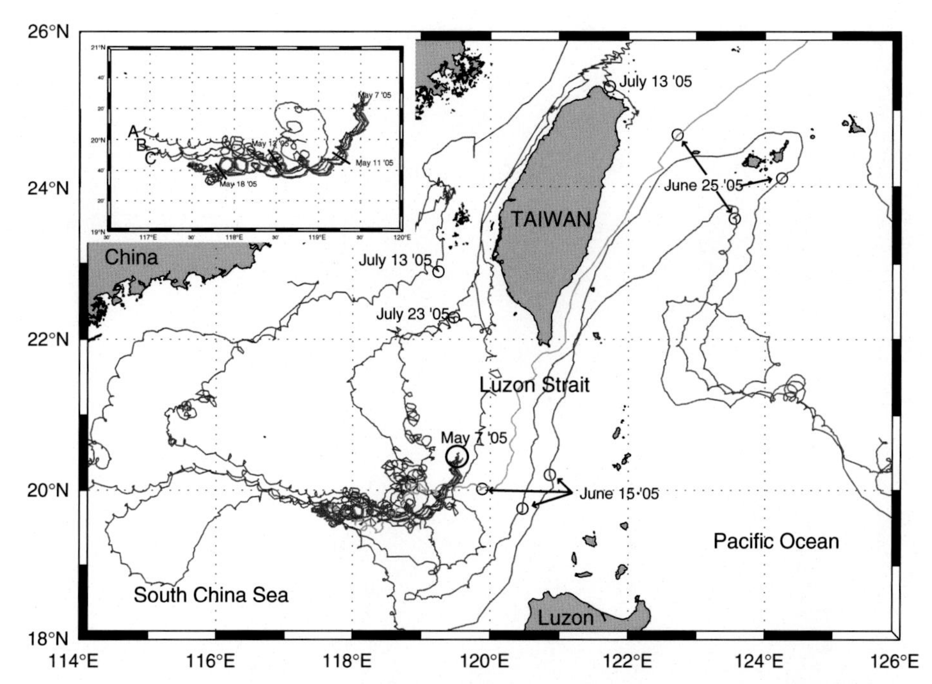

Plate 14 Trajectories of 8 drifters launched in May 2005 in the South China Sea west of the Luzon Strait. Blue (red and green) trajectories identify drifters that exited through the Taiwan Strait (Luzon Strait).

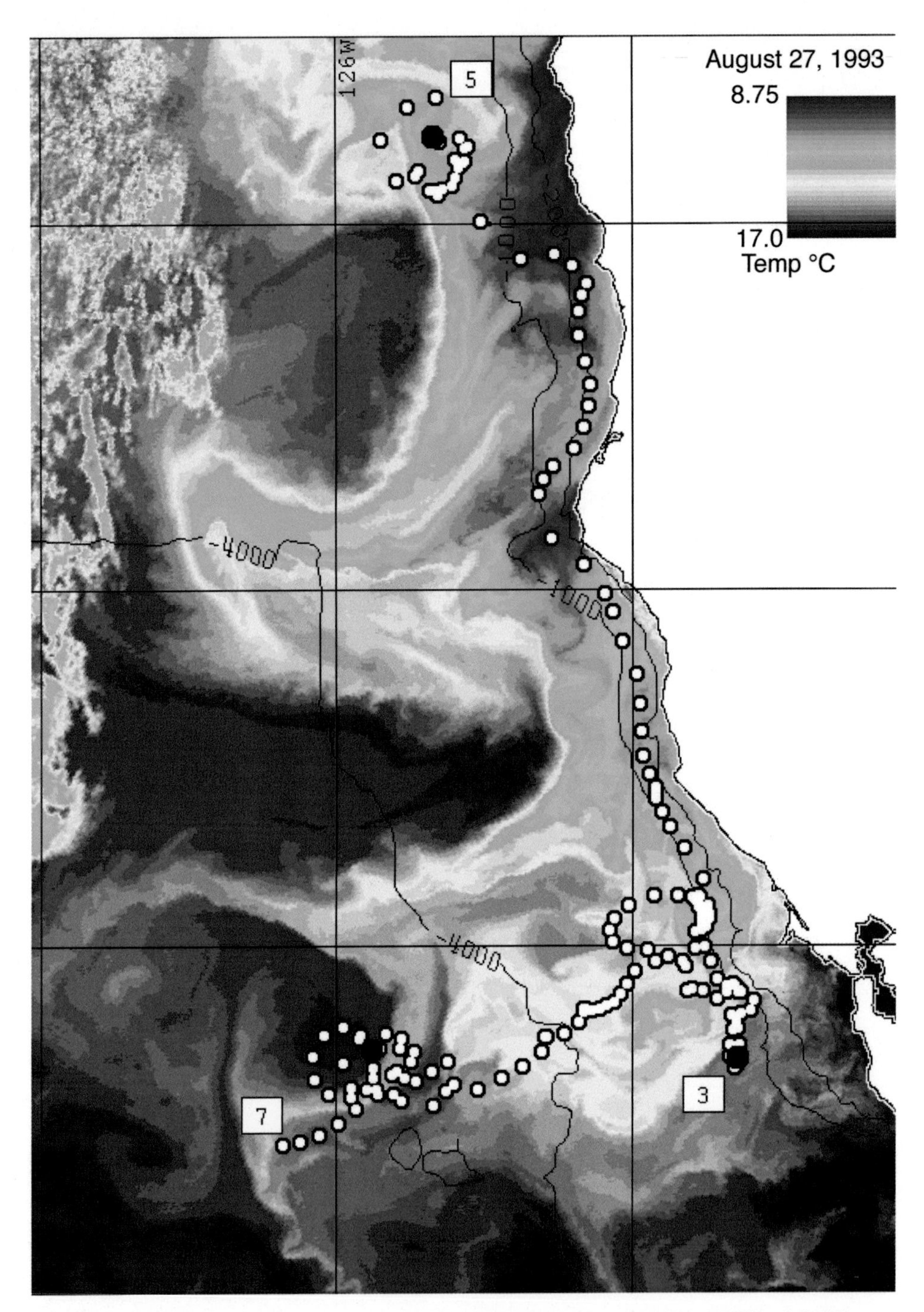

Plate 15 Trajectories of a triad of RAFOS floats (NPS5, NPS3, NPS7) launched west of S. Francisco, California in July 1993.

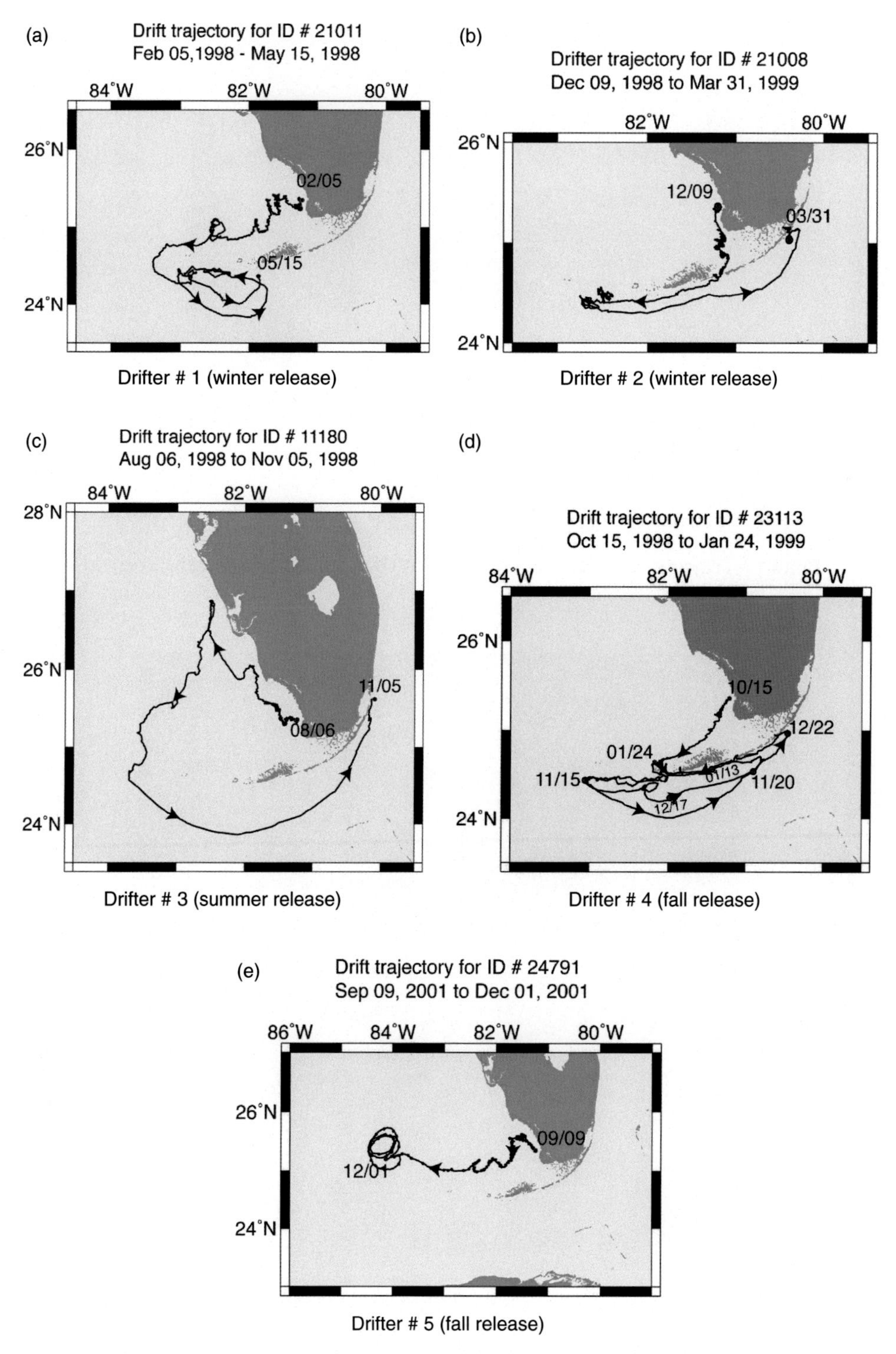

Plate 16 Trajectories of drifters launched from the western shelf of Florida in the framework of the Florida Circulation and Exchange Project (1995–2003). Drifters 1 and 2 have been launched during winter, drifter 3 during summer and drifters 4 and 5 during fall.

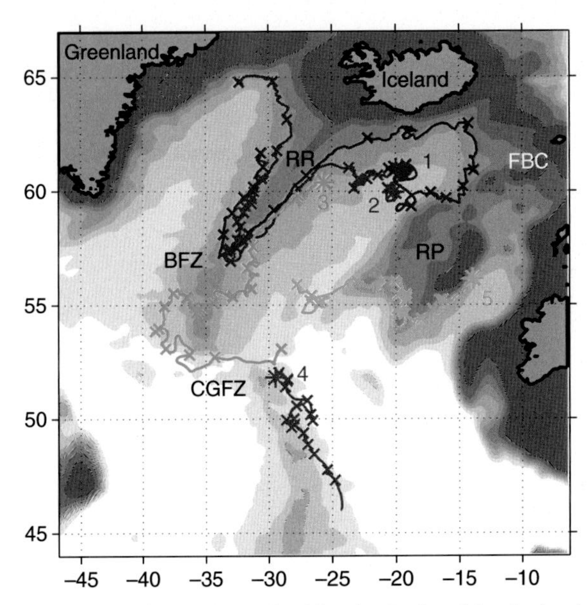

Plate 17 Trajectories of floats launched in the Iceland basin in the context of the Kiel SFB-460 project.

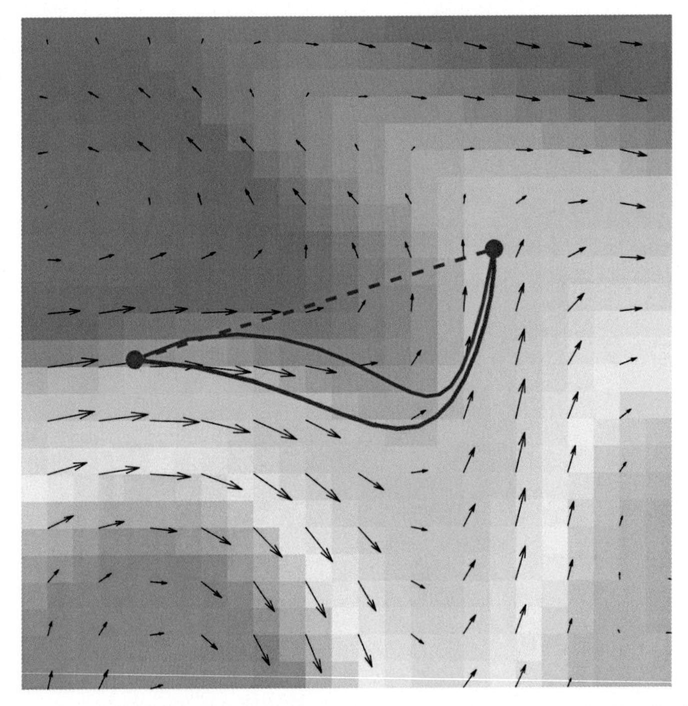

Plate 18 A drifter trajectory (red line) from a model-simulated circulation and time-averages of the simulated velocity (arrows) and pressure (color coded) fields. The blue lines are the drifter trajectory estimated using the active contour approach (see text), indicating the initial estimate (dashed line) and the estimate after 20 iterations (solid blue line).

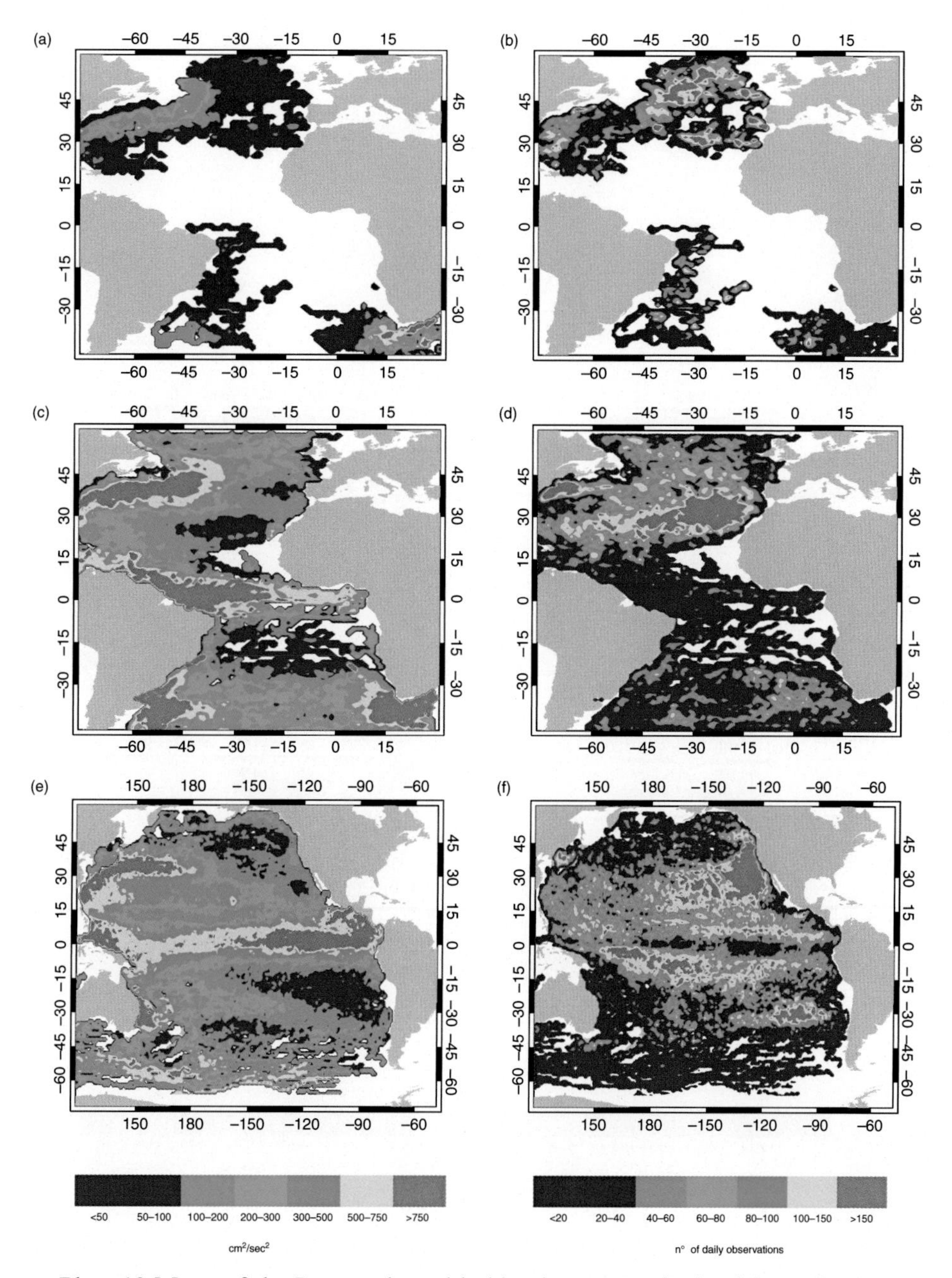

Plate 19 Maps of the Lagrangian eddy kinetic energy obtained from the AF (a), AD (c) and PD (e) datasets and number of daily observations per 1° square bin of the same datasets (b, d and f). The maps of the Lagrangian eddy kinetic energy (expressed in cm^2/sec^2) were obtained averaging over $1° \times 1°$ bins and assigning to each point the eddy kinetic energy of the trajectory.

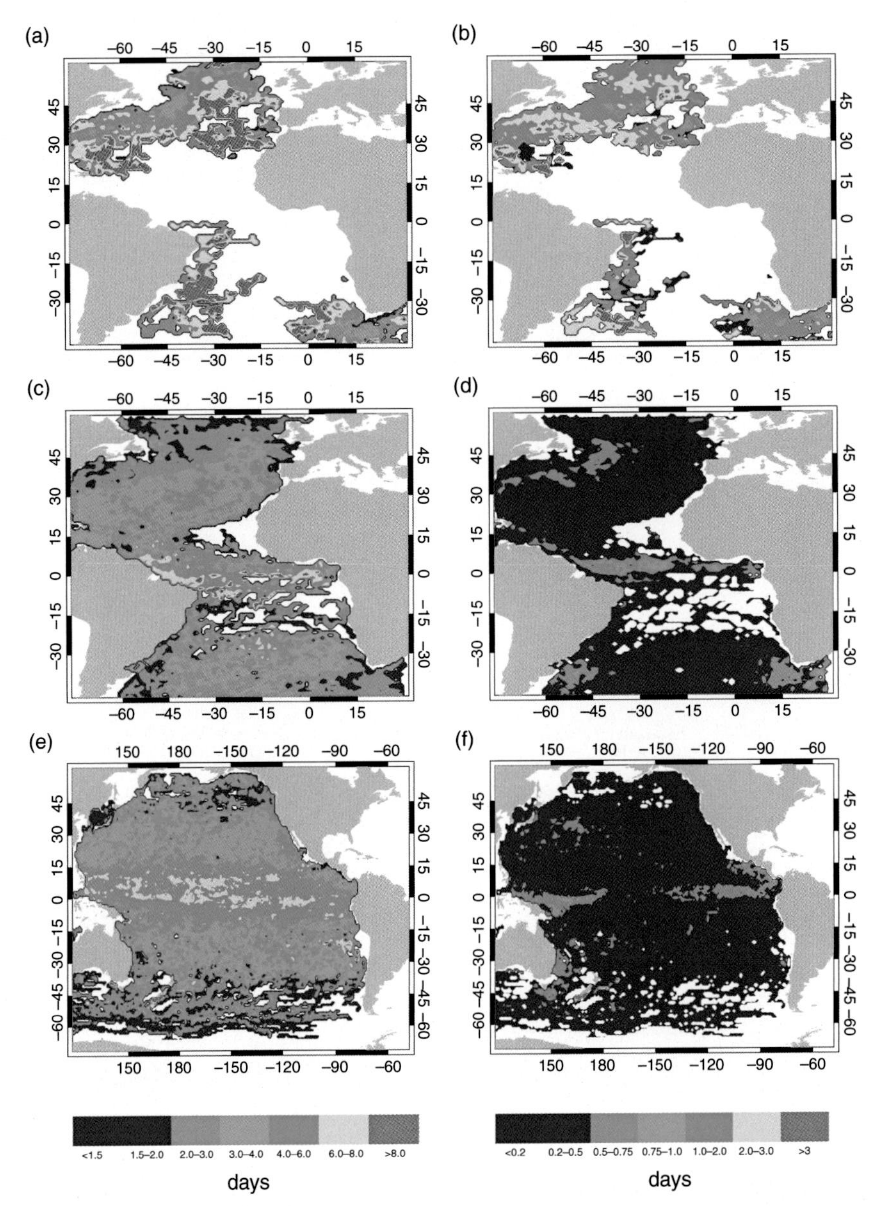

Plate 20 Maps of T_L and T_a of the AF (a, b), AD (c, d) and PD (e, f) datasets. The maps of the time scales (expressed in days) were obtained averaging over $1^\circ \times 1^\circ$ bins and assigning to each point the time scale computed on the entire trajectory.

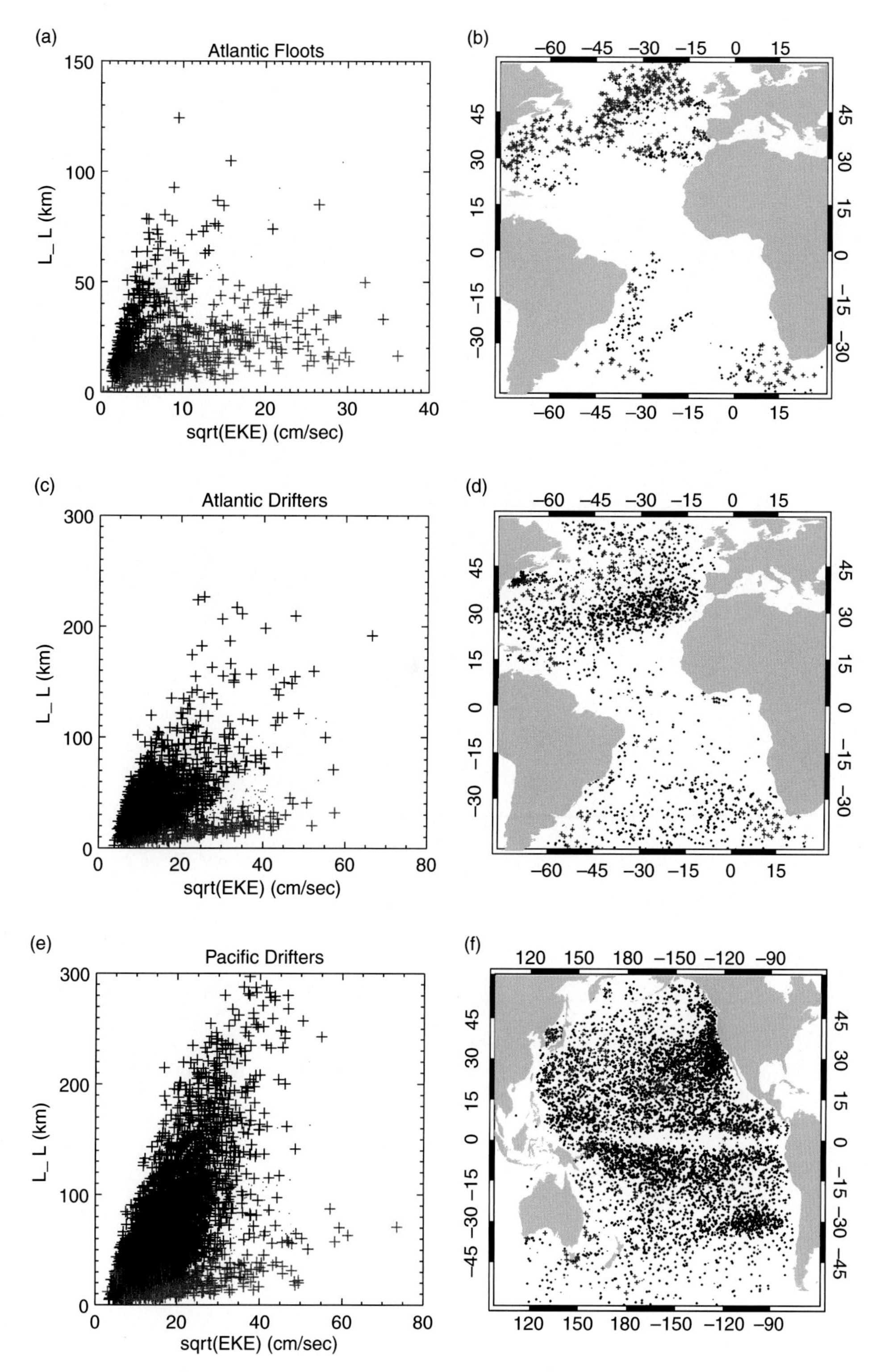

Plate 21 Plot of L_L vs. σ_U (panels a, c and e) and plot of the mean position of the trajectories (panels b, d and f) of AF, AD and PD datasets. The black crosses indicate floats and drifters with $y < 0.2$. The blue and red crosses are related to floats and drifters with $|y| = 1$, $x < 1$ and $|y| = 1$, $x > 1$, respectively.

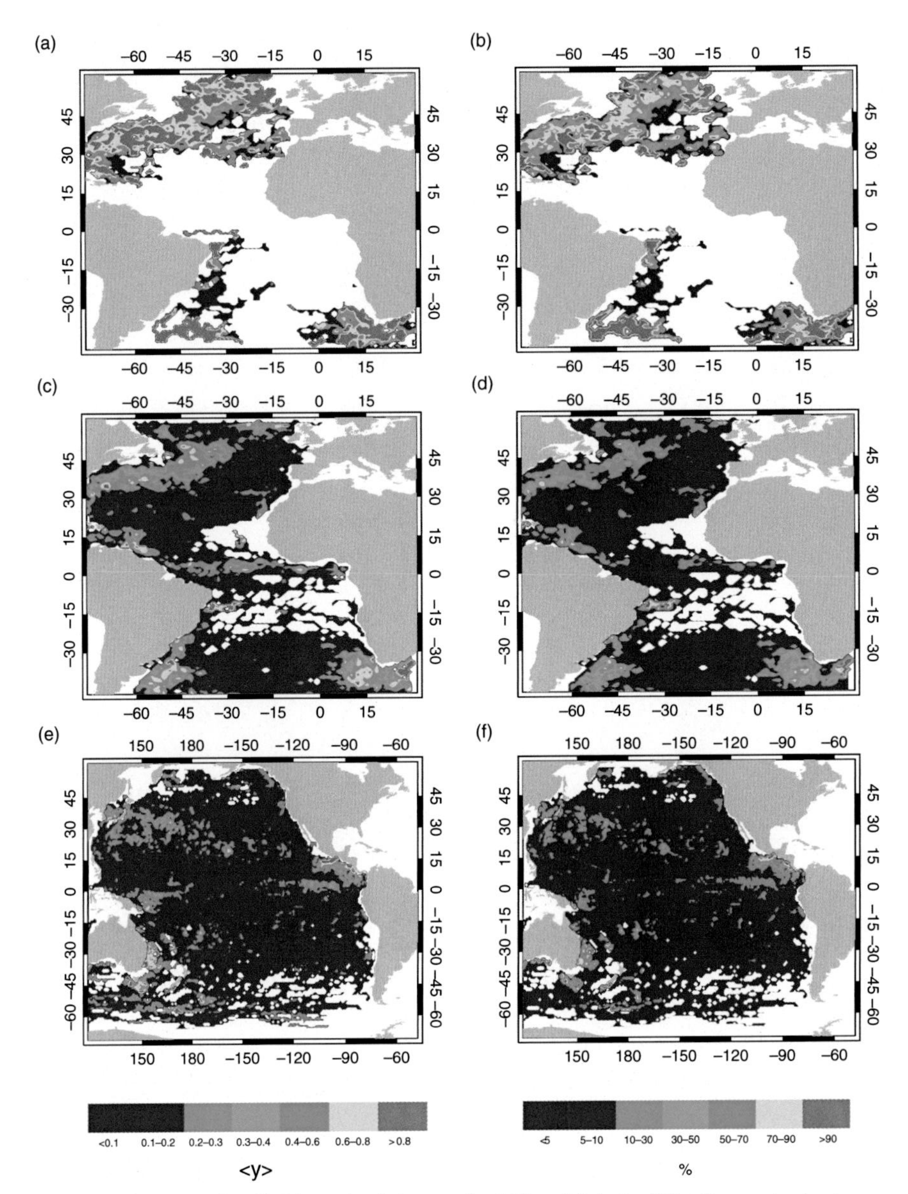

Plate 22 Maps of y (left) and of the probability (right) of finding a trajectory characterized by complex conjugates time scales (Classes III and IV) for the AF (a, b), AD (c, d) and PD (e, f) datasets. The probability map is computed considering, in a given bin, the ratio between the points of trajectories belonging to Classes III and IV and the sum of all the points.

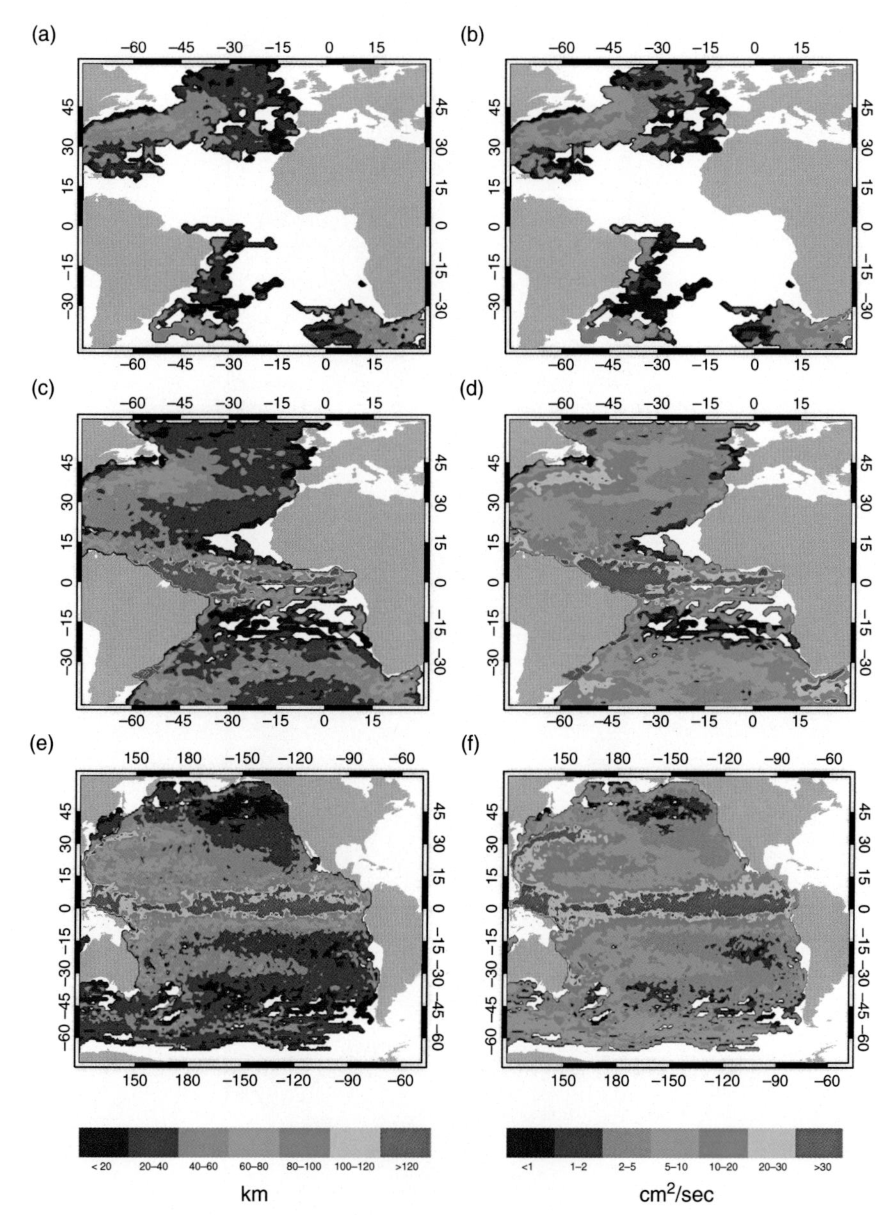

Plate 23 Maps of K_{app} (right) and of the Lagrangian length scale (left) for the AF (a, b), AD (c, d) and PD (e, f) datasets, respectively. The maps were obtained averaging over $1^{\circ} \times 1^{\circ}$ bins and assigning to each point the value of the trajectory. The Lagrangian length scale is expressed in km while for the the apparent diffusivity units are $10^7 \mathrm{cm}^2/\mathrm{sec}$.

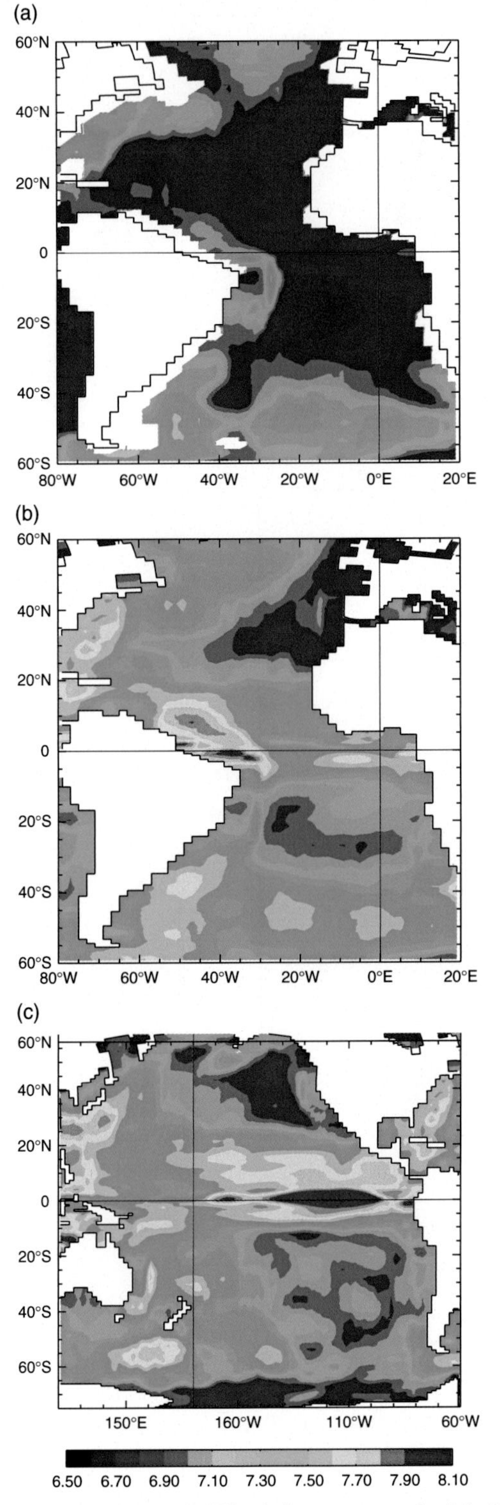

Plate 24 Annual mean horizontal diffusivity computed with OPA at 500 m in the Atlantic (a), at the surface in the Atlantic (b) and at the surface in the Pacific (c). The model has uniform longitudinal resolution of 2 degrees and variable meridional resolution ranging from 0.5 degrees in the tropics to 2 degrees in the polar regions. The base 10 logarithm of the value is shown; units are cm^2/sec. (*Courtesy of S. Calmanti and G. Madec.*)

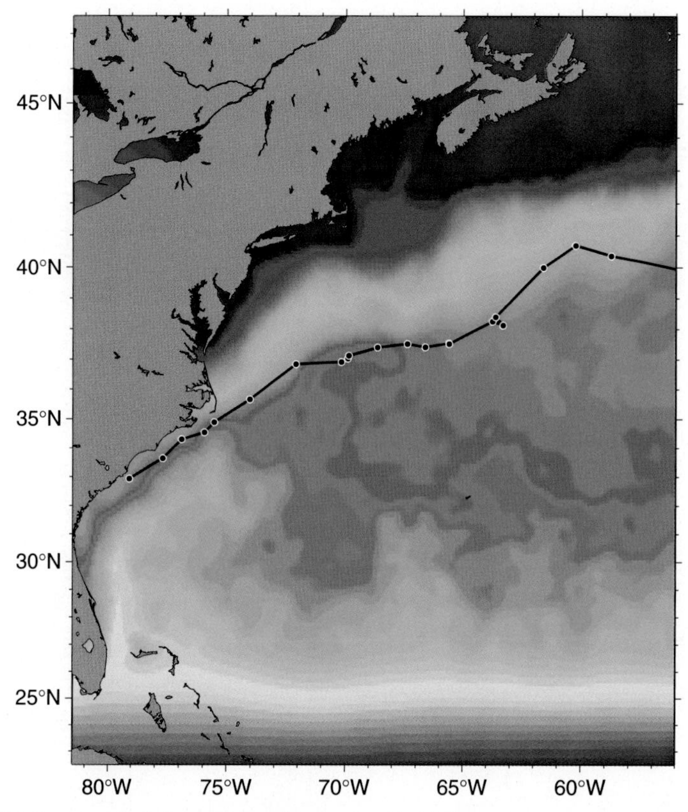

Plate 25 Track of right whale along the Gulf Stream over a several month period. Points denote 10 day progression of the whale to the east during the spring and early summer.

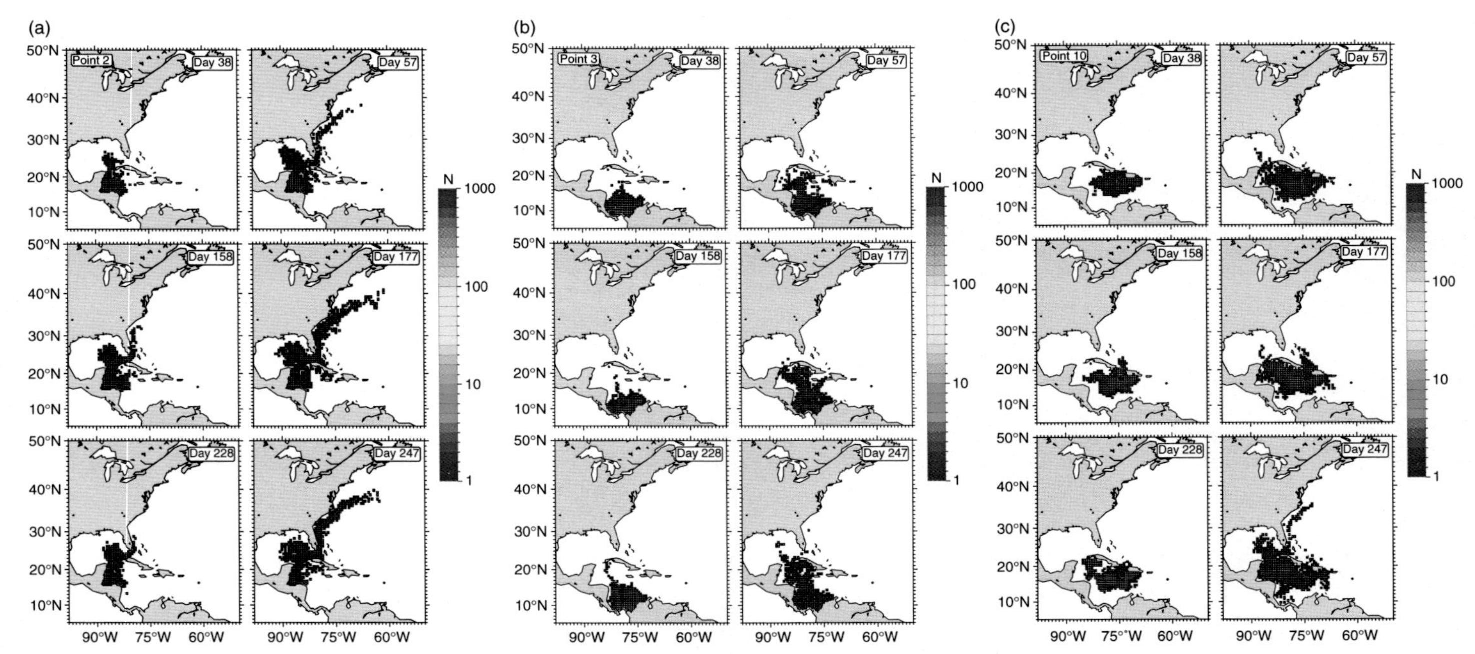

Plate 26 Drift simulations based on the same positions in Fig. 10.17 chosen from a random set of positions around the Caribbean rim. The clouds represent 1000 drifters launched at each point and followed for 24 and 42 days. Results are shown as particle densities on a grid. The densities involve the results of all five years of launches such that the density of particles divided by the 5000 total particle deployments is the probability of finding a particle in the grid box.

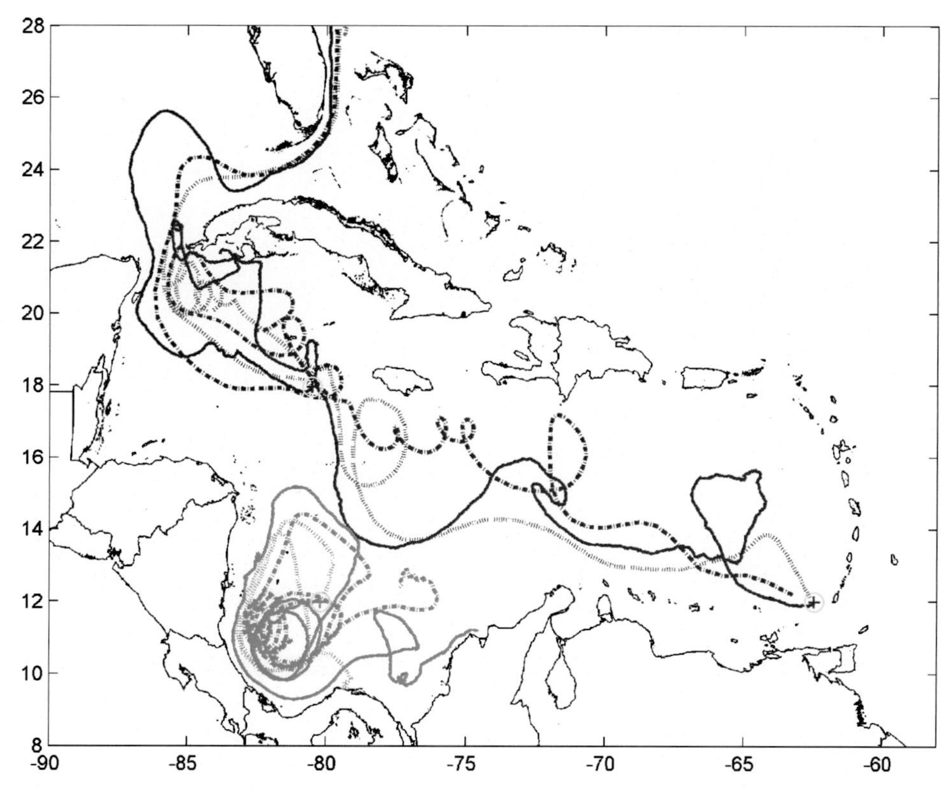

Plate 27 Comparison of drifter floats (solid lines) and simulated drifter tracks (dot and dashed lines) released in three different regions of the Caribbean Sea. Each color corresponds to an individual drifter release and each track is days duration. Simulations used output from the MICOM with wind forcing from ECMWF. Data for the real floats (solid lines) is from the Global Lagrangian Drifter Database available at: http://db.aoml.noaa.gov/cgibin/db/Bin/init_applet.x?gld + GLDKRIGGUI.class.

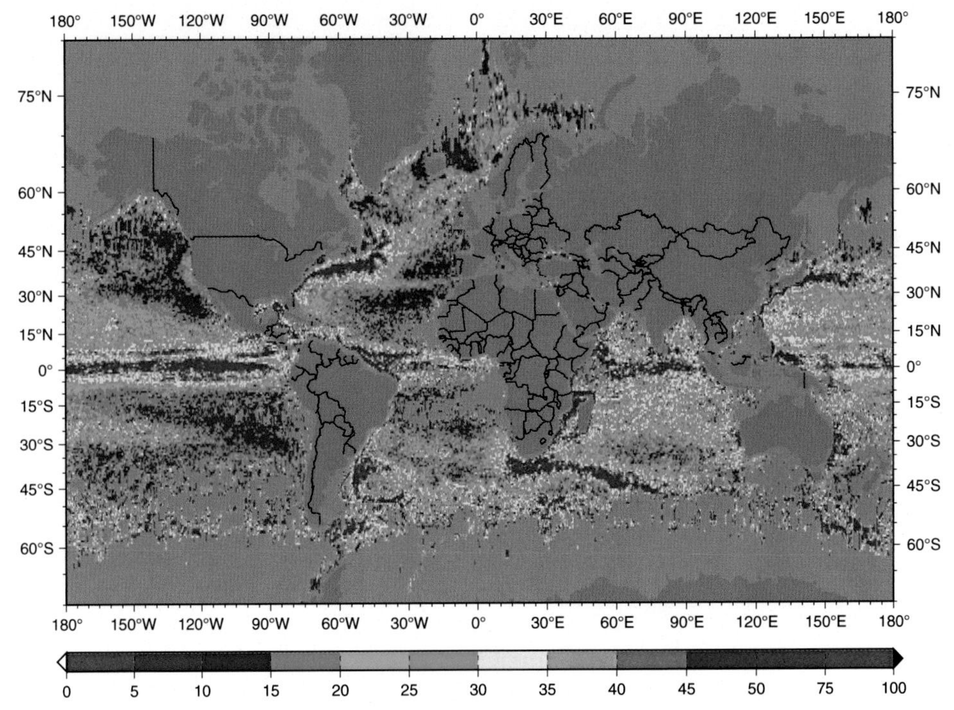

Plate 28 The average speed of sea surface currents, calculated from the climatological data base of the near-surface drifters, archived at AOML and detailed in Chapter 2 by Lumpkin and Pazos. The major ocean currents, along the western edge of ocean basins, and in both the equatorial and polar oceans are clearly visible in this global map. The average speed of the drifters is on the order of 10 cm/s in gyre interiors, 35 cm/s in regions close to the major currents and along the eastern boundaries of the ocean basins, and the average speed is on the order of 100 cm/s in the major ocean currents.

References

Bauer, S., M. S. Swenson, A. Griffa, A. J. Mariano, and K. Owens, 1998. Eddy-mean flow decomposition and eddy-diffusivity estimates in the tropical Pacific Ocean. *J. Geophys. Res.*, **103**, 30855–71.

Bennett, A. F. and B. S. Chua, 1999. Open boundary conditions for Lagrangian geophysical fluid dynamics. *J. Comp. Phys.*, **153**, 418–36.

Bretherton, F. P., R. E. Davis, and C. B. Fandry, 1976. Technique for objective analysis and design of oceanographic experiments applied to MODE-73. *Deep-Sea Res.*, **23** (7), 559–82.

Carter, E. F., 1989. Assimilation of Lagrangian data into a numerical-model. *Dyn. Atmos. Oceans.*, **13**, 335–48.

Carter, E. F. and A. R. Robinson, 1987. Analysis models for the estimation of oceanic fields. *J. Atm. Ocean. Tech.*, **4** (1), 49–74.

Chin, T. M., A. J. Mariano, and E. P. Chassignet, 1999. Spatial regression and multi-scale approximations for sequential data assimilation in ocean models. *J. Geophys. Res.*, **104** (C4), 7991–8014.

Christopherson, R. W., 2000. *Geosystems* (Fourth Edition), Englewood Cliffs, NJ: Prentice-Hall.

Evensen, G., 1994. Sequential data assimilation with a nonlinear quasi-geostrophic model using Monte Carlo methods to forecast error statistics. *J. Geophys. Res.*, **99** (C5), 10143–62.

Freeland, H. J. and W. J. Gould, 1976. Objective analysis of meso-scale ocean circulation features. *Deep-Sea Res.*, **23**, 915.

Gelb, A. (ed.), 1974. *Applied Optimal Estimation*. Cambridge, MA: MIT Press.

Ghil, M. and P. Malanotte-Rizzoli, 1991. Data assimilation in meteorology and oceanography. *Adv. Geophys.*, **33**, 141–266.

Halkin, D. and T. Rossby, 1985. The structure and transport of the Gulf-Stream at 73-degrees-W. *J. Phys. Oceanogr.*, **15**, 1439–52.

Hua, B. L., J. C. McWilliams, and W. B. Owens, 1986. An objective analysis of the POLYMODE local dynamics experiment. Part II: Streamfunction and potential vorticity fields during the intensive period. *J. Phys. Oceanogr.*, **15**, 506–22.

Ide, K. and M. Ghil, 1997a. Extended Kalman filtering for vortex systems. Part I: Methodology and point vortices. *Dynamics of Atmospheres and Oceans*, **27**, 301–32.

Ide, K. and M. Ghil, 1997b. Extended Kalman filtering for vortex systems. Part II: Rankine vortices and observing-system design. *Dynamics of Atmospheres and Oceans*, **27**, 333–50.

Ide, K., P. Courtier, M. Ghil, and A. Lorenc, 1997. A unified notation for data assimilation. *J. Metorolor. Soc. Japan*, **75**(1B), 181–9.

Ide, K., L. Kuznetsov, and C. K. R. T. Jones, 2002. Lagrangian data assimilation for point-vortex systems. *J. Turb.*, **53**.

Ishikawa, Y., T. Awaji, K. Akitomo, and B. Qiu, 1996. Successive correction of the mean sea surface height by the simultaneous assimilation of drifting buoy and altimetric data. *J. of Phys. Oceanogr.*, **26**, 2381–97.

Kalman, R. E., 1960. A new approach to linear filtering and prediction problems. *J. Basic Eng.*, **82**, 35–45.

Kamachi, M. and J. J. O'Brien, 1995. Continuous data assimilation of drifting buoy trajectory into an equatorial Pacific Ocean model. *J. Mar. Sys.* **6**, 159–78.

Kass, M., A. Witkin, and D. Terzopoulos, 1988. Snakes: Active contour models. *Int. J. Comput. Vision*, **1**, 321–31.
Kelly, K. A., 1989. An inverse model for near-surface velocity from infrared images. *J. Phys. Oceanogr.*, **19**, 1845–64.
Kuznetsov, L., K. Ide, and C. K. R. T. Jones, 2003. A method for assimilation of Lagrangian data. *Mon. Wea. Rev.*, **131**, 2247–60.
Lewis, J. M., 1982. Adaptation of P. D. Thompson's scheme to the constraint of potential vorticity conservation. *Mon. Wea. Rev.*, **110**, 1618–34.
Lewis, F. L., 1986. *Optimal Estimation*. New York: Wiley.
Mariano, A. J. and T. M. Chin, 1996. Feature and contour based data analysis and assimilation in physical oceanography. In *Stochastic Modelling in Physical Oceanography*, ed. R. J. Adler *et al.*, Cambridge, MA: Birkhäuser Boston, 311–42.
Mariano, A. J. and T. Rossby, 1989. The Lagrangian potential vorticity balance during POLYMODE. *J. Phys. Oceanogr.*, **19**, 927–39.
McWilliams, J. C. (1991). Geostrophic vortices. In *Nonlinear Topics in Geophysical Processes: Proceedings of International School of Physics "Enrico Fermi" Course 109*, ed. A. R. Osborne. Amsterdam: Elsevier, 5–50.
Mead, J. L. and A. F. Bennett, 2001. Towards regional assimilation of Lagrangian data: the Lagrangian form of the shallow water model and its inverse *J. Mar. Sys.*, **29**, 365–84.
Molcard, A., L. Piterbarg, A. Griffa, T. M. Ozgokmen, and A. J. Mariano, 2003. Assimilation of drifter positions for the reconstruction of the Eulerian circulation field. *J. Geophys. Res. Oceans*, **108**, 3056, doi:10.1029/2001JC001240.
Özgökmen, T. M., A. Molcard, T. M. Chin, L. I. Piterbarg, and A. Griffa, 2003. Assimilation of drifter observations in primitive equation models of midlatitude ocean circulation. *J. Geophys. Res. (Oceans)*, **108**, 3238, doi:10.1029/2002JC001719.
Pinardi, N. and A. R. Robinson, 1987. Dynamics of deep thermocline jets in the POLYMODE region. *J. Phys. Oceanogr.*, **17**, 1163–88.
Rossby, T., J. Price, and D. Webb, 1986. The spatial and temporal evolution of a cluster of SOFAR floats in the POLYMODE local dynamics experiment (LDE). *J. Phys. Oceanogr.*, **16**, 428–42.
Sorenson, H. W., 1970. Least-squares estimation: from Gauss to Kalman. *IEEE Spectrum*, **7**, 63–8.
Storvik, G., 1994. A Bayesian approach to dynamic contours through stochastic sampling and simulated annealing. *IEEE Transactions on Pattern Analysis and Machine Intelligence*, **16**, 976–86.
Thompson, P. D., 1969. Reduction of analysis error through constraints of dynamical consistency. *J. Appl. Meteorol.*, **8**, 738–42.

9

Observing turbulence regimes and Lagrangian dispersal properties in the oceans

VOLFANGO RUPOLO
ENEA, Roma, Italy

9.1 Introduction

Lagrangian instruments have been widely used in the last few decades to sample basic ocean properties, even in remote areas of the globe. Since transport properties depend on Lagrangian scales, an integrated analysis of the Lagrangian trajectories observed sparsely in space and time is fundamental for the understanding of the ocean transport properties. This analysis, however, is complex, due to the non-homogeneous character of the mesoscale structures and the existence of different regimes of dispersion.

Nowadays, Lagrangian data, at different depths and in all world ocean basins, are available through the WOCE (2002) archive. The first information that can be extracted from such data is a map of the mean currents and of the eddy kinetic energy, that is typically computed using the binning technique, where the float velocities are averaged over small spatial subregions (bins). This approach, that considers Lagrangian instruments as moving current meters, has often been exploited in the past in order to achieve a better description of the global oceanic circulation (e.g. McNally *et al.*, 1983; Richardson, 1983; Hofmann, 1985; Patterson, 1985; Davis, 1991a; Owens, 1991; Swenson and Niiler, 1996; Bauer *et al.*, 1998; Fratantoni, 2001; Bower *et al.*, 2002).

Lagrangian data have also been employed to study transport properties in the mesoscale range (e.g. Freeland *et al.*, 1975; Riser and Rossby, 1983; Rossby *et al.*, 1983; Colin de Verdière, 1983; Krauss and Böning, 1987; Figueroa and Olson, 1989; Zhang *et al.*, 2001; Bauer *et al.*, 2002). The objective of most of these studies was the characterization of the correlation properties of the particle velocities that were then used to derive a corresponding "diffusion coefficient." The framework of this analysis is Taylor's (1921) theory of *stationary* and *homogeneous* turbulence; moreover, dispersal properties

Lagrangian Analysis and Prediction of Coastal and Ocean Dynamics, ed. A. Griffa, D. Kirwan, A. Mariano, T. Özgökmen, and T. Rossby. Published by Cambridge University Press.

are proportional to the velocity spectrum at zero frequency, so quantifying low frequency variability is critical (Davis, 1991b). All this shows the intrinsic difficulty of this kind of analysis since the motion of a Lagrangian float is deeply affected both by the space non-homogeneities and by the nonstationarity of the Eulerian velocity field. Hence, averaging among long trajectories experiencing a mixture of different flow regimes can lead to misleading results.

Another important issue related to this topic is the study of the relationships between eddy kinetic energy and Lagrangian correlation scales. In fact the existence of one of the two regimes in which the time, or the space, correlation scale is nearly constant could be exploited to have a first estimate of diffusivity from the distribution of eddy energy.

The complicated nature of the Lagrangian dispersion has been evidenced also using simplified conceptual models. For instance, Elhmaidi *et al.* (1993) and Pasquero *et al.* (2001), considering the quasi-geostrophic 2D turbulence, have shown that to correctly model dispersal properties one should consider at least two regimes. Namely, distinguishing particles trapped in energetic coherent structures from those floating in a more quiescent turbulent background.

More recently, Berloff and McWilliams (2003) obtained improvements in the representation of dispersion using stochastic models in which the decorrelation parameters vary with statistical properties given by a probability distribution obtained from data.

To avoid a mixture of different regimes a first step is to try to identify, using a previous knowledge of the velocity field, some (large) geographical sub-domains that can be considered, in a first approximation, homogeneous. This approach has been utilized with good results in the analyses by Stammer (1997) based on altimetric TOPEX/POSEIDON data. On the other hand, Lagrangian experimental data may show a huge variability of behaviors even inside such sub-domains (see the work on the North Atlantic by Veneziani *et al.*, 2004, showing the coexistence of looping and non-looping trajectories in a same area). This is because the trajectories are measured at different times and in the turbulent flow.

A different approach, that will be explored in this chapter, consists in computing mean statistical properties considering only trajectories belonging to the same "dynamical class." Often the simple observation of the shape of the trajectory informs about the underlying dynamics of the flow much more than the statistical indices computed from the velocity time series. If using some quantitative index one could separate different classes of trajectories, representative of distinct dispersion regimes, the averaged statistical properties computed on such sub-domains could be useful first to investigate dispersion properties in these homogeneous classes and second to measure the quantitative relevance of these different regimes in the geographical space.

This quantitative index can be built using, as a paradigm of the observed ocean variability, idealized models that depend only on few parameters, such as the Lagrangian Stochastic Models (LSMs).

In general, a stochastic model can be "directly" employed to simulate Lagrangian statistical properties of the flow regime (e.g., Berloff and McWilliams, 2002b) or "inversely" used for extracting information about flow statistics (e.g., Griffa *et al.*, 1995). Here, we will focus our attention on the inverse use of a LSM, showing how it can be utilized as a tool to map qualitative and quantitative information about the nature of the flow in different geographical areas.

In particular, we will analyze sub-surface floats and surface drifters data, trying to detect how and where the presence of coherent structures influences the Lagrangian dispersion. The purpose of this work is also to provide a characterization of the dispersal properties of the main oceanic currents, believing that such kind of analysis can be useful to advance the representation of the eddy diffusivity in the Ocean General Circulation Models (OGCMs).

The application of stochastic models in oceanography has a long history. In this chapter we will only briefly review the hierarchy of the LSMs commonly used, referring the reader to the literature for technical details and applications. All the discussion of the present chapter will be done assuming that a float moves on a 2-dimensional surface and considering 1D LSMs with constant parameters. The first assumption is appropriate for many oceanic situations since particles move mainly on isopycnal surfaces. The second assumption is justified by the fact that in this chapter LSMs will be solely employed to define a model parameter useful to select, in the entire experimental dataset, different "classes" of trajectories.

This chapter is organized as follows. In Section 9.2, we define the statistical functions utilized in the chapter and we discuss the relationship between Lagrangian dispersal properties and velocity spectra, attempting to highlight distinct qualitative behaviors. In Section 9.3, we review the source of the Lagrangian variability and two distinct dynamical regimes of dispersion are introduced. In Section 9.4, we review the hierarchy of the commonly used LSMs. In Section 9.5, analyzing WOCE data, we select different classes of trajectories by means of a dimensionless index derived by the second-order LSM, and we show that these classes can be considered as representative of diverse regimes of dispersion. In Section 9.6, we discuss the results in terms of dispersal properties and, finally, in Section 9.7 we summarize the chapter content.

9.2 Lagrangian velocity spectra, velocity correlation function and diffusivity

In this section we shall introduce the statistical functions that are commonly employed when analyzing Lagrangian data, considering, for simplicity, the 1D space. The generalization to the 2D case is straightforward assuming the homogeneity of the velocity field. Given a velocity time series $u(t)$, its power spectrum (9.1), the velocity correlation function (9.2) and the structure function (9.5) are mathematically equivalent, but complementary descriptors. The power spectrum is defined as

$$P(\nu) = \lim_{T\to\infty} \frac{1}{T} A_T^*(\nu) A_T(\nu), \tag{9.1}$$

in which ν is the frequency, the asterisk indicates complex conjugation, $A(\nu) = \int_0^T u'(t) e^{-i2\nu\pi t} dt$ and $u'(t) = u(t) - \langle u(t) \rangle$, where the angle brackets means time average. The integral $\int_{-\infty}^{\infty} P(\nu) d\nu$ is equal to the eddy kinetic energy σ_U^2.

The correlation function $R(t)$ is defined as:

$$R(t) = \lim_{T\to\infty} \frac{1}{\sigma_U^2 \cdot T} \int_0^T u'(\tau + t) \cdot u'(\tau) d\tau. \tag{9.2}$$

The velocity autocorrelation function and the power spectrum are related by:

$$P(\nu) = 2\sigma_U^2 \cdot \int_0^\infty R(t) cos(2\pi\nu t) dt. \tag{9.3}$$

The Lagrangian integral time T_L is a measure of the time correlation of the velocity time series and is defined as:

$$T_L = \int_0^\infty R(t) dt. \tag{9.4}$$

Quantitative information about float dispersal, that is usually extracted from experimental Lagrangian data, is the particles displacement variance, or structure function,

$$S(t) = \langle |x(\tau + t) - x(\tau)|^2 \rangle. \tag{9.5}$$

The structure function is related to the velocity correlation function $R(t)$ and to the energy Lagrangian frequency spectrum $P(\nu)$ by the relations:

$$S(t) = 2\sigma_U^2 \cdot \int_0^t (t - \tau) \cdot R(\tau) d\tau, \tag{9.6}$$

$$S(t) = \frac{1}{\pi^2} \cdot \int_0^\infty \frac{P(\nu)}{\nu^2} \sin^2(\pi \nu t) \mathrm{d}\nu. \tag{9.7}$$

Taylor's (1921) result in form (9.6) and (9.7) is due to Kampé de Fériet (1939).

For small times $\sin^2(\pi\nu t) \sim (\pi\nu t)^2$ and equation (9.7) becomes (*ballistic regime*)

$$S(t) \sim t^2 \cdot \int_{-\infty}^{\infty} P(\nu)\mathrm{d}\nu = \sigma_U^2 \cdot t^2, \quad \textit{for} \ \ t \ll T_\mathrm{L}, \tag{9.8}$$

while for large times it can be shown (e.g. Krauss and Böning, 1987) that the other asymptotic behavior holds:

$$S(t) = P(0) \cdot t = 2\sigma_U^2 T_\mathrm{L} \cdot t, \quad \textit{for} \ \ t \gg T_\mathrm{L}. \tag{9.9}$$

For small times all frequencies equally contribute to particle displacements, while for large times the lower frequencies dominate and, if the velocity spectrum saturates at small frequencies, the Lagrangian diffusivity $K(t) = \frac{1}{2}\frac{\mathrm{d}}{\mathrm{d}t} S(t)$ asymptotes to the constant value (*random walk* or *Brownian regime*):

$$K(t) = \sigma_U^2 \cdot T_\mathrm{L}, \quad \textit{for} \ t \gg T_\mathrm{L}. \tag{9.10}$$

In a real trajectory, the variances of the velocity and acceleration time series are finite and consequently the following constraints on the spectral shape hold:

$$\int_{-\infty}^{\infty} P(\nu)\mathrm{d}\nu = \sigma_U^2 < \infty, \int_{-\infty}^{\infty} \nu^2 \cdot P(\nu)\mathrm{d}\nu = \textit{acceleration} \quad \textit{variance} = \sigma_A^2 < \infty. \tag{9.11}$$

This means that, possibly in a frequency range not observed by the experimental instrument, the logarithmic spectral slope of a spectrum $P(\nu) \propto \nu^{-\alpha}$ satisfies

$$\alpha > 3 \quad \textit{for} \ \ \nu \gg (T_\mathrm{L})^{-1}, \tag{9.12}$$

$$\alpha < 1 \quad \textit{for} \ \ \nu \ll (T_\mathrm{L})^{-1}. \tag{9.13}$$

The high frequencies asymptotic behavior (9.12), is physically due to the inertia of the Lagrangian particle that cannot be subjected to infinite accelerations, while the asymptotic behavior (9.13) establishes a limit in the growth of energy for small frequencies (large times), even in the unrealistic case of an unbounded domain.

Real oceanic spectra seldom exhibits a power-like behavior over a broad frequency (wave number) range; moreover the existence of a unique power law behavior in the entire frequency range is prevented by relations (9.11). Experimental spectra from Lagrangian floats (Colin de Verdière, 1983; Krauss and Böning, 1987; Rupolo *et al.*, 1996; Lumpkin and Flament, 2000) and from long time-series of altimetric TOPEX/POSEIDON data in extra tropical ocean (Stammer, 1997) are characterized by a low-frequency (wave number) plateau and by a decay in the mesoscale range with bi-logarithmic spectral slope $2 < \alpha < 3$. In regions characterized by strong mesoscale activity $\alpha > 3$ is often observed. It is easy to show (Rupolo, 1996) that for an idealized two-piece velocity spectrum, with a constant plateau for low frequencies, followed by a unique logarithmic spectral decay with slope α, the dispersal properties are dominated by low frequencies even for the shortest (and a fortiori for all) time lag for $\alpha > 2$.

In this chapter, we will mainly consider correlation and spectral properties, since in these frameworks it is easier to notice dissimilarities which indicate significant differences in the dynamics of the flow.

9.3 Variability of time and space scales: two different regimes of dispersion

Assuming that the particle speed remains correlated over a time T_{L}, the Lagrangian correlation length can be estimated as $L_{\mathrm{L}} = \sigma_U \cdot T_{\mathrm{L}}$. The evolution speed of the eddy field is $c_* = L_{\mathrm{E}}/T_{\mathrm{E}}$ (Lumpkin *et al.*, 2002), where T_{E} and L_{E} are the Eulerian counterparts of T_{L} and L_{L}. A qualitative view of the source of the variability of the Lagrangian motion is obtained considering the two limiting cases in which the Eulerian field evolves slowly ($\sigma_U > c_*$, strong spatial structure or *frozen turbulence* regime) and rapidly ($\sigma_U < c_*$, weak spatial structure or *fixed-float* regime) in the advective time scale.

In the frozen turbulence regime, the Lagrangian decorrelation is determined by the space variability of the Eulerian field: particles explore regions that are spatially decorrelated before the Eulerian velocity field has changed. In this case we have that

$$L_{\mathrm{L}} \sim L_{\mathrm{E}} \ll \sigma_U T_{\mathrm{E}} \tag{9.14}$$

or, equivalently,

$$T_{\mathrm{L}} \sim \frac{L_{\mathrm{L}}}{\sigma_U} \ll T_{\mathrm{E}}. \tag{9.15}$$

In the fixed-float regime, the float samples mesoscale fluctuations like a current meter. Before particles start exploring regions that are spatially decorrelated the time variability of the Eulerian field has already affected the particle motion. In this opposite limiting case, the Lagrangian decorrelation is exclusively determined by the time variability of the Eulerian velocity field and consequently we have

$$T_{\mathrm{L}} \sim T_{\mathrm{E}}, \tag{9.16}$$

or, equivalently,

$$L_{\mathrm{L}} \sim \sigma_U \cdot T_{\mathrm{E}} \ll L_{\mathrm{E}}. \tag{9.17}$$

In the fixed float regime the statistical properties of the Lagrangian and Eulerian time series are similar, while in the frozen turbulence regime Lagrangian and Eulerian correlation properties are intrinsically dissimilar and the frequency Lagrangian energy spectrum tends to the wave number Eulerian energy spectrum (Middleton, 1985). Inequalities (9.14)–(9.15) and (9.16)–(9.17) have to be interpreted in an order of magnitude sense, indicating that Lagrangian scales are expected to be (not much) smaller then the Eulerian counterpart.

The predominance of one of the two limiting cases is very unlikely in the real ocean. However, it is interesting to note their implications. If the oceanic velocity field were uniformly in a fixed-float, or frozen-field, regime it could be possible to use the more accessible Eulerian scales (e.g remote sensing data) as a proxy for Lagrangian scales. Moreover, if the Eulerian field selects the time and space scales independently of the kinetic energy we have in the frozen-field regime

$$L_{\mathrm{L}} \sim L_{\mathrm{E}} \sim const, \; K \sim \sigma_U^2 \cdot T_{\mathrm{L}} = \sigma_U \cdot L_{\mathrm{L}} \propto \sigma_U, \tag{9.18}$$

and in the fixed-float regime

$$T_{\mathrm{L}} \sim T_{\mathrm{E}} \sim const, \; K \sim \sigma_U^2 \cdot T_{\mathrm{L}} \propto \sigma_U^2. \tag{9.19}$$

Middleton (1985), studying several experimental datasets, has shown that the case $T_{\mathrm{L}}/T_{\mathrm{E}} < 1$ is the more generic one, while, in a recent study of experimental and numerical data in the North Atlantic, Lumpkin *et al.* (2002) have shown that neither a fixed-float or frozen-field regime characterizes the full set of observations. Nevertheless, results shown in the same paper as well as several other regional studies (Rossby *et al.*, 1983; Krauss and Böning, 1987; Figueroa and Olson, 1989; Brink *et al.*, 1991) indicate that for energetic mesoscale flows the regime $L_{\mathrm{L}} \sim const$, $K \propto \sigma_U$ seems to be the more appropriate. Finally, Hua *et al.* (1998), by means of numerical experiments, found that the geostrophic turbulence seems to be more consistent with the frozen turbulence regime.

9.4 Hierarchy of Markovian LSM

Brownian motion models the displacement of a particle which is subject to unknown forces acting on infinitesimal space and time scales. Contrastingly, Lagrangian motion in the sea is affected by a multitude of phenomena whose characteristic space and time scales are often comparable to the geographical extension sampled by the float and the observational period. This overlapping between scales has motivated the introduction of a hierarchy of LSM that takes into account the finite space and time size of the effect of the random variables on the Lagrangian dispersion.

Lagrangian velocity often is also characterized by oscillations of the autocorrelation function due to the spiraling of the fluid particles trapped in coherent structures. The ubiquitous presence of eddies in the ocean has then induced the introduction of stochastic models with "spin" (Borgas *et al.*, 1997; Reynolds, 2002; Veneziani *et al.*, 2004) in which the component of the horizontal velocity vector has a preferred sense of rotation.

In this section we will briefly review the main characteristics of the hierarchy of Markovian 1D LSM that are usually considered in the study of the dispersion in the ocean (for an exhaustive review see Pope, 1994; Griffa, 1996; and Berloff and McWilliams, 2002b). Furthermore, we will derive from the second-order LSM some analytical relations that will be used in the analysis of floats and drifters data (Section 9.5).

The order of a LSM, refers to the number of time scales that are incorporated in the model. In the following, the stochastic impulse is represented by a random increment $dw(t)$ from a normal distribution with zero mean and second-order moment $\langle dw \cdot dw \rangle = 2 \cdot dt$.

Markov 0 The 0 order LSM model is the random walk in which

$$\frac{dx}{dt} = U + \sqrt{2K} \cdot \frac{dw}{dt}, \tag{9.20}$$

where U is the mean velocity and K is the diffusion coefficient of the corresponding "advection–diffusion" equation for the probability density function P

$$\frac{\partial}{\partial t} P = -U \frac{\partial}{\partial x} P + \frac{\partial}{\partial t} (K \frac{\partial}{\partial t} P). \tag{9.21}$$

The typical time scale of the turbulent motion is infinitesimal, the particles variance grows linearly with time for all times, and the velocity field is characterized by a white noise spectrum (infinite energy).

Markov 1 In this model the Markov variable is the velocity and the incremental equations are:

$$\frac{\mathrm{d}x}{\mathrm{d}t} = U + u \tag{9.22}$$

$$\frac{\mathrm{d}u}{\mathrm{d}t} = -\frac{u}{T_v} + \sqrt{\frac{2\sigma_U^2}{T_v}} \cdot \frac{\mathrm{d}w}{\mathrm{d}t}. \tag{9.23}$$

The velocity obeys the classical Langevin equation and the particle conserves memory of its initial velocity for a time T_v. The velocity autocorrelation function decays exponentially ($R(t) = \mathrm{e}^{-\frac{t}{T_v}}$) with an e-folding time $T_v = T_\mathrm{L}$ (e.g. Griffa, 1996).

The associated Lorentzian spectrum of the velocity

$$P(\nu) \propto \frac{\frac{1}{T_v}}{\frac{1}{T_v^2} + \nu^2}, \tag{9.24}$$

is characterized by a plateau for frequencies lower than $1/T_v$ and it decreases as ν^{-2} for higher frequencies. In this model the variance of the acceleration is infinite.

Markov 2 In this model the random discontinuity is introduced in the acceleration field and the incremental equations are:

$$\frac{\mathrm{d}x}{\mathrm{d}t} = U + u \tag{9.25}$$

$$\frac{\mathrm{d}u}{\mathrm{d}t} = a - \frac{u}{T_v} \tag{9.26}$$

$$\frac{\mathrm{d}a}{\mathrm{d}t} = -\frac{a}{T_a} + \sqrt{2\sigma_U^2 \cdot \alpha_1\alpha_2} \cdot \frac{\mathrm{d}w}{\mathrm{d}t}. \tag{9.27}$$

These three equations are equivalent (Sawford, 1991) to the second-order stochastic process

$$\frac{\mathrm{d}^2u}{\mathrm{d}t^2} + \alpha_1\frac{\mathrm{d}u}{\mathrm{d}t} + \alpha_2 u = \sqrt{2\sigma_U^2 \cdot \alpha_1\alpha_2} \cdot \frac{\mathrm{d}w}{\mathrm{d}t} \tag{9.28}$$

with $\alpha_1 = (1/T_v + 1/T_a)$ and $\alpha_2 = 1/T_aT_v$.

When $T_a > 0$ the variances of both velocity and acceleration are finite; for $T_a \to 0$ relation (9.28) becomes a first-order equation and the model collapses to a first-order LSM.

This model has been introduced by Sawford (1991) to correct the small-scale behavior of the first-order LSM in the description of flows at high, but finite,

Reynolds number. In this context T_v is the time scale of the energy containing eddies and T_a is determined by the dissipation. In the mesoscale oceanic turbulence range the dissipation does not play a direct role and the acceleration time scale is simply another characteristic time scale of the flow representative of the correlation of the forcing (Griffa, 1996). The same model was subsequently developed to simulate the acceleration of a fluid particle in anisotropic and inhomogeneous flows by Pope (2002).

The correlation function of the velocity field obtained integrating (9.25)–(9.27), is given by

$$R(t) = \frac{\beta_2 e^{\beta_1|t|} - \beta_1 e^{\beta_2|t|}}{\beta_2 - \beta_1}, \tag{9.29}$$

where

$$\beta_{1,2} = \frac{-\alpha_1 \pm \sqrt{\alpha_1^2 - 4\alpha_2}}{2} \tag{9.30}$$

are roots of $z^2 + \alpha_1 z + \alpha_2 = 0$, and are equal to $\beta_1 = -1/T_v$, $\beta_2 = -1/T_a$.

It is easy to show (Berloff and McWilliams, 2002b), that the following relation holds

$$\frac{1}{T_v T_a} = \frac{\sigma_A^2}{\sigma_U^2}, \tag{9.31}$$

where σ_U^2 and σ_A^2, represent the velocity and acceleration variances.

The Lagrangian correlation time T_L, obtained integrating (9.29), is equal to

$$T_L = T_v \cdot \left(1 + \frac{T_a}{T_v}\right). \tag{9.32}$$

Depending on the sign of the radicand in (9.30) we have the two cases:

$$T_v, T_a \in R \quad for \quad \alpha_1^2 \geq 4\alpha_2$$

$$T_v, T_a = a \pm ib \in C \quad for \quad \alpha_1^2 < 4\alpha_2.$$

The appearance of a complex conjugates pair of time scales is symptomatic of a missing but dynamically important spatial dimension. Reynolds (2002) has shown the correspondence between multi-dimensional $(n-1)$th-order LSM and one-dimensional nth-order LSM. In the parameter range in which $T_v, T_a \in C$, the Markov-2 model is then equivalent to a two-dimensional model Markov-1 with "spin," where the non isotropy of the velocity field is expressed by a "handedness" about the vertical axis to simulate spiraling particles motion (Borgas *et al.*, 1997).

For $t \sim 0$, $R(t)$ defined by (9.29) is characterized by a quadratic behavior as is expected in flow with finite accelerations; for $t >> T_a$ the correlation function asymptotes to the exponential form characteristics of the first-order LSM.

The corresponding velocity spectrum $P(\nu)$ can be expressed as

$$P(\nu) = \frac{2\alpha_1\alpha_2\sigma_U^2}{\left[\alpha_2 - 4\pi^2\nu^2\right]^2 + 4\pi^2\alpha_1^2\nu^2}, \tag{9.33}$$

and has a relative maximum, for $\nu_0 = \frac{1}{8\pi^2}(2\alpha_2 - \alpha_1^2)$, only if $\alpha_1^2 < 2\alpha_2$, i.e. when T_v, $T_a \in C$ and when the imaginary part exceeds the real one ($b > a$). It is interesting to note that for $2\alpha_2 < \alpha_1^2 < 4\alpha_2$ (i.e. when T_v, $T_a \in C$ with $b < a$) the correlation function (9.29) oscillates but the velocity spectrum (9.33) has no peak.

Relation (9.33) can be written in terms of the two time scales T_v and T_a in the following way:

$$P(\nu) = \frac{2\sigma_U^2 \cdot (T_v + T_a)}{\left(1 - 4\pi^2\nu^2 T_a T_v\right)^2 + (T_a + T_v)^2 4\pi^2\nu^2}. \tag{9.34}$$

This spectrum is characterized by a plateau for low frequencies and it decreases in the intermediate frequency range with a variable logarithmic slope ranging from −2 to −4. Rescaling both $P(\nu)$ and ν with the Lagrangian correlation time $T_L = T_v + T_a$

$$\hat{P}(\hat{\nu}) = \frac{P(\hat{\nu})}{T_L}, \quad \hat{\nu} = \nu \cdot T_L, \tag{9.35}$$

we can express the "bi-normalized" power spectrum $\hat{P}(\hat{\nu})$ as :

$$\hat{P}(\hat{\nu}) = \frac{2}{\left(1 - \frac{4\pi^2\hat{\nu}^2 \cdot y}{(1+y)^2}\right)^2 + 4\pi^2\hat{\nu}^2} \quad \textit{for} \quad T_a, T_v \in R \quad \textit{with} \quad y = \frac{T_a}{T_v} \tag{9.36}$$

and

$$\hat{P}(\hat{\nu}) = \frac{2}{\left[1 - \pi^2\hat{\nu}^2 \cdot (1 + x^2)\right]^2 + 4\pi^2\hat{\nu}^2} \quad \textit{for} \quad T_a, T_v \in C \quad \textit{with} \quad x = \frac{b}{a}. \tag{9.37}$$

Expression (9.36) collapses to (9.37) for $\alpha_1^2 \to 4\alpha_2$ (i.e. $T_a \to T_v$, $y \to 1^-$).

In panel d of Fig. 9.7 are plotted analytical spectra (9.36) and (9.37) for different values of y and x. The motivation for considering bi-normalized

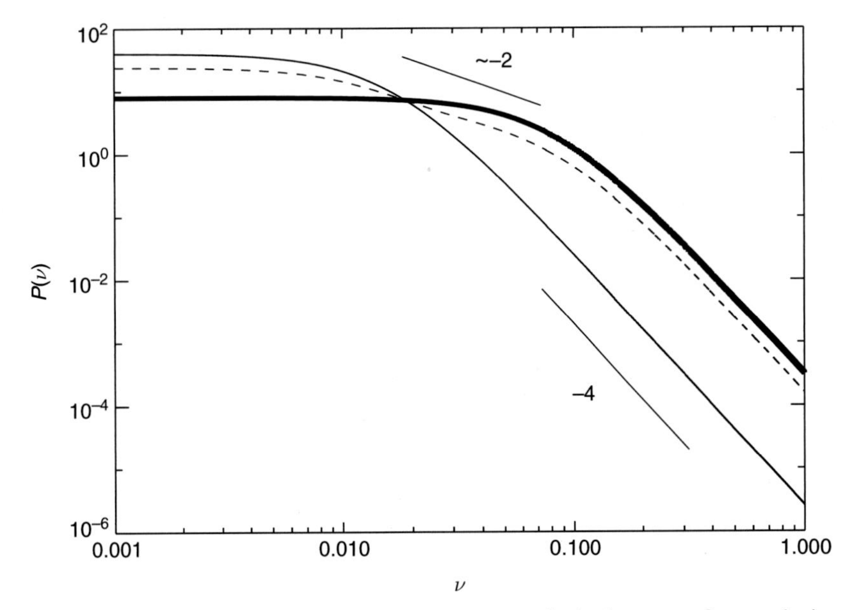

Figure 9.1 The continuous lines represent two analytical spectra from relation (9.33) with $T_v = T_a = 2$ days and $T_v = T_a = 10$ days (bold line). The average between these two spectra (dashed line) shows a "spurious" spectral decay with a bi-logarithmic slope ~ 2. The frequency ν is expressed in $(\text{days})^{-1}$.

spectra is given by the fact that the average of dimensional spectra characterized by the same ratio $y = T_a/T_v$ but with different values of the two time scales, hides the functional form of the individual spectra. This is shown in Fig. 9.1 where a spurious ~ -2 behavior in the intermediate frequency range results from the average of two curves in which this kind of spectral decay is not present. This is obviously true also for the velocity correlation and structure functions.

This simple observation suggests that, using the second-order LSM as a paradigm for the oceanic dispersion, the ratio between acceleration and velocity time scales $y = T_a/T_v$ can be utilized to separate the Lagrangian trajectories in homogeneous categories.

In the next section we will use this approach in the analysis of experimental Lagrangian data.

9.5 Data analysis

All the data analyzed were downloaded from the WOCE (2002) archive. In particular, we consider all the surface drifters data in the Atlantic and Pacific Oceans and all the subsurface SOFAR and RAFOS floats moving at a nominal depth between 400 and 1000 m in the Atlantic Ocean.

Drifters and floats positions were mostly recorded every 6 and 24 hours. When a different recording time step was used, velocities were interpolated to have a 6-hour and daily observation for the drifters and floats dataset, respectively. Gaps in the time series smaller than four consecutive points were linearly interpolated. When larger gaps were found the original trajectory was divided in more pieces.

To fully resolve the mesoscale frequency range, and to minimize the effects of the non-homogeneity of the velocity field, we perform spectral analysis on pieces of trajectories of 128 days. We have 960, 2397 and 6770 such pieces for the Atlantic Floats (AF), Atlantic Drifters (AD) and Pacific Drifters (PD) datasets, for a total of 1 296 256 daily observations (see Tables 9.1, 9.2 and 9.3 and panels b, d and f of Fig. 9.2 for a map of daily observations per 1° square bin).

In the following we will show several maps of parameters computed from the trajectories since they highlight the marked inhomogeneous character of the oceanic mesoscale. All these maps were obtained by averaging, on a grid of $1^\circ \times 1^\circ$, between all the points falling in a given bin, and assigning to each point the value computed for the trajectory. This technique allows us to map quantities computed along the entire trajectory (as correlation parameters). On the other hand, the non-locality of the considered values can affect the results in case of trajectories spanning a very large geographical area.[1]

For each trajectory we compute the Lagrangian eddy kinetic energy $\sigma_U^2 = \frac{1}{2}\left(\sigma_{u_1}^2 + \sigma_{u_2}^2\right)$, where $\sigma_{u_i}^2 = \frac{1}{N-1}\sum_{j=1}^{N}[u_i(j) - \overline{u_i}]^2$, $\overline{u_i} = \frac{1}{N}\sum_{j=1}^{N} u_i(j)$, and N is the number of points of the trajectory. The resulting maps (panels a, c and e of Fig. 9.2) show that maxima of Lagrangian kinetic energy are associated to the main currents: the equatorial bands in both basins, the Gulf Stream, the Agulhas currents, the Brazil–Malvinas confluence in the Atlantic Ocean, and the Kuroshio and the East Australia currents together with the area West of Central America in the Pacific, coherently with previously published results (e.g, Stammer, 1997; Lumpkin *et al.*, 2002).

In Figs. 9.3 and 9.4 we show the total, zonal and meridional velocity spectra and velocity correlation functions for the three datasets. These quantities are obtained by averaging on a mixture of diverse regimes distributed geographically and thus no dynamical significance should be attached to them. Nevertheless some considerations can be made:

[1] Usually, geographical maps with Lagrangian data are constructed considering "pseudo-Eulerian" averages obtained using only values of data in the given bin.

Table 9.1 *List of SOFAR and RAFOS experiments. Principal Investigator, Institution, geographical area interested and number of 128 days long trajectories for each experiment.*

Experiment	P.I.	Institution	Area	N°	First year	Float type
PreLDE	Rossby	URI	North West North Atl.	42	1975	S
LDE	Rossby	URI	North West North Atl.	14	1978	S
Gulf Stream	Richardson, Price, Schmitz, Owens, Rossby	WHOI URI	North West North Atl.	2	1979	S
Gulf Stream Recirculation	Richardson, Price, Schmitz, Owens, Rossby	WHOI URI	North West North Atl.	39	1980	S
Site L	Owens, Price	WHOI	North West North Atl.	49	1982	S
TOPOGULF	Ollitraut	IFREMER	East North Atl.	79	1983	S
Eastern Basin	Price, Richardson	WHOI	East North Atl.	6	1984	S
Newfoundland Basin	Owens	WHOI	North West North Atl.	16	1986	S
Iberian Basin	Zenk, Boebel	IFM	East North Atl.	14	1990	R
Brazil Basin	Boebel, Zenk	IFM	West South Atl.	128	1992	R
North Atl. Meddies	Richardson	WHOI	East North Atl.	11	1993	R
Nort Atl. Current	Rossby, Prater	URI	North West North Atl.	65	1993	R
Meddy Seeding	Bower	WHOI	East North Atl.	7	1993	R
W. Bound. Curr.	Bower	WHOI	North West North Atl.	32	1994	R
EUROfloat	Gould, Zenk, Williams, Speer, Cantos	Various	East North Atl.	29	1995	R
ACCE	Bower	WHOI	North East North Atl.	242	1996	R
ACCE	Rossby, Prater, Carr	URI	North East North Atl.	65	1997	R
CANIGO	Ambra, Kantos, Knoll	Various	East North Atl.	33	1997	R
KAPEX	Boebel, Rossby	UCT URI	South East South Atl.	87	1997	R

Table 9.2 *Number of original drifter trajectories in the Atlantic and Pacific Oceans.*

	79	80	81	82	83	84	85	86	87	88	89
Atl.	–	–	–	–	–	–	–	–	–	–	20
Pac.	40	72	23	32	35	29	31	71	153	285	324

	90	91	92	93	94	95	96	97	98	99	00
Atl.	54	62	115	225	226	260	317	358	384	473	445
Pac.	363	426	575	630	589	621	732	513	468	629	655

Table 9.3 *Number of 128 days long trajectories for the three datasets*

	No of trajectories	No of days
AF	960	122880
AD	2397	306816
PD	6770	866560

(i) the subsurface floats are less energetic than the surface drifters. The latter ones show a well-defined anisotropy, probably due to the β-induced zonal amplification (Lacasce, 2000);
(ii) the Lagrangian meridional velocity spectra are similar to the generic shape evidenced by Stammer (1997) analyzing Topex/Poseidon altimetric data and are characterized by a low frequency plateau followed by a spectral decay;
(iii) even if subsurface floats data were mostly recorded in areas where the motion is characterized by a massive presence of coherent structures, like the Gulf Stream and Agulhas currents or the Mediterranean outflow area, no striking differences (except the anisotropy) are observed in the shape of the spectra and velocity correlation functions between the floats and drifters datasets, nor in the spectral slope decay or in the oscillatory behavior of the correlation function.

In this section, we will not discuss the anisotropy of the spectral and transport properties; we shall consider total energy spectra and correlation functions (average of the meridional and zonal components) and we will focus on a way to select homogeneous sub-sets of trajectories, following the suggestions given by the simple relations developed in Section 9.4 and computing, for each trajectory, the ratio $y = T_a/T_v$.

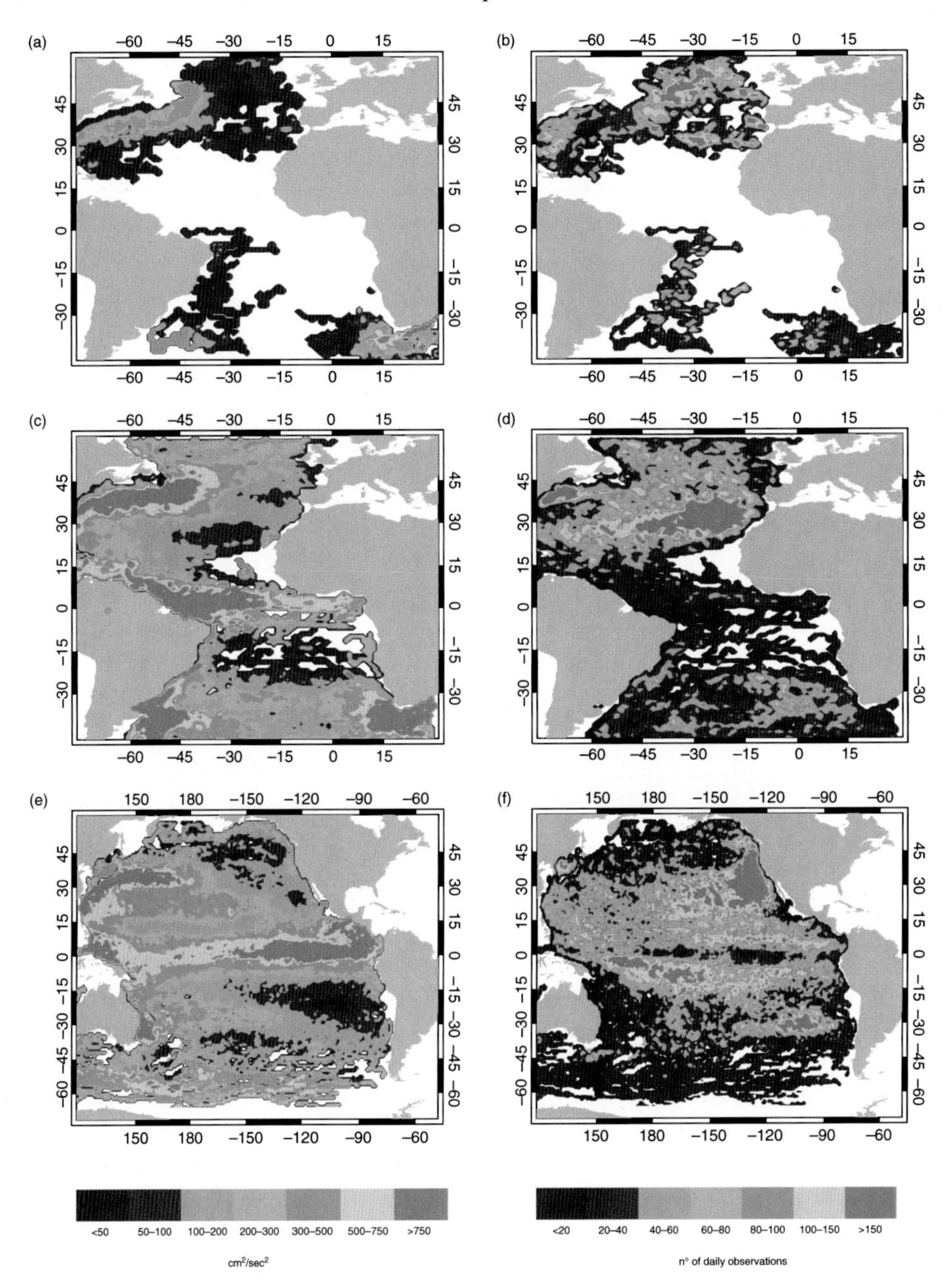

Figure 9.2 Maps of the Lagrangian eddy kinetic energy obtained from the AF (a), AD (c) and PD (e) datasets and number of daily observations per 1° square bin of the same datasets (b, d and f). The maps of the Lagrangian eddy kinetic energy (expressed in cm^2/sec^2) were obtained averaging over $1° \times 1°$ bins and assigning to each point the eddy kinetic energy of the trajectory. See Plate 19 for color version.

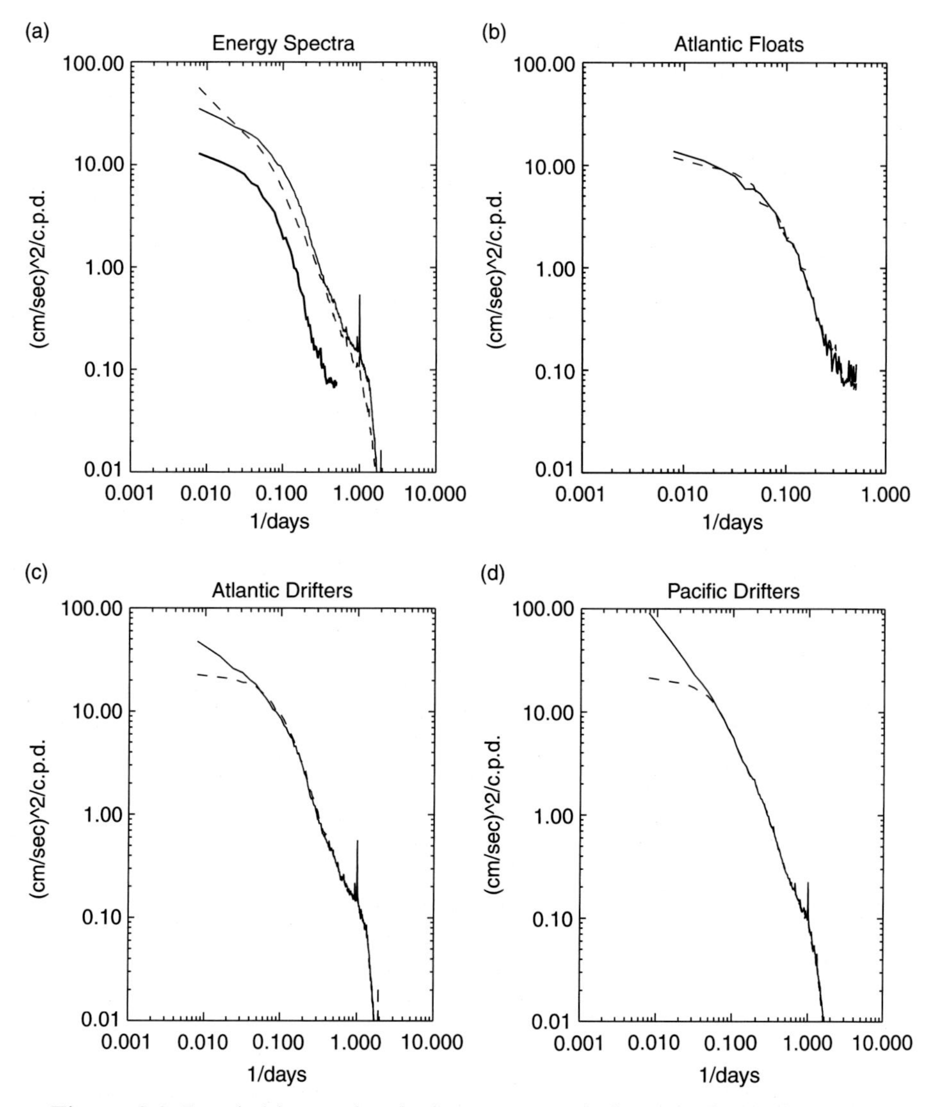

Figure 9.3 Panel (a): total velocity spectra of the AF (bold line), AD (continuous line) and PD (dashed line) datasets. Panels (b), (c) and (d): zonal (continuous line) and meridional (dashed line) velocity spectra for the AF, AD and PD datasets, respectively.

The estimate of the Lagrangian correlation time scale is obtained from Equation (9.4) integrating in the first positive lobe the zonal and meridional components of the velocity correlation (9.2) and then considering T_L as the semi sum of the two values. This "first zero crossing technique," even if questionable, is commonly used (for a discussion of the several techniques employed in the literature see Lumpkin *et al.*, 2002). The estimates of

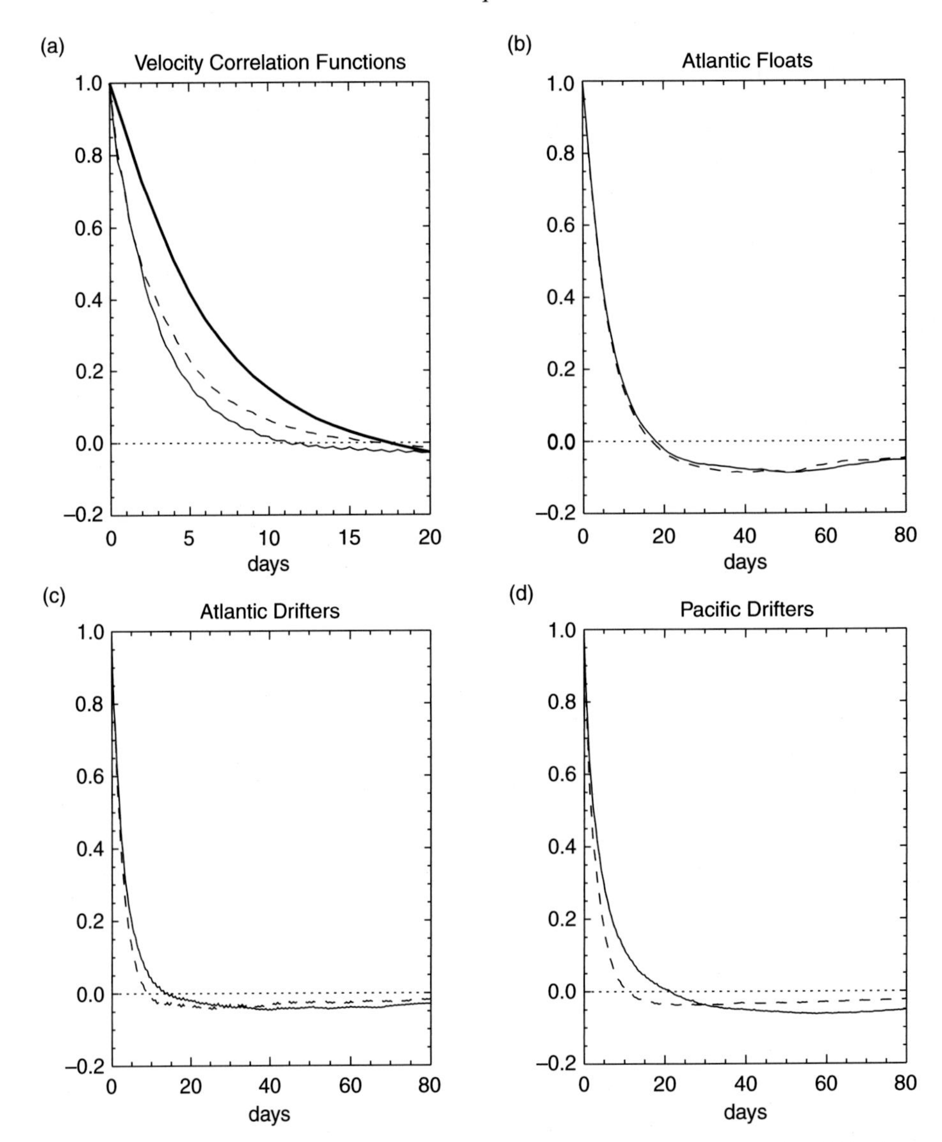

Figure 9.4 Panel (a): total velocity correlation function of the AF (bold line), AD (continuous line) and PD (dashed line) datasets. Panels (b), (c) and (d): zonal (continuous line) and meridional (dashed line) velocity correlation function of the AF, AD and PD datasets, respectively. The zero value is indicated by dotted lines.

T_v and T_a are obtained, for each trajectory, using relations (9.31) and (9.32), which give

$$T_v = \frac{T_{\mathrm{L}} + \sqrt{T_{\mathrm{L}}^2 - 4\frac{\sigma_U^2}{\sigma_A^2}}}{2}, \quad T_a = \frac{T_{\mathrm{L}} - \sqrt{T_{\mathrm{L}}^2 - 4\frac{\sigma_U^2}{\sigma_A^2}}}{2}, \tag{9.38}$$

where, given the accelerations time series $a_i(j)$, the variance $\sigma_A^2 = \frac{1}{2}\left(\sigma_{a_1}^2 + \sigma_{a_2}^2\right)$ is computed as follows

$$\sigma_{a_i}^2 = \frac{1}{N-2}\sum_{j=1}^{N-1}[a_i(j) - \overline{a_i}]^2, \quad \text{with} \quad \overline{a_i} = \frac{1}{N-1}\sum_{j=1}^{N} a_i(j). \tag{9.39}$$

When $T_{\mathrm{L}}^2 < 4\sigma_U^2/\sigma_A^2$ (that is equivalent to $\alpha_1^2 < 4\alpha_2$, in the notation of Section 9.2), we have that $T_v, T_a = a \pm ib \in C$, $|T_v| = |T_a|$ and $|y| = 1$.

In Fig. 9.5 we show, for the three datasets, the probability density functions with their cumulative for T_{L} (panels a and b) and T_a (panels c and d), while in Fig. 9.6 we map their geographical distribution. As expected (see also Table 9.4), the AF dataset shows, for both T_{L} and T_a, a distribution shifted toward higher values.

The geographical distribution shows a general tendency to have larger values of T_{L} in the less energetic interior of the oceans in opposition to the energetic currents (the Gulf Stream, the Agulhas currents in the Atlantic, and the regions East of Australia and West of Central America in the Pacific) that are characterized by having shorter decorrelation velocity time scales (the opposite is true for the acceleration time scale). However, this general tendency is not observed in the Kuroshio current (the same result was observed by Stammer, 1997), in the area east of Brazil and in the central Pacific equatorial band.

As a first exercise, we compute the mean experimental "bi-normalized" velocity spectra, obtained rescaling, for each trajectory, $P(\nu)$ and ν with the Lagrangian correlation time T_{L}, and then we look for the best fit with the analytical expressions (9.36) and (9.37). To achieve that, we choose the value of y (when $T_v, T_a \in R$, (9.36)) or x (when $T_v, T_a \in C$ (9.37)) that minimizes the distance

$$s = \frac{1}{N}\sum_{i=1}^{N}\left[P_{\mathrm{exp}}(\hat{\nu}_i) - P_{\mathrm{an}}(\hat{\nu}_i)\right]^2, \tag{9.40}$$

where P_{exp} and P_{an} designate experimental and analytical spectra, and the sum is performed till $\hat{\nu}_i = 1$. The results are shown in Fig. 9.7 and the best analytical

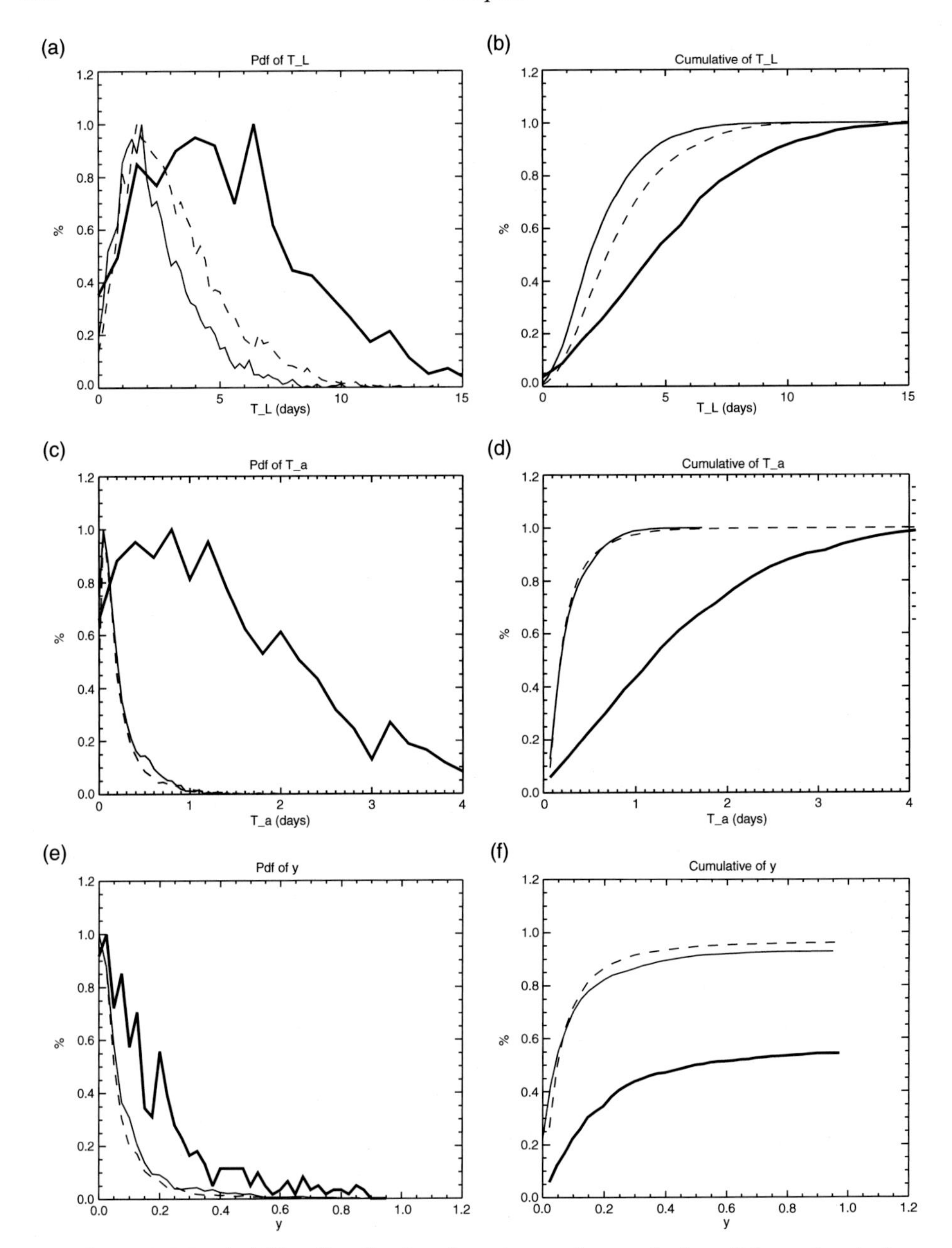

Figure 9.5 Probability distribution function and cumulative of the estimated values of T_L (panels a and b), T_a (panels c and d) and y (for $y < 1$, panels e and f). Bold, thick and dashed lines refer to the AF, AD and PD datasets, respectively. The Pdfs are normalized with their maximum value.

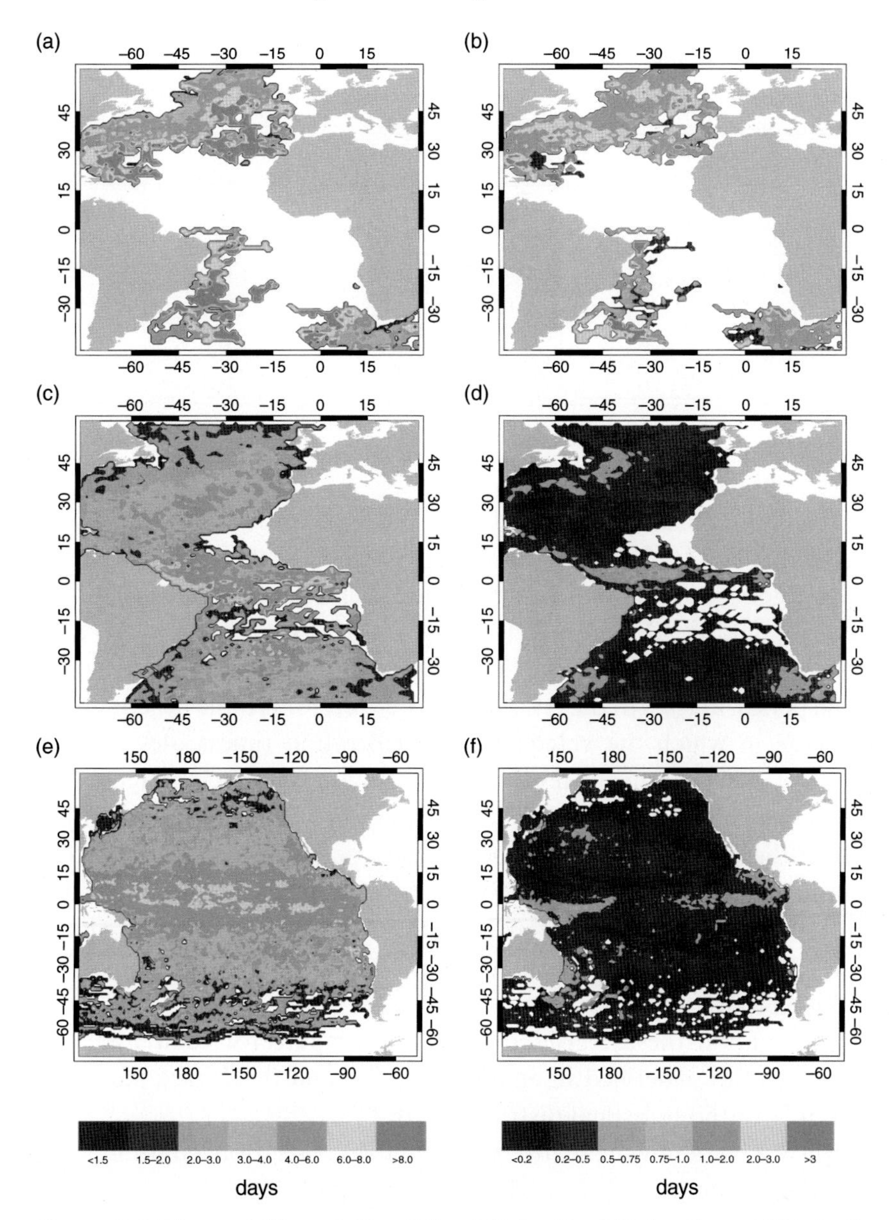

Figure 9.6 Maps of T_L and T_a of the AF (a, b), AD (c, d) and PD (e, f) datasets. The maps of the time scales (expressed in days) were obtained averaging over $1^\circ \times 1^\circ$ bins and assigning to each point the time scale computed on the entire trajectory. See Plate 20 for color version.

Table 9.4 *Mean value and standard deviation of* T_L, T_v, T_a *(expressed in days) and* y. *The number in parenthesis is the median of the relative pdf.*

	T_L	T_v	T_a	y
AF	6.5 ± 3.3 (4.80)	4.8 ± 3.2 (3.20)	1.6 ± 1.1 (1.20)	0.6 ± 0.4 (0.52)
AD	3.2 ± 1.6 (2.00)	3.0 ± 2.2 (2.20)	0.3 ± 0.2 (0.15)	0.2 ± 0.3 (0.05)
PD	4.0 ± 2.8 (2.80)	3.7 ± 2.8 (3.20)	0.3 ± 0.3 (1.20)	0.1 ± 0.2 (0.05)

approximation (dashed line) is obtained for $y = 0.1$ and $y = 0.2$ in the AD and PD datasets, and for $|y| = 1$, $x = 1.4$ for the AF dataset. The analytical fits are relatively good, at least in the intermediate frequency range. The result for the subsurface dataset AF confirms (see e.g Griffa, 1996 and Rupolo *et al.*, 1996) that the first-order LSM is not suitable for the representation of subsurface float dispersal, at least in the Atlantic. Moreover, for AF the best fit is obtained in a range in which the velocity and acceleration time scales are complex conjugates. This demonstrates that this technique clearly highlights the main characteristics of this dataset, i.e. the massive presence of looping trajectories in which the memory time and the rotation period are of the same order of magnitude (e.g., Richardson, 1993; Veneziani *et al.*, 2004). Contrastingly, the small value of $y = T_a/T_v \sim 0.1$–0.2 for the AD and PD datasets shows that surface drifters dispersal is reasonably well represented, especially for large times, with a first-order LSM.

Building upon these considerations we compute for each trajectory the indices y and/or x defined in equations (9.36) and (9.37) (see panels e and f of Fig. 9.5 for the statistical distribution of y).

In Fig. 9.8, we show the scatter plot of the decorrelation length $L_L = \sigma_U \cdot T_L$ versus the square root of the eddy kinetic energy σ_U (panels a, c and e for the three datasets). The black crosses indicate values of drifters and floats characterized by having $y < 0.2$, while the blue and red crosses are relative to the floats and drifters whose time scales lie in the complex conjugate space with, respectively, $x < 1$ and $x > 1$. For small y the decorrelation length is approximately linearly dependent on σ_U while, when the time scales become complex, the dependence on the eddy kinetic energy is definitely weaker and for large σ_U, L_L is nearly constant. Moreover, plotting the mean position of the trajectories belonging to these three sub datasets (panels b, d and f) it is possible to see that

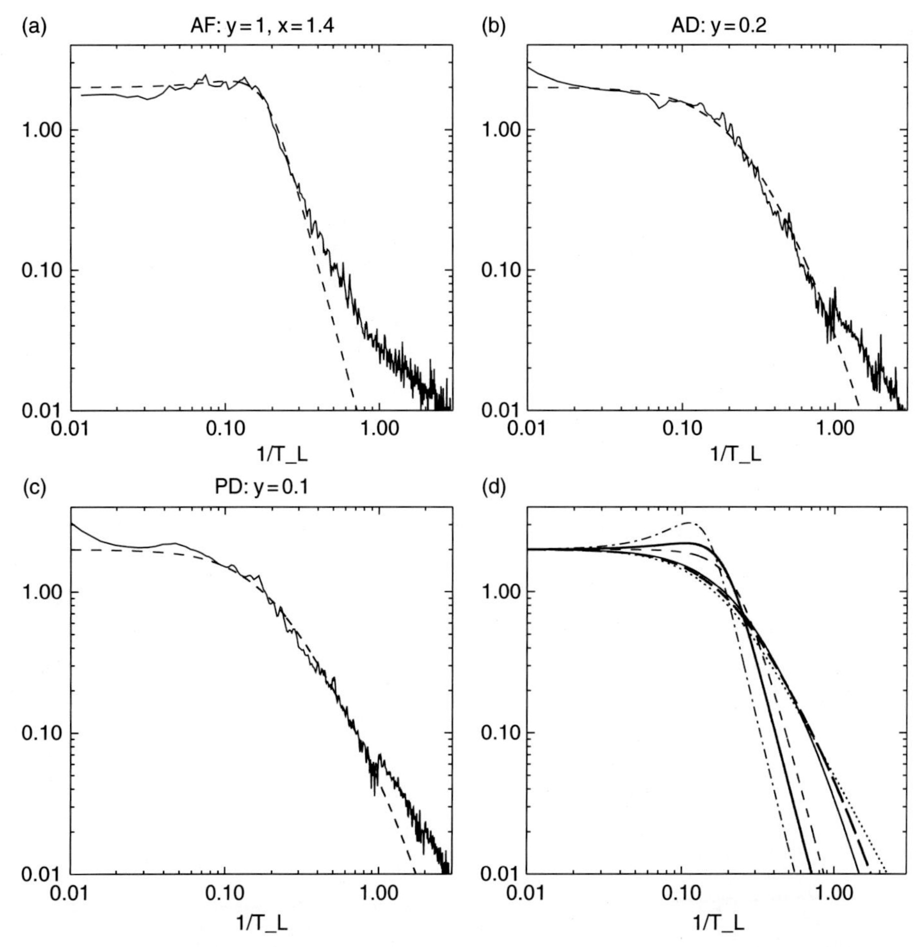

Figure 9.7 Experimental "bi-normalized" energy spectra (continuous line) of the AF, AD and PD datasets (panels a, b and c). The dashed lines represent the "best fit" from the second LSM analytical spectrum ((9.36) and (9.37)). Axes are dimensionless. In panel d are reported the 3 best fits for AF (bold line), AD (thick line) and PD (thick dashed line). The dotted, dashed and dashed-dotted lines represent the analytical expressions (9.36) and (9.37) with $y = 0$, $|y| = 1$, $x = 1$ and $|y| = 1$, $x = 2$, respectively.

the trajectories that have complex conjugates time scales (blue and red points) are mainly located in the regions characterized by strong currents. Recalling the definitions of Section 9.2, we interpret these results affirming that the ratio $y = T_a/T_v$ can be considered as an useful index to discriminate between the "frozen turbulence" and "fixed-float" regime.

To get a direct feeling about the qualitative difference between trajectories characterized by diverse values of the ratio $y = T_a/T_v$, we plot in Fig. 9.9 typical

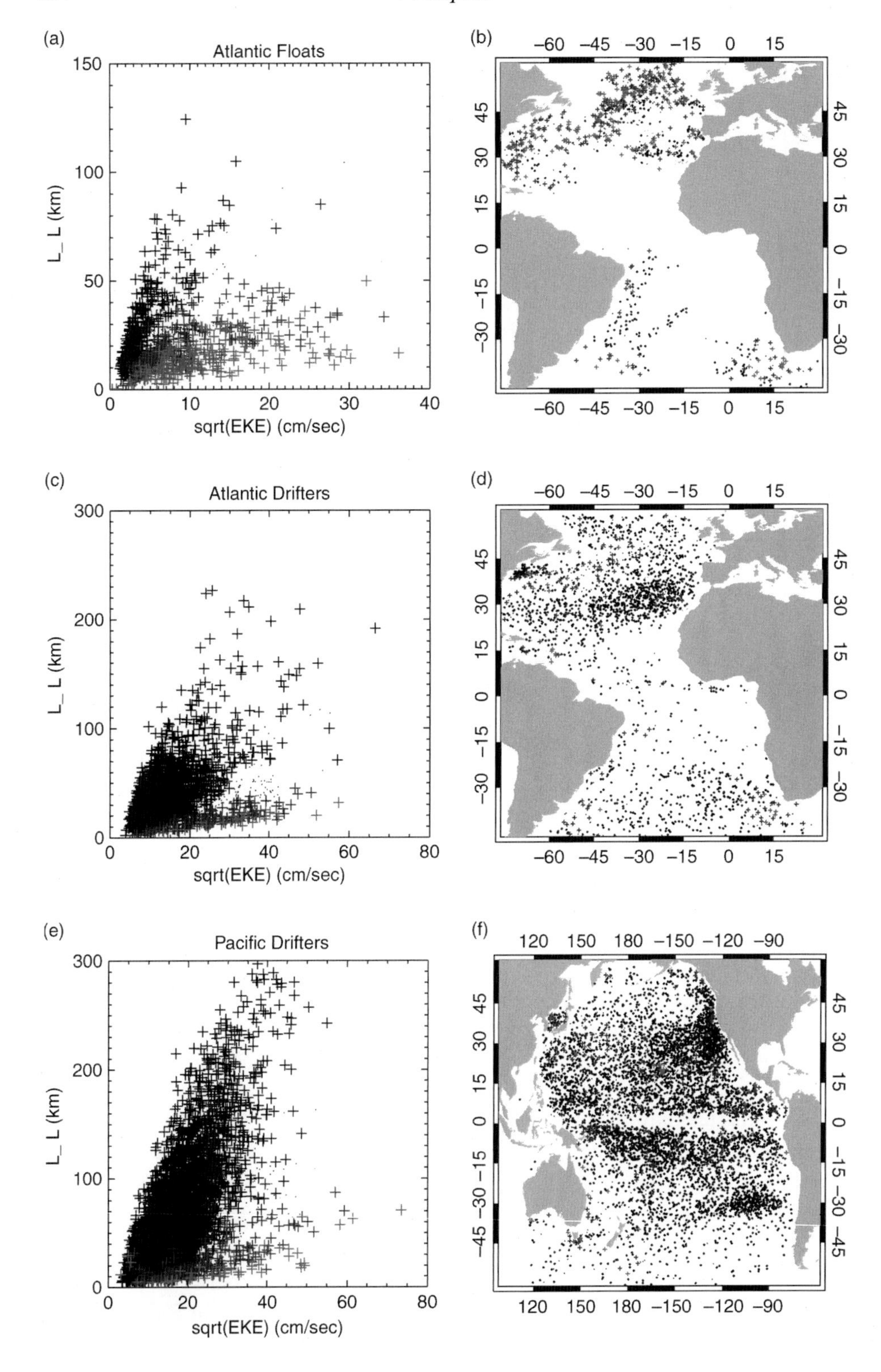
(a)
Atlantic Floats
L_ L (km)
sqrt(EKE) (cm/sec)
(b)
(c)
Atlantic Drifters
L_ L (km)
sqrt(EKE) (cm/sec)
(d)
(e)
Pacific Drifters
L_ L (km)
sqrt(EKE) (cm/sec)
(f)

trajectories of the four classes defined by $y < 0.2$ (Class I) , $0.4 < y < 0.8$ (Class II), $|y| = 1$ $x < 1$ (Class III) and $|y| = 1$ $x > 1$ (Class IV).

The trajectories belonging to the two extreme Classes I and IV are considerably dissimilar; in Class I are grouped floats and drifters characterized by large-scale meandering influenced by high frequency variability. In Class IV trajectories are clearly influenced by the presence of coherent structures and very often they rapidly whirl, trapped inside an eddy with a well-defined length scale (looping behavior). This is partly true also for the trajectories characterized by $|y| = 1$ and $x > 1$ (Class III), even if in this case the length scale of the loops is not well defined and often the floats/drifters seem to jump between eddies of different size. Finally, floats/drifters of the Class II ($0.4 < y < 0.8$) have trajectories showing intermediate characteristics and, meandering around large-scale structures, they did not show the influence of high frequency motions.

In the presence of coherent structures, the trajectory of a particle can display two distinct behaviors. Particles trapped in the coherent eddies quickly rotate, while the rest of the particles meander in the less energetic turbulent background and experience only sporadic looping behavior.

Richardson (1993) has introduced a simple criterion of defining a "looper" as a trajectory that undergoes at least two consecutive loops in the same direction. Veneziani *et al.* (2004), analyzing subsurface floats in the North West Atlantic, found that in this area the percentage of "looping" floats varies from 15% to 40%. They also propose a quantitative criterion, based on the mean angular velocity of the trajectory, to define "looping" trajectories.

To look at all the trajectories analyzed in this work (10 127!) is a demanding task;[2] we have the feeling that the method proposed here is capable of identifying looping trajectories, but in the following we will call "turbulent" the trajectories belonging to Classes III and IV.

It is crucial to note the different relative importance (Table 9.5) of these classes in the three datasets. While in AF almost half of the trajectories (46%)

Figure 9.8 Plot of L_L vs. σ_U (panels a, c and e) and plot of the mean position of the trajectories (panels b, d and f) of AF, AD and PD datasets. The black crosses indicate floats and drifters with $y < 0.2$. The blue and red crosses are related to floats and drifters with $|y| = 1, x < 1$ and $|y| = 1, x > 1$, respectively. See Plate 21 for color version.

[2] An inventory (postscript file) of all the trajectories analyzed here divided per classes is available at clima.casaccia.enea.it/staff/rupolo/trajectorybook/index.html

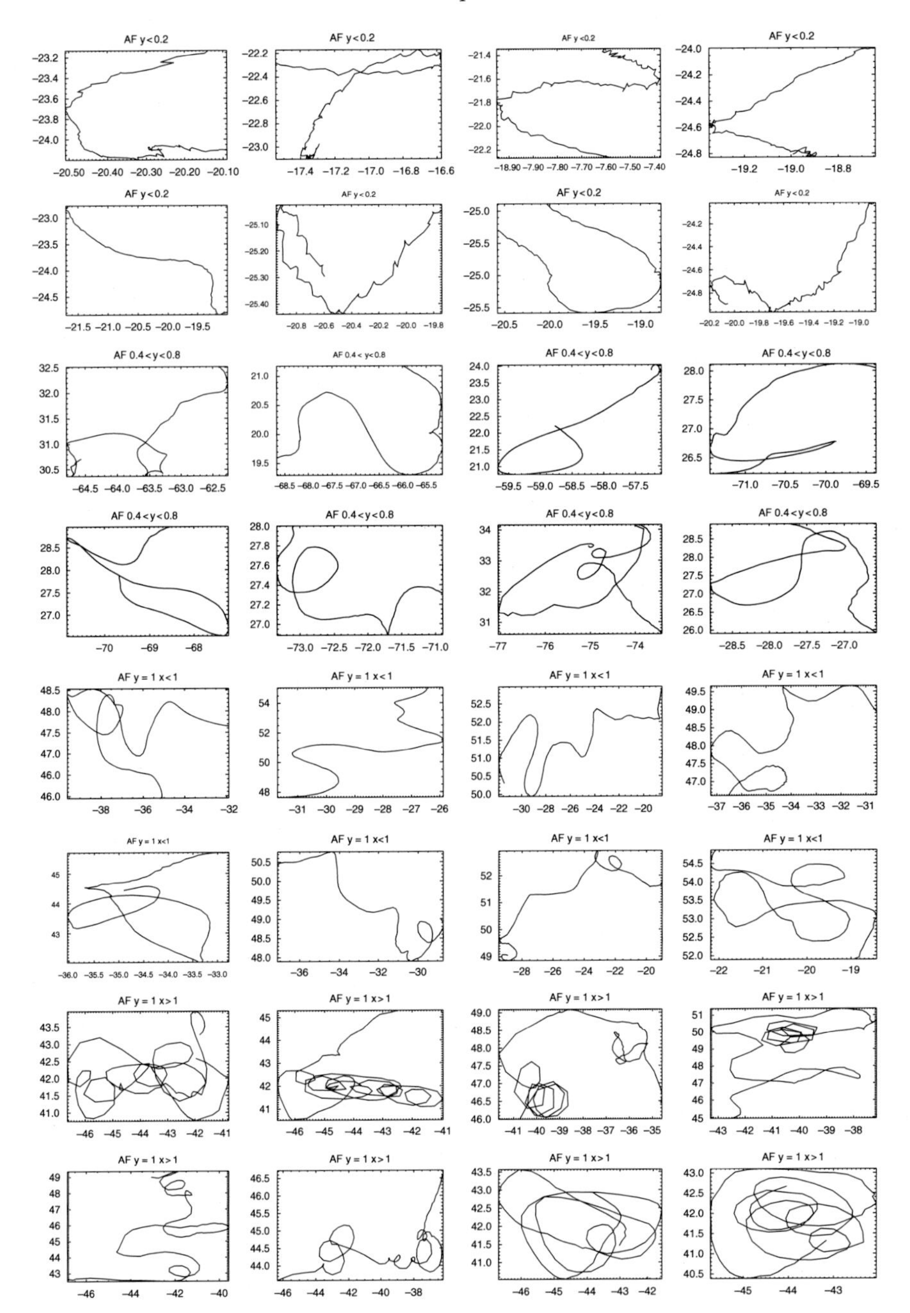

Figure 9.9a AF dataset. Typical trajectories belonging to, (from top to bottom) Class I ($y < 0.2$, the two upper rows), Class II ($0.4 < y < 0.8$, rows 3 and 4), Class III ($|y| = 1, x < 1$, rows 5 and 6) and Class IV ($|y = 1, x > 1$, rows 7 and 8).

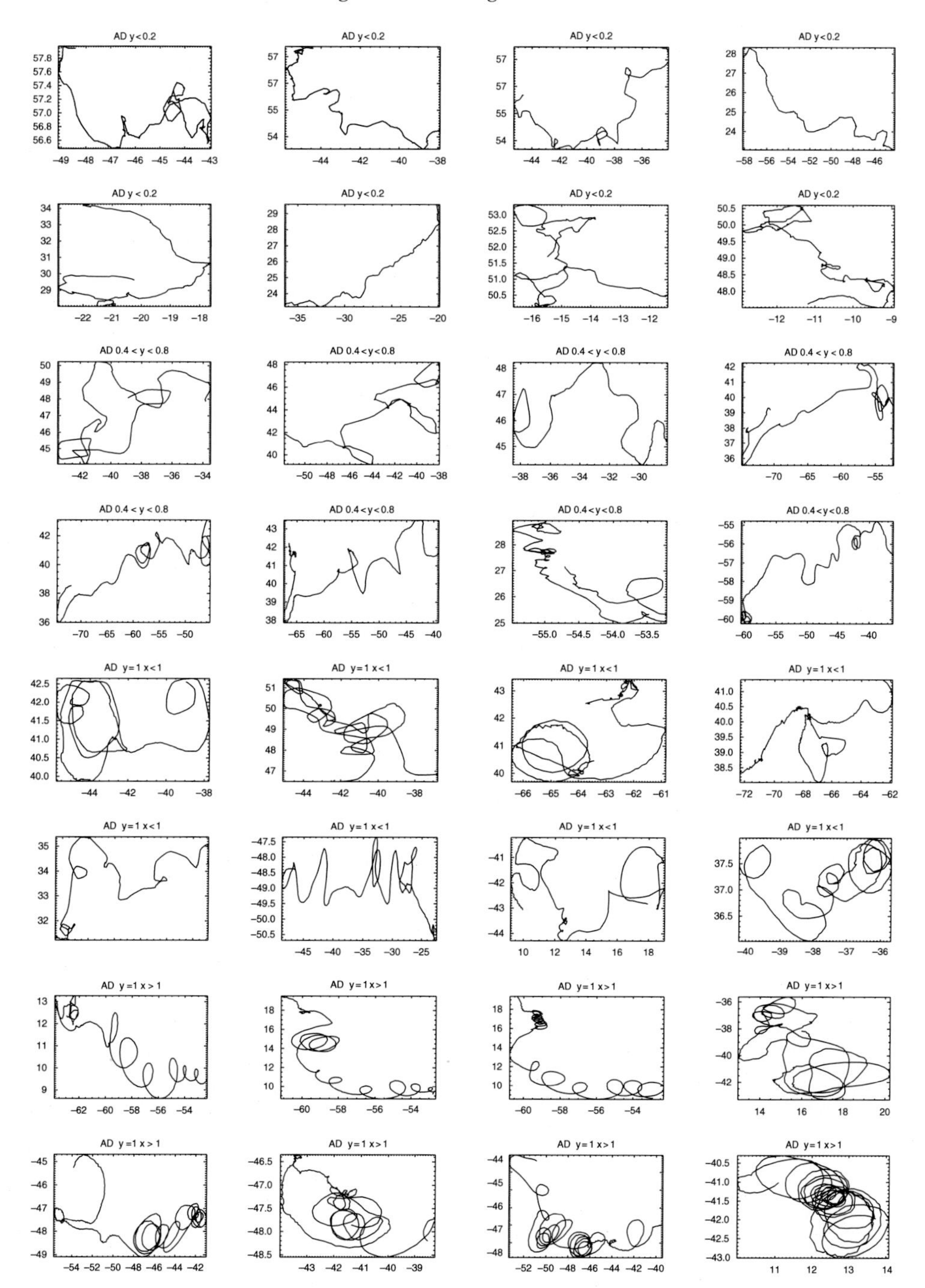

Figure 9.9b Same as Fig. 9a for dataset AD.

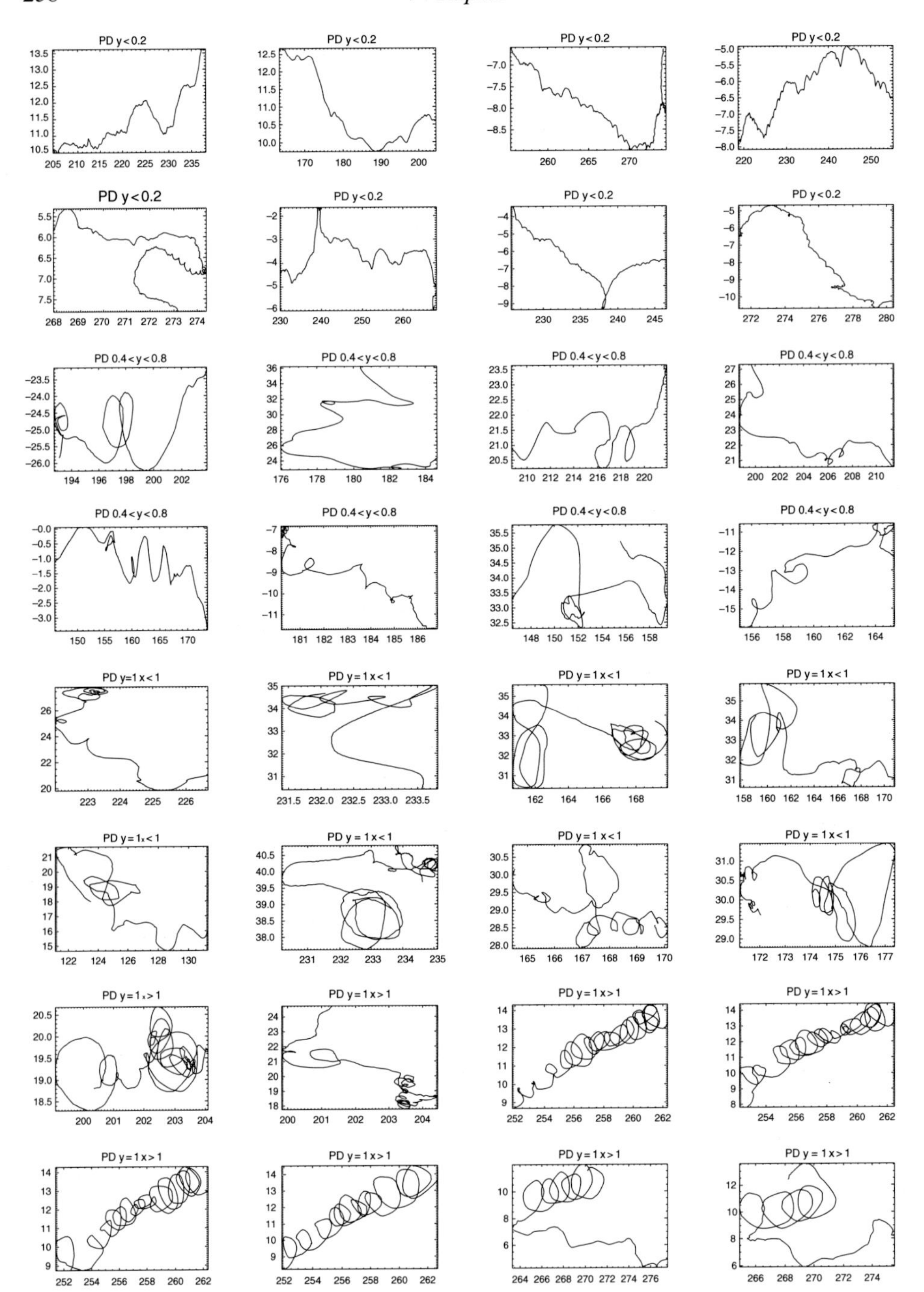

Figure 9.9c Same as Fig. 9a for dataset PD.

Table 9.5 *Mean value and standard deviation of* y *(Classes I and II) and* x *(Classes III and IV). The lower values indicate the number of trajectories and the relative percentage of each class.*

	Class I $y < 0.2$	Class II $0.4 < y < 0.8$	Class III $\|y\| = 1, x < 1$	Class IV $\|y\| = 1, x > 1$
AF	(y) 0.09 ± 0.05 327, **34%**	(y) 0.6 ± 0.1 63, **7%**	(x) 0.6 ± 0.2 246, **26%**	(x) 1.4 ± 0.3 194, **20%**
AD	(y) 0.06 ± 0.05 1922, **80%**	(y) 0.5 ± 0.1 89, **4%**	(x) 0.6 ± 0.3 147, **6%**	(x) 1.1 ± 0.1 26, **1%**
PD	(y) 0.06 ± 0.02 5838, **86%**	(y) 0.5 ± 0.1 165, **2%**	(x) 0.5 ± 0.2 219, **3%**	(x) 1.2 ± 0.2 49, **1%**

Table 9.6 *Mean value and standard deviation of the correlation time* T_L *(expressed in days) of the different classes of trajecories for the three datasets.*

	Class I $y < 0.2$	Class II $0.4 < y < 0.8$	Class III $\|y\| = 1, x < 1$	Class IV $\|y\| = 1, x > 1$
AF	9.2 ± 3.0	6.0 ± 2.4	5.4 ± 2.0	3.0 ± 1.3
AD	3.6 ± 1.6	1.9 ± 0.6	1.5 ± 0.4	1.1 ± 0.2
PD	4.3 ± 2.0	2.2 ± 0.9	2.1 ± 1.4	1.8 ± 2.4

belong to the "turbulent" Classes III and IV, in AD and PD the great majority of drifters (80% and 86%, respectively) is influenced by high frequency motions (Class I). Some minor difference is observed between PD and AD, where there is a slightly major occurrence of trajectories of Classes III and IV (4% and 7%, respectively). Finally, coherent with previous observations (e.g. Krauss and Böning, 1987) and results from 2D turbulence studies (e.g. Babiano *et al.*, 1987; Pasquero *et al.*, 2001), "turbulent" trajectories are characterized by smaller correlation times and greater energetic contents (see Tables 9.6 and 9.7).

In Fig. 9.10, we show the dimensional (panels a, c and e) and bi-normalized (b, d and f) energy spectra for the four classes previously defined. Dimensional AF spectra of the floats belonging to the Classes III and IV do not show any peak. On the contrary, the presence of a peak in the spectra of Class IV of AD and PD, shows that these (few) drifters loop with a period of about 5–10 days.

Table 9.7 *Mean value and standard deviation of the eddy kinetic energy* σ_U *(expressed in* cm^2/sec^2*) of the different classes of trajectories for the three datasets.*

	Class I $y < 0.2$	Class II $0.4 < y < 0.8$	Class III $\lvert y \rvert = 1, x < 1$	Class IV $\lvert y \rvert = 1, x > 1$
AF	28 ± 44	61 ± 56	132 ± 178	190 ± 215
AD	256 ± 324	804 ± 565	720 ± 532	1172 ± 698
PD	273 ± 282	440 ± 324	648 ± 596	1127 ± 1085

In the three datasets, the bi-normalized version of the spectra of each class has a very similar shape that highlights, in terms of spectral slope and presence of peak, the characteristics of the distinct regimes. This is evident even looking at the correlation functions of each class plotted in Fig. 9.11; when the time axis is normalized by T_L (panels b, d, and f) they display a very similar shape. In particular, it is worth noting the clear oscillations in the Class IV and the smooth behavior for small times for the Classes II, III and IV, as expected for trajectories characterized by a finite variance of acceleration. Nevertheless, some minor differences can be observed, as the presence of a negative lobe for the velocity correlation functions of Class I of AF. These, apparently small, differences may have major influence when considering the structure function, that is related to the velocity correlation function by the Taylor relation (9.6). Finally, we may notice that there are some discrepancies between the expected theoretical behavior and the computed normalized velocity correlation functions, as for example the weak oscillatory behavior for functions belonging to Class III or the value of their integral that, for definition, should be equal to 1. For instance, the integral of the "normalized" correlation functions is about 0.6 for all the classes belonging to the PD dataset. This is not surprising, since all the analysis is based upon the estimate of the correlation time by means of the "first zero crossing" technique.

On the other hand, we think that the coherence of the results presented here confirms the validity of the approach based upon the screening of the trajectories by means of a dimensionless index, derived from the simple second-order LSM. The extreme classes defined by this index are characterized by different shapes of the trajectories and by different spectral and correlation properties. In particular, the dependence of the decorrelation length on the eddy kinetic energy is very dissimilar, suggesting that the ratio $y = T_a/T_v$ is an useful guide to discriminate between "frozen turbulence" and "fixed-float" regimes.

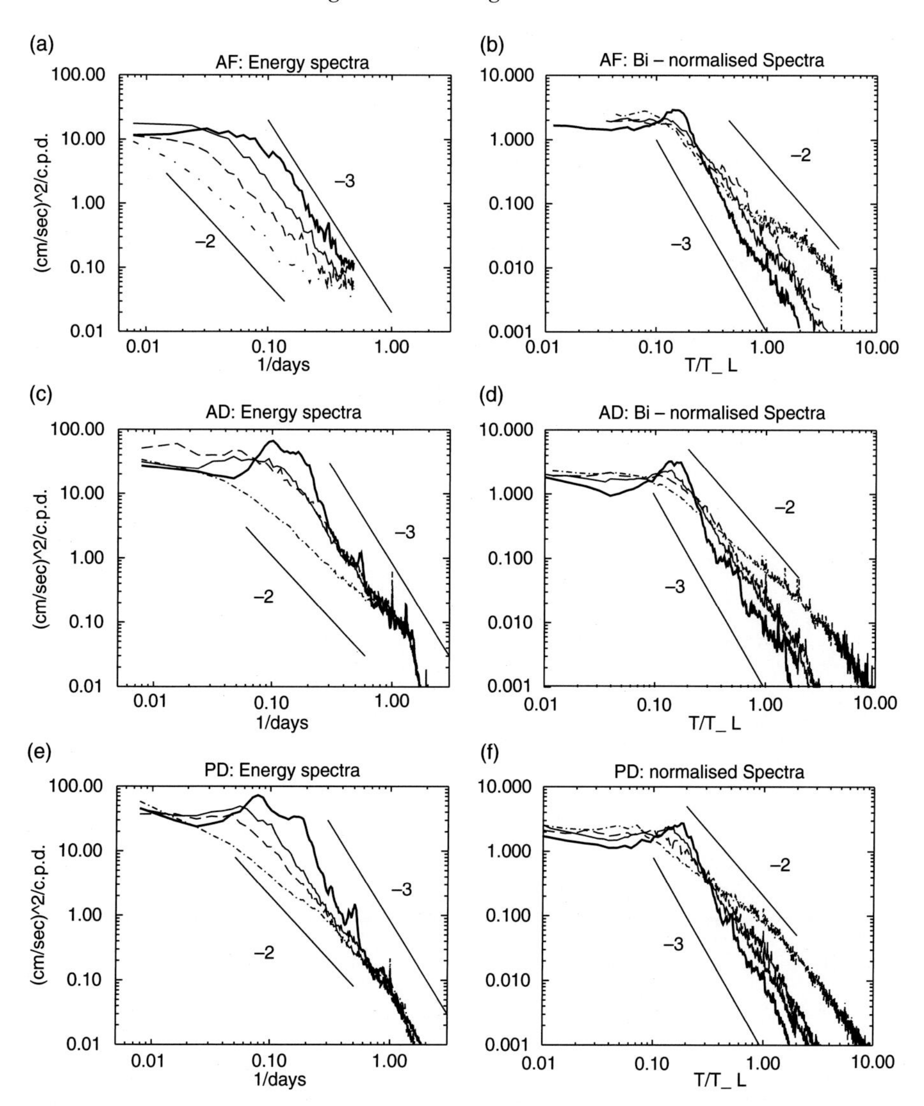

Figure 9.10 Dimensional (left column) and bi-normalized (right column) velocity spectra of AF (upper row), AD (central row) and PD (lower row) datasets. The dashed-dotted and the dashed lines represent mean spectra computed considering floats/drifters belonging to Classes I and II ($y < 0.2$, and $0.4 < y < 0.8$). The thin and thick continuous lines represent spectra computed considering floats/drifters belonging to Classes III and IV ($|y| = 1$, $x < 1$ and $|y| = 1$, $x > 1$, respectively).

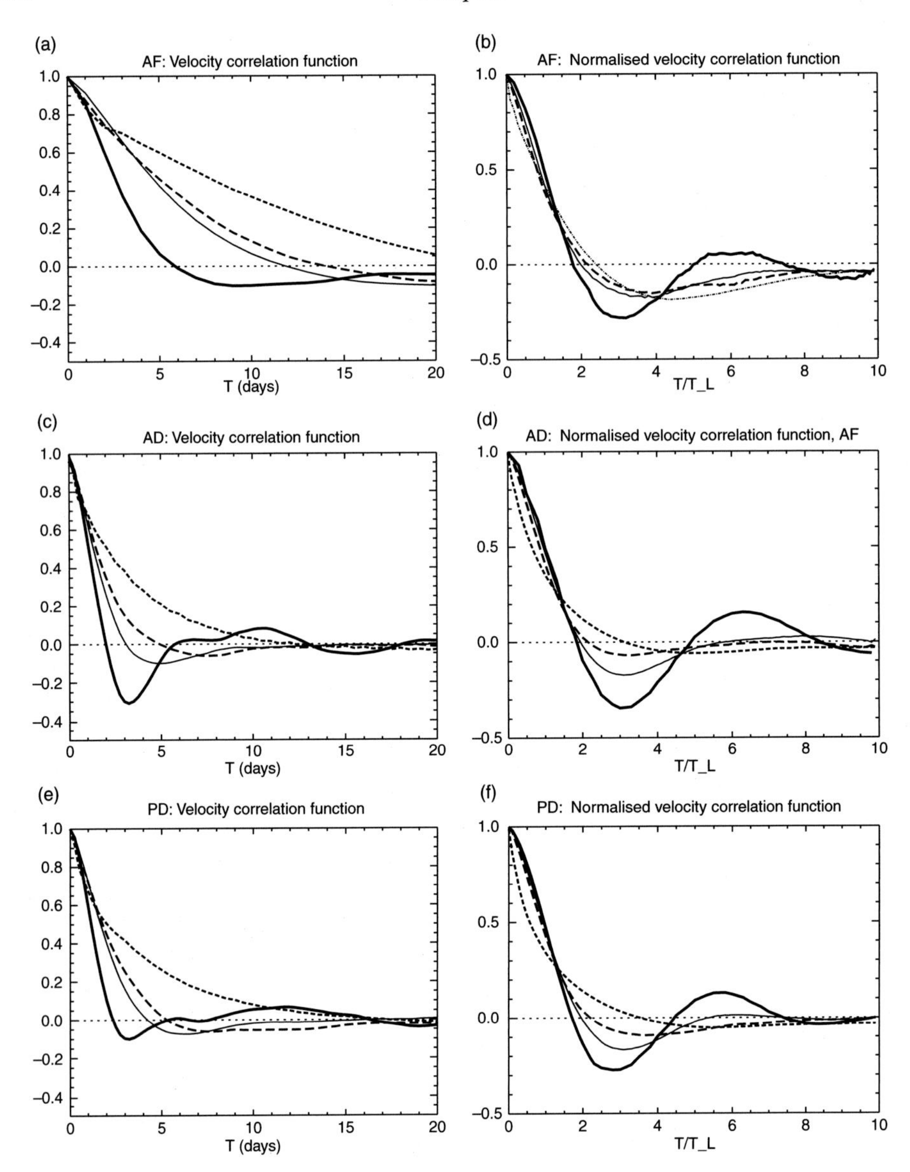

Figure 9.11 Velocity correlation functions vs. time (left column) and with the time axis normalized by T_L (right column) for AF (upper row), AD (central row) and PD (lower row) datasets. The dashed-dotted and the dashed lines represent mean correlation functions computed considering floats/drifters belonging to Classes I and II ($y < 0.2$, and $0.4 < y < 0.8$). The thin and thick continuous lines represent mean correlation functions computed considering floats/drifters belonging to Classes III and IV ($|y| = 1, x < 1$ and $|y| = 1, x > 1$ respectively). The zero value is indicated by dotted lines.

9.6 Turbulence regimes and dispersal properties

The division of trajectories in different classes allows us to estimate the role of the diverse regimes in the particles' dispersion, both in time and in geographical space.

We first consider, for the PD dataset, the dimensionless structure function, obtained integrating the Taylor relation (9.6) without the dimensional value σ_U^2, and rescaling time by T_L. In this way, we get a qualitative view of the influence of the coherent structures (thin and thick continuous lines in panel a of Fig. 9.12) that appears to be more important only for times smaller than 4–5 T_L. In the same figure, it is interesting to note the diverse behavior in the transition from the quadratic (ballistic motion) to the linear (random walk) regime. Obviously, to get information on the real dispersal properties, one has to look at the dimensional structure function. The population of the AD and PD datasets is largely dominated by trajectories belonging to Class I. Consequently, we find it more interesting to concentrate on the sub-surface floats in the Atlantic (panels b, c and d of Fig. 9.12). In this dataset, the population of "turbulent" trajectories is rather consistent (see Table 9.5) and the contribution of floats belonging to Classes III and IV appears to be more important (>50%) for all times (Fig. 9.12, panel d). The relevance of the trajectories characterized by a strong looping behavior (Class IV) rapidly decreases (thick continuous line of panel d) from 30% to 20% in the first 20 days, and the corresponding structure function converges to random walk more rapidly than for the other classes. This random walk regime is probably related to the motion of the eddy, in which the trajectory is embedded, in the turbulent background, as occurs in 2D turbulence (Bracco *et al.*, 2000; Pasquero *et al.*, 2001). Conversely, the contribution of trajectories belonging to Class I grows in the first 40 days (dashed-dotted line, panel d), while for larger times the relative structure function saturates, showing a sub-diffusive behavior.

This analysis was performed on Lagrangian data covering huge geographical areas; the role of the different regimes in different locations can be estimated by mapping the bin averaged values of the ratio $y = T_a/T_v$ (panels a, c and e of Fig. 9.13). In the same figure, we also show the maps of the probability of finding a trajectory belonging to the Classes III and IV.

The AF (panels a and b) dataset is the most homogeneous; the mean value of y is everywhere relatively large. The highest values (>0.8) and the highest probabilities (>80%) of finding "turbulent" trajectories are observed in the Gulf Stream and Agulhas currents, in the Brazil–Malvinas Current and, not uniformly, in the Mediterranean outflow area.

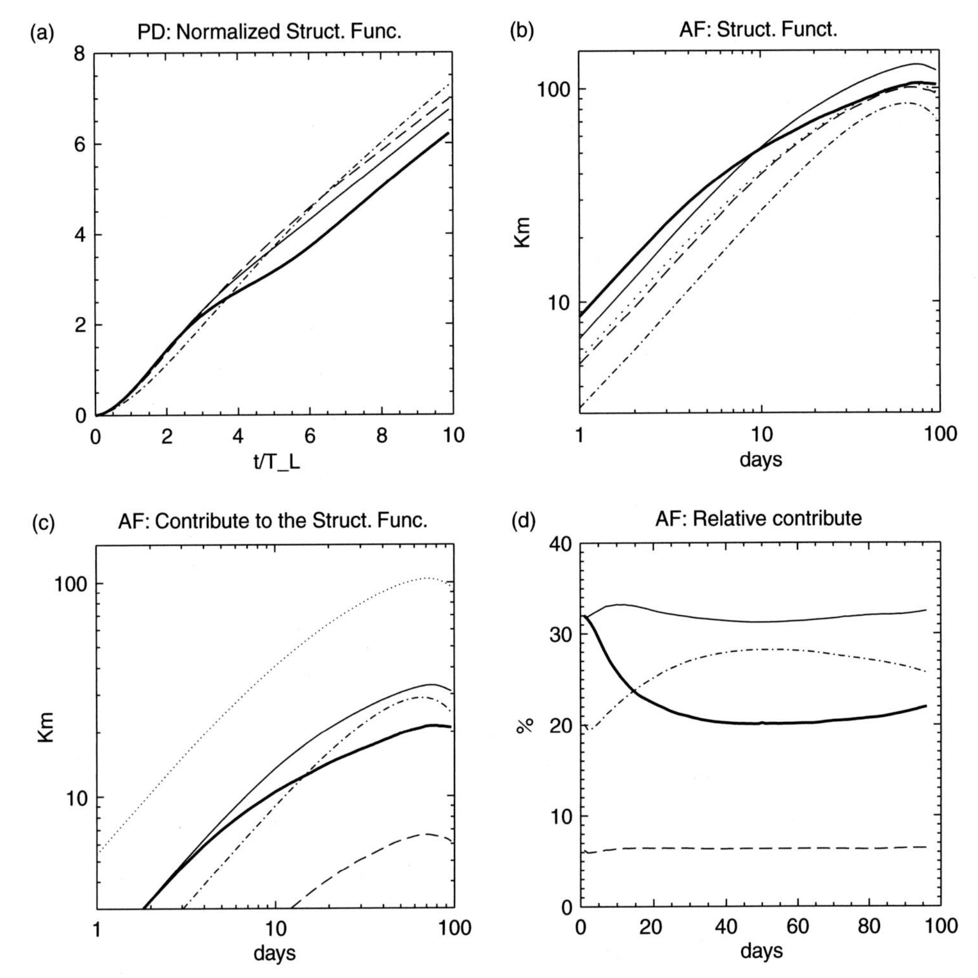

Figure 9.12 Panel (a): Normalized structure function for the four classes of trajectories for the PD dataset. Panel (b): Structure functions for the entire AF dataset (dashed line) and for the four classes of trajectories. Panel (c): Structure function for the entire AF dataset (dashed line) and contribution of the four classes of trajectories. The contribution of each class is computed multiplying the relative structure function by the percentage representative of the population. Panel (d) : Relative contribute to the structure function of the four classes of trajectories, expressed in percentage. Note that the four classes do not represent the entire dataset. In all panels dashed-dotted, dashed, thin continuous and thick lines represent Class I, II, III and IV, respectively.

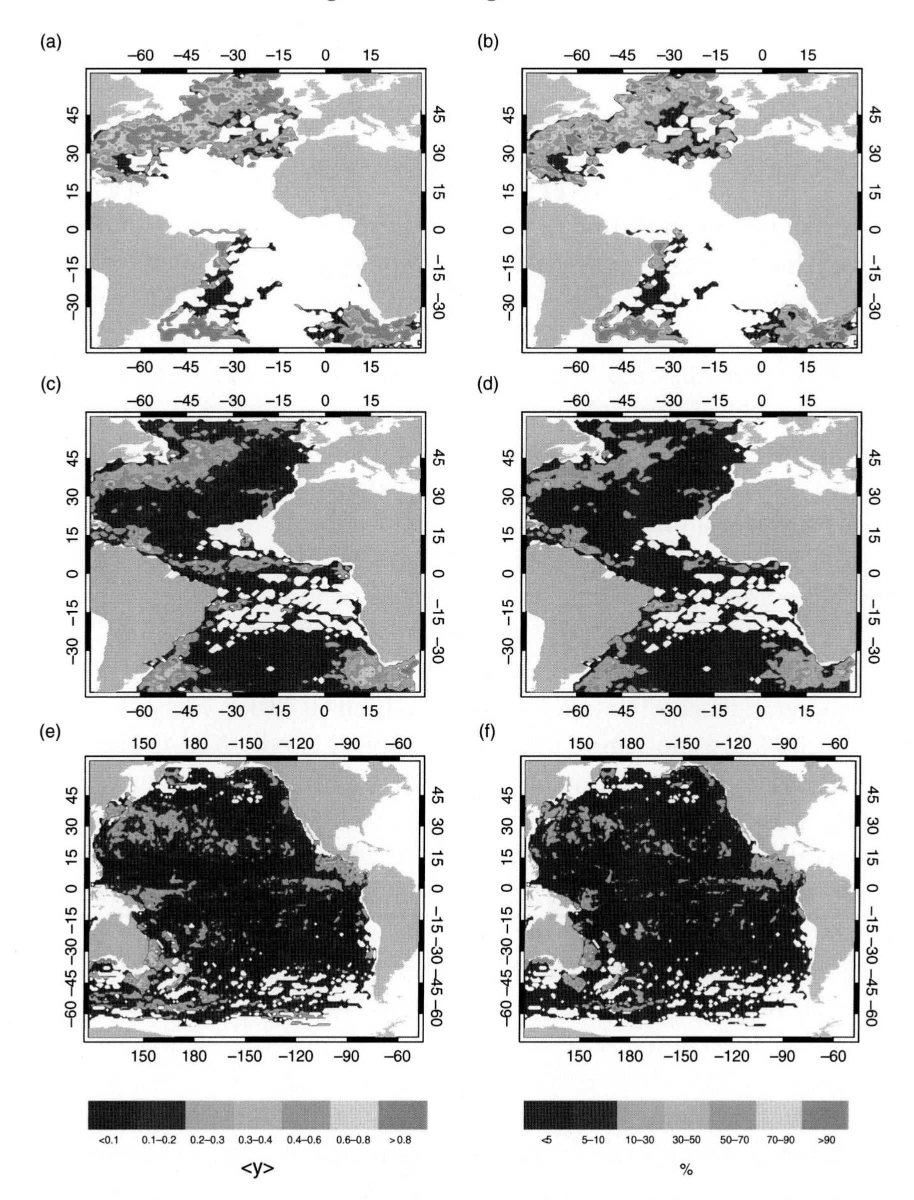

Figure 9.13 Maps of y (left) and of the probability (right) of finding a trajectory characterized by complex conjugates time scales (Classes III and IV) for the AF (a, b), AD (c, d) and PD (e, f) datasets. The probability map is computed considering, in a given bin, the ratio between the points of trajectories belonging to Classes III and IV and the sum of all the points. See Plate 22 for color version.

With the exception of this latter area, the same regions are evidenced in the AD dataset (panels c and d), where relatively high values of y are observed also, in the central equatorial band and near to the northern coasts of South America.

The Pacific Ocean, at least at the surface, behaves differently; in a quiescent background one can distinguish two turbulent areas east and south east of Australia and west of Central America; moreover a large tongue with intermediate values of y extends forward south east from the Kuroshio area. Interestingly, relatively high values of y are observed also in the Japan Sea.

We have seen that the time and space scales of ocean variability do not support the eddy diffusivity model (Markov 0, see Section 9.4). Moreover, the Lagrangian diffusivity is not defined (Section 9.2) if the velocity spectrum does not saturate at low frequencies, as for instance the zonal component of the velocity spectra for the AD and PD datasets (Fig. 9.3, Section 9.5). Nevertheless, the use of an eddy diffusivity model is common in OGCMs. Consequently, in the maps of Figure 9.14 we plot the Lagrangian correlation length and the apparent diffusivity $K_{\mathrm{app}} = \sigma_U^2 \cdot T_{\mathrm{L}} = \sigma_U \cdot L_{\mathrm{L}}$ (Stammer, 1998; Lumpkin *et al.*, 2002) that can be considered as a proxy of the Lagrangian eddy diffusivity K.

The Lagrangian correlation length is less than 60–80 km everywhere, except than in the Kuroshio region and in the tropical bands. K_{app} is higher in the main energetic current paths but, consistently with the previous results, the maximum values are observed in the area characterized by high energy and intermediate and low values of y, such as the Kuroshio region and the equatorial area. On the contrary, where "turbulent" trajectories are predominant, the same energetic content gives rise to a lower value of K_{app}.

It is interesting, at this point, to synthesize the information obtained from the maps presented in this chapter. Essentially, we have observed that the main energetic systems in the Atlantic and Pacific Oceans can be divided into two categories.

The first is characterized by a Lagrangian correlation *time* decreasing with the eddy kinetic energy and is massively populated by "turbulent trajectories." To this category belong the Gulf Stream, Agulhas and the East Australia currents, together with the subsurface area of the Mediterranean outflow and the region west of Central America.

The second category is characterized by a Lagrangian correlation *length* increasing with the eddy kinetic energy and is populated by trajectories having a low value of the ratio $y = T_a/T_v$. The energetic equatorial bands and the Kuroshio current belong to this category.

The analysis performed here indicates that in these two categories prevails the "frozen-turbulence" and "fixed-float" regimes, respectively.

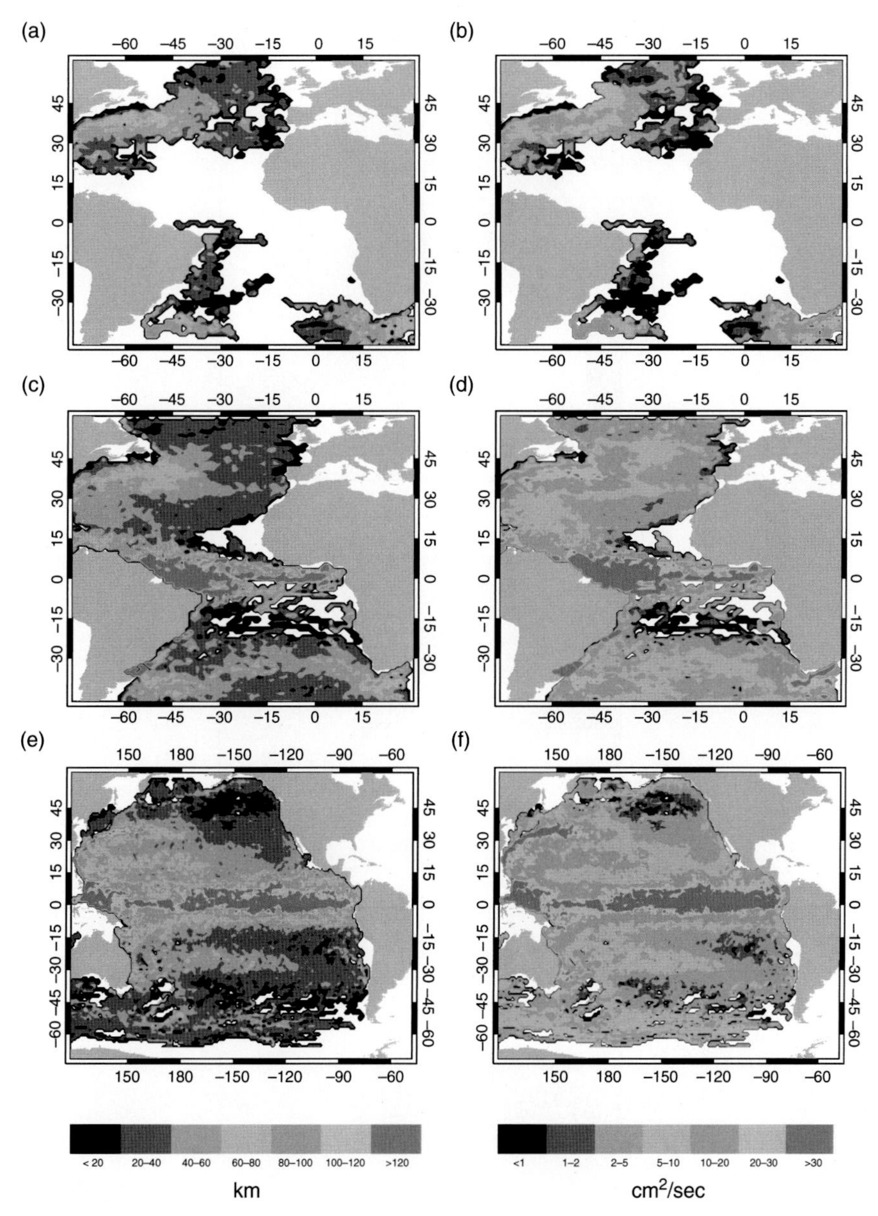

Figure 9.14 Maps of K_{app} (right) and of the Lagrangian length scale (left) for the AF (a, b), AD (c, d) and PD (e, f) datasets, respectively. The maps were obtained averaging over $1^\circ \times 1^\circ$ bins and assigning to each point the value of the trajectory. The Lagrangian length scale is expressed in km while for the the apparent diffusivity units are 10^7 cm^2/sec. See Plate 23 for color version.

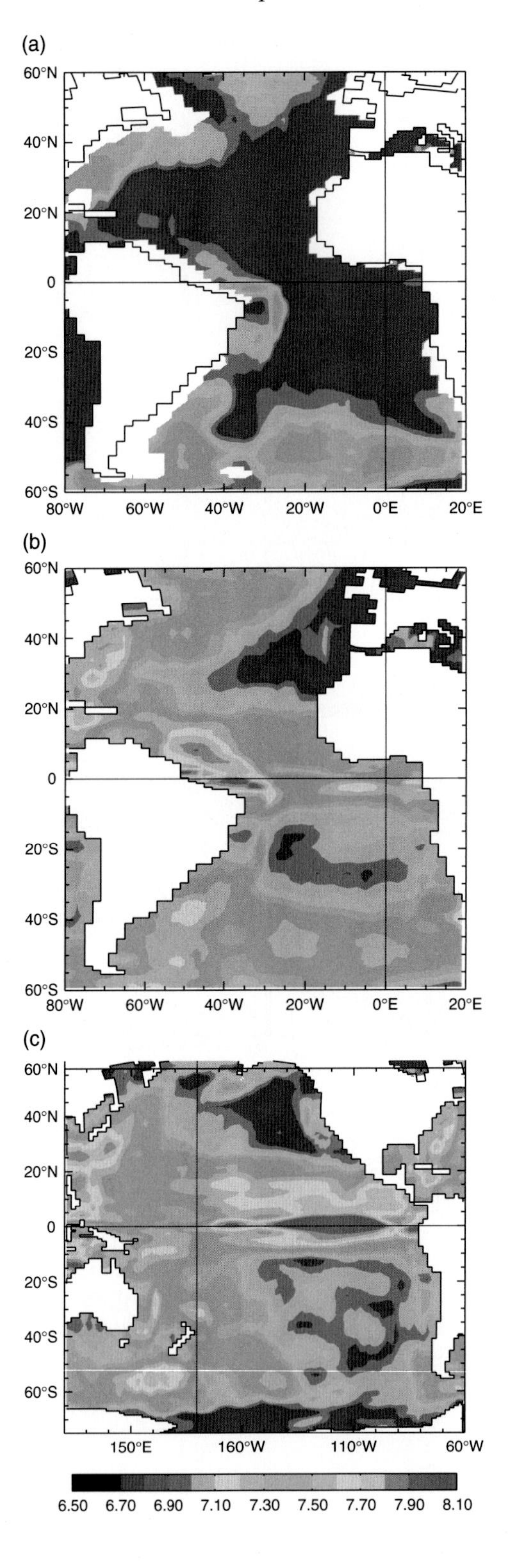
(a)
60°N
40°N
20°N
0
20°S
40°S
60°S
80°W
60°W
40°W
20°W
0°E
20°E
(b)
60°N
40°N
20°N
0
20°S
40°S
60°S
80°W
60°W
40°W
20°W
0°E
20°E
(c)
60°N
40°N
20°N
0
20°S
40°S
60°S
150°E
160°W
110°W
60°W
6.50
6.70
6.90
7.10
7.30
7.50
7.70
7.90
8.10

We believe that this kind of a geographical characterization of the main oceanic currents can be useful to advance the representation of the eddy diffusivity in the OGCMs since, in these two regimes, the Lagrangian diffusivity varies differently according to the Lagrangian eddy kinetic energy.

In the last few years a large effort has been devoted by the scientific community to improve the parameterization of eddy diffusivity. This effort was mainly focused on the attempt of mimicking, in coarse resolution models, the effect of baroclinic instability by means of a diffusion coefficient (or an eddy-induced velocity) proportional to some index representative of the isopycnal slope. The justification for this approach is that baroclinic instability is the main mechanism of energy extraction and is the primary source of eddies (e.g. Treguier *et al.*, 1997, Visbeck *et al.*, 1997 and Bryan *et al.*, 1999). Nevertheless, in the oceans exist also regions characterized by large eddy activity in the absence of strong density gradients. Stammer (1997) has shown (with altimeter data) that in the vicinity of the East Australia Current, the Brazil–Malvinas confluence and the Agulhas Retroflection, a strong eddy field is associated with weak mean baroclinic currents. This suggests that the scenario in which the source of eddies is the kinetic, rather than the potential, energy cannot be totally neglected, and possibly a complete parameterization of eddies should take into account complementary approaches. We found strong indications that in the same regions the Lagrangian diffusivity is proportional to the square root of the eddy kinetic energy. This shows how Lagrangian data, that directly record both kinetic energy and the existence of eddies, may help to indicate in which geographical area a particular parameterization is appropriate.

As an example of this methodology, we mention the parameterization of the eddy diffusivity proposed by Babiano *et al.* (1987) that has been applied to a regional OGCM by Rupolo *et al.* (2003). In that case, the parameterization of Babiano *et al.* results into a scaling of the eddy diffusivity coefficient with the square root of the eddy kinetic energy. Such parameterization has been implemented on a global OGCM (Calmanti, personal communication) and the spatial distribution for the eddy diffusion coefficients resembles the structure that are observed in the present analysis (Fig. 9.15). It should be mentioned that

---

Figure 9.15 Annual mean horizontal diffusivity computed with OPA at 500 m in the Atlantic (a), at the surface in the Atlantic (b) and at the surface in the Pacific (c). The model has uniform longitudinal resolution of 2 degrees and variable meridional resolution ranging from 0.5 degrees in the tropics to 2 degrees in the polar regions. The base 10 logarithm of the value is shown; units are cm^2/sec. (*Courtesy of S. Calmanti and G. Madec.*) See Plate 24 for color version.

the parameterization of Babiano *et al.* does not account for the contribution of baroclinic instability to the diffusivity fields. Rather, it describes non-local effects of geostrophic turbulence and the results discussed in this chapter suggest that (and where) it could be considered as a tool to improve OGCM results.

9.7 Summary and concluding remarks

In this chapter, we have briefly reviewed the relationship between correlation functions and dispersal properties and a hierarchy of 1D LSMs. The order of a LSM, refers to the number of timescales that are incorporated in the model, and the analytical expression of the velocity spectrum of the second-order LSM can be expressed in terms of the ratio between the velocity and acceleration time scales T_v and T_a. We have shown that the index $y = T_a/T_v$ can be useful to rationalize the analysis of real Lagrangian trajectories and to avoid a mixture of different regimes, when averaging quantities. We analyzed subsurface floats data in the Atlantic basin and surface drifters data in the Atlantic and Pacific Oceans, and observed that trajectories having respectively a small value of y, and a complex conjugates pair of time scales, belong to two extreme classes. Trajectories belonging to these classes are characterized by dissimilar shapes, correlation and dispersal properties. In particular, the Lagrangian correlation length varies according to the energy very differently, indicating that this index can be an useful tool to discriminate between the two limiting cases in which the spatial structure is strong (frozen-turbulence regime) or weak (fixed-float regime). On the other hand, the spectral and correlation properties of trajectories belonging to the same class are similar.

These observations allow us to estimate the role of the diverse regimes in the particles' dispersion, both in time and in the geographical space. The dimensionless form of the structure function shows that coherent structures, when present, prevail in the Lagrangian dispersion only for time smaller than 4–5 T_L. The same analysis, performed in the dimensional space, has led to the characterization of the role of the coherent structures in the dispersion in the subsurface (400–1000 m) in the Atlantic basin.

The subsurface flow in the Atlantic Ocean above the main thermocline turns out to be definitively the most turbulent; almost half of the floats have complex conjugates time scales and very often display a clear looping behavior.

At the surface, drifters having a low ratio $y = T_a/T_v$ largely dominate. Nevertheless, one can identify in the Atlantic extended areas (Agulhas Current and the west boundary currents) in which a relatively high percentage of the trajectories display turbulent behavior (Fig. 9.13). In the Pacific Ocean turbulent trajectories are confined to two hot spots (east of Australia and west of Central America) and partly in a large tongue extending toward south east from the Kuroshio area.

In section 9.6, we discussed these findings trying to highlight how they can be exploited to test parameterization of eddy diffusivity in OGCMs.

Finally, we want to conclude with two remarks concerning some technical details concerning the analysis of Lagrangian data.

In this chapter, we adopted the approach of calculating quantities along each trajectory and then averaging among the resulting statistics. This technical procedure, that allows us to maximize the amount of data when mapping Lagrangian scales, was already used by Lumpkin *et al.* (2002) and partly tested by Berloff *et al.* (2002a). An alternative approach consists in binning Lagrangian velocities in the physical space and then to compute statistics in the obtained Eulerian map. In this case, when dealing with time correlation properties, one has first to compute a mean correlation function and then to estimate the correlation time. We believe that both of these technical procedures are acceptable, and their different use can highlight different aspects of the information contained in the Lagrangian data. The statistical analysis of extremely variable and non-stationary data, such as Lagrangian trajectories, is intrinsically difficult and requires a large spectrum of experimental approaches.

The last consideration refers to the estimate of the acceleration time scale, directly from the trajectory. This computation obviously depends on the (different) interpolations and smoothing procedures applied to the original data. We believe that this dependence can affect the relative population of the Classes I and II, leaving unaltered the estimate of the number of buoys strongly influenced by the presence of coherent structures (Classes III and IV).

Acknowledgments

Drifter and float data were obtained from the WOCE Data Products Committee. I would like to thank R. Iacono and M. Gwyn-Fawke for stimulating discussions. I'm also grateful to two anonymous referees that helped to improve this chapter. This work was partly supported by the National Program "Ambiente Mediterraneo" funded Italian Ministry for Research (MIUR) and by the EU project MFSTEP (contract EVK3-CT-20-00075).

References

Babiano, A., C. Basdevant, P. Le Roy, and R. Sadourny, 1987. Single particle dispersion, Lagrangian structure function and Lagrangian energy spectrum in two-dimensional incompressible turbulence. *J. of Mar. Res*, **45**, 107–31.

Bauer, S., M. S. Swenson, A. Griffa, A. J. Mariano, and K. Owens, 1998. Eddy mean flow decomposition and eddy-diffusivity estimates in the tropical Pacific Ocean.1. Methodology. *J. of Geoph. Res.*, **103**, 30855–71.

Bauer, S., M. S. Swenson, A. Griffa, A. J. Mariano, and K. Owens, 2002. Eddy mean flow decomposition and eddy-diffusivity estimates in the tropical Pacific Ocean. 2. Results. *J. of Geoph. Res.*, **107**, 3154, doi:10.1029/2001JC000613.

Berloff, P., J. McWilliams, and A. Bracco, 2002a. Material transport in oceanic gyres. Part I: Phenomenology, *J. of Phys. Oceanogr.*, **32**, 797–830.

Berloff, P. and J. McWilliams, 2002b. Material transport in oceanic gyres. Part II: Hierarchy of stochastic models, *J. of Phys. Oceanogr.*, **32**, 797–830.

Berloff, P. and McWilliams, 2003. Material transport in oceanic gyres. Part III: Randomized stochastic models, *J. of Phys. Oceanogr.*, **33**, 1416–45.

Borgas, M., T. Flesch, and B. Sawford, 1997. Turbulent dispersion with broken reflectional symmetry. *J. of Fluid Mech.*, **332**, 141–56.

Bower, A. S., B. LeCann, T. Rossby, W. Zenk, J. Gould, K. Speer, P. L. Richardson, M. D. Prater, and H. M. Zhans, 2002. Directly measured mid-depth circulation in the northeastern North Atlantic Ocean, *Nature*, **419**, 603–7.

Bracco, A., J. Lacasce , C. Pasquero, and A. Provenzale, 2000. The velocity distribution of barotropic turbulence. *Phys. Fluid*, **12**, 2478–88.

Brink, K., R. Beardsley, P. Niiler, M. Abbott, A. Huyer, R. Samp, T. Stanton, and D. Stuart, 1991. Statistical properties of near-surface flow in the California coastal transition zone. *J. of Geoph. Res.*, **96**, 14693–706.

Bryan, K., J. Dukowicz, and R. Smith, 1999. On the mixing coefficient in the parameterisation of bolus velocity. *J. of Phys. Oceanogr.*, **29**, 2442–56.

Colin de Verdière A., 1983. Lagrangian eddy statistics from surface drifters in the eastern North Atlantic. *J. of Mar. Res.*, **41**, 375–98.

Davis, R. E., 1991a. Observing the general circulation with floats. *Deep Sea Res.*, **38** (Suppl.), 531–71.

Davis, R. E., 1991b. Lagrangian ocean studies. *Annu. Rev. Fluid Mech.*, **23**, 43–64.

Elhmaidi, D., A. Provenzale, and A. Babiano, 1993. Elementary topology of two-dimensional turbulence. *J. of Fluid Mech.*, **242**, 655–700.

Figueroa, H. A. and D. B. Olson, 1989. Lagrangian statistics in the South Atlantic as derived from SOS and FGGE drifters. *J. of Mar. Res.*, **47**, 525–46.

Fratantoni, D. M., 2001. North Atlantic surface circulation during the 1990s observed with satellite-tracked drifters. *J. of Geoph. Res.*, **106**, 22067–93.

Freeland H. J., P. B. Rhines, and H. T. Rossby, 1975. Statistical observations of the trajectories of neutrally buoyant floats in the North Atlantic. *J. of Geoph. Res.*, **33**, 383–404.

Griffa A., K. Owens, L. Piterbarg, and B. Rozovskij, 1995. Estimates of turbulence parameters from Lagrangian data using a stochastic particle model. *J. of Mar. Res.*, **53**, 371–401.

Griffa A., 1996. Applications of stochastic particle models to oceanographic problems. In *Stochastic Modelling in Physical Oceanography*, ed. R. Adler, P. Muller, and B. Rozovskii. Cambridge, MA: Birkhäuser Boston.

Hofmann, E. E., 1985. The large-scale horizontal structure of the Atlantic Circumpolar Current from FGGE drifters. *J. of Geoph. Res.*, **90**, 7087–97.

Hua, B. L., J. McWilliams, and P. Klein, 1998. Lagrangian accelerations in geostrophic turbulence. *J. Fluid. Mech.*, **366**, 157–76.

Kampé de Fériet, J., 1939. Les fonctions aléatoires stationnaires et la théorie statistique de la turbulence homogène. *Ann. Soc. Sci. Bruxelles*, **59**, 15–194.

Krauss, W. and C. W. Böning, 1987. Lagrangian properties of eddy fields in the northern North Atlantic as deduced from satellite-tracked buoys. *J. of Mar. Res.*, **45**, 252–91.

Lacasce, J. H., 2000. Floats and f/H. *J. Mar. Res.*, **58**, 61–95.

Lumpkin, R. and Flament, P., 2000. Lagrangian statistics in the central North Pacific. *J. of Mar. Syst.*, **24**, 141–55.

Lumpkin, R., A. M. Treguier, and K. Speer, 2002. Lagrangian eddy scales in the Northern Atlantic Ocean. *J. of Phys. Oceanogr.*, **32**, 2425–40.

McNally, G. J., Patzert, W. C., Kirwan A., and D. Vastano, 1983. The near surface circulation of the North Pacific using satellite tracked drifting buoys. *J. of Geoph. Res.*, **88**, 7507–18.

Middleton, J. F., 1985. Drifter spectra and diffusivities. *J. of Mar. Res.*, **43**, 37–55.

Owens, W. B., 1991. A statistical description of the mean circulation and eddy variability in the NW Atlantic using SOFAR floats. *J. of Phys. Oceanogr.*, **28**, 257–383.

Pasquero, C., A. Provenzale, and A. Babiano, 2001. Parameterization of dispersion in two dimensional turbulence. *J. of Fluid Mech.*, **439**, 279–303.

Patterson, S. L., 1985. Surface circulation and kinetic energy distributions in the Southern Hemisphere oceans from FGGE drifting buoys. *J. of Phys. Oceanogr.*, **15**, 865–84.

Pope, S. B., 1994. Lagrangian PDF methods for turbulent flows. *Annu. Rev. Fluid Mech.*, **26**, 23–63.

Pope, S. B., 2002. A stochastic Lagrangian model for acceleration in turbulent flows. *Phys. of Fluids.*, **14**, (7), 2360–75.

Reynolds, A. M., 2002. On Lagrangian stochastic modelling of material transport in oceanic gyres. *Physica D*, **172**, 124–38.

Richardson, P. L., 1983. Eddy kinetic energy in the North Atlantic from surface drifters. *J. of Geoph. Res.*, **88**, 4355–67.

Richardson, P. L., 1993. A census of eddies observed in North Atlantic SOFAR float data. *Progress in Oceanography*, **31**, 1–50.

Riser, S. and H. T. Rossby, 1983. Quasi-Lagrangian structure and variability of the subtropical western North Atlantic circulation. *J. of Mar. Res.*, **41**, 127–62.

Rossby, H. T., S. Riser, and A. Mariano, 1983. The western North Atlantic – a Lagrangian view-point. *Eddies in Marine Science*, ed. A. Robinson. Berlin: Springer-Verlag.

Rupolo, V., B. L. Hua, A. Provenzale, and V. Artale, 1996. Lagrangian spectra at 700 m in the western North Atlantic. *J. of Phys. Oceanogr.*, **26**, 1591–607.

Rupolo, V., A. Babiano, V. Artale, and D. Iudicone, 2003. Sensitivity of the Mediterranean circulation to horizontal space-time-dependent tracer diffusivity field in a OGCM. *Nuovo Cimento*, **26 C** 4, 387–415.

Sawford, B. L., 1991. Reynolds number effects in Lagrangian stochastic models of turbulent dispersion. *Phys. Fluid A*, **3**(6), 1577–86.

Stammer, D., 1997. Global characteristics of ocean variability estimated from regional TOPEX/POSEIDON altimeter measurements. *J. of Phys. Oceanogr.*, **27**, 1743–69.

Stammer, D., 1998. On eddy characteristics, eddy transport and mean flow properties. *J. of Phys. Oceanogr.*, **28**, 727–39.

Swenson, M. S. and P. P. Niiler, 1996. Statistical analysis of the surface circulation of the California current. *J. of Geoph. Res.*, **101**, 22631–45.

Taylor, G. I., 1921. Diffusion by continuous movement. *Proc. Lon. Math. Soc.*, **20**, 196–212.

Treguier, A., I. Held, and V. Larichev, 1997. On the parameterisation of quasigeostrophic eddies in primitive equation models. *J. of Phys. Oceanogr.*, **27**, 567–80.

Veneziani, M., A. Griffa, A. M. Reynolds, and A. Mariano, 2004. Oceanic turbulence and stochastic models from subsurface Lagrangian data for the north-west Atlantic Ocean. *J. of Phys. Oceanogr.*, **34**, 1884–1906.

Visbeck, M., J. Marshall, and T. Haine, 1997. Specification of eddy transfer coefficient in coarse-resolution ocean circulation model. *J. of Phys. Oceanogr.*, **27**, 381–402.

WOCE Data Products Committee, 2002. *WOCE Global Data: Subsurface Floats and Surface Velocity Data, Version 3.0*, WOCE Report No. 180/02, WOCE International Project Office Southampton, UK.

Zhang, H. M., M. D. Prater, and H. T. Rossby, 2001. Isopycnal Lagrangian statistics from the North Atlantic Current RAFOS float observations. *J. of Geoph, Res.*, **106**, 13817–36.

10

Lagrangian biophysical dynamics

DONALD B. OLSON

Rosenstiel School of Marine and Atmospheric Science, University of Miami, Miami, Florida, USA

The use of a particle-following or Lagrangian perspective to follow the dynamics of life in the sea is explored from a variety of perspectives. The discussion begins with the consideration of the energetics of marine organisms and a demonstration that mean field models fail to adequately describe the life of large marine fishes in the sense that they require sizable, >100–1000 X aggregation of prey over the average biomass density in the ocean. In place of a mean field model in time a structured population model where populations are dependent on space, time, age, and their metabolism is derived. Having introduced the structured model it is then argued that it is impractical to use such a model except in a Lagrangian frame. Methods for coupling these models in a Lagrangian description of the marine environment are then discussed. This section of the manuscript ends with an appraisal of the amount of spatial aggregation required to support large pelagic fishes such as swordfish and tunas. The second portion of the paper goes on to provide examples of trajectories in different marine environments including boundary currents, mesoscale eddy fields and fronts, and the coastal environment. An emphasis on the dynamics of trajectories at various trophic levels provides insights on aggregation mechanisms and rates.

The last two sections introduce methods of modeling population structure with Lagrangian trajectories. The first looks at the oceanographic connectivity of populations in the Caribbean based on a Lagrangian model embedded in a global circulation model (GCM) of the North Atlantic. The simulations are compared with observed drifter trajectories and available morphological and genetic information on a reef fish. The results suggest that the drift patterns and our knowledge of regional currents are consistent with the observed distribution of fish morphometric and genetic traits. The methods discussed can be modified to include active movement in the target animals if the behavior behind these movements were better known. Other factors may,

Lagrangian Analysis and Prediction of Coastal and Ocean Dynamics, ed. A. Griffa, D. Kirwan, A. Mariano, T. Özgökmen, and T. Rossby. Published by Cambridge University Press.

however, be important in reinforcing the pattern set by possible pathways for recruitment. The final model is a simplified one using representative trajectories with mixing to simulate long-term population dynamics in the ocean. Examples include consideration of genetics in ocean gyres and the spread of coral species in the Caribbean following the closure of the Panamanian Isthmus around three million years ago.

10.1 Introduction

While it is not important to most readers of this book to define the term Lagrangian in general, it is worthwhile stating it here, both for the reader whose interest is in learning basic Lagrangian or particle following methods, and to introduce a slight modification in driving forces over those in the fluid itself. Biologically it makes more sense to follow an individual organism than it does to follow a population. However, for various reasons it is easier to track the latter, just as in fluid dynamics where we are following a continuum construct tracking a water parcel instead of a water molecule. One might argue, however, that the history of individual organisms is more relevant than the statistical dynamics involved with the role of individual water molecules in the world. Therefore what follows is an argument that the details of individual trajectories of organisms are one important aspect of understanding their population dynamics. The life paths of organisms are fundamental features of ocean ecosystems and study of ecosystems at some levels demand a Lagrangian perspective. For the moment, this does not mean that all of the dynamics have to be resolved at the individual scale in the sense of an Individual Based Model (IBM), but that for many purposes even aggregated population models have advantages within Lagrangian implementations. In approaching these ideas it is helpful to introduce a set of working hypotheses.

(1) The events that determine the fate of organisms are inherently tied to the structural elements that define life histories, such as age, past nutrient availability, and reproductive opportunities, therefore it is essential to follow individuals or subpopulations with similar histories.

This sets up the argument for using an approach that uses the Lagrangian, i.e. individual following, frame of reference. This is typically referred to as an IBM (Craig *et al.*, 1979; Huston *et al.*, 1988; DeAngelis and Gross, 1992; Judson, 1994). An early review of the possibilities in tracking individual organisms can be found in the May 1993 Transactions of the American Fisheries Society (Van Winkle *et al.*, 1993). It is interesting that none of these eight models are explicitly spatial. One focus here is methods of doing this. In many cases the number of individuals that actually make up a population render an IBM impractical. The alternative of following subpopulations of Lagrangian particles is discussed in Dutkiewicz *et al.* (1993) and Olson and Hood (1994). Finally, the new techniques in tracking animals using acoustic or light-sensor based positioning (Block *et al.*, 1997; Brill *et al.*, 1999; Lutcavage *et al.*, 2000; Block *et al.*, 2001; Gunn and Block, 2001; Davis and Stanley, 2001; Graves *et al.*, 2001; Brill *et al.*, 2002; Schaefer and Fuller, 2002) and genetics for deriving connections between populations make Lagrangian perspectives a priority.

The second hypothesis concerns the environmental features that correspond to the life cycles of marine organisms.

(2) Oceanic life is often closely governed by the interaction between physical, chemistry and biology in specific features such as fronts and eddies where the biological processes are enhanced. These enhancements and the trophic cascades they allow are fundamental to the existence of the observed ocean ecosystem.

At this point it is important to introduce a definition of the term front. The term comes from meteorology where most readers are familiar with the passage of weather fronts and their effect on our life styles. To formally define them as regions of maximal temperature gradient in space (Palmen and Newton, 1969; Olson 2002) seems to leave behind the total impact they have on humans and the rest of nature. This problem becomes even greater when one considers the free drifting or planktonic life style common in the ocean environment (see Hitchcock and Cowen, Chapter 11 this volume). Here, the focus will not be restricted to plankton, but will involve the influence of active swimming behavior on the trajectories in marine organisms. The current analysis will also forego a discussion of the small-scale interactions involved with turbulent enhancement of foraging (see Rothschild and Osborne, 1988; Yamazaki *et al.*, 2002). Instead the focus will be on the mesoscale and larger space and time scales (>10 km/>1 day) that correspond to the daily ambit of a fish and larger invertebrates. A major conclusion, as supported below, is that full treatment of the complexities of marine life at these scales demands a Lagrangian perspective.

In our current life experience, even if we are responding to storms while at sea, the view is on a grid. The experience is very different without our current knowledge of weather data on a grid. For a free drifting organism or a sailing ship of old, events such as hurricanes or frontal passages are encountered without foreknowledge. This is not to imply that either older mariners or plankton lack adaptations that increase their survival rates in the presence of certain ocean environments, but that there is a wide range of behaviors, involving modifications of particle-based, Lagrangian behavior, that are crucial to the organization and function of many oceanic ecosystems. Here the goal is to understand these adaptations and marine ecosystems as complex adaptive systems (Levin 1999, 2002).

The discussion begins with a consideration of the needs of large, upper trophic level organisms to be able to live in the average ocean. Specifically, the bioenergetics of large pelagic fish is considered. These energetic arguments focus on spatial descriptions of the ocean environment and on the inherent patchiness involved in it (Haury *et al.*, 1977). An argument is made that enhanced production at various trophic levels make features such as fronts a fundamental contributor to the overall productivity of the ocean. Following a description of flows and behavior of fauna in frontal systems, a set of models is introduced. Both the nature of physical flows and the problem of working with models that treat the structure of real populations in terms of age, metabolism, or genetics make the Lagrangian reference frame the only practical one in which to formulate these models. A discussion of the means of doing this is followed by several examples.

The chapter is organized into four main sections starting with a description of marine population dynamics from a structured perspective and then a discussion of placing structured populations into ocean ecosystems using a Lagrangian perspective. These sections are followed by a description of actual trajectories of organisms in various marine ecosystems including eastern and western boundary currents, mesoscale features, and coastal ecosystems. This third section also explores the impact of animal behaviors on the dynamics of these different regimes. The last section introduces two methods of using Lagrangian simulations to model oceanic environments. The applications include a consideration of dispersal of organisms across the Caribbean and the contribution this makes to recruitment dynamics in benthic species. A final model is used to explore structured population dynamics over evolutionary time scales. An example of the spread of coral species in the Caribbean over the last several million years is used to consider the long-term use of Lagrangian techniques. The paper ends with a short set of conclusions on use of Lagrangian techniques in understanding marine life.

10.2 Describing the dynamics of marine populations

10.2.1 The average fish: mean field models

Mean field models are introduced to explore the energetic needs of the average large pelagic fish such as a tuna or a billfish, filling the top position in an open ocean food web. This involves the calculation of the energetic needs of the fish, the average primary productivity of the ocean, and then finally the area of ocean needed to support the fish. This non-spatial approach will be compared to the results of following organisms through space. The analysis begins with the computation of the energetic needs of several species.

The energetics of an animal such as a fish depends on three basic factors: (1) the engestion of food and the efficiency of its conversion through metabolic pathways, (2) the cost of maintenance of both the basal metabolism and activities of the animal, and (3) the energy required for growth and reproduction. The second factor is typically known as catabolism. The last factor, involving addition of biomass to either the individual or their offspring, is typically called anabolism. If the goal is to document the biomass of a population of a given species in the averaged sense, the analysis begins with a measure of the population in terms of numbers or density (numbers per area), N_i where the i indicates the species involved. Summing over this index therefore allows consideration of an entire ecosystem in principle. In the mean field model biomass is given by the product of N_i and the average weight of the individuals, $< W_i >$, where the angle brackets denote a spatial average. In the mean field model of the ecosystem, an average over the environment reduces the analysis to the spatial mean dynamics, the biomass (B_i) of an individual species is $< B_i > = N_i < W_i >$. In the spatially averaged view of a population, a given area with some average carrying capacity must maintain $< N_i >> 0$. Statistically as argued below this might not be the case if it is the correlation between fluctuations in carrying capacity and population that leads to successful mean populations. This leads to the problem of solving for the amount of forage it takes to support this biomass.

The energetics of an organism can be written in terms of its food requirements (R_e) and the food ration it receives (E, engestion),

$$R_e = G + M + (1 - \epsilon)E,$$

where G is the growth of the individual or reproduction (anabolism), M is the metabolism (catabolism) including the basal metabolism, metabolic costs of growth, digestion, reproduction, and activity. Finally the last term represents the loss of energy through egestion of fecal material. Therefore ϵ is the assimilation efficiency. In fisheries science the growth of fish with age is often treated in terms of length because this is an easily and accurately measurable variable. Here, however, the analysis will be done in terms of weight (W_i) that can be directly converted to energy terms. The growth of individuals then can be written as,

$$\mathrm{d}W_i/\mathrm{d}t = G = \epsilon E - \mu.$$

For assimilation (ϵE) exceeding the metabolic costs (μ), growth is positive ($\mathrm{d}W_i/\mathrm{d}t > 0$). Population dynamics can be divided into two states dependent on the sign of G. When E is sufficient to make G positive the organism is in the well-fed state. When G is negative the organism is challenged. This can be part of the life history as induced by increases in μ and expulsion of gametes during reproduction, or in the case of insufficient forage to maintain basal metabolism and crucial behaviors such as foraging, leading to starvation. A key issue in the overall range of a species is the portion of space that allows enough growth to allow reproduction, the reproductive range, versus the overall range that allows the species to escape starvation. Before proceeding with a Lagrangian description of these ranges, it is important to introduce a fuller description of an organism's biology in terms of the structure of its population with respect to metabolism (using W_i as a proxy), aging, and genetics.

10.2.2 Structured population models

Governing factors in the dynamics of populations with respect to population structure are age (a, ontogenetic effects), metabolic requirements (W, physiological influences), and adaptive capability (genetic or behavioral; see Metz and Diekmann, 1986; Murray, 1993). For example, changes in weight, $W_i(a, t)$, with either age (a) or time (t) can modify mortality as well as fecundity rates. On longer temporal scales changes in populations are structured by their genetic traits and these are modified over time through selection, drift, or mutation to produce changes in the very nature of the population and its interactions with the environment. It is the complexity of carrying information concerning the structuring variables that leads to the conclusion that it is important to follow individuals if one wants to understand the basic nature of life.

The physiological, ontogenetic, and behavioral components of a species' life history plays an important role in the manner in which these organisms interact with their environment. The demographics of a population are a reflection of the past history of the population in terms of food resources, thermal conditions that set the physiology, aging, and behavior of the population. The history of the environmental conditions that the population has encountered therefore is said to structure the population, i.e. determine its demographic character. In a planktonic organism these histories in a given local population are differentiated through mixing that causes various subpopulations with different temporal and spatial histories to co-populate an area. The problem is then to construct a model that can incorporate the different histories within a local population and keep track of the various variables that structure these subpopulations. The problem becomes even more complicated when the organism has significant motive capabilities and travels on its own through the ocean environment based on the behavioral clues available to it.

Structure can be defined as variations within individuals that make up a population. Here a local population is defined in terms of the density, N_i, the number of individuals per unit area or volume where i denotes the subpopulation with a similar history. The problem is then to express the factors that lead to changes in N_i in space or time. These factors can be separated into the influence of the physical environment, here advection and mixing by ocean currents and the thermal conditions, and biotic factors that control a population's interaction with both its food supply and its predators. Perhaps in organisms like copepods the most important interaction, however, is the interaction of physiological status and stage progression with a combination of physical and biological conditions they experience. Mathematically, the dynamics of the populations are expressed in terms of an expansion of the total temporal derivative with respect to the various factors that structure the population. The general form becomes an expansion of N_i in terms of time, t, spatial location, $\vec{x}$, the somewhat plastic stages of organisms such as copepods, stage (s_i), or age (a) in the population, and the population's physiological status as denoted here by the variable, W_i. Adding a mortality term, m, this expansion becomes,

$$\mathrm{d}N_i(t, a, w, \mathbf{x}) = \frac{\partial N_i}{\partial t}\,\mathrm{d}t + \frac{\partial N_i}{\partial \vec{x}}\,\mathrm{d}\mathbf{x} + \frac{\partial N_i}{\partial a}\,\mathrm{d}a_i + \frac{\partial N_i}{\partial W_i}\,\mathrm{d}W_i = -mN_i\,\mathrm{d}t,$$

where the velocity that a population is carried through the environment, $\mathbf{V} = \mathrm{d}\mathbf{x}/\mathrm{d}t$, population aging a, and the population variations with respect to individual weight have been defined in terms of the partial derivatives in the previous equation.

Fluid dynamicists will recognize the material derivative in the first two partial expressions although here one might also include mobile behavior on the part of the population. So the total or material derivative is the Lagrangian description of the population. The age or stage term gives the equation the form of McKendrick–Von Foerster (McKendrick, 1926; Von Foerster, 1959). In the case of straight age dependence a progresses linearly with time (t) from time of birth. In taxa such as copepods and the even more variable arthropods, the euphausiids, age is not a good variable, rather the stages(s) of development and their dependence on forage and temperature and salinity in setting stage duration must be considered. For most fish and mammals age is linear with time although many of the parameters that govern population dynamics vary as a function of age making the governing equations non-autonomous. An example of this would be allometric growth and saturation with age.

The dependence of a population on its genetic structure can also be considered (Webb, 1985; Olson *et al.*, 2001). This can be done by linking the parameters a, G, and m and their dependence on environmental parameters including competing or predatory species, temperature and salinity to variable sets of genotypes within a population. Of special concern here is the nature of ocean currents and mixing and the role that behavior plays in interactions with a population and selection of particular phenotypes associated with this underlying genetic structure.

It is worth distinguishing the changes in the properties of the equation if $m = m(N)$, i.e. the equation is nonlinear in N_i. Of course, it is assumed that in most cases m will depend on the expansion variables, cf. $m = m(a, W(t))$. This problem has both initial conditions, $N(a, 0)$, $W(a, 0)$, and a source condition representing reproduction in the population. The latter can be written in terms of an integral of a fecundity function (f) over age,

$$N(0,t) = \int_0^\infty f(a, W, \gamma_f)\, da.$$

Here fecundity is assumed to be a function of age, physiological variables (W), and an effective sex ratio for the population (γ_f). Notice that the initial condition and the birth rate at $t = 0$ do not necessarily match. In this case the solution will have initial transients that must be solved for.

There are various means of seeking solutions to the McKendrick–Foerster equation (cf. Webb, 1985; Murray, 1993). For now consider the linear problem, $m \neq m(N)$. One of the most straightforward methods is to introduce a cohort function for a population born at a given time, i.e. $N_c(t) = N(a_c + t)$ for $t \geq t_c$. The trajectory of the age of a cohort in time is a straight line as long as aging is linear or stage duration $D = 1$. The problem can then be solved for a individual cohort by integration along the characteristic as shown in Fig. 10.1. This solution for a given cohort is then

$$N_c(t) = N_c(t_c) \exp\left[-\int_{t_c}^{t} m(t')\, dt'\right].$$

This can then be extended to an entire range of cohorts to give the general solution,

$$N(a,t) = \begin{cases} N(0, t - a_c)\exp[-\int_0^a m(t')dt], & t > a \\ N(a - t, 0)\exp[-\int_{a-t}^{a} m(t')dt'], & t \leq a. \end{cases}$$

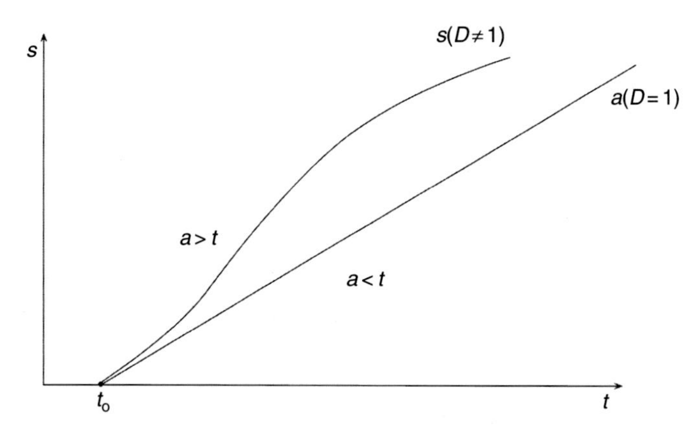

Figure 10.1 The characteristics of populations versus time in the McKendrick–von Foerester model for organisms undergoing linear aging, $a(D=1)$, as occurs in many vertebrates and stage, $s(D\neq 1)$, based transitions as occur in many invertebrates such as copepods where the aging trajectory depends on the physical conditions encountered as part of their life history such as temperature impacts on development and growth (McLaren *et al.*, 1989 and Huntley and Lopez, 1992). Here D is the duration factor for a life stage.

For simple choices of m these integrals can be solved. Another method for the long term limit on the population, $t \gg a_{\max}$, is to make use of similarity solutions of the form $N(a,t) = \bar{N}(a)e^{\gamma t}$. Upon substitution back into the original equation and some work on the birth condition this leads to viable solutions. Examples of similarity solutions are given in Murray (1993). The degree that these solutions can be found depends on the exact nature of the mortality, m.

10.2.3 Mortality as a function of condition

The simplest choice of parameterization for mortality is to assume it is constant or perhaps a constant function of age, $m(a)$. In general, however, one assumes that the mortality in a marine population depends on the condition of the stock in terms of its metabolic status, i.e. $W(a,t)$. In the structured model of Ault and Olson (1996) this is assumed to involve a linear increase in mortality as an individual's weight deviates from its optimal weight, $W_o(a)$. This has subsequently been applied to several fisheries models (Ault *et al.*, 1999, 2003). In the spatial context both the constant m and linearly varying $m(W(a))$ are somewhat problematic since they lead to the conclusion that there is a finite probability of finding some finite population everywhere in a domain as long as it is viable in some portion of it (see Olson and Hood, 1994). While one can

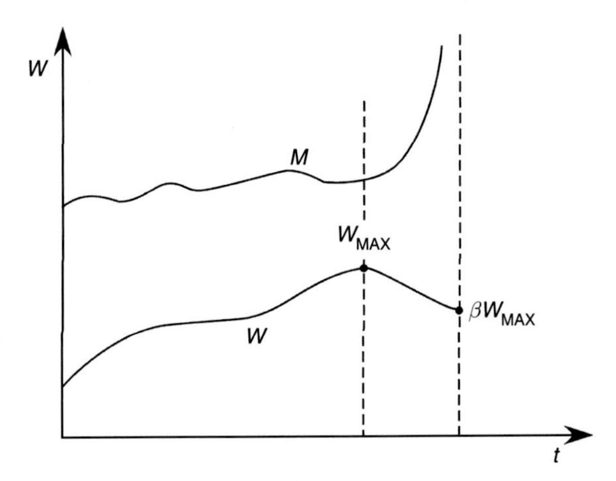

Figure 10.2 Mortality (M) as a function of condition (W) in populations based on the Olson *et al.* (2005) increase in mortality under suboptimal foraging situation. In the starvation model of Olson *et al.* the mortality rate goes to infinity effectively killing a population on an unsuccessful foraging trajectory. The Ault and Olson (1996) model in contrast leads to e-folding decreases with an appreciable impact of population dynamics, but an expectation that there is still a finite probability of finding a living organism on any trajectory.

just set that population effectively to zero at some very low concentration or probability density, to actually get regions with no population a modified mortality must be introduced.

Here the mortality formulation of Olson *et al.* (2005) is introduced. This involves consideration of the actual metabolism of an organism (Fig. 10.2) and its influence on mortality under starvation conditions. This model also introduces inter-specific interactions through predation and competition between the ith and jth populations N_i, N_j. The influence of fisheries mortality, $F(N_i, N_j)$ is also explicitly considered. Here the dependence is on multiple species interactions within a fishery that include mortality in the target species and by-catch. The equations follow from the general form introduced above with a population equation for N_i

$$\begin{aligned} \mathrm{d}N_i/\mathrm{d}t &= \partial N_i/\partial t + \partial N_i/\partial a + \nabla \cdot (\vec{V} N_i) \\ &= f(N_i(a > a_m)) - m_i(W_i, N_j) - F_i(N_i, N_j) \end{aligned}$$

and an equation for the population's weight, W_i,

$$\mathrm{d}W_i/\mathrm{d}t = E(K_i, N_j) - \mu_i(N_i, W_i, A_i),$$

where $E(K_i, N_j)$ is a function that describes the engestion and assimilation of food based on the carrying capacity of the environment, K_i, the availability of prey and the amount of competition. The function $\mu(N_i, W_i, A_i)$ describes the metabolic costs (catabolism) for the species in terms of its population density, weight, and activity (A_i). The weight gain is typically referred to as the anabolism. Note that this nomenclature differs slightly from von Bertalanffy's (1968) definition which sets the rate of weight gain equal to the difference between anabolism and catabolism. The author thinks the usage here is more accurate in a physiological sense. In practice as long as one is consistent with the use of the formulation this difference is semantic. Here the metabolism is treated as a single variable, W_i, but a more general formulation discussed below would include storage of energy in terms of lipids in addition to the protein that forms the body frame of individual organisms.

The population and weight equations are coupled through the dependence of fecundity, f, and mortality on weight and the effect of competition on population size in the metabolic equation. Reproduction is viewed here as a loss of mass (W_i) in the adults and should realistically involve an increase in the activity related costs. For the mortality it is possible to consider two possible states. These are the well-fed state where food is not limiting and the resultant natural mortality is set by predation on N_i and a background mortality $m_o(a)$. The second state occurs under starvation, $E = 0$. There are various responses that an organism takes to cope with the lack of food (see Davidson, 1994). Here the use of stored lipids and then the consumption of the protein framework of the individual, i.e. W, are considered. Both of these depend upon the total mass the body has attained and therefore the past history of the organism's foraging success. Since the metabolism of ectothermal organisms also is very dependent on temperature this also depends on the past environmental history of the creature. The formulation introduced by Olson *et al.* (2005) considers the effect that burning proteins from the body has on mortality. As starvation proceeds an organism can feed its metabolism, μ, by metabolizing its own structures up to a point where this becomes lethal. This typically is some fraction of the total mass initially attained prior to the onset of starvation, W_{max}, such that the lethal limit is $W_i^o = aW_{i,max}$. The result also depends on the rate of weight loss that determines the time that the individual has to adapt to starvation conditions. The effect on the mortality in the population equation can then be written

$$m = m_o + [\dot{W}/(W - W^o)]^-.$$

Here the raised minus sign denotes an operator such that the term in brackets is zero for $dW_i/dt = \dot{W}_i > 0$ and equal to the absolute value of the quantity inside the bracket when $\dot{W}_i < 0$. Under this starvation condition the mortality climbs to infinity as W_i approaches W_i^o and the population dies in a finite amount of time. The consequences of this type of mortality will be explored more fully below, but first the importance of a Lagrangian approach to application of this model to the real ocean is discussed.

10.3 Lagrangian implementation of structured models

10.3.1 Coupling the model to the rest of the ecosystem

Structured models with their inherent dependence on the past history of the organisms are most conveniently posed in a Lagrangian frame that follows either individuals' or groups of individuals' trajectories through the environment (Olson *et al.*, 2001). To demonstrate this, consider the calculation of the population dynamics above on a fixed Eulerian grid. First a Lagrangian model of a population across a space of $m \times m$ dimension will require $\rho \times m \times m$ particles where ρ will be greater than one in order to get the same resolution in space, but will not be much larger than one. The particle field must also resolve the (n) age classes in the population so the final number of particles is $2n \times \rho \times m \times m$. Here the factor of two arises from the fact that there is an equation for both N_i and W_i. Additional organisms included in the model increase the number of variables accordingly. The equivalent Eulerian computation demands $[2n \times (m \times m)]^2$ variables since every grid point needs the historical contribution to its current population from every other grid point across the age spectrum. Since $\rho \ll (m \times m)^2$ for a typical model simulation the Eulerian model is the most expensive undertaking. This does not mean that a given model will make use of only Lagrangian techniques.

Almost all of the available physical models currently in operation are Eulerian based. Therefore implementation of a Lagrangian structured population model will typically be done in the context of an Eulerian grid that provides the physical environmental information. This will include the velocity field and the thermodynamic setting in terms of temperature and salinity. The determination of the Lagrangian trajectories for planktonic organisms can be accomplished by computing them from the Eulerian velocity distribution. In the examples below this is done using a fourth-order Runga–Kutta scheme at full spatial resolution but with subsampled time steps from the Eulerian model. This higher order integration scheme is employed to accurately capture trajectories in circular eddies from model fields at one day intervals. Simpler methods can be utilized when the trajectories are computed within numerical models' integration (see Figueroa and Olson, 1994). In order to take into account the subgrid scale diffusion in the physical model an autocorrelated random walk is added to the Lagrangian model (see Dutkiewicz *et al.*, 1993; Olson and Hood, 1994; Cowen *et al.*, 2000; Idrisi *et al.*, 2004). In the case of free swimming nekton their behavior needs to be added to the model. This involves assessing their swimming ability and its energetic cost to the activity (A_i) term above. Behavioral considerations treated to date in models include kinesis or behavior modification based on conditions sensed at the current time (Humston *et al.*, 2001, 2004) and the effects induced by schooling behavior in the presence of turbulence (Flierl *et al.*, 1999). Finally, for attempts to treat a fuller picture of the entire ecosystem, models might be cast in a hybrid system where part of the biological calculation is done in the Lagrangian frame and part on the Eulerian grid of the physical model.

The basic model for an underlying ecosystem can be pictured to consist of nutrients, *Nu*, phytoplankton, *P*, and zooplankton, *Z*, where the latter might consist of an age structured population. Typically, these variables are expressed in terms of nitrogen in units of (mol/kg). The total population of zooplankton is further divided into stages such that, $Z_T = \sum_i Z_i$, consists of i stages of development. The zooplankton can further be written in terms of the mean individual weight at a given stage, W_i (or in micro-moles of nitrogen per individual), times the population density, N_i (numbers) in keeping with the structured model introduced above. In many efforts, however, the zooplankton is lumped into one pool. In the biogeochemical literature, where the interest is in nutrient and carbon cycling, multiple forms of nutrient, a detrital pool (*De*) and sinking are typically considered explicitly (Riley *et al.*, 1949; Fasham *et al.*, 1990; Olascoaga *et al.*, 2005). The models take into account the amount of a nutrient currency such as the nitrogen in each pool. These models can be run within the physical models themselves (Eulerian coordinates) as shown in various efforts (see Flierl and McGillicuddy, 2002). The structured model introduced above can then be run with this type of model representing the lower trophic levels on the grid they advect through. The choice of model and whether to put the lower trophic levels into the physical model depends on the application and the nature of the overall modeling frame available. The simulations below are based on a set of transportable codes that are designed for implementation in any physical/biological model with a minimum of effort.

At this point the basic biological model and its Lagrangian implementation have been introduced. It is now time to apply these to the ocean environment. This treatment will start with an assessment of the actual bioenergetics at various trophic levels in the marine ecosystem. This is followed by a characterization of types of trajectories and then an analysis of outcomes along these.

10.3.2 Bioenergetics across trophic levels

Application of the theory above to marine population involves obtaining estimates of the basal metabolisms, the cost of swimming, and the energy transfer between trophic levels. Here the strategy is to start at the upper levels and estimate the energy requirements in the upper trophic levels. The basal metabolisms are dependent on the total body mass, and are typically parameterized with a power function. The latter takes the form

$$\mu = \alpha W^{\beta}$$

where the metabolism is measured in kJ/day. The coefficients α and β depend on the type of organism with lower taxa having lower values in both. Added to this is the cost of activity, A. These involve many aspects of an organism's behavior. Here the focus is on swimming costs. The arguments based on this will underestimate the total cost, but provide a minimum energetic cost for making a living in the marine ecosystem.

The expenditure of energy involved with motion is estimated based on work done against drag for nekton and the potential energy needed to lift zooplankton on their diel migrations added to the base costs of running an organism's metabolism. Here a set of examples is considered in order to assess the different energetics related to representative taxa and their paths through the ocean environment. An example of an IBM simulation with full bioenergetics for larval fish can be found in Hinckley *et al.* (1996). Another very innovative work is the spatially explicit consideration of striped bass (*Morone saxatilis*) in Chesapeake Bay by Brandt and Kirsch (1993). Implementation of such models involves an evaluation of primary productivity, the processing of energy up the food chain, and finally a look at the energetics of the top of the oceanic trophic system.

10.3.3 The energetics of pelagic fish and the average ocean

In order to assess the status of large pelagic fish in the ocean one can consider their energetics and the energy available to them as a forage base. This basically follows the form introduced above, but makes use of some additional details concerning the actual physiology of the fish. There are two approaches that can be taken to estimate the energetics of a fish. The first involves actually measuring the oxygen consumption of the animal to assess its metabolism and monitoring its food intake. This has been done for skipjack (*Katsuwonus pelamis*) and yellowfin (*Thunnus albacares*) by Kitchell *et al.* (1978). The

Kitchell *et al.* work is introduced here as an introduction to fisheries' energetics using the Wisconsin model. The results from this earlier work will be compared to the more recent appraisals of tuna energetics as reviewed in Korsmeyer and Dewar (2001). The work on tunas includes extensive lab and field work that is not currently possible with the billfish. The only alternative approach is to measure the growth of fish through time and infer the energy costs entailed. The first method is of course more accurate, but it is typically impractical with most large pelagics. Here the second method is used to obtain estimates on swordfish (*Xiphas gladius*).

Work on tunas by Kitchell *et al.* (1978) uses oxygen consumption measurements to estimate the weight specific metabolism of the fish under starvation conditions. This is then corrected for metabolic costs missed by this oxygen measurement due to the cost of digestion or Standard Dynamic Action (SDA; Davidson, 1994). They also report on measurements of tuna maximum consumption and come up with total metabolic cost of maintaining the fish. They then compare the resulting allometric curves relating energy to the fish's weight to the maximum consumption rate of the tunas to argue that fish in the wild operate at two to three times their maintenance metabolism (Kitchell *et al.*, 1978). The result then is then the expected energy cost of the fish and its maximum rate of energy gain through foraging as a function of the tunas' mass. Converted to measures of total fish mass the basal metabolism for maintenance is $\mu = 122W^{1.016}$ kJ/day for the skipjack and $\mu = 122W^{0.80}$ kJ/day for the yellowfin. The maximum consumption rates are $C = 1050W^{0.70}$ kJ/day for the skipjack and $C = 1050W^{0.65}$ kJ/day for the yellowfin. These fish reach weights of 20 kg and 200 kg respectively. The Kitchell *et al.* (1978) work shows consumption (C) and the weight-specific metabolic curves converging at these maximum weights suggesting a consistent picture of the tunas' life history. While this convergence of possible forage and energetic needs seems to suggest an evolutionary constraint, it is not clear how this limit is reached since the fish compete in the ocean on the basis of individual fish, i.e. there is not a weight-specific limit. The curves based on multiplying the weight specific rates by the weight of the fish (W_i) are used here. On a fish basis the consumption (C) and minimum energy curves never intersect (Fig. 10.3). Here these early estimates of tuna metabolism will be used to provide a minimum envelope for the requirements of these animals. Newer work allows some checks on this earlier assessment.

The synthesis of Korsmeyer and Dewar (2001) and the literature within provide a check on the earlier Kitchell *et al.* (1978) work and provide a more refined estimate of the metabolic costs of being a tuna. The more recent work of Brill (1979, 1987), introducing new techniques involving surgical or

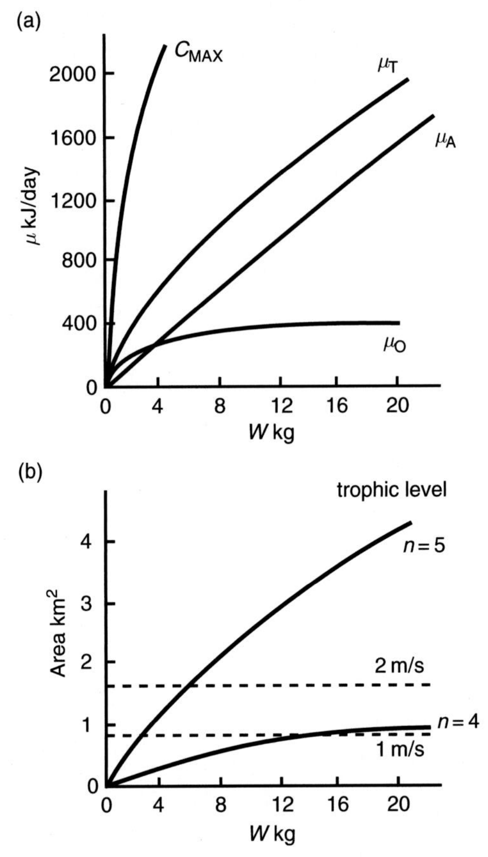

Figure 10.3 (a) The metabolic costs for a yellowfin tuna swimming at 1 m/s plotted as a function of weight using Brill (1987) results on μ_o and swimming costs from Korsmeyer and Dewar (2001). The estimated cost of activity is μ_A, and μ_T is the total energy cost. The curve labeled C_{MAX} is the maximum consumption rate from Kitchell *et al.* (1978). Energy conversions are taken from Bernie and Levy (1988) using the oxygen mass from Pauling (1970). The lower curve is the estimate of μ_o, the next curve up the estimate of swimming cost and the next curve the total estimate of minimum cost for the fish to live in the ocean. (b) Area in square kilometers required to maintain the minimal energy balance (a) in a yellowfin tuna as a function of the fish's weight, W. The lower curve is the estimated minimum forage area needed assuming resources are evenly distributed. The curves assume that fish are foraging at the fourth ($n = 4$) or fifth ($n = 5$) trophic level in an oligotrophic sea with a primary production of 1.6 kJ/m^2/day. The upper curve shows the areas needed to maintain maximum consumption on the part of the fish. See text for details.

drug-induced torpor as a method of assessing the true basal metabolism of tunas (Standard Metabolic Rate, SMR), provides another cross check on the earlier conclusions. Gooding *et al.* (1981) revisit the methods reviewed in Kitchell *et al.* in newer experimental settings. The data on skipjack are not consistent between these works, with Brill's methods (skipjack $\mu_o = (145.4 \pm 9.6) W^{0.563}$ kJ/day) giving much lower slopes for the metabolic curves than either the Kitchell *et al.* (1978) or Gooding *et al.* (1981) results. The yellowfin data, however, are more consistent and provide a useful place to make the conclusion of this work.

Using the Brill (1979, 1987) estimates of yellowfin Standard Metabolic Rate (SMR), or resting state metabolism ($\mu_o = (101.2 \pm 9.5) W^{0.573}$ kJ/day), it is possible to produce a minimum floor for tuna energetics. This work involving immobilized tuna provides a minimum in energy expenditure and therefore a floor for the calculation here (Fig 10.3). Added to this is the cost of swimming at a nominal speed necessary to provide a yellowfin with enough oxygen ($\sim$1 m/s, Kitchell *et al.*, 1978). Then subtracting the swimming costs given in Korsmeyer and Dewer (2001) provides the total minimum estimated energy costs for yellowfin as a function of weight. These are the minimum in the sense that they do not include a correction for SDA, non-aerobic metabolism associated with burst swimming, growth or reproduction. Estimates in Korsmeyer and Dewar (2001) suggest that the non-aerobic and growth alone roughly double the actual costs. Note that these minimum costs fall far below the maximum consumption curve from Kitchell *et al.* (1978) feeding studies. This is consistent with the idea that tunas are designed for rapid movement and are capable of massive consumption rates. The question becomes how the ocean obliges with the needed resources.

To answer the question of how the ocean resources are allocated to tunas, the ability of a yellowfin to harvest the ocean environment is estimated by allowing the fish to swim through an oligotrophic sea. Assuming that yellowfin are between the fourth and fifth trophic level and a maximum trophic efficiency of 20% (Parsons *et al.*, 1988 estimate efficiencies for transfer of energy between trophic levels to be between 10 and 20%), and a uniform distribution of resources Fig. 10.4 gives an estimate of the area a tuna needs to cover to obtain its daily ration. Assuming a search radius of 5 m for a total search across track of 10 m and a velocity of 1 m/s a fish can cover 0.86 km^2 in a day. Therefore at fourth trophic level fish can harvest their environment successfully up to a weight around 15 kg, while at fifth level appropriate minimum energy requirements are not met beyond the two kilogram size. Tuna, however, are not solitary animals, but travel in schools that can number hundreds to thousands of fish. While a school extends its coverage of the environment above that available to the senses of a single fish, even with the entire school in a line the net result is the same. For other school geometries there is a net loss of efficiency. Therefore if a single fish can just meet its metabolic requirements, a school must share, the individual portion is equal to the schools daily grazing area divided by the number of fish it contains. Based on these minimal resource considerations the conclusion is that tunas require an aggregation of resources by a factor of 100 to 1000-fold in order to exist in an oligotrophic ocean. That is schools subsist on foraging on other schools. The large pelagics are therefore adapted to seek out aggregations of prey and then efficiently consume them (Brill, 1996; Essington *et al.*, 2001). These adaptations peak with the endothermic (warm blooded) bluefin tuna (Carey and Lawsen, 1973).

The problem is not whether a single fish can actually realize this type of area, but that both of these species are typically organized into schools that can number 100 to 1000 fish. This means that a small school containing 100 fish would have to cover 100 km^2/day. Now considering a school of 100 1 kg fish packed with the fairly loose rate of a fish per 5 m comfort radius (Flierl *et al.*, 1999) in a circular array the school would need to swim at 1.2 m/s continuously to obtain the necessary ration. This is above the swimming rate of 0.7 m/s that is needed to meet the oxygen requirements of the fish (Magnuson and Weininger, 1978). Assuming the scaling from the laboratory measurements on small fish pertains to 100 kg fish, a school of 100 would have to swim over 900 m/s to obtain sufficient forage in a homogeneous ocean. While the scaling to these large fish is problematic (Brill, 1979) it is not expected to be off by more than a factor of two. So having neglected social costs involved with schooling and the energy costs of predator avoidance and reproduction, it seems that these tunas are approaching the minimum metabolic requirements for foraging the mean field ocean. In order to accommodate these latter activities, however, it is important that both the different trophic levels that make up the food webs utilized by these fish aggregate and that the tunas themselves have behaviors that integrate cues to finding these and modify their trajectories to adequately harvest them.

The swordfish are solitary in terms of their forage behavior, but with their greater weight and high-energy life-style also have a challenging problem in terms of energetics. The energy cost for swordfish is worked out from growth rates and estimates of their metabolic costs (Withers, 1992; Diana, 1995). Although these numbers do not include the physiological measurements available for the tunas they seem to be consistent with an overall expectation for large pelagic predators. A calculation similar to the one above for tunas suggests that swords require between 1.3 and 15.7 km^2 per day (Radlinski, per. comm.). Again this can be attained at a reasonable cruising speed and optical foraging range, for example 1 m/s and 10 m optical response gives a coverage of only 0.9 km^2/day, but adding reproductive and predator avoidance costs suggest that aggregation swordfish and their forage resource must be important aspects of their life history.

Since it is obvious that these populations of tunas and other large pelagics such as swordfish exist and have flourished to the extent that we routinely exploit them, the problem becomes one of understanding the trajectories that lead to both the necessary prey aggregations and the requisite behavior on the part of the large pelagic predators that allows them to successfully utilize them. As outlined in the theoretical issues for the basic structured models above, it is the details of the trajectories provided by the ocean circulation and the modification of these related to behavior of both the predator and prey that govern the final dynamics of the ecosystem. Following the arguments above on the problem of viewing these trajectories and the dynamics along them in a Lagrangian frame, the final issue to explore is the specification of the predator/prey trajectories that lead to the observed ecosystem.

10.4 Trajectories of marine life

10.4.1 Types of movement in the ocean

The final problem in this review is that of documenting the actual trajectories that allow marine organisms to successfully utilize the ocean environment. As demonstrated by the energetic considerations above, these trajectories must not only provide access to the physical environment that is conducive to survival in terms of temperature, salinity, and light, but there they must also provide access to nutrient resources or forage. In biological oceanography it is typical to classify organisms into several groups.

- Plankton: free drifting organisms carried by ocean currents.
  - Holoplankton: organisms that spend their entire life as plankton. These include both phytoplankton (plants) and zooplankton (animals).
  - Meroplankton: animals or plants that spend only a portion of their life cycle as plankton. These include the larvae of many bottom dwelling (benthic) organisms and many forms that become free swimming forms or nekton as they grow.
- Nekton: free swimming organisms.
  - Micro-nekton: smaller organisms with ability to move well especially in the vertical, but with little ability to navigate horizontally.
  - Pelagics: free swimming organisms with the ability to range across the ocean typically with the ability to navigate.

There is some overlap between these classifications since as expected there is a gradation in the abilities of organisms to move themselves through the environment. For example, even many uni-cellular phytoplankton have the capacity of modifying their depth by either modulating their buoyancy or actively propelling themselves in the vertical using flagella. These and the small copepods that make up a large portion of the zooplankton then modify the action of horizontal currents on their distribution. The combination of differential growth and aggregation through these vertical migrations acts to produce patch structure in the lower trophic levels in the ocean. This patchiness in turn acts as a factor in patterning the behavior at higher trophic levels to eventually produce the ecosystem that has to support the energetic balance calculated above.

The discussion here begins with true plankton and their trajectories through various marine environments (see Hitchcock and Cowen, Chapter 11 this volume). For these organisms trajectories correspond to fluid trajectories themselves. Since the oceans are approximately incompressible there is no change in the concentration of these forms following the flow except for the increases associated with reproduction. Trajectories that lead to this population growth then depend on the introduction of nutrients along these trajectories and the light environment that is sampled. For example, trajectories in an upwelling system can carry a dilute number of phytoplankton cells in a nutrient-replete water mass upward in the light field. This then leads to a plankton bloom. Zooplankton entrained into this upwelled fluid can then make use of the increase in primary production to grow and reproduce. For the schematic trajectories (Fig. 10.4) to produce a sustainable ecosystem some of the trajectories must take portions of the plankton community back into the waters feeding into the upwelling to act as seeds for the plankton bloom. This can occur through turbulent mixing, but the closure within an upwelling cell can be greatly enhanced through the introduction of a very small amount of vertical motion on the part of the plankton community.

10.4.2 Eastern boundary currents and upwelling

Differential motion relative to fluid parcels occurs across a large range of organisms that are considered planktonic. A correctly timed change in vertical position relative to the flows in an upwelling system can greatly enhance the response of the ecosystem to the physically forced system (Fig. 10.4). The basic nature of upwelling systems is upward motion providing a source of limiting nutrients into the surface layers as first described by Nathansohn (1906). The problem becomes positioning of organisms to make use of this nutrient source. This is accomplished by one of the dominant upwelling phytoplankton types, the diatoms, by developing a differential sinking pattern through the successive generations of cells (Round *et al.*, 1990; Sumich, 1999). This involves size fractionation in the diatoms silica frustules (skeleton) as they reproduce

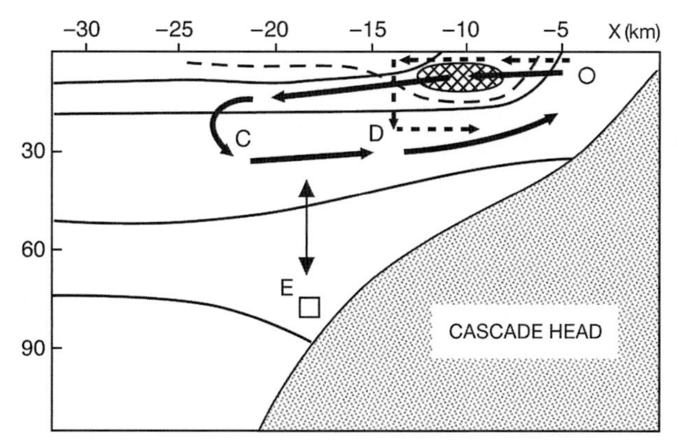

Figure 10.4 Schematic of the Oregon/California upwelling system showing the trajectories of diatoms (dashed line), copepods (circle and solid line) in the secondary cell induced by the upwelling dynamics. The figure also indicates the behavioral driven concentration of coastal euphausiids (square, vertical line) to make use of this system. The solid contours are density surfaces and the dashed and hatched contours the chlorophyll distribution with the cross hatching indicating the maximum. This in turn patterns the behavior of larger predators such as hake, tunas and blue whales to make use of the California current ecosystem.

through cloning. This occurs because half of each overlapping frustule is refilled upon cell division from the inside making the smaller side create a smaller cell half each division. This can continue until the cells are up to 25% smaller than the original cells created under sexual reproduction. The latter resets the population, but does not typically occur during bloom conditions. Since essentially all diatoms are denser than sea water (Round *et al.*, 1990) the result of sinking and differential vertical mixing in the turbulent field is expected to sort the diatoms vertically in the water column as it extends offshore in an upwelling system. Eventually it is the larger portion of the size spectrum that sinks to depth and enters the fluid being carried back on shore. These cells can either complete the circuit themselves, or produce spores to finish the loop by sexually reproducing. This is one example of the means phytoplankton can adapt to the details of fluid trajectories in upwelling systems. Equally important to the ecosystem is succession between phyto-planktors as the fluid is carried off shore (Parsons *et al.*, 1988). It is not clear to what degree phytoplankton use sinking to their advantage or whether the deep chlorophyll maximum found off the coast is due to sinking, adaptation, and/or just subduction associated with the offshore physical advection (see Bienfang *et al.*, 1981, 1982; Washburn *et al.*, 1991; Kadko *et al.*, 1991). Hoffmann *et al.* (1991) use Lagrangian simulations to consider the dynamics of upwelling off California. Phytoplankton behavior will be further explored below in terms of the mechanisms at work in various ecosystems. The California ecosystem is further structured by adaptive behavior on the part of the secondary producers in the zooplankton.

Similar mechanisms for differentially positioning populations of copepods within upwelling zones are documented by Peterson *et al.* (1978) among others. Peterson's work on *Calanus marshallae* in the upwelling off Oregon demonstrates the ontogenetic (age-related) behaviors that allow this copepod to exploit the upwelling system (Fig. 10.1). This small herbivore comes up into the newly upwelled fluid on the coast as a reproductively capable adult. This places the young hatched nauplii in the outset of the offshore advected phytoplankton bloom. Their concentration reaches its maximum in the phytoplankton biomass maximum in the outer upwelling front (Peterson *et al.*, 1978; Olson, 2002). The older copepods are then carried offshore until they reach a stage where they sink back into the deeper onshore advected fluid. These then molt to become adults just below the coastal upwelling. This behavior with age then provides trajectories over the life cycle in *C. marshallae* and provides means of utilizing the upwelling system in the California Current. These patterns in life cycles and behavior continue into the higher trophic levels.

At the level of the micro-nekton, there are a large number of primitive shrimp-like euphausiids that are found in the California Current system (Brinton, 1962). Many of these are just following trajectories that take them into the coastal regime from the North Pacific. In some species such as *Nematoscelis difficus* the presence of these species involves simple advection into the system with possible reproduction, but not a behavior that really makes use of the inner upwelling system (Reid *et al.*, 1978; Olson and Hood, 1994). *Euphassia pacifica*, on the other hand, while found across the entire boreal North Pacific seems to have a definite vertical migration behavior that allows it to utilize the thin strip under the coastal upwelling (Mackas *et al.*, 1997; Fig. 10.4). The apparent behavior that allows these animals to maintain themselves in this zone is diel vertical migrations between the upper levels with their equatorward, and seaward flow and the reversed flows at depth. This is apparently sufficient to maintain thick concentrations of *E. pacifica* along the fringe of the North American Pacific coast. Vertical migration between the undercurrent and upper level flows provides a mechanism for retaining *E. pacifica* populations along the coast in proximity to the upwelling resources. This is an example of the type of trajectory modification relative to the ocean circulation that allows upper trophic levels to exist in the ocean. These behaviors are repeated or modified in the upper trophic levels where active migration along systems by adults becomes important. This becomes the case in fishes such as hake (*Merluccius productus*, Mackas *et al.*, 1997) who migrate along the west coast of North America to spawn in regimes that provide increased survival chances for their young while still maximizing the adult foraging experience. Therefore it is the complicated behaviors that modify the already complicated ocean currents (Fig. 10.5, see Brink *et al.*, 1991) to produce the spatial structure of food webs in settings such as the California Current. Before looking at the actual details of behaviors at various trophic levels, it is worthwhile considering the nature of the other major coastal upwelling systems in the western boundary currents.

10.4.3 Western boundary currents

Many of the same attributes might be expected in western boundary currents, but with modifications that take into account the different ecosystems. The overall upwelling in western boundary currents is equivalent to that in their eastern boundary counterparts (Olson, 2001). While the tendency for the eddy field and the boundary current structure to disperse offshore under the influence of beta drift in the eastern boundary currents stretches these environments seaward as mentioned above, the beta driven compression of western boundary currents through western intensification reduces the area of upwelling to a narrow ribbon (see Olson, 2002). Along coast transit times are seasonal or longer in the eastern boundary currents, while they involve time scales of weeks in the Gulf Stream or Kuroshio front (Fig. 10.6). The trajectories that arise for these meandering fronts have been quantified using floats by Bower and Rossby (1989) and Bower (1991). The Lagrangian time scales in eastern boundary currents then are much longer than the biological time scales even at higher trophic levels, while in western boundary currents only the phytoplankton time scales are faster than the physical ones. It is difficult to reconstruct a diagram like that in Fig. 10.4 for a western boundary current because of the rapid downstream advection. The evidence for interactions between trophic levels is also hard to document in these systems (Nakata, 1990; Olson, 2001, 2002). While the trajectories shown above for eastern boundary conditions do not necessarily insure a match between an organism's path and its nutritional resources, it is harder to see these necessary matches in western boundary current trajectories. Observations such as those of *Nannocalanus minor* in a Gulf Stream meander by Ashjian and Wishner (1993) show that copepods can respond quickly to phytoplankton blooms driven by the meandering and reproduce. The fate of the newly produced eggs in the very patchy meander environment is less certain (Olson, 2002). The Gulf Stream fronts as they extend eastward off the coasts separate eutrophic and oligotrophic waters and involve considerable mixing of these

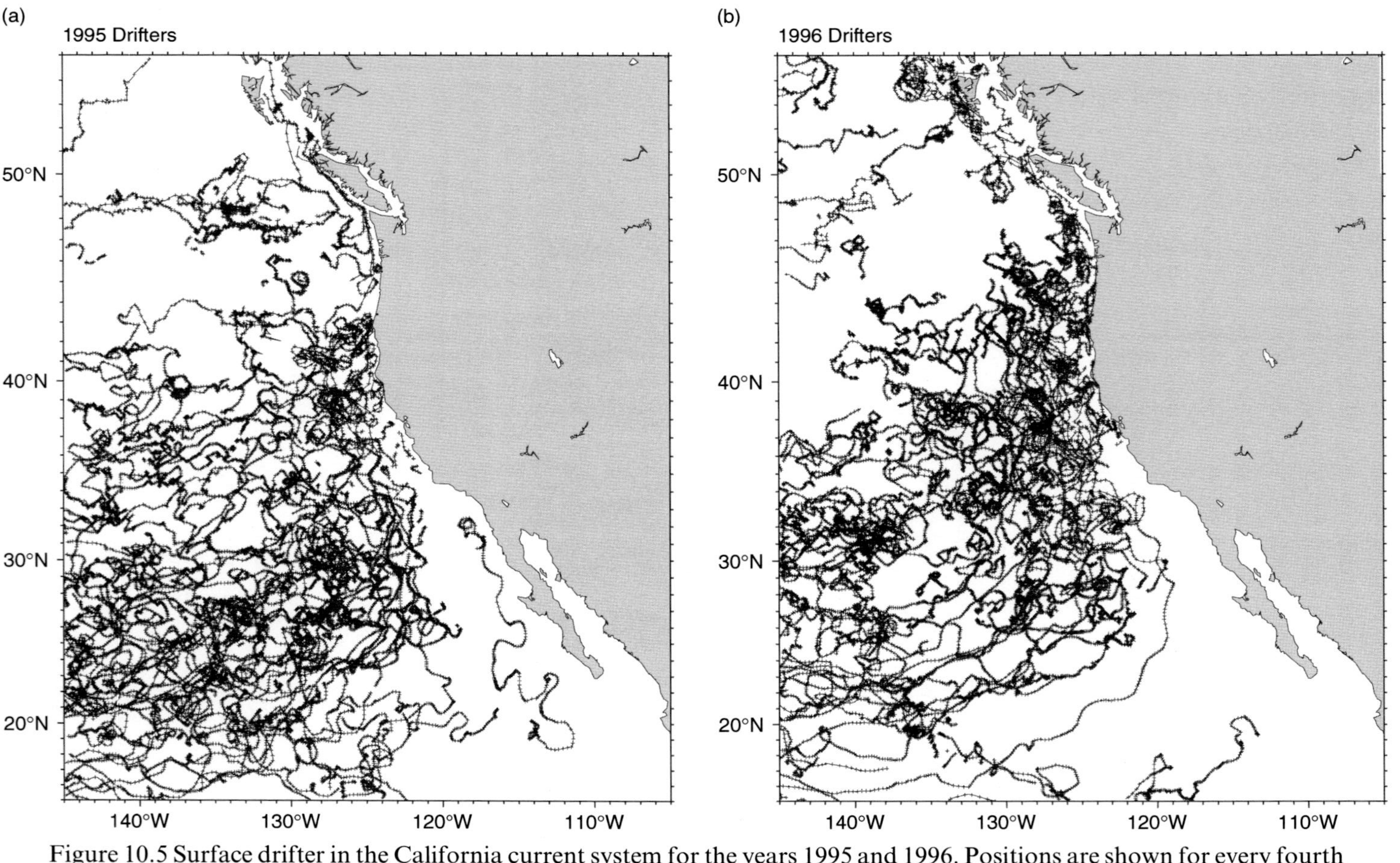

Figure 10.5 Surface drifter in the California current system for the years 1995 and 1996. Positions are shown for every fourth day. While the different deployments of units in the two years make it hard to make a complete comparison, it is possible to discern some variations between years. A major issue in relationship to Fig. 10.4 and the discussion in the text is the complex nature of the two-dimensional flow revealed in these surface constrained trajectories.

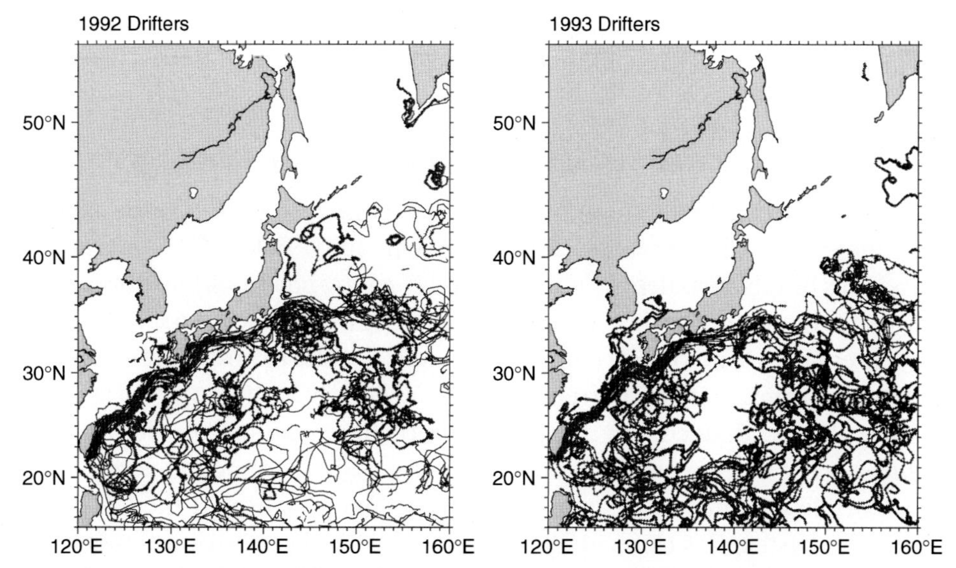

Figure 10.6 Surface drifters from the Kuroshio for 1992 and 1993. The scale of the plots and the plotting interval (4 days) is the same as in Fig. 10.5 to allow an inter-comparison of a western and eastern boundary current domain. To the author's knowledge there were no concentrated launches in the Kuroshio, so the concentration of trajectories there depicts the concentration of surface confined particles into the strong western boundary current front.

waters. While eastern boundary currents also involve lateral mixing of water masses, the interactions between different biomes is less intense. In the western boundaries it is probable that there is considerable loss of organisms that are mixed into environments in which they are not competent to take either the physical or biological challenge. It is interesting that the same biogeographic indicator organisms, transitional or temperate, occupy both the eastern boundary currents and the poleward sides of the western boundary current extensions (Brinton, 1962; Reid *et al.*, 1978). From the large-scale physical point of view it is difficult to explain the existence of endemic species inhabiting a part of the ocean circulation with unidirectional flow between two lethal environments (Reid *et al.*, 1978; Olson and Hood, 1994). The secret of maintaining populations in these zones lies in the nature of the mesoscale field and its ability to introduce transition zone species back to the west (Wiebe and Flierl, 1983; Olson, 1991). The opportunistic nature of these fauna is borne out in the fact that they invade and dominate both the cold, cyclonic eddies (rings) produced when temperate waters are extruded equatorward across the boundary current and the warm, initially oligotrophic anticyclones that are formed on the current's poleward side. It is the details of the fluid trajectories in proximity to these eddies or rings that plays a crucial role in the overall ecology of these organisms (Flierl, 1981; Flierl and McGillicuddy, 2002).

10.4.4 Lagrangian dynamics at the mesoscale

When sampled at high enough spatial and temporal scales the oceans are dominated by the presence of a turbulent eddy field. The Lagrangian dynamics of this turbulence defines the daily environment of most organisms far better than large-scale climatologies (see Figs. 10.5 and 10.6). Turbulence by its very nature involves a complex cascade of scales in both time and space. For the purposes of the current discussion, however, the interest is in the mesoscale as defined by the local baroclinic radius of deformation ($Rd = \sqrt{g'h}/f$ where g' is the reduced gravity, h the thermocline depth, and f the Coriolis parameter; Olson, 1991). In the inner portion of the California upwelling system this can be 10 km or less, while in the subtropical side of the Gulf Stream or Kuroshio it is around 50 km. It is this scale that dominates both an Eulerian and a Lagrangian description of eddies. As important to the Lagrangian view of eddies, however, is the intensity of the flow relative to the motion of the eddy itself through the fluid (Flierl, 1981). If the translation of the eddy through the fluid either due to large-scale advection or self-induced Rossby-drift (Nof, 1981; Flierl and McGillicuddy, 2002), c, is less than the swirl velocity, $v \sim (g'/f)\nabla h$, in an eddy the core fluid becomes trapped and moves with the eddy (Flierl, 1981; Dewar and Flierl, 1985). This retention of a portion of foreign habitat on the opposite sides of the western boundary current in rings provides an interesting natural experiment where organisms compete to dominate the slowly changing environment in the ring. In most cases the evidence shows the temperate or transition species out-competing their subtropical and boreal counterparts to dominate the ring cores (Wiebe and Flierl, 1983; Wiebe *et al.*, 1985). Rings drift westward on the earth due to the changes in the Coriolis acceleration across them. This is a nonlinear manifestation of the tendency for planetary waves to have a westward phase velocity (Nof, 1981; Flierl and McGillicuddy, 2002). Biologically this westward motion and the tendency for rings either to accumulate along the western boundary or to rejoin the western boundary current has the effect of bringing elements of the transition fauna and flora back to the western end of the biogeographic zone. Therefore rings provide a mechanism for maintaining the transition zone populations against the uni-directional mean circulation.

In addition to providing a means of closing the transitional biogeographic distributions in the oceans, rings and their surrounding fronts provide an important ecological interaction site where frontal enhancement can contribute to the aggregation of prey needed to support large pelagics such as swordfish. The mechanisms involved with frontal enhancement are discussed in detail by Olson (2002). They consist of upward transport of nutrients and phytoplankton blooms, aggregation of photo- or geo-tactic organisms as they are swept into the front (Olson and Backus, 1985), and active accumulation of fauna in the front in response to the increased forage base. Observations of the mid-water or mesopelagic, lanternfish *Benthosema glaciale* in ring 82B by Olson and Backus (1985) show that aggregation through frontogenesis can create increases in population levels in ring fronts of 100 to 1000 fold over the surrounding environment. These aggregations took place over a period of nearly 50 days. Estimates based on models of frontal dynamics and measurements in other frontal zones suggest that ring fronts aggregate organisms more slowly than the western boundary currents themselves, but produce a persistent enhanced environment that can last for months to years. In the case of Kuroshio warm core rings, which can persist for years in the large zone separating the Kuroshio from the Oyashio, rings pattern the major fishing efforts. Pacific saury (*Cololabis saira*) are a boreal species that avoids the ring cores and are aggregated in the cold streamers that the rings pull out of the Oyashio front (Saitoh *et al.*, 1986; Sugimoto and Tameishi, 1992). At the same time flying squid (*Ommastrephes bartrami*) and skipjack tuna (*Katsuwonus pelamis*) are found in aggregations on the inner portion of the ring front in the modified subtropical waters. Mackerel (*Scomber japonicus*) and Pacific bluefin tuna (*Thunnus thynnus*) are also found in the frontal bands associated with ring passage (Sugimoto and Tameishi, 1992). Similar patterns are found in the distribution of long-line fishing effort on swordfish around warm-core rings in the Gulf Stream (Podesta *et al.*, 1993). While the enhancement involved with the western boundary current fronts and their associated rings can easily provide 100 to 1000 fold increases in forage concentration, this seems to still be somewhat marginal when compared to the energy requirements derived above. One therefore expects even more aggregation to occur at finer scales within these fronts. Observations suggest that these submesoscale aggregations exist at the scale of a kilometer or so (Yamamoto and Nishizawa, 1986). The evidence for these aggregations is heavily biased towards the smaller nekton, while it is in the higher trophic levels that further aggregation must make up the difference in the energy requirements. The evidence for further aggregation at these intermediate trophic levels comes from the fisheries for species such as saury and squid and from acoustic surveys (Arnone *et al.*, 1990; Olson *et al.*, 1994). The details of the spatial distributions and movements involved with higher trophic level interactions remains very poorly known.

10.4.5 Trajectories and behavior

Before further consideration of other examples of biophysical interactions it is worthwhile exploring the detailed nature of trajectories in eastern and western boundary currents and their mesoscale eddy fields. Of major interest is the manner in which trajectories connect different portions of the flow field, and the locations of expected lower trophic response that is conducive for successful foraging at the higher levels. There is also the question of the behavioral traits and the cues that allow nekton to fully utilize the environment and find the necessary aggregations of prey. Before considering the effects of biological behavior it is worthwhile discussing the classification of trajectories of passive particles in realistic oceanic flows. The goal is to provide the means of classifying the structure of trajectories in the Lagrangian frame (see Oswatitsch, 1959).

For the physical flow field the trajectories can be quantified by considering the eigenvalues of the Jacobian of the velocity field or strain matrix (Perry and Fairlie, 1974). The shear matrix is comprised by the gradients in the velocity field, $\nabla\mathbf{V}$. For a two-dimensional surface flow with velocity components u, v these gradients can be written out in terms of the normal deformation, $N = \partial u/\partial x - \partial v/\partial y$; the shear deformation, $S = \partial u/\partial y + \partial v/\partial x$; the divergence, $D = \partial u/\partial x + \partial v/\partial y$; and the vorticity, $\zeta = \partial v/\partial x - \partial u/\partial y$. The shear matrix then can be written as

$$\begin{bmatrix} D+N & S-\zeta \\ S+\zeta & D-N \end{bmatrix}.$$

The eigenvalues for this matrix are

$$\lambda = -\left[p \pm \sqrt{p^2 - 4q}\right]$$

where $p = -[(N+D)+(N-D)]$ and $q = [(N+D)(N-D)-(S+\zeta)\ (S-\zeta)]$. These eigenvalues define the geometry of the trajectories in the flow field. For real eigenvalues there are stable and unstable nodes and saddle points, while the complex eigenvalues lead to foci that again can be stable or unstable in the sense that flow is into or out of them (Fig. 10.7). Of major importance are the regions where λ is purely imaginary and the flow elliptical. These regions are separated from surrounding flow by a stagnation point $\mathbf{V} = 0$. Upon appropriate transformation to follow the trajectory of individual closed vortices or eddies, the analysis reveals the ability of these systems to transport portions of ocean habitat relatively intact (Flierl, 1981; Olson, 1991). When the eigenvalues are purely real they also lead to stagnation points separating portions of the flow through a hyperbolic stagnation point. Here the patterns

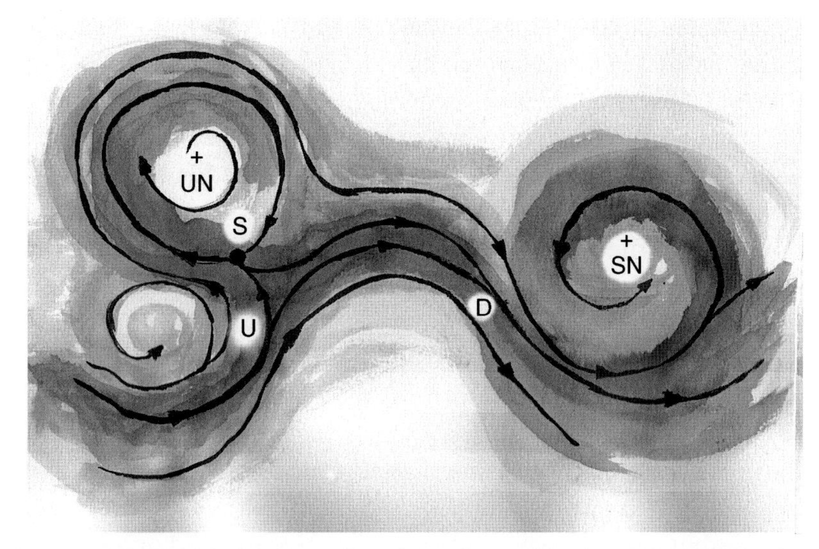

Figure 10.7 Categorization of trajectories using local two-dimensional analysis of the shear matrix to infer both the nature of the horizontal flow, but also its interaction with the vertical velocity field, i.e. zones of upwelling and downwelling. The solid lines are trajectories while the ink washes are meant to depict the flow of a tracer in the domain. The letters denote the various identifiable areas in the flow; UN indicates an unstable node; SN denotes a stable node with a flow into it, U indicates upwelling and D denotes downwelling in the meander trough. The solid point at the bottom of the anticyclonic vortex is a septrix point that acts to isolate the flow in the vortex from the surrounding flow. This is an important aspect of the succession dynamics discussed above. The basic description of this flow is based on meanders in the Kuroshio or Gulf Stream extension (Olson *et al.*, 1994), but the concepts behind the classification are applicable to the wider nature of oceanic flows.

in the ocean ecosystem induced by the flows are first explored followed by a discussion of adding active swimming behavior to the problem.

The combination of possibilities involved with the solutions for flows (Fig. 10.7) introduce a set of conditions that act to pattern life in the ocean. The geometries involved are shown in Fig. 10.7. The example envisions a synoptic example that can be found in any western boundary current extension. Similar flows are found in eastern boundary currents, although the amplitude of flows differs significantly. The major change involves the amplitude of mesoscale eddies with weaker eddies having smaller trapped areas (Flierl, 1981) and weaker impacts of lifting fluid as they move with less eddy induced productivity enhancement (McGillicuddy *et al.*, 1999; Flierl and McGillicuddy, 2002). In other venues the geometry of flows is crucial for biological retention in regions as explored further below. It is possible to picture a complementary matrix for the one above that includes the terms that govern the motion of fish through the biophysical cues that make up their environment. This could then be used to categorize types of swimming trajectories in a manner similar to the physical flows. Here the role of flow geometry and behavior in producing the level of aggregation required by the energetic analysis above is explored.

Although it is difficult to observe directly, the ability of organisms to swim can be implemented within the same framework. While the behavior of individual fish may be hard to picture in terms of vector field variables, one can introduce similar terms to define density dependent actions such as schooling behavior. These can then be coupled to the physical fields to consider such behaviors in realistic oceanic turbulence as suggested by Flierl *et al.* (1999). Due to its prevalence in the ocean, an important aspect to be considered is the act of schooling. If as suggested in the calculations above resources are limited to the extent that aggregation of prey must involve 100 fold or higher increases in prey density, why do many upper trophic level organisms organize themselves in schools. The inherent low-order property of a school is that it has to share forage. This introduces a very different population dynamics (Cosner *et al.*, 1999) that leads to ratio dependence in the population interaction terms. Trade-offs involve increased security either on the individual (low chance of predation) or group level (scare tactics). The school also provides a wider taxis sweep of the environment as discussed in the case of tuna. Before any further consideration of upper trophic levels it is important to document aggregation in the primary and lower trophic levels.

10.4.6 Biological enhancement and aggregation dynamics

The calculations above suggest that there is an upward increase in the degree of aggregation versus foraging ability in marine taxa as one goes up the trophic system. Even in conservative computations the level of increase has to be in the order of 100 fold or more. The question is then the mechanisms that might lead to this level of aggregation of resources. The most commonly addressed enhancement mechanism in marine systems is the upwelling of nutrients into the photic zone and subsequent phytoplankton blooms. This is manifest in the difference in primary production between eutrophic waters and the oligotrophic open ocean environment. This only provides a factor of at most 200 in terms of the mean field gradients in primary production (Parsons *et al.*, 1988; Valiela, 1984). As discussed above in the context of the California Current, movement of phytoplankton cells relative to the waters they are suspended in may play a role in redistributing them in ways that are adaptive for them and the herbivores that rely upon them. The question of relative motion in the phytoplankton has been under discussion for a long time (Gran, 1915; Brandt, 1920; Allen, 1932; Smayda, 1970). There is still debate on how cells sink. Quoting Smayda (1970), Mann and Lazier (1996) have a long discussion over the point of whether or not cells sink in concert with Stokes Law. The law states that spherical objects at low Reynolds number will sink with a velocity, $w_s = \frac{2}{9} g' r^2 / v$, where $g' = g(\rho - \rho_c) \,/\, \rho$ and ρ_c is the cell density, ρ the water density, v is the viscosity, and g the gravitational acceleration. Mann and Lazier (1996) consider the non-spherical nature of phytoplankton as a reason for the slower fall rates recorded in Smayda (1970), but it seems more likely to this author that the failure to follow a Stokes curve involves size related changes in cell density. The problem is the difficulty in measuring ρ_c (Smayda, 1970). In the case of the diatoms in the upwelling system introduced above one should expect the ratio of cell sap to frustrule to change as the cells become smaller. With these changes the outcome in terms of cell trajectories in Fig. 10.4 becomes uncertain. Considering a cell 100 μm in diameter with $v = 10^{-6}\,\mathrm{m^2/s}$ and a sinking rate of 5 m/day (see Sverdrup *et al.*, 1942), the Stokes equation can be used to estimate a density contrast of $(\rho_c/\rho) = 2.7 \times 10^{-3}$. This is approximately four times the density contrast between the surface and the 50 m level in the upwelling system (Fig. 10.4) suggesting that sinking is not strongly controlled by the physical stratification on the depth scale of the euphotic zone. If one assumes that the density contrast is constant throughout the cloning process the resulting spread of the cells with a 10% decrease in cell size per generation produces a spread of a population between 50 m for the large initial diatom to 23 m for the level of the shallowest and smallest clone after ten days (Fig. 10.8). The spread of the population allows the diatom to sample a broad set of depths and for the larger, deeper population, a chance to make use of the shoreward flow back into the upwelling (Fig. 10.4).

Even if one expects local spatial/temporal enhancements above the mean field values it is clear that additional mechanisms must be involved. As summarized in Olson (2002) these mechanisms include the interaction of geotaxis (affinity to the ocean surface) and phototaxis (behavior tied to light levels) that accumulate organisms under flow environments with negative D, i.e. convergence, as experienced in fronts (Fig. 10.9). Observations in ring fronts (Olson and Backus, 1985) and riverine plume fronts (Govoni *et al.*, 1989; Govoni and Grimes, 1992) suggest that the sweeping action of fronts with these behaviors can lead to aggregations that amount to 100–1000 fold increases. In the take over by temperate species of ring cores discussed above Olson (1986) argues that these frontal processes are important for explaining

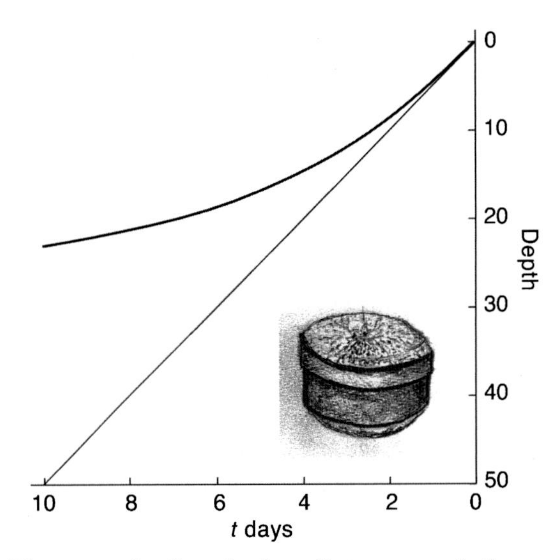

Figure 10.8 The spread of a cloning diatom population assuming 10% decrease in size per generation for the smaller frustrule (see inset for diatom geometry) and constant size of the larger. The lower curve tracks the larger fraction while the upper tracks the upper level of the population that is made up of ever smaller cells. The dynamics assume constant density ratio and Stokes sinking velocities. See text for model details.

the Wiebe *et al.* (1985) observations. See Olson (2002) for a summary of the basic mechanisms and rate estimates.

If one includes reproduction in response to increases in primary production (see Ashjian, 1993) the net effect of an increase at the lowest trophic level is magnified. However, most observations (Nakata, 1990; Ashjian, 1993; Ashjian and Wishner, 1993) imply a great deal of patchiness in the secondary trophic level response. Models of lower trophic responses to meandering can be found in Flierl and Davis (1993) and Flierl and McGillicuddy (2002). The question becomes how upwelling enhancement of primary production interacts with the behavioral aggregation of zooplankton and micronekton. For predator/prey interactions, a major issue with higher trophic levels is the means that mobile nekton use to find the necessary resource aggregations.

The issue of the nekton involves a higher level of Lagrangian dynamics in the sense that it involves both the ocean circulation and its imprint on the distribution of prey, but also the degree that individual or school trajectories are determined by specific sets of behavioral cues. While swimming strongly modifies the trajectories of nektonic animals, their pathways are still influenced by the flows in the ocean. The mechanisms by which oceanic currents set

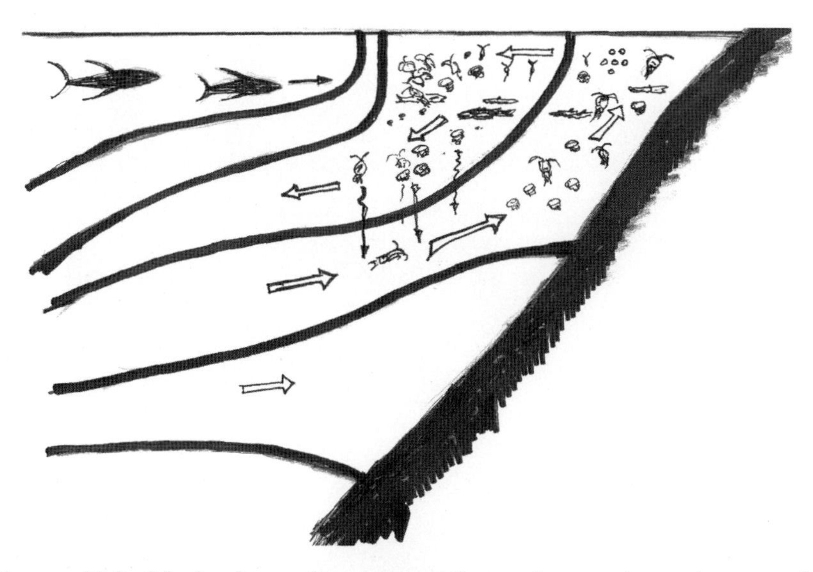

Figure 10.9 Mechanisms for aggregation of organisms in oceanic environments. The figure depicts increased production upon the introduction of nutrients into the light field through upwelling, behavioral aggregation in intermediate plankton and micro-nekton in frontal domains with simple photo- or geo-taxis, and active searching leading to aggregation in highly mobile nekton. For a fuller discussion of these mechanisms see Olson (2002).

up cues determine the navigation of large free swimming organisms. The use of cues varies between various taxa, and can include visual, acoustic, and sensations such as temperature. The first two of these can lead to taxis in the sense that they provide the ability to sense conditions at a distance. Due to the optical properties of seawater compared to acoustic properties, visual cues are less effective than acoustic ones. Typically visual sensing is limited to ranges of tens of meters or less while some marine mammals may communicate over ranges that encompass entire ocean basins. Our knowledge of the use of acoustics in pelagic fishes is rudimentary, but it is conceivable that a large pelagic such as a swordfish can even hear prey or their con-specifics foraging behavior at fairly great distances. Tactile senses of conditions in sea water such as salinity or temperature provide local information. By modifying swimming behavior in response to thermal preferences nekton can move along thermal features such as fronts (Humston *et al.*, 2001; Olson, 2002). Such movements can become very effective migration tools such that long coherent fronts as those in the Gulf Stream or Kuroshio become pathways in the sea (Olson and Podesta, 1988; Humston *et al.*, 2001). The mechanisms that concentrate forage in these features make behaviors that maintain large pelagics within

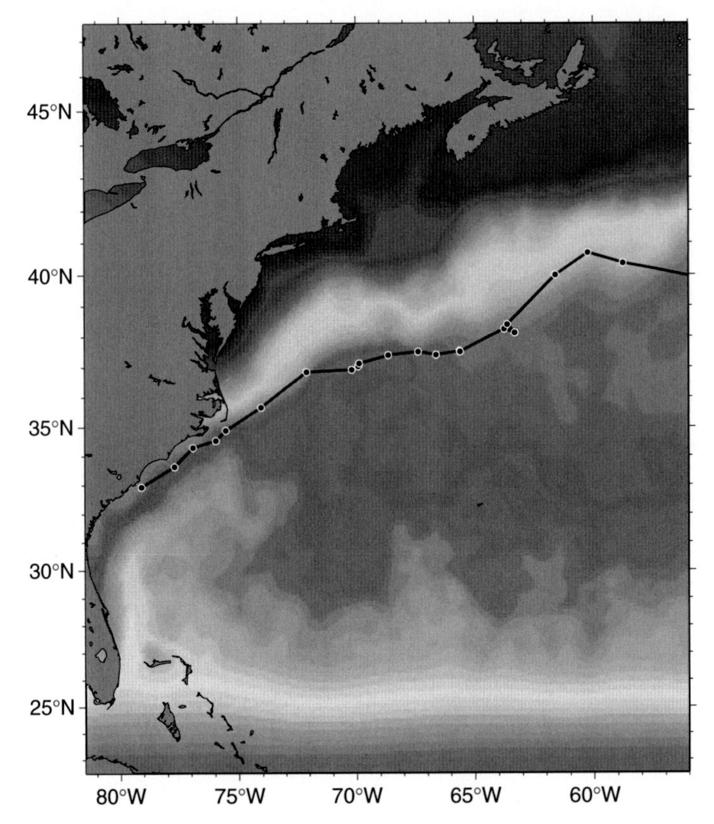

Figure 10.10 Track of right whale along the Gulf Stream over a several month period. Points denote 10 day progression of the whale to the east during the spring and early summer. See Plate 25 for color version.

them selective in an evolutionary sense. This stems from the increased survival in organisms that can optimize contact with adequate forage while still avoiding predation.

An example of the use of the Gulf Stream front by a northern right whale (*Eubalaena glacialis*) is seen in its trajectory in Fig. 10.10. This whale was found with a net caught on her flukes in the waters off Georgia. A satellite transponder was placed in the netting. The author was contacted by investigators at the New England Aquarium to assist them in deciding if the satellite tag had come off the whale and was free drifting or not. The trajectory is straight down the Gulf Stream front as seen in the satellite thermal chart. The progression down the front however is extremely slow compared to the surface currents expected in the Gulf Stream. The conclusion is that the whale is maintaining its position in the front and actually swimming against it. Presumably this behavior involves using the front as a foraging site. The trip out into the North Atlantic eventually took the right whale to an old whaling ground off the Azores where these whales are known to congregate in summer months. As fall approached the whale returned to the US east coast along the Gulf Stream front. The question becomes one of how the animal keeps itself in the front. There are a number of cues that can be used to keep in the front including detection and navigation along the temperature gradient, optical cues tied to the large increase in light attenuation across the northern side of the front, and finally the aggregated forage resources along the front. Similar patterns show up in sperm whale (*Physeter macrocephalus*) and swordfish (*Xiphias gladius*) that congregate along the Gulf Stream front in their foraging for squid (*Illex spp.*) along the front (Waring *et al.*, 1993; Olson, 2002) and its topographic partner, the shelf break front (Stillwell and Kohler, 1985). The search for swordfish along the Gulf Stream makes its front a discernable feature in charts of fishing effort by the US long-line fleet (Podesta *et al.*, 1993).

The evidence then suggests that the dynamics of the fronts provides the necessary aggregation factor required to assure the survival of large pelagic predators. Olson (2002) calculated the time scales involved with creation of frontal aggregations based on either geo- or phototaxis in the presence of frontogenesis. Active frontogenesis of course does not take place continuously along the Gulf Stream front, but is associated with the phase of meanders (Hitchcock *et al.*, 1993; Olson *et al.*, 1994; Olson, 2002). One would also expect that the aggregation time scale proceeds up the foodchains with primary production increases being followed by secondary responses and behavioral aggregation under frontogenesis and finally the arrival and aggregation of larger predators. Podesta *et al.* (1993) tried to find hints of this in the long-line operations in the Gulf Stream but could not. More recent efforts with better records from the Hawaiian long-line fleet suggest time scales from days to weeks (Polovina *et al.*, 2001). There is one more factor that leads to aggregations of prey capable of sustaining large pelagics in frontal zones like the Gulf Stream: migration.

The squid that make up a major portion of the diet of both the sperm whales and swordfish are migratory and have an annual lifespan. They attain most of their growth in the highly productive shelf waters off New England and eastern Canada. *Illex illecebrosus* spawns in the Gulf Stream front between Florida and Newfoundland (ODor and Dawe, 1998) and then migrate poleward onto the shelves as far north as the Labrador Sea where they feed before making a return journey into the Gulf Stream to spawn as adults. Since they die after spawning their single year turn around and the long migration makes them quite different from the swordfish that have a similar migration pattern, except for the use of the shelf environment, but are long lived therefore distributing their genes through a number of year classes instead of the single one (see O'Dor, 1998; Bakun and Csirke, 1998). The squid spread their migrations between what O'Dor (1998) calls micro-cohorts that spawn at different times. In the case of *I. illecebrosus* this involves two protracted migrations that correspond to spring and fall blooms in the Gulf Stream. The similar short-fin squid in the Kuroshio (*Todarodes pacificus*) is divided into three subpopulations with winter migrations in both the Kuroshio and the Sea of Japan and additional summer and autumn migrants in the latter. The eggs and larval squid then are carried northwards in these systems by the currents. A major factor in this is a constraint that egg hatching demands access to the warmer waters (O'Dor, 1998). For the swordfish the squid represent forage importation from the rich northern waters that are below their temperature threshold of around 10°C (Carey and Robison, 1981). The squid resource may also play a role in swordfish spawning which occurs along the southeastern coast of the USA (Govoni *et al.*, 1999). The spatial distribution in both predator and prey allows for accumulation of biomass while the aggregation needed for spawning provides enhanced foraging for the predator. The spatial/temporal spacing between adults and young introduced by equatorward migration in the adults in both squid and swordfish also separates populations in a way that allows cannibalism to be a successful life strategy in both the squid and the fish. For an analysis of cannibalism in space/time see Olson *et al.* (2005). Before going on to other settings it is worthwhile considering modification of the trajectories in pelagic nekton tied to schooling.

10.4.7 Schooling dynamics

Schooling behavior is common in the ocean and can involve hundreds of individuals aggregated into tight social groups that move as one. Theoretically schooling aids in foraging in that a larger group has better search and capture capabilities and in survival by intimidating potential predators and increasing the probability of survival for individuals under attacks. Schooling also increases swimming efficiency through interactions between individual swimmers (see Vogel, 1994). On the negative side schooling demands sharing of forage thereby increasing costs as noted above (Cosner *et al.*, 1999). There is considerable debate on the evolution of schooling in fish. Wilson (2000) suggests that schooling derives from the nomadic life strategy of fishes, although many non-nomadic fish such as groupers and snappers definitely school. The degree that schooling in fish such as tunas are random wanderings or fine-tuned migrations on definable behavioral cues (Humston *et al.*, 2001) deserves further consideration. At a macroscale schooling fish move with a single trajectory, but at fine scale schooling introduces additional motions tied to maintenance of the school and its interactions with other organisms and oceanic flows. One issue that arises is the limits to school size from the perspective of forage needs and actual maintenance of the school in the presence of oceanic turbulence (Flierl *et al.*, 1999). While the Flierl *et al.* work introduces the mechanisms behind schooling as an interaction of a group of individuals in a turbulent physical environment, the important aspects of foraging dynamics in schools where the fish are both predators and prey still remain to be addressed.

Schooling can also completely modify the range of fishes under conditions where it is obligate and decimated populations begin schooling with fish with different migration affinities (Bakun, 2001, 2005; Bakun and Cury, 1999). The ideas in these works suggest that schooling becomes a trap forcing individuals with different habitat affinities to join the majority of fish in a school and then ending up outside its preferred habitat. Bakun argues that this sets the response of pelagic fish populations to predation, climate shifts, and fishing. In order to explore these ideas further it is important to address quantitatively the nature of the trajectories involved and the underlying dynamics behind migrations in the pelagic domain.

10.4.8 Trajectories of organisms and recruitment

An important factor in fisheries science is the process through which individuals enter the population that is actually fished. This is termed recruitment to the fishery and usually occurs at a given age. It is important to understand recruitment pathways in order to minimize mortality in the pre-recruits that might create a recruitment failure. Perhaps the best example of the latter is the impact that dams without fish ladders have on salmon in various rivers in the northwestern USA (Harden, 1996). In general the linkages are less obvious than those in salmon to the extent that there is considerable doubt as to where fish recruit from. A major goal in fisheries management is to insure that the take of a fishery does not threaten future populations. In many populations the total range over which a population exists is larger than the reproductive range in which they actually successfully reproduce. One approach then is to identify and protect the reproductive range. A case in point occurs in the spiny lobster (*Panulirus argus*) in Florida waters. This species has meroplanktonic larvae with a competency period (viable time span) in the plankton of up to a year and can recruit from sites as far away as Brazil. Lobsters spawned in Florida waters meanwhile are probably largely dispersed into the North Atlantic. One argument is that fishermen should be allowed then to catch all of the lobsters in Florida since they do not contribute to recruitment! Without going into issues like the other ecosystem impacts that would result in fishing out a species from a region, it is clear that the trajectories that can carry lobster recruits to Florida are an issue. There also needs to be more attention to local closed circulations that can lead to retention and self recruitment in an area (Yeung and Lee, 2002; Paris *et al.*, 2002). It is also important to introduce the other factors that govern recruitment.

While the focus here is on transport, there are other factors that determine recruitment to a given location. If the probability of successful recruitment to a region i having been spawned at site j is P_{ij}, then this probability can be written as a product of the probability of success in each of the processes involved. These include the probability of being spawned in the first place at site j, or P_{si}; the probability of successful transport from i to j, written $P_{\mathrm{T}ij}$; and finally the probability that the act of settlement into the habitat at site i is successful, P_{Ri}. Finally, if the interest is in recruitment into a fishery there a final factor, the probability of surviving to a catchable age at site i, P_{ai}. Our interest here is focused on the transport issues that along with spawning are relatively well known. The importance of the transport in determining population connectivity has prompted considerable debate (Roberts, 1997; Cowen *et al.*, 2000; Armsworth, 2002; James *et al.*, 2002). Much more effort needs to be paid to the entire suite of probabilities before an understanding of recruitment is achieved. In particular the coupling of the transport between sites to the probability of successful settling requires more careful consideration. It is at this stage smaller scale dynamics involving tides and internal wave fields may play an important role (Shanks, 1983, 1988; Pineda, 1991, 1994; Leichter *et al.*, 1996; Forward and Tankersley, 2001).

Before introducing methods for attempting reconstructions of individual trajectories between spawning sites and potential future territory it is worthwhile examining the types of trajectories in those outlined above that can contribute to the problem. The problem is defined as finding trajectories that contribute to the introduction of recruits from the place in which they are spawned to their new homes. The first issue is to determine what sites are connected to other sites and which sites are separated by closed trajectories or isolated by saddle points.

10.4.9 Some examples of Lagrangian problems in the coastal ocean

Here, in short vignettes , a number of relevant marine ecosystem problems are introduced from a Lagrangian prospective. These include the nature of coastal regimes including both benthic and demersal species (bottom dwelling and bottom associated). The systems considered are bank circulations and coastal plumes driven by riverine and estuarine outfalls. Respectively these provide areas that are retentive, elliptical trajectories, and connective along coast flows with saddle points that connect individual riverine/estuarine outfalls with the continental shelf environment. As discussed below the latter includes transitions in trajectories that depend on the nature of the estuary dynamics and the outfall plume response to the transport out of the estuarine system. These dynamics and the tendency for enclosed (elliptical) flows over topographic highs produce patterns similar to that described above for a meandering boundary current extension.

Flows over banks on continental shelves create closed circulations tied to the Taylor constraints as discussed above and the rectification of tidal forcing that tends to contribute to enhanced anticyclonic flows around these features. The nature of these circulations shows up in the basic life cycle of cod, *Gadus morhua*, on Georges Bank and elsewhere in their North Atlantic habitat where spawning occurs in an upstream portion of the Bank such that the larval fish are placed in an optimal location with respect to both retention on the Bank (Fig. 10.11, Lough and Manning, 2001) and presumably forage potential. This apparent adaptation of local populations to individual features provides a means for sustaining these local populations. Not all cod and haddock populations, however, are tied to topographic banks. Some are associated with broad shelf areas such as the Acto-Norwegian cod in the Barents Sea and island systems such as found in Iceland. An important aspect of the total dynamics of these species involves the concept of interconnected local populations or metapopulations that have distinct demographic interrelationships on population time scales and the degree to which they are more distantly interconnected by occasional or very long time scale gene flow (evolutionary time scale). In the case of the cod there is strong evidence both in terms of tagged fish and genetics that the populations from Norway to Georges Banks involve populations consisting of distinct metapopulations, but that are interrelated in terms of gene flow (Jensen, 1972; Mork *et al.*, 1985). It is, however, not clear whether this inter-population flux is capable of maintaining populations. A significant contributor to the level of interconnection is the freshwater coastal plumes that dominate much of the high latitude North Atlantic (Chapman and Beardsley, 1989). In the western North Atlantic the freshwater that sets up the shelf trapped southerly flow comes mostly from the Arctic down the East Greenland current and then is augmented by Arctic waters coming through the Canadian Archipelago and riverine inputs from the St. Lawrence and rivers around the Gulf of Maine. A similar buoyancy driven current carries waters northward along the Norwegian shelf. This originates

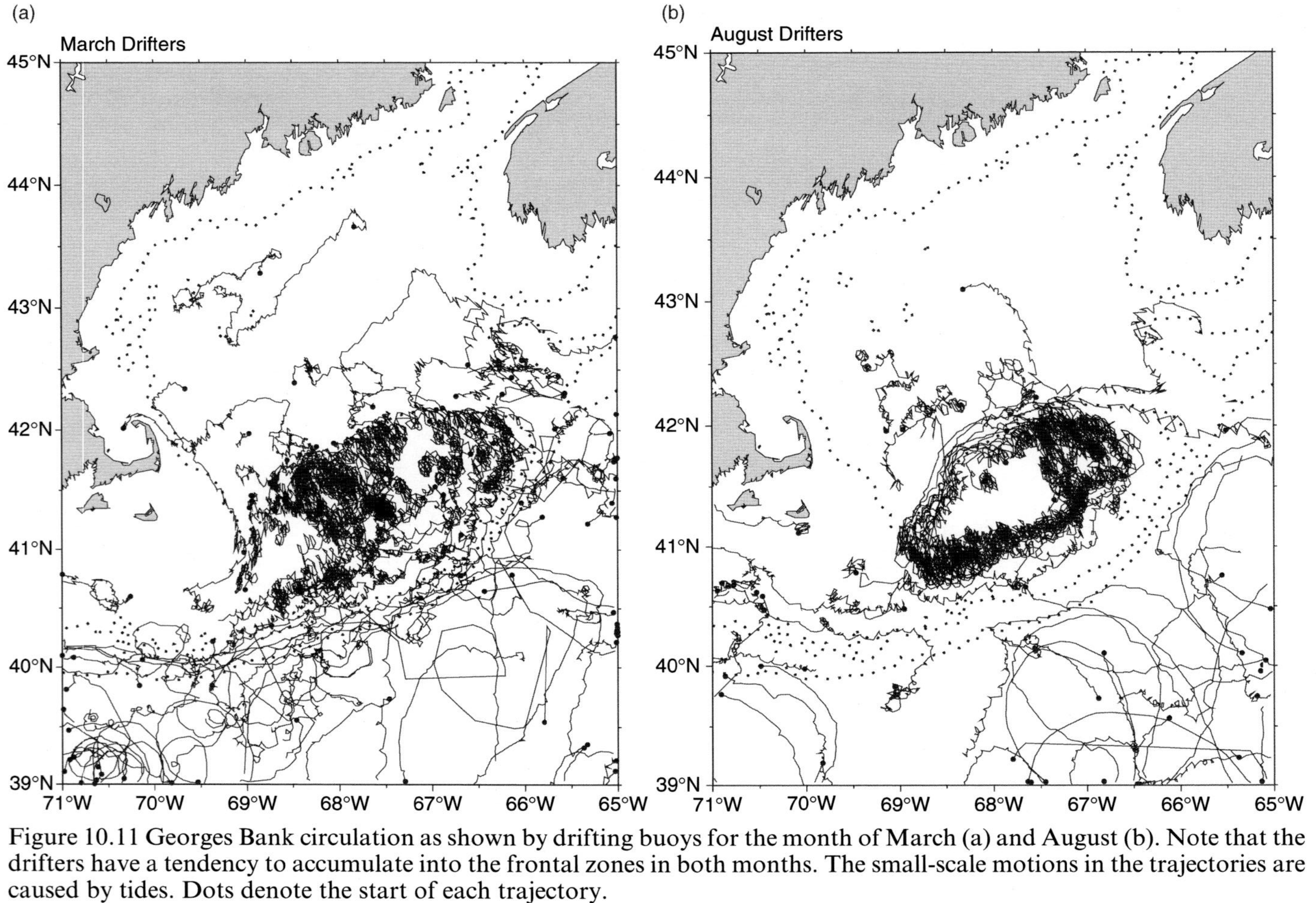

Figure 10.11 Georges Bank circulation as shown by drifting buoys for the month of March (a) and August (b). Note that the drifters have a tendency to accumulate into the frontal zones in both months. The small-scale motions in the trajectories are caused by tides. Dots denote the start of each trajectory.

from the Baltic outflow and is augmented from the various fjord systems. This flow eventually crosses into the Barents Sea where its characteristic may play a role in the life history of the Arcto-Norwegian cod stock (Nakken, 1994). The northern extent of cod is defined by the presence of polar waters and winter sea ice. To the south, the determinants of cod range are not as clear. On the west the last major stock is that on Georges Bank although individuals can be found southward in the Mid-Atlantic Bight. There are cod stocks that extend into the shelf seas of Europe down to the English Channel. The southern boundaries are strongly correlated with temperature and the end of terrains created during the last ice age. A major aspect in both the eastern and western extent of Atlantic cod involves the changes in conditions along these buoyancy driven current systems that interconnect their demersal/bank based habitats and the degree that the coastal fronts isolate them against offshore advection of their young and isolation from predators.

To understand the habitat for Atlantic cod from a trajectory perspective there are three scales that become important. At the largest scales there are the cyclonic gyres that occupy the ocean interiors (Jonsson, 1994) and the coastal circulations discussed above. These introduce the basic shelf break and bank scale flows that along with tidal force modifications set up the individual habitats for most cod stocks. The final piece in the dynamics for cod in the current context is the age-dependent behavioral aspects that allow them to use these habitats successfully. The first two scales define the domain of individual stocks and their interconnections (Fig. 10.11a). The last aspect, however, is the crucial one that potentially sustains cod populations within the context of the larger scale issues (Fig. 10.11b). Therefore on a global basis the freshwater plumes along the coastlines and banks that comprise most of cod habitat provide conduits a few radius of deformation wide that delimit the habitat edge and may or may not provide pathway for adult cod migrations. For juvenile cod the prograde fronts with density surfaces impinging on the topography provide a partial barrier that might protect them from being swept into the open ocean. Indeed in the case of Georges Bank off New England the adults migrate to spawn on the northeast peak of the bank thereby placing their new offspring well within the anticyclonic circulation and away from the shelf edge front where they would be vulnerable to lethal transport events associated with Gulf Stream warm-core rings impinging on the shelf front and breaking it (Flierl and Worblewski, 1985; Myers and Drinkwater, 1989). While the hypothesis that rings might influence cod year classes was not borne out in the observations, there are many bank fish larvae that are entrained offshore by rings as shown in the latter publication. The absence of any larval cod in the streamers of shelf water pulled off in ring/shelf interaction events suggests that the control of trajectories by cod through their life cycle is well adapted to maximize retention in the shelf circulation. More recent studies associated with the Global Ocean Ecosystem program (GLOBEC) Georges Bank program provide additional insight into the mechanisms that isolate cod and haddock from the threat posed by rings.

Lagrangian observations and models of Georges Bank associated with GLOBEC verify and further quantify the general patterns of spawning and around bank flow (Werner *et al.*, 1993; Naimie *et al.*, 2001). The new observations and insights also provide a definitive mechanism that adaptively isolates young cod from offshore advection while presumably offering them a prey aggregation (Davis, 1984) that is capable of maintaining the needs of growing schools of cod. Lough and Manning (2001) present data and a conceptual model of young cod using behavior to maintain themselves in the inner bank tidal fronts. The scale of these features is approximately 10 km on a bank that is almost 200 km across (Fig. 10.12). They are advected regularly on and off the central bank by the semidiurnal M2 tide (Loder

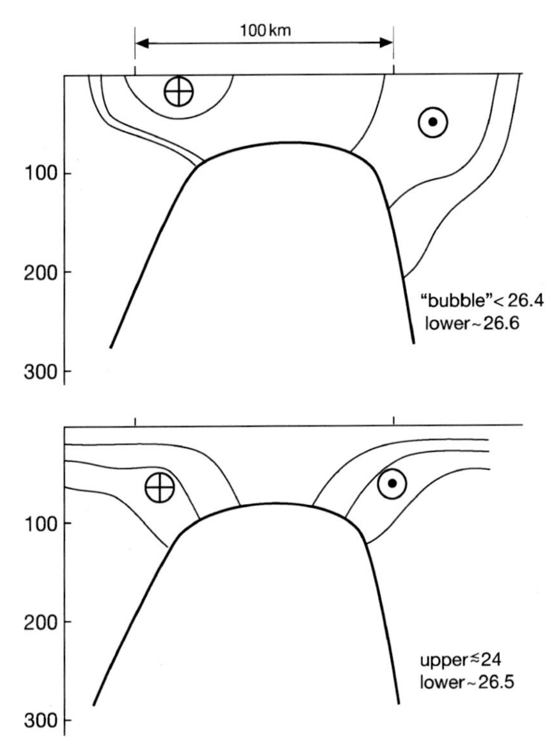

Figure 10.12 Depiction of the frontal domains on Georges Bank for winter and late summer conditions. The contours depict isopycnals. The range is noted at the lower right in each panel.

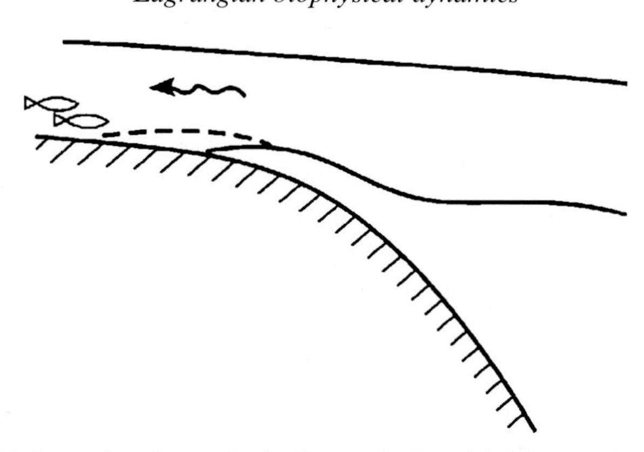

Figure 10.13 Sweeping dynamics in fronts during tidal fluctuations in frontal position. The fish avoiding the conditions on the other side of the front are swept into the center of the bank on each incoming tide. Differential behavior such as this provides a mechanism for an organism to maintain position within critical physical trajectories as detailed in Fig. 10.7.

and Wright, 1985; Lough and Manning, 2001), but a simple combination of photo- and geotaxis with possible interactions with associated prey aggregation in the front suggests that these fronts both provide an optimal environment and protection for ring driven entrainment into the hostile offshore environment (Fig. 10.13).

Similar questions concerning pathways that fauna use in coastal environments arise in the interconnections along coastlines in general. While subpolar demersal species such as cod depend on deep fjords, glacially produced banks and shelf break oceanic flows, temperate to tropical areas involve very different coastal systems. These systems have well-mixed and partially stratified estuaries where freshwater enters semi-enclosed embayments flooded by the increase of sea level associated with the end of the last ice age. In todays ecosystems the drowned river valleys and the associated continental shelf environments allow broad new niches. The time scales involved with the expansion of these are discussed below. Here the current function of these systems is the issue. Estuaries with their strong boundary constraints are often thought of in terms of a two-dimensional (along-center and depth) system. Actual estuaries involve considerable cross sectional adjustment and in well-mixed systems depend on the details of the turbulent mixing itself in three dimensions. Actual trajectories of individuals within these systems are furthermore determined by behavior in relationship to water chemistry and flow. For example, the affinity to the bottom substrate during portions of the tidal cycle seems to provide an important rectified migration tool for shrimp in estuarine systems in Biscayne Bay (Ault *et al.*, 1999). Similar links between behavior allows larval menhaden to find their way into the North Carolina sounds that form their nursery grounds (Forward *et al.*, 1999; Hare *et al.*, 1999). In the case of sibling crab species in the Chesapeake system, the blue (*Callinectes sapidus*) and red (*Ovalipes ocellatus*) crabs achieve reproductive isolation by alternatively spawning on the shelf and allowing the estuarine circulation to carry the young into the embayment and the opposite strategy, i.e. spawning in the bay and having the young reintroduced to the shelf using behavior that keeps them in the outgoing waters (Epifanio and Garvine, 2001; Welch and Forward, 2001; Forward *et al.*, 2004). The dynamics of the freshwater outfall onto the continental shelf also provides pathways that bound trajectories to the coast and provide possible linkages between estuarine systems distributed spatially along coastlines.

As they are introduced to the shelf, freshwaters from estuaries whether they are embayments or rivers undergo adjustment under the influence of the Earth's rotation (Garvine, 1986). This adjustment in the time-dependent case is governed by the propagation of Kelvin waves along the coast (Gill, 1982) and the across shelf relaxation to the Rossby radius of deformation (R_d) as introduced above. In the case where the freshwater outfall is low enough to allow Kelvin waves to form a down-coast plume (Fig. 10.14), the result is a retrograde front bounding a R_d wide flow down-shore with the coast on the right in the northern hemisphere and to the left in the southern hemisphere. The result is a uni-directional bundle of trajectories taking freshwater outfall plumes equatorward on the western sides of oceans and poleward on eastern boundaries. This directionality to circulations places interesting constraints on populations that use coastlines and these plumes. Before providing some examples, it is important to introduce the issue of outfall transports that exceed the capacity of the linear Kelvin/Rossby adjustment models transport. A linearly adjusted coastal plume transport can be estimated using an initial value problem that pictures the introduction of a given outflow, $Q(\mathrm{m}^3/\mathrm{s})$ onto a shelf. Linear adjustment can accommodate a flow that fills a volume R_d wide and a flow that balances the Kelvin waves' ability to fill the plume, $c = \sqrt{g'h}$. While this is somewhat simplistic in the face of mixing, wind and tidal forcing, the bounding estimate for the linear plume is $g'h^2/2$ where the factor of one half involves the average of the plumes depth (Fig. 10.14b). Additional mixing only decreases the rates of along coast flows while counter plume wind fields can actually reverse along coast transport (Munchow and Garvine, 1993). Therefore the use of the river plume model only provides a template of the coastal connections possible. The directional bias introduced by the Earth's rotation produces Lagrangian trajectories that extend down coast with the coast to the right in the northern hemisphere and to the left in the southern. It

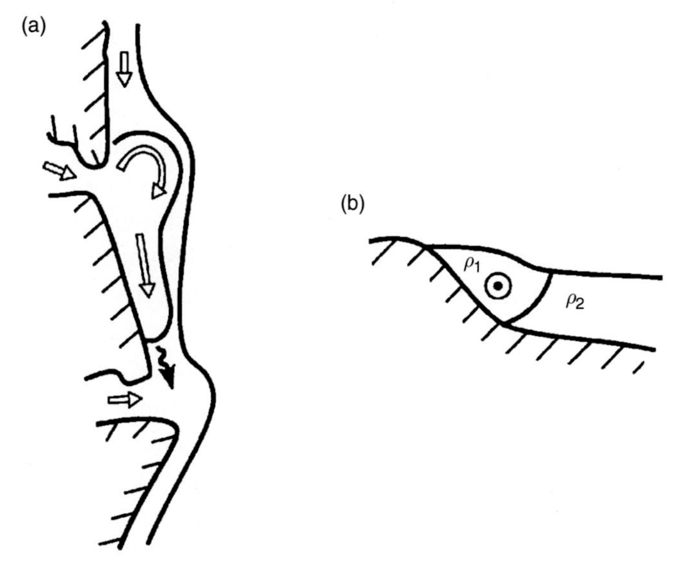

Figure 10.14 Schematics of a riverine plume on a continental shelf. (a) The horizontal plan view of a plume propagating with the coastline to the right in the northern hemisphere. (b) The cross sectional geometry of a typical riverine plume as a two-layer system with densities ρ_1 and ρ_2.

is important to consider also the situation where the river or estuary outfall exceeds the ability of the linear Kelvin adjusted plume to accommodate a coastal outfall. The first suggestion is that the extra transport produces an anticyclonic eddy at the mouth of the estuary that accumulates the excess fluid. The elliptical trajectories in the eddy act as a retention mechanism, concentrating organisms at the mouth of a super-critical estuary. Observations in Chesapeake Bay suggest that such eddies exist and do concentrate organisms (Hood *et al.*, 1999). These eddies evolve in their own right eventually spilling out on to the continental shelf or moving along the coast themselves. This can involve transport back against the adjustment tendency as the vortex sees its image in the coast (McCreary *et al.*, 1997). Finally, in stratified estuarine systems the flow is two-layer and affinity to either the fresher upper waters or the more saline deep waters with age allows various species to use the estuary as a nursery ground by having their young ride in on the deeper flow. Other organisms have the opposite strategy and place their young in the outflowing waters. This deposits them in the plume or eddy depending on Q. These organisms with adults in the embayments may be using the coastal plume to provide linkage between bays in their progeny.

10.5 Modeling trajectories of marine organisms

10.5.1 Building a simulation model

As outlined in the introduction there are a number of reasons for attempting to provide models for the movement of marine organisms ranging from fisheries to human health. There are a number of Eulerian models for lower trophic levels, most of which in recent publications are designed to address marine biogeochemistry (Fasham *et al.*, 1990). For problems involving more complicated ecosystem issues and for longer-term simulations of interactions between ocean biology and chemistry Lagrangian methods become the preferred or only option. A calculation of the efficiency of Lagrangian methods as compared to fixed grid computations is found above. Here a set of methods for implementing a Lagrangian simulation is discussed.

The basic simulation problem involves predicting trajectories that connect different portions of a marine ecosystem. Typically one wishes to understand populations at a site and therefore how the chosen site receives immigrants from the larger ecosystem. It is therefore a natural inclination to attempt to construct trajectories by reversing time and back computing the source locations. Unfortunately this is not generally possible. Both the stochastic problem discussed below and the Eulerian advection/diffusion equation (Cantrell and Cosner, 2003) are not reversible. Technically the reversal of time in the equations is not supported on the data (Padron, 2003). On a more experience-based level this is the reason it is impossible to stir cream back out of a coffee. Here a forward problem for calculating connections is demonstrated. The example will choose random spawning sites and compute arrivals across a spatial domain. Without specific information on behavior the simulations assume surface layer, planktonic or passive drift. For examples of incorporating behavior in a model see Flierl *et al.* (1999) or Humston *et al.* (2001). In the current context it is only important to specify the ocean currents.

There are several sources for physical velocity fields that can be used to estimate the action of ocean circulation on the motion of organisms. These vary considerably in terms of reliability and resolution. There are trade-offs between using constructions of the ocean circulation based on traditional hydrographic measurements constrained by limited Eulerian current meter measurements of currents at specific points; the available inventory of actual Lagrangian measurements based on drifters and floats; and the results of various model simulations of ocean currents. Ultimately the best results, given our current state of knowledge, will involve of an inter-comparison of these three sources and a careful consideration of the results that they imply. Since none of the direct or indirect measurements of currents are complete enough to specify a current field adequate to describe advection in the plankton in most of the world's oceans, most analyses are forced to make use of models. The problem then becomes verifying the model's realism against the available measurements. Here one such verification for the Inter-American Seas (IAS, Caribbean and Gulf of Mexico) will be discussed followed by an application of a model suite to simulate the spread of meroplanktonic organisms in the region. First, however, the components of the models used are briefly described.

The model fields used to represent the ocean circulation are from a set of simulations using the Miami Isopycnal Coordinate Ocean Model (MICOM) North Atlantic simulations of Paiva *et al.* (1999). The drift simulations are completed using an updated version of the particle following or Lagrangian model used in Cowen *et al.* (2000) to consider connectivity issues in the eastern Caribbean. The model makes use of daily velocity fields from five years of a MICOM simulation forced by European Center for Medium Range Forecasting (ECMWF) six hourly winds (Fig. 10.15). The MICOM model is a numerical model of the Atlantic Ocean between 24° S and approximately 70° N. It has a horizontal grid of 1/12 of a degree and 19 layers in the vertical. More details on the model can be found in Paiva *et al.* (1999). The Lagrangian model uses the MICOM velocity field to provide the deterministic portion of a particle's trajectory. To this a stochastic velocity component is added to represent subgrid scale mixing (Dutkiewicz *et al.*, 1993; Griffa, 1996). The subgrid diffusion is chosen based on the 1/12 degree grid scale and the compilation of scale-dependent Lagrangian diffusivities based on the dye studies of Murty (1976). The turbulence model of Dutkiewicz *et al.* (1993) involves an autocorrelated random walk, i.e. a random walk with a memory time scale T and a diffusivity, from the Murty (1976) relationships. The model has been used in a number of applications including the original work and further considerations of meandering currents in Olson and Hood (1994) within the context of the MICOM model. The basic code has also been implemented in the Princeton Ocean model with varying applications from Cuban rafters to oil spills (Mooers *et al.*, 2002; Mooers and Maul, 1998). A simplified version of the model was used in the dispersal calculations in Cowen *et al.* (2000) and Olson *et al.* (2001). The current simulations use the original Dutkiewicz code with its fourth-order Runga–Kutta integration and a new random number generator (Kloeden and Platen, 1995). The MICOM velocities have been interpolated to a cartesian grid while maintaining the 1/12° velocity patterns. Arrival codes are those originally developed by Cowen *et al.* (2003).

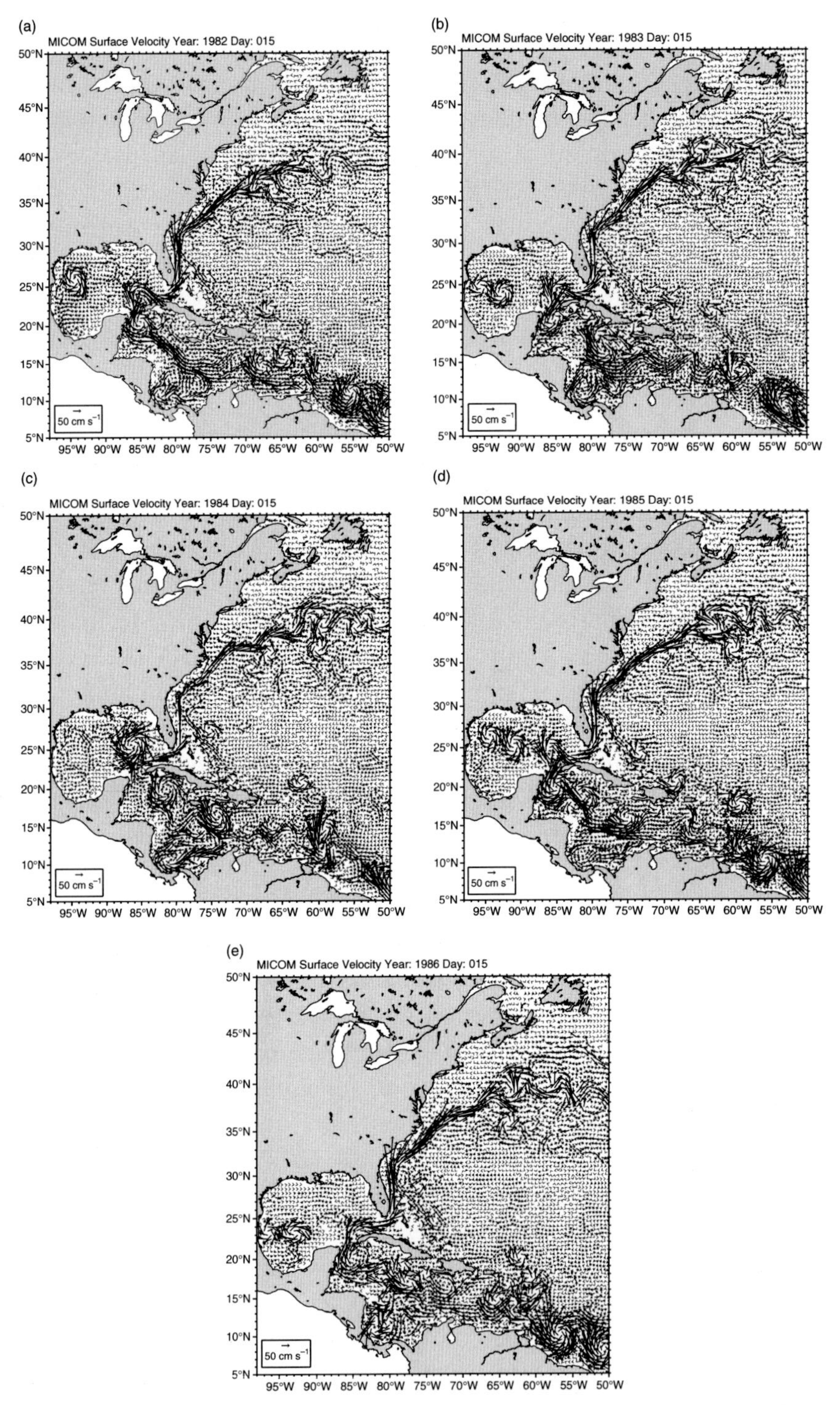

Figure 10.15 An example of the Miami Community Ocean Model (MICOM) velocity fields in the Caribbean. The plots show subsampled velocity vectors for day 15 (January 15th) for each of five model years (1982–1986). See text for discussion and comparison with observations.

The runs involve 1000 particles launched at around 100 randomly chosen points around the outside of the Caribbean from Barbados to Cuba. Simulations are run for 24 day and 42 day periods in three different seasons (Julian days 15–57; 135–177; 205–247) and for each of five years from the early 1980s (ECMWF wind-driven MICOM runs (1982–1986)). Run durations correspond to the average larval duration and the upper range for the Caribbean (Hood and Zastrow, 1993). While it would be naive to expect any of these years to correspond to the actual ocean on those dates, it is reasonable to believe that the model captures the magnitude and basic structure of the mean circulation and the mesoscale eddy field and some aspects of their interannual variability. Since the goal is to explore a full set of possibilities the behavior of the particles are not correlated. See Griffa *et al.* (2004) for methods for individual simulations that maintain subgrid scale connections for application in oilspill tracking and other applications that demand simulation of individual events as opposed to computation of dispersal probabilities. Here the goal is to estimate the total set of possible outcomes given the various MICOM fields (Fig. 10.15). To test the assumption that the combination of the MICOM simulations and the subgrid scale turbulence code actually approximate the real system the results are compared with observed drifters in the region.

The basic information available to verify the MICOM model is the degree to which the model reproduces the surface salinity field, the transport through various straits in the IAS, and the surface velocity patterns measured by drifters deployed across the region. The surface drifter data involves all of the available World Ocean Circulation Experiment style surface drifters deployed in the region between 1988 and 1999 (Fig. 10.16a, b). The drifters have a 10 m tall 3D holey-sock set at a mean depth of 15 m and have been carefully tested to validate their water following capabilities (see Niiler, 2001). The transport estimates are based on studies involving a variety of shipboard and mooring work described in Wilson and Johns (1997) in the straits and passages. The salinity field comparison involves comparing the T/S observations described in Hansel *et al.* (2004) with those produced in the model. The detailed description of the inter-comparison between the model and observations will appear in a manuscript in preparation.

10.5.2 Oceanographic connectivity in the Greater Caribbean: an example

As an example of using Lagrangian methods to explore the biology of an area, the problem of the connections between various portions of the Greater Caribbean or Inter-American Sea (IAS) is considered. The Bahamas, for

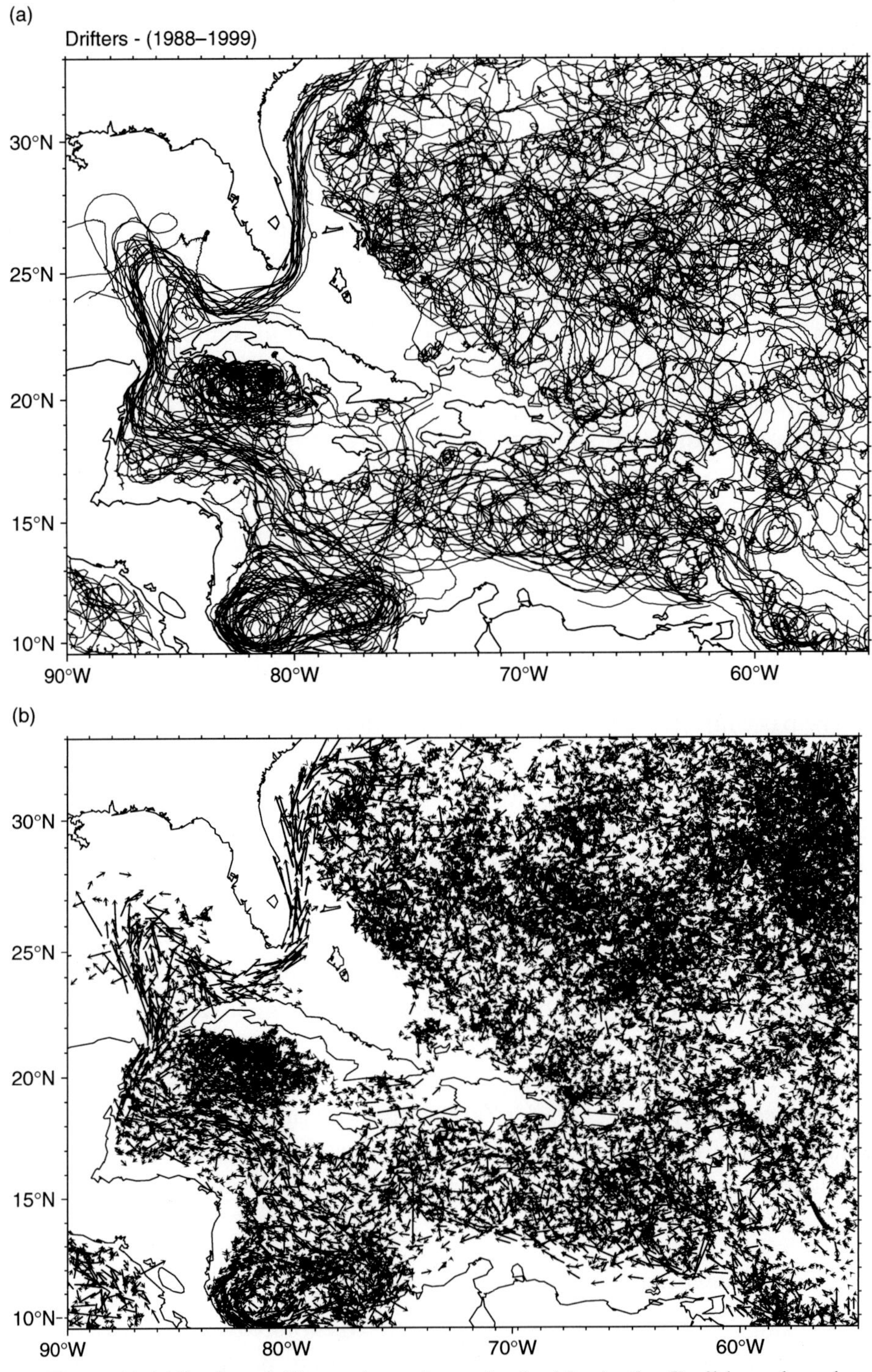

Figure 10.16 Surface drifter trajectories and velocities in the Caribbean based on WOCE standard drifters. The first panel shows the drifter trajectories while the second shows daily average velocities.

example, sit at the outer edge of the IAS and share many, if not all, of the marine benthic species with the region. The Bahamas and the east coast of Florida represent the northernmost range of many of these species with the exception of the small island of Bermuda (Fig. 10.16). These three regions can be connected to the rest of the Caribbean through the western boundary current, the Gulf Stream or by flows to the north of the Greater Antilles and into the Bahamas from the east. Within the Caribbean itself there is a broad flow from east to west that is driven by the North Equatorial Current (Mooers and Maul, 1998) and an additional component from the south brought across the equator as part of the North Brazil Current and its rings (Johns *et al.* 1990; Fratantoni *et al.*, 1995). Most of this flow goes on to form the boundary current although a portion is entrained into the semi-closed Columbian gyre (Figs. 10.15, 10.16). This gyre is connected to the rest of the Caribbean by a septrix point connecting elliptical trajectories in the gyre to the cross Caribbean flow off the coast of Nicaragua. North and west of this point the flow is variable as it crosses the Cayman basin and intensifies along the Belize and Mexican coast to form the Gulf Stream. This current sweeps through the Straits of Yucatan and forms a loop in the Gulf of Mexico before coursing through the Straits of Florida and back into the North Atlantic (Figs. 10.15, 10.16). The other route to the Bahamas is from the east on the northern side of the Greater Antilles (Puerto Rico, Hispanola, and Cuba). One question is the degree that these waters are from the open subtropical gyre or from the south where Caribbean fauna can be introduced. The flow through the passages separating these islands is also of interest in terms of regional connections. The historical view (see Nof and Olson, 1983) is that the Mona Passage between Puerto Rico and Hispanola is shallow and unimportant, while the major connection is a flow from the Atlantic into the Caribbean through the Windward Passage between the latter island and Cuba. Here a set of Lagrangian simulations within the MICOM model will be used to consider the features that make up oceanographic connectivity for the surface layer plankton in the Caribbean.

The species distributions between the Greater Caribbean differs somewhat but all of the region hosts populations of conch (*Stombus gigas*), spiny lobster (*Panulirus argus*), and a large number of tropical fish ranging from those found in grass beds, to reefs and out into the deep, pelagic ocean. While there are a broad set of fish that are of acute interest because of fisheries concerns, groupers, billfish in particular, the discussion here will concentrate on a small reef fish whose genetic and morphological attributes have recently been charted (Taylor and Hellberg, 2003). Conch and lobster are of interest because of their commercial importance and the large difference in the time their larval forms spend in the plankton. *Stombus gigas* larvae spend a limited time in the plankton with the maximum time estimated at around two weeks. Spiny lobsters, on the other hand, are competent meroplanktors for up to a year. The cleaner goby (*Gobiosoma evelynae*) studied by Taylor and Hellberg (2003) has a short meroplanktonic stage with time scales (∼20 days) similar to conch. Their work details several color morphs that form subpopulations around the Caribbean. These populations also demonstrate genetic differentiation in their mitochondrial DNA. In considering the simulations below the differences between long and short planktonic periods in possible connections and the possible disconnections for some species across the Caribbean are the points of interest.

Simulations for Lagrangian drift across the Caribbean in the MICOM model are shown in Figs. 10.17 and 10.18. An inter-comparison with actual drifter trajectories is given in Fig. 10.19. Figure 10.17 shows the drift of particles originating at a single point at various time lags. Figure 10.17 by comparison shows the outcome of similar simulations, but in different years to highlight the inter-annual variations in drift outcomes. Notice that breaks in all years occur at the inferred barriers in the Taylor and Hellberg work. Interconnections to the western boundary are more variable with good connection between the Jamaica area and the coast of Mexico and even Florida occurring in one year and not the other (Fig. 10.20). The difference between these two years involves the nature of the fronts and eddy fields in the Cayman basin. In the first case there is a coherent front across the basin that provides trajectories linking the central Caribbean to the Mexican and Florida coasts within weeks. In the other year the Cayman basin is dominated by eddies that make the transport to these western areas diffusive and slow (Fig. 10.21). These factors create variability in the connectivity, defined by the probability of a successful arrival of meroplanktonic larvae across portions of the domain, with time. This addition of temporal variability broadens the debate between Roberts (1997) and Cowen *et al.* (2000) over recruitment across ocean spaces. Here recruitment is defined in the traditional fisheries sense as successful arrival to our ability to catch (sense) the animal. The scenario implied in Figs. 10.20a, 10.21a is more in keeping with Roberts picture of well connected regimes, while the other year renders dispersion patterns that amplify the Cowen *et al.* (2000) conclusion.

The simulations of the Caribbean only provide a starting place for understanding the workings of its marine ecosystem. The overall results suggest that the details of the mesoscale eddy field are important determinants of initial eddy spread. There is evidence of local retention in most of the sites considered and in most areas a small but significant chance of making a coherent transit in

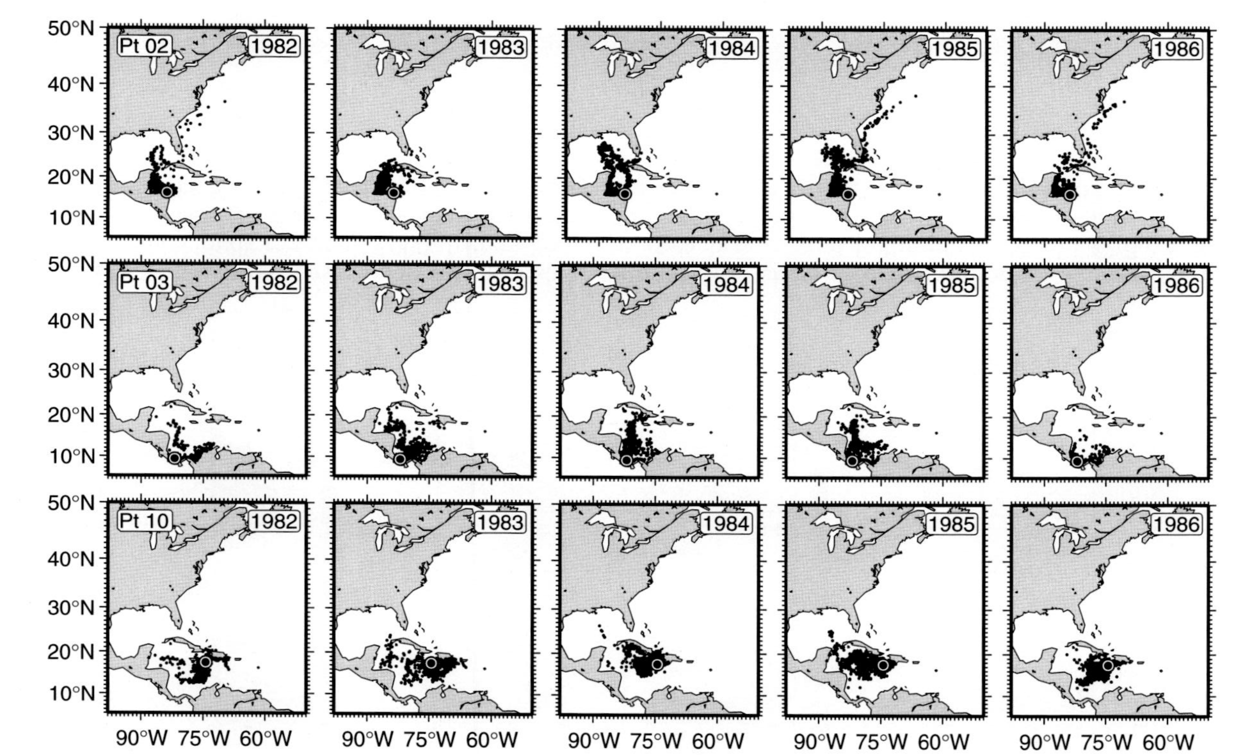

Figure 10.17 Drifter simulations using the MICOM model with the turbulent subcode discussed in the text. Drifter locations are shown after 42 days in the simulation for the indicated starting points off Nicaragua (a), Panama (b) and Hispanola (c). Each simulation involves 1000 particles in each of five different model years 1982–1986.

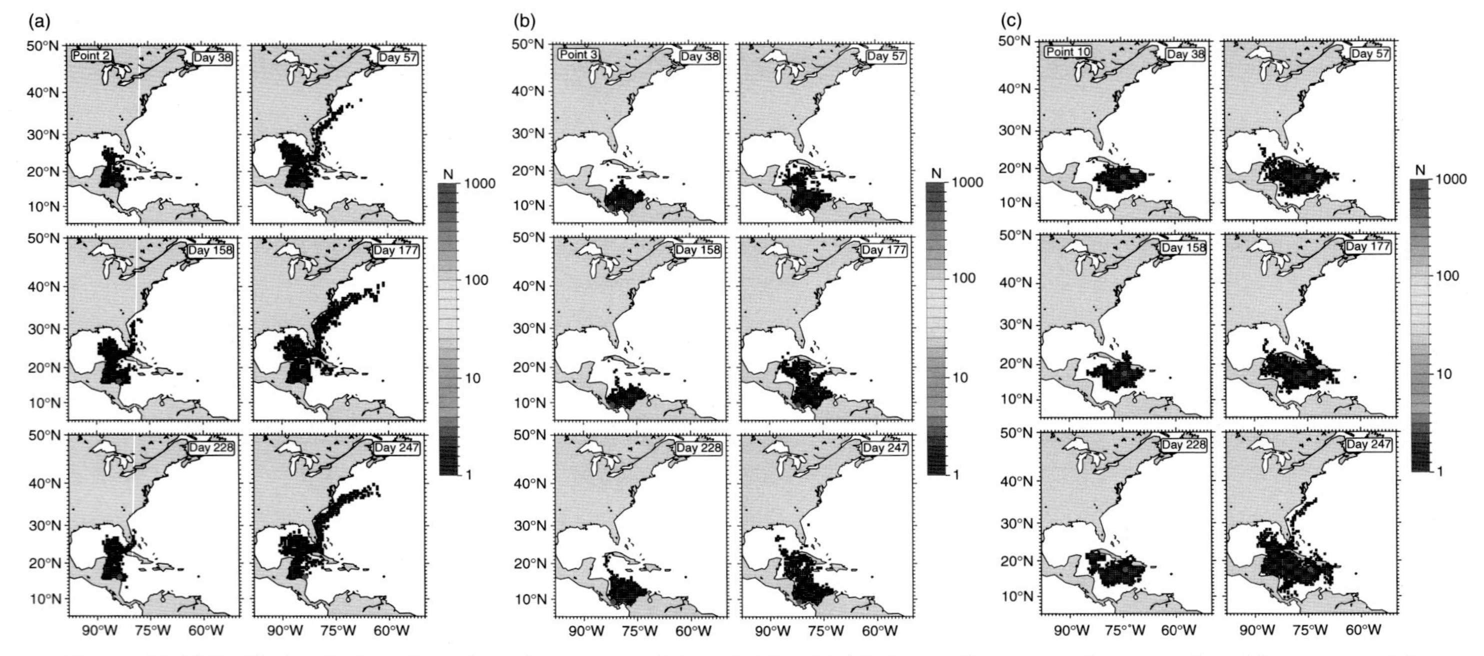

Figure 10.18 Drift simulations based on the same positions in Fig. 10.17 chosen from a random set of positions around the Caribbean rim. The clouds represent 1000 drifters launched at each point and followed for 24 and 42 days. Results are shown as particle densities on a grid. The densities involve the results of all five years of launches such that the density of particles divided by the 5000 total particle deployments is the probability of finding a particle in the grid box. See Plate 26 for color version.

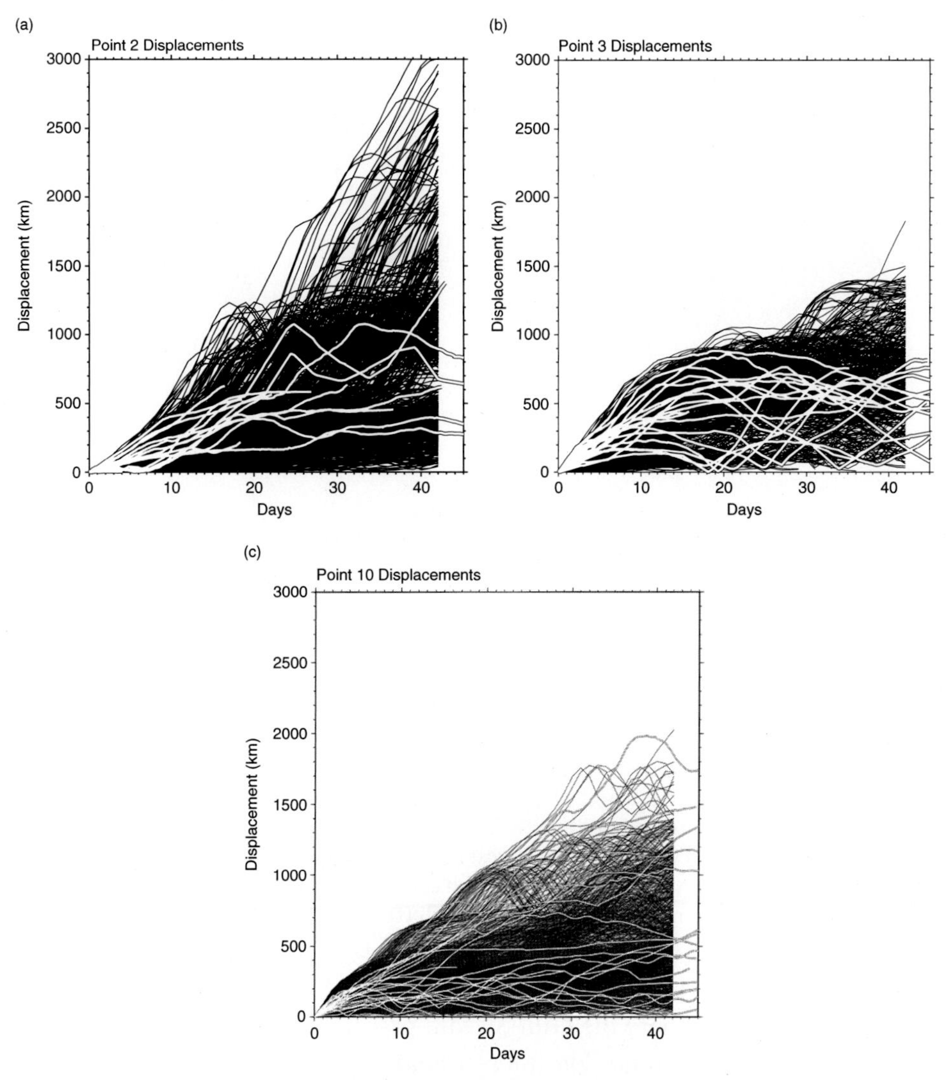

Figure 10.19 Dispersion diagrams for available drifters in the Caribbean and the simulated drifters for three different points in Figs. 10.17 and 10.18. Point 2 is off Nicaragua, point 3 off Panama and point 10 off Hispanola. Actual drifters are superimposed on the simulated drifters in the light cored curves. Note that in general the curves are compatible and there is even confirmation of the rare long-range dispersion in one of the drifter plots. The low number of actual drifter observations, however, makes it impossible to statistically verify the arrival conclusions.

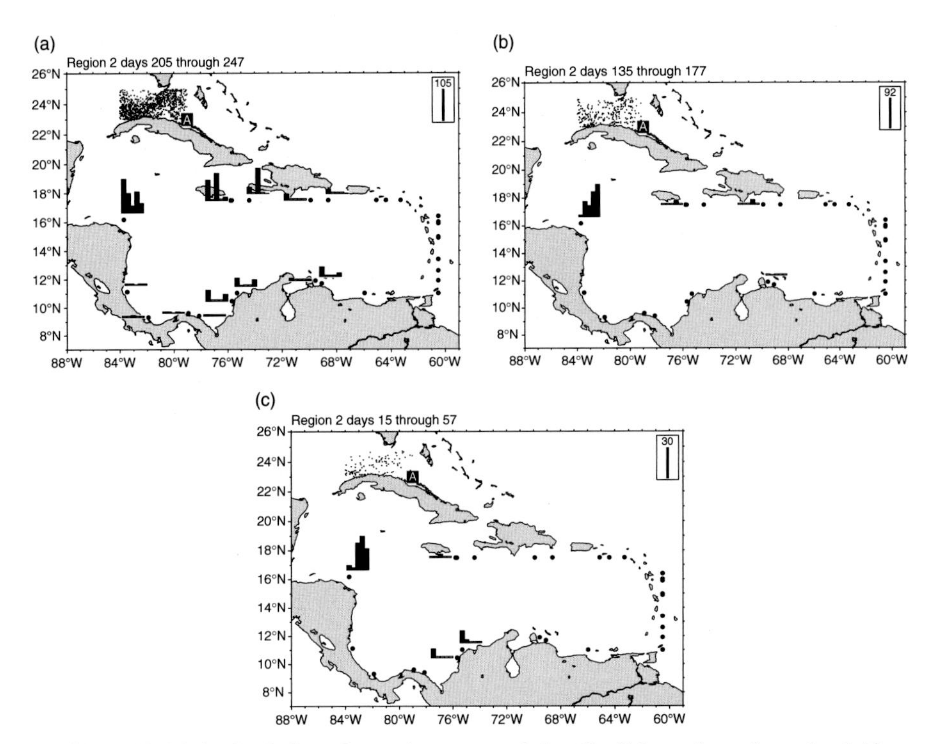

Figure 10.20 Arrival plots for points around the Caribbean based on the drift simulations. Each arrival location is shown by a set of particles at the end of their drift. The contribution of each of the random locations to these arrivals is displayed as histograms at each launch point for each of the five years of simulation. Points without any successful arrivals in any of the five simulated years are not shown. The level of total arrivals over the space is shown by the scale at the upper right hand side of the charts.

an oceanic front across a wide expanse of space. This is especially true in the western Caribbean where the Gulf Stream provides some very long pathways out into the North Atlantic. The inter-annual differences in the simulations are larger than the seasonal ones over most of the area. In particular this is evident in changes in the western Caribbean where there are large differences in dispersion across the Cayman Sea between Jamaica and the Mexican coast and between Jamaica and the Bahamas through the Windward Passage. Finally, the reader is reminded that probabilities of arrival from these simulations provide only one of the probabilities needed to estimate the probability of recruitment. The probability of successful settlement involves details of population dynamics at a local scale.

Details of individual trajectories, sensory clues, habitat availability, and predation set the final probability of a successful settlement. Consideration of reef fish larvae at Barbados by Paris *et al.* (2002) suggests that retention

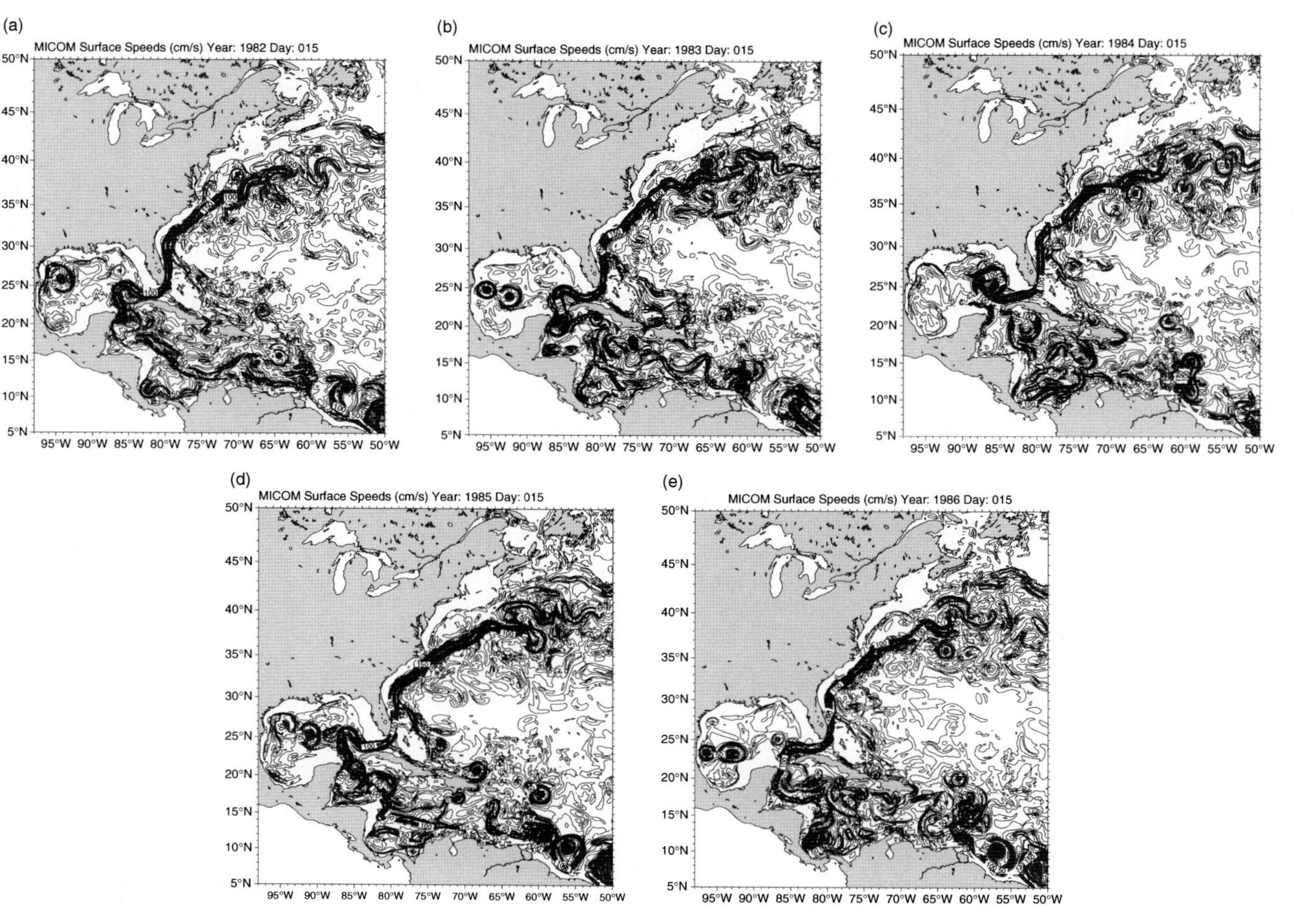

Figure 10.21 Contours of current speed in selected MICOM periods showing the variations in frontal patterns across the Caribbean in the model. Continuous strips of high velocity connecting different regions indicate frontal connections between the different areas. Notice the difference between the two years with strong interconnections in the western Caribbean in 1982 and low connections in 1986.

around islands is the major determinant of recruitment to an individual island's ecosystem. In other words the important component in these ecosystems for many creatures is self-recruitment making the inter-island connectivity a secondary issue at the population level. At the scale of individual islands the role of behavior becomes more prominent (Armsworth, 2001; Paris *et al.*, 2002). Behavior can be as simple as adaptive vertical redistribution of larvae (Yeung and McGowen, 1991; Yeung *et al.*, 1993; Paris *et al.*, 2002) or it may involve directed horizontal swimming (Armsworth, 2001). These in turn are strongly influenced by the nature of the flow on the scale of individual islands and reefs.

10.5.3 Retention on topography: a final look

Retention gyres were discussed above in relationship to the Georges Bank problem. In the Caribbean problem they also play an important role. The problem in the current analysis is the fact that while the MICOM model does produce closed elliptical circulations around many of the islands, with its 25 m depth limit and its lateral boundary conditions, it is unlikely that it adequately depicts the coastal flows. These flows typically consist of anticyclonic, Taylor-caps around the topography. These are retrograde in terms of their topographic fronts that intercept the bottom similar to those above in Fig. 10.12. Observations of eddies attached to sea mounts and islands are common (Bogard *et al.*, 1997; Cowen and Castro, 1991; Paris *et al.*, 2002; Beckmand and Mohn, 2002). There are also commonly wakes of eddies downstream of islands (Lobel and Robinson, 1986). The dynamics of both the trapped eddies and those in the wakes, however, are but poorly known. It is clear that seamounts (Rogers, 1994) and islands typically have enhanced biomass over their surroundings. With significant retention in the around island flows, they therefore may make important contributions to the maintenance of local populations. This does not necessarily explain total range of populations or genetic connections. Therefore, it is important to understand both the nature of local retention, i.e. elliptical local trajectories, and the degree that these regions are connected across the inclosing septrix point. Dispersion between islands depends upon both the degree of transport across the bounding flow and then the connections between islands. The simulated trajectories above perhaps provide reliable estimates of the latter.

10.5.4 Simplified models for longer-term population dynamics

The example above using the MICOM model in an attempt to simulate the complicated aspects of a regional circulation and its relationship to its

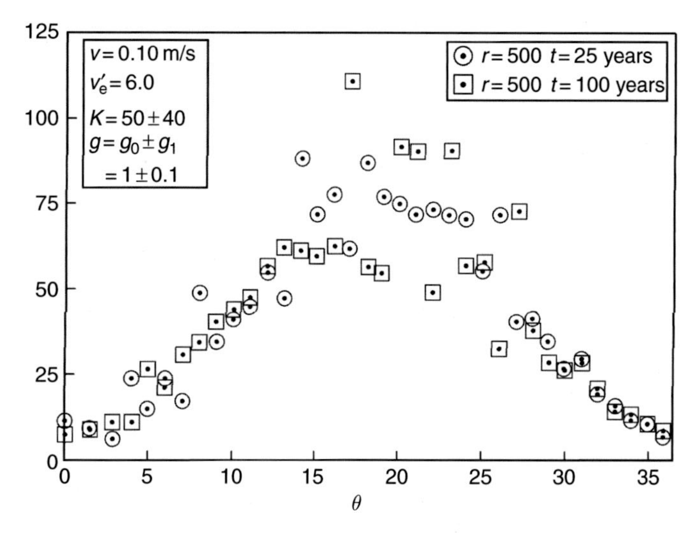

Figure 10.22 An example of simulations in a simple circular gyre in a Lagrangian based model. The carrying capacity is a sinusoid with a maximum of $K = 90$ in the center of the diagram and a low of $K = 10$ in the southern gyre. Note that many particles are exceeding the carrying capacity at the 100 year point leading to extinction events (circles). The gyre here has a radius of 500 km and a mean advection of 0.10 m/s. Most of the hundred years of simulations have patterns similar to the squares which are at 25 years.

ecosystems represents one means of using Lagrangian methods to consider the marine ecosystem. To explore the role of the more complex aspects of interactions between trajectories and population structure, a simple model system that uses specified mean trajectories with superimposed mixing along them is introduced (Fig. 10.22; Olson *et al.*, 2001). The simplest case is a model of a closed gyre as a mean flow around a loop. Particles follow the mean flow and respond to ecosystem variables and are mixed via that same autocorrelated mixing scheme introduced above. Populations interact along the specified trajectories in the presence of the stochastic mixing and the imposed environmental parameters that influence the development of the structured populations. The code was originally written with the idea of applying it to the issue of genetic structure (Olson *et al.*, in revision). A significant question posed by a reviewer of the first manuscript was what happens without any genetics. This question leads back to Hutchinson's (1961) paradox of the plankton, i.e. why there are so many species in samples of the plankton. For example, Venrick (1982, 1986) finds hundreds of species of phytoplankton in repeated samples north of Hawaii. Similar anomalous numbers appear in the central gyre zooplankton at these sites (McGowen and Walker, 1985). One explanation is that each species has its own niche somewhere and is found at the sites north of Hawaii because of diffusion. See Olson and Hood (1994) for a basic argument based on logistic dynamics and formal means of proving the necessary conditions for persistence of such mixed populations in Cantrell and Cosner (2003). Here simple logistic models with fairly restricted trajectories are considered.

A simple set of logistic models within the framework of the very simple flow around a circular mean trajectory with mixing and local competitive interactions in the presence of variable carrying capacity as discussed in Olson *et al.* (2001) allows a method of considering basic population dynamics in a Lagrangian frame. Consider a logistic model that involves the interaction of subpopulations on particles with local competition around the specified trajectory. The dynamics of populations on each particle is governed by

$$\mathrm{d}N_i/\mathrm{d}t = r_i N_i(1 - b_{ij}N_j/K_i(\theta)),$$

where the rate coefficients r_i is set by each subpopulation's abilities, the interaction term, b_{ij} in the simulations occurs over an increment of the total trajectory, $\delta\theta$, defined as the competitive arc. Here the time derivative includes advection and mixing around the gyre.

At least under logistic dynamics, persistence depends on gyre size, carrying capacity gradients, and mixing rates. Given a large enough gyre and mixing that does not allow species with lower r to be extinguished, species will coexist on a variable carrying capacity (K). In essence the Lagrangian models on a loop suggest that in planktonic or even mobile animals the definition of niche in terms of persistence depends on a complicated combination of parameters even in the over-simplistic logistic model when realistic spatial dynamics are considered. For gyres with radii larger than 500 km populations on all particles persist for 100 years in the model. While smaller gyres have extinctions that eliminate populations on approximately half the 750 particles on a time scale that is proportional to their radius and therefore the rate at which populations are swept through the gradient in K. A gyre with a radius of 50 km undergoes extinctions in less than a year or approximately ten circulation times around the gyre at 0.10 m/s. The extinctions take place in the high carrying capacity regions when particles with high populations overwhelm the K (Fig. 10.22). Extinctions start in a 500 km gyre at around 100 years or approximately 100 circulation time scales. The population dynamics of this simple gyre example is just one problem that can be approached in this way. Olson *et al.* (2001; in revision) explore population genetics in the same model. Simple specified mean trajectory models with mixing can be used to explore long-term population dynamics in the ocean.

As another example of the types of problems that could be addressed using these methods consider the circulation in the Greater Caribbean and its changes since the Pliocene when the Isthmus of Panama closed isolating the basin from the Pacific. This closing of the passage into the Pacific is thought to represent the formation of the modern Gulf Stream and the climate shifts leading into the Pleistocene ice-ages by reorganizing the flow of heat and salt in the North Atlantic (Stanley, 1995; Stanley and Ruddiman, 1995). The result in the Caribbean was a set of faunal changes including a large number of extinctions (McNeill *et al.*, 1997; Budd *et al.*, 1994). Here the circulations in the Pliocene and today are described by a simple set of trajectories isolating the basic components of the flows (Fig. 10.23). A particle-based model like the gyre example can then be run on these trajectories to reconstruct the Caribbean since around 3 million years ago. The introduction of particles from one trajectory to another is then accomplished by applying a probability of exchange at branch points in the flows such as the splitting of the Caribbean Current to enter the Cayman or Columbian basins. The model above could then be scaled up to run a model that actually considers evolutionary time scales and explicitly allows for the changes in the ocean connectivity that occurred when the isthmus closed. Such a computation has one last effect, however, to take into account. That is successful arrival on the part of a meroplanktor.

A case in point is the appearance of the branching coral *Acropora palmata* in the Caribbean (McNeill *et al.*, 1997). *A. palmata* appears in the geological record around 3.6 to 2.6 Ma (McNeill *et al.*, 1997). It is found in the southern Caribbean then for around a million years before it suddenly spreads to the northern Caribbean, Florida and the Bahamas. In considering Fig. 10.23 the question becomes what led to this final spread in *A. palmata*? From the figure and our knowledge of western boundary currents in today's oceans a proto-Gulf Stream is expected to have existed in the Pliocene sea. While it might not provide the pathway it does today from the Caribbean to Florida it is unlikely that there was no connectivity. A hypothesis is that *A. palmata* did not spread to the northern regions because it could not successfully settle there in the face of the coral communities that were already established. That is the trajectories were there but it could not successfully compete on arrival in these waters. The suggestion is that it took the first ice age related lowering of sea level to erase the playing field that allows *A. palmata* to finally fill the northern niches. Genetic information on mantis shrimp in the Indonesian archipelago suggests that there is little connection between islands that are clearly well connected in terms of observed current trajectories (Barber *et al.*, 2000). The best explanation for this is that it is very difficult for larvae to

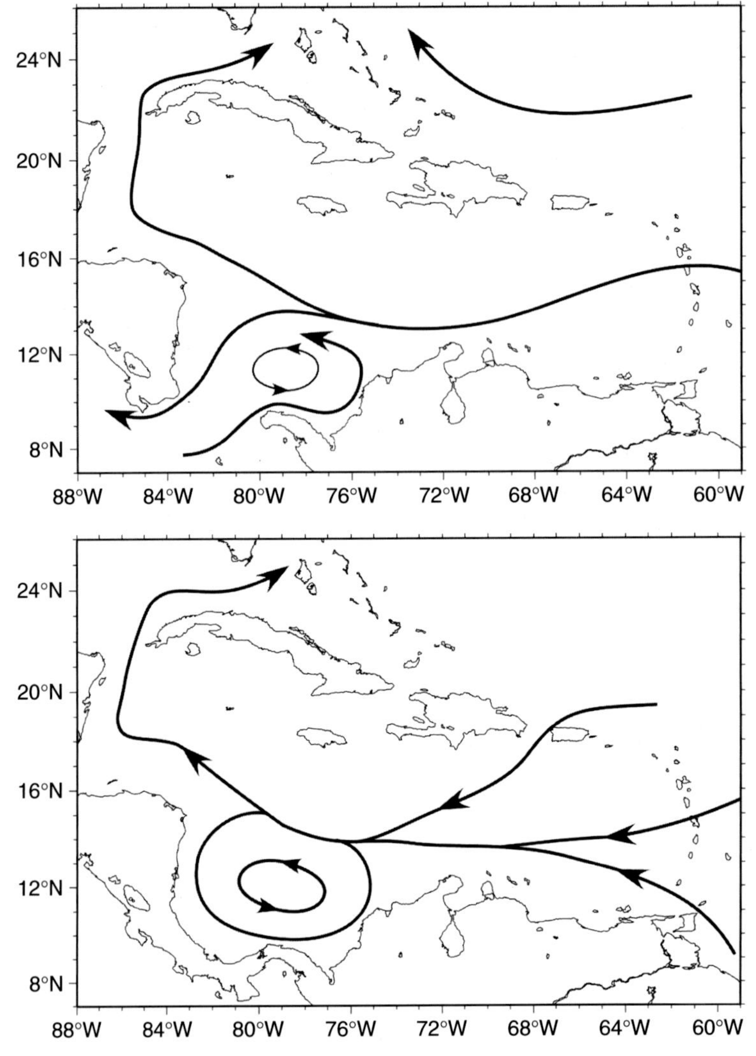

Figure 10.23 Trajectories for proposed simulations of the impact of closing the Panamanian isthmus and its impact on life in the Caribbean. The upper panel depicts conditions approximately three million years before present when the Caribbean was open to the Pacific. This weakened the Gulf Stream system and the subtropical gyre. The isthmus closed subsequently leading to the current situation in the lower panel.

actually successfully settle. Similar genetic distances are measured in *Acropora cervicornis* in the Caribbean by Vollmer and Palumbi (2004). The problem of arrival in the cleaner gobies studied by Taylor and Hellberg (2003) may also be an important part of the story. A color morph that is not found commonly on a reef will be hard pressed to settle in a region where large fish do not view it as a servant. Sexual exclusion might also be playing a role in maintaining these patterns.

10.6 Conclusions

The discussion above is meant to convince the reader of the utility of using the Lagrangian perspective to understand life in the marine environment. In particular following individuals or portions of populations as they move through the ocean allows a direct appreciation of the factors that lead to their aggregation based on physics and behavior. The energetic calculations imply that understanding these dynamics in detail is crucial for population dynamics between various trophic levels. The picture that arises is a system where there is ten-fold or more aggregation at each trophic level in a food chain. If the goal is to understand fisheries it is important to think of these aggregations and how different types of fishing gear interact with them. In particular, this might be a very important area for finding means of reducing by-catch in fisheries, i.e. catching undesirable species from an economic or conservation basis in a fishery. Since many fisheries make use of this tendency to aggregate up a food chain, a better understanding of the aggregation process in time and space may provide important methods for managing fisheries' effort. Finally, the examples shown above provide ways of analyzing the structure and the interconnections (or disconnections) in the marine environment. Understanding recruitment across a marine area like the Caribbean requires the specification of the trajectories involved and should in the long run include the behavioral aspects that modify the physical transport and dispersal.

New tools such as animal tracking technologies and genetics provide rich possibilities in gaining a further understanding of marine ecosystems. The model tools discussed here provide a complementary set of methods for interpreting both the tagging and genetic results. An important issue in point is the difference between population level and genetic connectivity. In principle one individual per generation can supply enough gene flow to maintain a link between populations. On the population level, however, it takes at bare minimum two individuals to connect two regions. In practice population connectivity demands far larger numbers than this. In the cases studied above the act of aggregating and the existence of rare but persistent large-scale connections such as provided by fronts supply means for successful population connectivity. In the spectrum of life in the oceans there are examples of connections being based on larval drift and other examples where it is adult migration that keeps the species connected across its habitat.

The scope of models presented also allows an interesting frame for reconsidering the predator–prey problem. The last model of populations competing in a model gyre becomes even more interesting when populations interact by eating each other. One assumes that extinctions become more frequent and change markedly with the type of group dynamics that are used in the sense of Cosner *et al.* (1999).

Acknowledgments

This work was supported on two Biocomplexity grants from NSF (OCE-998150 and OCE-0119976) and the NSF funded Florida Straits billfish project (OCE-0136132). The Caribbean drift simulations are part of a Sea Grant funded effort to study spiny lobster populations (NA16RG-2195). Earlier work on the simulation model and the structured population models was supported by the Office of Naval Research. Current support from ONR on Lagrangian dynamics is also gratefully acknowledged (N000140310439). Geoff Samuels and Jean Carpenter helped to produce the graphics. Modeling discussions with A. Griffa, C. Paris, C. Cosner and S. Cantrell contributed to the effort. D. McNeill and A. Clement contributed to the discussion of long-term changes in the Caribbean and ways of modeling them. J. Ault and L. Farmer provided help with the bioenergetics model as did the MSC 371 reading class.

References

Allen, W. E., 1932. Problems of flotation and deposition of marine plankton diatoms. *Trans. Amer. Micr. Soc.*, **51**, 1–7.

Armsworth, P. R., 2001. Directed motion in the sea: efficient swimming by reef fish larvae. *J. Theor. Biol.*, **210**, 81–91.

Armsworth, P. R., 2002. Recruitment limitation, population regulation, and larval connectivity in reef fish metapopulations. *Ecology*, **83**, 1092–104.

Arnone, R. A., R. W. Nero, J. M. Jech, and I. DePalma, 1990. Acoustic imaging of biological and physical processes within Gulf Stream meanders. *EOS*, **71**, 982.

Ashjian, C. J., 1993. Trends in copepod species abundances across and along a Gulf Stream meander: evidence for entrainment and detrainment of fluid parcels from the Gulf Stream. *Deep-Sea Res.*, **40**, 461–82.

Ashjian, C. F. and K. F. Wishner, 1993. Temporal and spatial changes in body size and reproductive state in *Nannocalanus minor* (copepoda) females across and along the Gulf Stream. *J. Plankton Res.*, **15**, 67–98.

Ault, J. S. and D. B. Olson, 1996. A multicohort stock production model. *Trans. Amer. Fish. Soc.*, **125**, 343–63.

Ault, J. S., J. Luo, S. G. Smith, J. E. Serafy, J. D. Wang, G. A. Diaz, and R. Humston, 1999. A spatial dynamic multistock production model. *Can. J. Fish. Aquat. Sci.*, **56** (Suppl.), 4–25.

Ault, J. S., J. Luo, and J. D. Wang, 2003. A spatial ecosystem model to assess spotted seatrout population risks from exploitation and environmental changes. In *Biology of the Spotted Seatrout*, ed. S. A. Bortone. Boca Raton: CRC Press, 267–96.

Bakun, A., 2001. School-mix feedback: a different way to think about low frequency variability in large mobile fish populations. *Prog. Oceanogr.*, **49**, 485–511.

Bakun, A., 2005. Seeking an expanded suite of management tools: implications of rapidly-evolving adaptive response mechanisms (such as school-mix feedback). In *Proceed. World Conference on Sustainable Fisheries*, ed. N. Erhardt, *Bull. Mar. Sci.*, **76**, 463–84.

Bakun, A. and J. Csirke, 1998. Environmental processes and recruitment variability. *Squid Recruitment Dynamics* FAO Tech. Pap. 376, ed. P. G. Rodhouse, E. G. Dawer, and R. K. O'Dor.

Bakun, A. and P. Cury, 1999. The school-trap: a mechanism promoting large-amplitude out-of-phase population oscillations in small pelagic fish species. *Ecol. Letters*, **2**, 349–51.

Barber, P. H., S. R. Palumbi, M. K. Erdmann, and M. K. Moosa, 2000. Biogeography: A marine Wallaces line? *Nature*, **406**, 692–3.

Beckman, A. and C. Mohn, 2002. The upper ocean circulation at Great Meteor Seamount Part II: Retention potential of the seamount induced circulation. *Ocean Dynamics*, **52**, 194–204.

Bernie, R. M. and M. N. Levy, 1988. *Physiology*. St. Louis: C. V. Mosby.

Bienfang, P., 1981. Sinking rates of heterogeneous, temperate phytoplankton populations. *J. Plankton Res.*, **3**, 235–53.

Bienfang, P., J. Szyper, and E. Laws, 1982. Sinking rate and pigment responses to light-limitation of a marine diatom: Implications to dynamics of chlorophyll maximum layers. *Oceanol. Acta*, **6**, 55–62.

Block, B. A, J. E. Keen, B. Castillo, H. Dewar, E. V. Freund, D. J. Marcinek, R. W. Brill, and C. Farwell, 1997. Environmental preferences of yellowfin tuna (*Thunnus albacares*) at the northern extent of its range. *Mar. Biol.*, **130**, 119–32.

Block, B. A., H. Dewar, S. B. Blackwell, T. D. Williams, E. D. Prince, C. J. Farwell, A. Boustany, S. L. H. Teo, A. Seitz, A. Walli, and D. Fudge, 2001. Migratory movements, depth preferences, and thermal biology of Atlantic bluefin tuna. *Science*, **293**, 1310–14.

Bogard, S. J., A. B. Rabinovich, P. H. LeBlond, and J. A. Shore, 1997. Observations of seamount-attached eddies in the North Pacific. *J. Geophys. Res.*, **102**, 12, 441–12, 456.

Bower, A. S., 1991. A simple kinematic mechanism for mixing fluid parcels across a meandering jet. *J. Phys. Oceanogr.*, **21**, 173–80.

Bower, A. S. and H. Rossby, 1989. Evidence of cross-frontal exchange processes in the Gulf Stream based on RAFOS float data. *J. Phys. Oceanogr.*, **19**, 1177–90.

Brandt, K., 1920. Ueber den Stoffwechsel im Meere. 3 *Abhandllung. Wissenschaftliche Meeresuntersuchungen*, Abteilun Kiel, Neue Folge **8**, 185–429.

Brandt, S. B. and J. Kirsch, 1993. Spatially explicit models of striped bass growth potential in Chesapeake Bay. *Trans. Amer. Fish. Soc.*, **122**, 845–69.

Brill, R. W., 1979. The effect of body size on the standard metabolic rate of skipjack tuna, *Katsuwonus pelamis. Fish. Bull.*, **77**, 494–8.

Brill, R. W., 1987. On the standard metabolic rates of tropical tunas, including the effect of body size and acute temperature change. *Fish. Bull.*, **85**, 25–35.

Brill, R. W., 1996. Selective advantages conferred by the high performance physiology of tunas, billfishes, and dolphin fish. *Comp. Biochem. Physiol.*, **113A**, 3–15.

Brill, R. W., B. A. Block, C. H. Boggs, K. A. Bigelow, E. V. Freund, and D. J. Marcinek, 1999. Horizontal movements and depth distribution of large adult yellowfin tuna (*Thunnus albacares*) near the Hawaiian Islands, recorded using ultrasonic telemetry: Implications for the physicological ecology of pelagic fishes. *Mar. Bio.*, **133**, 395–408.

Brill, R. W., M. Lutcavae, G. Metzger, P. Bushnell, M. Arendt, J. Lucy, C. Watson, and D. Foley, 2002. Horizontal and vertical movements of juvenile bluefin tuna (*Thunnus thynnus*) in relation to oceanography conditions of the western North Atlantic, determined with ultrasonic telemetry. *Fish. Bull.*, **100**, 155–67.

Brink, K. H., R. C. Beardsley, P. P. Niiler, M. Abbott, A Huyer, S. Ramp, T. Stanton, and D. Stuart, 1991. Statistical properties of near-surface flow in the California coastal transition zone. *J. Geophys. Res.*, **96**, 14693–706.

Brinton, E., 1962. The distribution of Pacific euphausiids. *Bull. Scripps Inst. Oceanogr.*, **8**, 51–270.

Budd, A. F., T. A. Stemann, and K. G. Johnson, 1994. Straitigraphic distributions of genera and species of Neogene to recent Caribbean reef corals. *J. of Paleo.*, **68**, 951–77.

Cantrell, R. S. and C. Cosner, 2003. *Spatial Ecology via Reaction-Diffusion Equations.* Chichester: Wiley.

Carey, F. G. and K. D. Lawsen, 1973. Temperature regulation in free-swimming bluefin tuna. *Comp. Biochem. Physiol.*, **44**, 375–92.

Carey, F. G. and B. H. Robison, 1981. Daily patterns in the activities of swordfish, *Xiphias gladius*, observed by acoustic telemetry. *Fish. Bull.*, **79**, 277–92.

Chapman, D. C. and R. C. Beardsley, 1989. On the origin of shelf water in the Middle Atlantic Bight. *J. Phys. Oceanogr.*, 19, 384–91.

Cosner, C., D. L. DeAngelis, J. S. Ault, and D. B. Olson, 1999. Effects of spatial grouping on the functional response of predators. *Theor. Pop. Bioi.*, **56**, 65–75.

Cowen, R. K. and L. R. Castro, 1991. Relation of coral reef fish larval distributions to island scale circulation around Barbados, West Indies. *Bull. Mar. Sci.*, **54**, 228–44.

Cowen, R. K., K. M. M. Lwiza, S. Sponaugle, C. B. Paris, and D. B. Olson, 2000. Connectivity of marine populations: Open or closed? *Science*, **287**, 857–9.

Cowen, R. K., C. B. Paris, D. B. Olson, and J. L. Fortuna, 2003. The role of long distance dispersal versus local retention in replenishing marine populations. *J. Gulf Caribbean Res.*, **14**, 129–37.

Craig, R. B., D. L. De Angelis, and K. R. Dixon, 1979. Long- and short-term dynamic optimization models with application to the feeding strategy of the Loggerhead shrike. *Amer. Nat.*, **113**, 31–51.

Davidson, V. L., 1994. Integration of metabolism. In *Biochemistry*, ed. V. L. Davidson and D. B. Sittman. Philadelphia: Harwal Pub., 511–21.

Davis, C. S., 1984. Interaction of a copepod population with the mean circulation of Georges Bank. *J. Mar. Res.*, **42**, 573–90.

Davis, T. L. O. and D. A. Stanley, 2001. Vertical and horizontal movements of southern bluefin tuna (*Thunnus maccoyii*) in the Great Australian Bight observed with ultrasonic telemetry. *Fish. Bull.*, **100**, 448–65.

DeAngelis, D. L. and L. J. Gross, 1992. *Individual-based models and approaches in ecology: populations, communities, and ecosystems.* New York: Chapman and Hall.

Dewar, W. K. and G. R. Flierl, 1985. Particle trajectories and simple models of tracer transport in coherent vortices. *Dyn. Atmos. Oceans*, **9**, 215–52.

Diana, J. S., 1995. *Biology and Ecology of Fishes*. Traverse City, MI: Cooper Pub. Group.

Dutkiewicz, S., A. Griffa, and D. B. Olson, 1993. Particle diffusion in a meandering jet. *J. Geophys. Res.*, **98**, 16487–500.

Epifanio, C. E. and R. W. Garvine, 2001. Larval transport on the Atlantic continental shelf. *Estuar. Coast. Shelf. Sci.*, **52**, 51–77.

Essington, T. E., J. F. Kitchell, and C. J. Walters, 2001. The von Bertalanffy growth function, bioenergetics, and the consumption rates of fish. *Can. J. Fish. Aquat. Sci.*, **58**, 2129–38.

Fasham, M. J. R., H. W. Ducklow, and S. M. McKelvie, 1990. A nitrogen-based model of plankton dynamics in the oceanic mixed layer. *J. Mar. Res.*, **48**, 591–639.

Figueroa, H. A. and D. B. Olson, 1994. Eddy resolution versus eddy diffusion in a double gyre GCM. Part I: The Lagrangian and Eulerian Description. *J. Phys. Oceanogr.*, **24**, 371–86.

Flierl, G. R., 1981. Particle motions in large amplitude wave fields. *Geophys. Astrophys. Fluid Dyn.*, **18**, 39–74.

Flierl, G. R. and C. S. Davis, 1993. Biological effects of Gulf Stream meandering. *J. Mar. Res.*, **51**, 529–60.

Flierl, G. and D. McGillicuddy, 2002. Mesoscale and submesoscale physical-biological interactions. In *The Sea*, Vol. 12, ed. A. R. Robinson, J. J. McCarthy, and B. J. Rothschild. New York: John Wiley and Sons, 113–86.

Flierl, G. R. and J. W. Worblewski, 1985. The possible influence of warm core rings upon shelf water larval fish distributions. *Fish. Bull.*, **83**, 313–30.

Flierl, G., D. Grunbaum, S. Levin, and D. B. Olson, 1999. From individuals to aggregation: the interplay between behavior and physics. *J. Theoretical Bio.*, **196**, 397–454.

Forward, R. B. Jr., M. C. DeVries, R. A. Tankersley, D. Rittschof, W. F. Heuler, J. S. Burke, J. M. Welch, and D. E. Hoss, 1999. Behaviour and sensory physiology of Atlantic menhaden larvae, *Brevoortia tyrannus*, during horizontal transport. *Fish. Oceanogr.*, **8** Suppl., 37–56.

Forward, R. B. Jr. and R. A. Tankersley, 2001. Selective tidal-stream transport of marine animals. *Oceanogr. Mar. Biol. Ann. Rev.*, **39**, 305–53.

Forward, R. B. Jr., J. H. Cohen, R. D. Irvine, J. L. Lax, R. Mitchell, A. M. Schick, M. M. Smith, J. M. Thompson, and J. I. Venezia, 2004. Settlement of blue crab *Callinectes sapidus* megalopae in a North Carolina estuary. *Mar. Ecol. Prog. Ser.*, **269**, 237–47.

Fratantoni, D. M., W. E. Johns, and T. L. Townsend, 1995. Rings of the North Brazil Current: Their structure and behavior inferred from observations and a numerical simulation. *J. Geophys. Res.*, **100**, 10,633–54.

Garvine, R. W., 1986. The role of brackish plumes in open shelf waters. In *The Role of Freshwater Outflow in Coastal Marine Ecosystems*, ed. S. Skreslet. Berlin: Springer-Verlag, 47–65.

Gill, A. E., 1982. *Atmosphere-Ocean Dynamics*. New York: Academic Press.

Gooding, R. M., W. H. Neill, and A. E. Dizon, 1981. Respiration rates and low-oxygen tolerance limits in skipjack tuna, *Katsuwonus pelamis*. *Fish. Bull.*, **79**, 31–48.

Govoni, J. J. and C. B. Grimes, 1992. The surface accumulation of larval fishes by hydrodynamic convergence in the Mississippi River plume front. *Cont. Shelf. Res.*, **12**, 1265–76.

Govoni, J. J., D. E. Hoss, and D. R. Colby, 1989. The spatial distribution of larval fishes about the Mississippi River plume. *Limnol. Oceanogr.*, **34**, 178–87.

Govoni, J. J., B. W. Stender, and O. Pashuk, 1999. Distribution of larval swordfish, *Xiphias gladius*, and probable spawning off the southeastern United States. *Fish. Bull*, **98**, 64–74.

Gran, H. H., 1915. The plankton production in northern European waters in the spring of 1912. *Bull. Plankt. Cons. Int. Explor. Mer.*, 7–142.

Graves, J. E., B. E. Luckhurst, and E. D. Prince, 2001. An evaluation of pop-up satellite tags for estimating post release survivial of blue marlin (*Makaira nigricans*) from a recreational fishery. *Fish. Bull.*, **100**, 134–42.

Griffa, A. 1996. Application of stochastic paricle models to oceanographic problems. In *Stochastic Modelling in Physical Oceanography*, ed. R. Adler, P. Muller, and B. Rozovskii. Cambridge, MA: Birkhäuser Boston, 113–28.

Griffa, A., L. I. Piterbarg, and T. Ozgokmen, 2004. Predictability of Lagrangian particle trajectories: Effects of smoothing of the underlying Eulerian flow. *J. Mar. Res.*, **62**, 1–35.

Gunn, J. and B. Block, 2001. Advances in acoustic, archival, and satellite tagging of tunas. *Fish Physiol.*, **19**, 167–224.

Hansell, D. A., N. R. Bates, and D. B. Olson, 2004. Excess nitrate and nitrogen fixation in the North Atlantic. *Mar. Chem.*, **284**, 243–65.

Harden, B. 1996. *A River Lost*. New York: Norton.

Hare, J. A., J. A. Quinlan, F. E. Werner, B. O. Blanton, J. J. Govoni, R. B. Forward, L. R. Settle, and D. E. Hoss, 1999. Larval transport during winter in the SABRE study area: results of a coupled vertical larval behaviour-three-dimensional circulation model. *Fish. Oceanogr.*, **8** Suppl., 57–76.

Haury, L. R., J. A. McGowan, and P. H. Wiebe, 1977. Patterns and processes in the time-space scales of plankton distributions. In *Spatial Pattern in Plankton Communities*, ed. J. H. Steele. New York: Plenum Press, 277–428.

Hinckley, S., A. J. Hermann, and B. A. Megrey, 1996. Development of a spatially explicit, individual-based model of marine fish early life history. *Mar. Ecol. Prog. Ser.*, **139**, 47–68.

Hitchcock, G. L., A. J. Mariano, and T. Rossby, 1993. Mesoscale pigment fields in the Gulf Stream observation in a meander crest and trough. *J. Geophys. Res.*, **98**, 8425–45.

Hoffmann, E. E., K. S. Halstrom, J. R. Moisan, D. B. Haidvogel, and D. L. Mackas, 1991. Use of simulated drifter tracks to investigate general transport patterns and residence times in the coastal transition zone. *J. Geophy. Res.*, **96**, 15041–52.

Hood, E. D. and C. E. Zastrow, 1993. Ecosystem- and taxon-specific dynamic and energetics properties of larval fish assemblages. *Bull. Mar. Sci.*, **53**, 290–335.

Hood, R. R., H. V. Wang, J. E. Purcell, E. D. Houde, and L. W. Harding, Jr., 1999. Modeling particles and pelagic organisms in Chesapeake Bay: Convergent features control plankton distributions. *J. Geophys. Res.*, **104**, 1223–44.

Humston, R., J. Ault, M. Lutcavage, and D. B. Olson, 2001. Schooling and migration of large pelagics relative to environmental cues. *Fish. Oceanogr.*, **9**, 136–46.

Humston, R., D. B. Olson, and J. S. Ault, 2004. Behavioral assumptions in models of fish movement and their influence on population dynamics. *Trans. Amer. Fish. Soc.*, **133**, 1304–28.

Huntley, M. E. and M. D. G. Lopez, 1992. Temperature-dependent production of marine copepods: a global synthesis. *Amer. Nat.*, **140**, 201–42.

Huston, M. A., De Angelis, D. L., and W. M. Post, 1988. New computer models unify ecological theory. *BioScience*, **38**, 682–91.

Hutchinson, G. E., 1961. The paradox of the plankton. *Amer. Nat*, **95**, 137–45.

Idrisi, N., M. J. Olascoaga, Z. Garraffo, D. B. Olson, and S. L. Smith, 2004. Mechanisms for emergence from diapause of *Calanoides carinatus* in the Somali current. *Limnol. Oceanogr.*, **49**, 1262–8.

James, M. K., P. R. Armsworth, L. B. Mason, and L. Bode, 2002. The structure of reef fish metapopulations: modelling larval dispersal and retention patterns. *Proc. R. Soc. Lond. B*, **269**, 2079–86.

Jensen, A. C., 1972. *The Cod*. New York: T. Y. Crowell Co.

Johns, W. E., T. N. Lee, F. Schott, R. Zantopp, and R. Evans, 1990. The North Brazil Current Retroflection: seasonal structure and eddy variability. *J. Geophys. Res.*, **95**, 22, 103–20.

Judson, O. P., 1994. The rise of the individual-based model in ecology. *Trends Ecol. Evol.*, **9**, 9–14.

Jonsson, S., 1994. Cyclonic gyres in the North Atlantic. In *Cod and Climate Change, ICES Mar. Sci. Symp.*, **198**, 287–91.

Kadko, D. C., L. Washburn, and G. Jones, 1991. Evidence of subduction within cold filaments of the Northern California coastal transition zone. *J. Geophys. Res.*, **96**, 14, 909–26.

Kitchell, J. F., W. H. Neill, A. E. Dizon, and J. J. Magnuson, 1978. Bioenergetic spectra of skipjack and yellowfin tunas. In *The Physiological Ecology of Tunas*, ed. G. D. Sharp and A. E. Dizon. New York: Academic Press, 357–68.

Kloeden, P. E. and E. Platen, 1995. *Numerical Solution of Stochastic Differential Equations*. Berlin: Springer-Verlag.

Korsmeyer, K E. and H. Dewar, 2001. Tuna metabolism and energetics. *Fish Physiol.*, **19**, 35–78.

Leichter, J. J., S. R. Wind, S. L. Miller, and M. W. Denny, 1996. Pulsed delivery of subthermocline waters to Conch Reef (Florida Keys) by internal tidal bores. *Limnol. Oceanogr.*, **41**, 1490–1501.

Levin, S. A., 1999. *Fragile Dominion*. Cambridge, MA: Perseus Pub.

Levin, S. A., 2002. Complex adaptive systems: Exploring the known, the unknown and the unknowable. *Bull. Amer. Math. Soc.*, **40**, 3–19.

Lobel, P. S. and A. R. Robinson, 1986. Transport and entrapment of fish larvae by ocean mesoscale eddies and current in Hawaiian waters. *Deep-Sea Res.*, **33**, 483–500.

Loder, J. W. and D. G. Wright, 1985. Tidal rectification and frontal circulation on the sides of Georges Bank. *J. Mar. Res.*, **43**, 581–604.

Lough, R. G. and J. P. Manning, 2001. Tidal-font entrainment and retention of fish larvae on the southern flank of Georges Bank. *Deep-Sea Res.* II, **48**, 631–44.

Lutcavage, M. E., R. W. Brill, G. B. Skomal, B. C. Chase, and J. Tutein, 2000. Tracking adult North Atlantic bluefin tuna (*Thunnus thynnus*) in the northwestern Atlantic using ultrasonic telemetry. *Mar. Bio.*, **137**, 347–58.

Mackas, D. L., R. Kieser, M. Saunders, D. R. Yelland, R. M. Brown, and D. F. Moore, 1997. Aggregation of euphausiids and Pacific hake (*Merluccius productus*) along the outer continental shelf off Vancouver Island. *Can. J. Fish. Aquat. Sci.*, **54**, 2080–96.

Magnuson, J. J. and D. Weininger, 1978. Estimation of minimum sustained speed and associated body drag of scombrids. In *The Physiological Ecology of Tunas*, ed. G. D. Sharp and A. E. Dizon. New York: Academic Press, 313–38.

Mann, K. H. and J. R. N. Lazier, 1996. *Dynamics of Marine Ecosystems*. Malden, MA: Blackwell Sci.

McCreary, J. P., Jr., S. Zhang, and S. R. Shetye, 1997. Coastal circulations driven by river outflow in a variable-density 1 1/2 layer model. *J. Geophys. Res.*, **102**, 15,535–54.

McGillicuddy, D., R. Johnson, D. Siegel, A. Michaels, N. Bates, and A. Knap, 1999. Mesoscale variations of biogeochemical properties in the Sargasso Sea. *J. Geophys. Res.*, **104**, 13381–94.

McGowan, J. A. and P. W. Walker, 1985. Structure in the copepod community of the North Pacific Central Gyre. *Ecol. Monogr.*, **49**, 195–226.

McKendrick, A. G., 1926. Application of mathematics to medical problems. *Proc. Edinb. Math. Soc.*, **44**, 98–130.

McLaren, I. A., J.-M. Sevigny, and C. J. Corkett, 1989. Temperature-dependent development in *Pseudocalanus* species. *Can. J. Zool.*, **67**, 559–564.

McNeill, D. F., A. F. Budd, and P. F. Borne, 1997. Earlier (late Pliocene) first appearance of the Caribbean reef-building coral *Acropora palmate*: Stratigraphic and evolutionary implications. *Geology*, **25**, 891–4.

Metz, J. A. J. and O. Diekmann, eds., 1986. *The Dynamics of Physiologically Structured Populations*. Berlin: Springer-Verlag.

Mooers, C. N. K. and G. A. Maul, 1998. Intra-American Sea Circulation. In *The Sea* Vol. 11, ed. A. R. Robinson and K. H. Brink. New York: John Wiley and Sons, 183–208.

Mooers, C. N. K., L. Gao, W. D. Wilson, W. E. Johns, K. D. Leaman, H. E. Hurlburt, and T. Townsend, 2002. Initial concepts for IAS-GOOS. In *Operational Oceanography Implementation at the European and Regional Scales*, ed. N. C. Flemming, S. Vallerga, N Pinardi, H. W. A. Behrens, G. Manzella, D. Prandle, and J. H. Stel. Amsterdam: Elsevier Science B. V., 379–90.

Mork, J., N. Ryman, G. Stahl, F. M. Utter, and G. Sundnes, 1985. Genetic variation in Atlantic cod (*Gladus morhua* L.) throughout its range. *Can. J. Fish. Aquati. Sci.*, **42**, 1580–7.

Munchow, A. and R. W. Garvine, 1993. Buoyancy and wind forcing of a coastal current. *J. Mar. Res.*, **51**, 293–322.

Murray, J. D., 1993. *Mathematical Biology*. New York: Springer-Verlag.

Murty, C. R., 1976. Horizontal diffusion characteristics in Lake Ontario. *J. Phys. Oceanogr.*, **6**, 76–84.

Myers, R. A. and K. Drinkwater, 1989. The influence of Gulf Stream warm core rings on recruitment of fish in the northwest Atlantic. *J. Mar. Res.*, **47**, 635–56.

Naimie, C. E., R. Limeburner, C. G. Hannah, and R. C. Beardsley, 2001. On the geographic and seasonal patterns of near-surface circulation on Georges Bank from real and simulated drifters. *Deep-Sea Res.* II, **48**, 501–18.

Nakata, K., 1990. Abundance of nauplii and protein synthesis activity of adult female copepods in the Kuroshio front during the Japanese sardine spawning season. *J. Oceanogr. Soc. Japan.*, **46**, 219–29.

Nakken, O., 1994. Causes of trends and fluctuations in the Arcto-Norwegian cod stock. *ICES Mar. Sci Symp.*, **198**, 212–28.

Nathansohn, A., 1906. Uber die Bedeutung vertikaler Wasserbewegung fur die Produktion des Planktons. *Abhandlungen der Koniglichen Sachsischen Gesellschaft der Wissenschaften zu Leipzig*, Mathematische-Physische Klasse **29**, 355–441.

Niiler, P. P., 2001. The world ocean surface circulation. In *Ocean Circulation and Climate*, ed. G. Siedler, J. Church, and J. Gould. San Diego: Academic Press, 193–204.
Nof, D., 1981. On the beta-induced movement of isolated baroclinic eddies. *J. Phys. Oceanogr.*, **11**, 1662–72.
Nof, D. and D. B. Olson, 1983. On the flow through broad gaps with application to the Windward Passage, *J. Phys. Oceanogr.*, **13**, 1940–56.
O'Dor, R., 1998. Squid life-history strategies. In *Squid Recruitment Dynamics*, FAO Tech. Pap. 376, ed. P. G. Rodhouse, E. G. Dawer, and R. K. O'Dor, 233–54.
O'Dor, R. and E. G. Dawe 1998. *Illex illecebrosus*. In *Squid Recruitment Dynamics*, FAO Tech. Pap. 376, ed. P. G. Rodhouse, E. G. Dawer, and R. K. O'Dor, 77–104.
Olascoaga, M. J., N. Idrisis, and A. Romanou, 2005. Biopysical isopycnic-coordinate modeling of plankton dynamics in the Arabian Sea. *Ocean Modelling*, **8**, 55–80.
Olson, D. B., 1986. Lateral exchange within Gulf Stream warm-core ring surface layers. *Deep-Sea Res.*, **33**, 1691–704.
Olson, D. B., 1991. Rings in the ocean. *Ann. Rev. Earth and Planet. Sci.*, **19**, 283–311.
Olson, D. B., 2001. Biophysical dynamics of western transition zones. *Fish. Oceanogr.*, **10**, 1–18.
Olson, D. B., 2002. Biophysical dynamics of ocean fronts. In *The Sea*, Vol. 12, ed. A. R. Robinson, J. J. McCarthy, and B. J. Rothschild. New York: John Wiley and Sons, 187–218.
Olson, D. B. and R. H. Backus, 1985. The concentrating of organism at fronts: a cold-water fish and a warm-core ring. *J. Mar. Res.*, **43**, 113–37.
Olson, D. B. and G. P. Podesta, 1988. Oceanic fronts as pathways in the sea. In: *Signposts in the Sea*, ed. W. F. Herrnkind and A. B. Thistle. Tallahassee, FA: Florida State Univ. Press, 1–15.
Olson, D. B. and R. R. Hood, 1994. Modelling pelagic biogeography. *Prog. Oceanogr.*, **34**, 161–205.
Olson, D. B., G. L. Hitchcock, A. J. Mariano, C. J. Ashjian, G. Peng, R. W. Nero, and G. P. Podesta, 1994. Life on the edge: marine life and fronts. *Oceanography*, **7**, 52–60.
Olson, D. B., C. Paris, and R. Cowen, 2001. Lagrangian biological models. *Encyclopedia of Ocean Sci.*, 1438–43.
Olson, D. B., C. Cosner, S. Cantrell, and A. Hastings, 2005. Persistence of fish populations in time and space as a key to sustainable fisheries. *Bull. Mar. Sci.*, **76**, 213–32.
Oswatitsch, K., 1959. Physikalische Grundlagen der Stromungslehre. *Hankbuch der Physik*, **8**, 1–124.
Padron, V., 2003. Effect of aggregation on population recovery modeled by a forward-backward pseudoparabolic equation. *Trans. Amer. Math. Soc.*, **356**, 2739–56.
Paiva, A. M., J. T. Hargrove, E. P. Chassignet, and R. Bleck, 1999. Turbulent behavior of a fine mesh (1/12°) numerical simulation of the North Atlantic. *J. Mar. Sys.*, **21**, 307–20.
Palmen, E. H. and C. W. Newton, 1969. *Atmospheric Circulation Systems*. New York: Academic Press.
Paris, C. B., R. K. Cowen, K. M. M. Lwiza, D. P. Wang, and D. B. Olson, 2002. Objective analysis of three-dimensional circulation in the vicinity of Barbados, West Indies: implication for larval transport. *Deep-Sea Res.*, **49**, 1571–90.
Parsons, T. R., M. Takahashi, and B. Hargrave, 1988. *Biological Oceanographic Processes*. Oxford: Pergamon Press.

Pauling, L., 1970. *General Chemistry*. New York: Dover.
Perry, A. E. and B. D. Fairlie, 1974. Critical points in flow patterns. *Adv. Geophys*, **18B**, 299–316.
Peterson, W. T., C. B. Miller, and A. Huchinson, 1978. Zonation and maintenance of copepod populations in the Oregon upwelling zone. *Deep-Sea Res.*, **26A**, 467–94.
Pineda, J., 1991. Predictable upwelling and shoreward transport of planktonic larvae by internal tidal bores. *Science*, **253**, 548–51.
Pineda, J., 1994. Internal tidal vores in the nearshore: warm-water fronts, seaward gravity currents and the onshore transport of neustonic larvae. *J. Mar. Res.*, **52**, 427–58.
Podesta, G. P., J. A. Browder, and J. J. Hoey, 1993. Exploring the association between swordfish catch and thermal fronts on the U. S. longline grounds in the western North Atlantic. *Cont. Shelf Res.*, **13**, 253–77.
Polovina, J. J., E. Howell, D. R. Kobayashi, and M. P. Seki, 2001. The transition zone chlorophyll front, a dynamic global feature defining migration and forage habitat for marine resources. *Prog. Oceanogr.*, **49**, 469–83.
Reid, J. L., E. Brinton, A. Fleminger, E. L. Venrick, and J. A. McGowan, 1978. Ocean circulation and marine life. In *Advances in Oceanography*, ed. H. Charnock and G. E. R. Deacon. New York: Plenum Press, 65–130.
Riley, G. A., H. Stommel, and D. F. Bumpus, 1949. Quantitative ecology of the plankton of the western North Atlantic. *Bull. Bingham Oceanogr. Coll.*, **12**, 1–169.
Roberts, C. M., 1997. Connectivity and management of Caribbean coral reefs. *Science*, **278**, 1454–7.
Rogers, A. D., 1994. The biology of seamounts. *Mar. Bio*, **30**, 305–50.
Rothschild, B. J. and T. R. Osborn, 1988. Small-scale turbulence and planktonic contact rates. *J. Plankton Res.*, **10**, 465–74.
Round, F. E., R. M. Crawford, and D. G. Mann, 1990. *The Diatoms: Biology Morphology of the Genera*. New York: Cambridge University Press.
Saitoh, S. I., S. Kosaka, and J. Iisaka, 1986. Satellite infrared observations of Kuroshio warm-core rings and their application to study of Pacific saury migration. *Deep-Sea Res.*, **33**, 1601–40.
Schaefer, K. M. and D. W. Fuller, 2002. Movements, behavior, and habitat selection of bigeye tuna (obesus) in the eastern equatorial Pacific, ascertained through archival tags. *Fish. Bull.*, **100**, 765–88.
Shanks, A. L., 1983. Surface slicks associated with tidally forced internal waves may transport pelagic larvae of benthic invertebrates and fishes shoreward. *Mar. Ecol. Prog. Ser.*, **13**, 311–15.
Shanks, A. L., 1988. Further support for the hypothesis that internal waves can cause shoreward transport of larval invertebrates and fish. *Fish. Bull.*, **86**, 703–14.
Smayda, J. J., 1970. The suspension and sinking of phytoplankton in the sea. *Oceanogr. Mar. Biol. Ann. Rev.*, **8**, 353–414.
Stanley, S. M., 1995. New horizons for paleontology, with two examples: the rise and fall of the Cretaceous supertethys and the cause of the modern ice age. *J. Paleo.* **69**, 999–1007.
Stanley, S. M. and W. F. Ruddiman, 1995. Neogene ice age in the North Atlantic Region: climatic changes, biotic effects and forcing factors. In *Effects of Past Global Change on Life*. Washington: National Academy Press, 118–33.
Stillwell, C. E. and N. E. Kohler, 1985. Food and feeding ecology of the swordfish *Xiphias gladius* in the western North Atlantic Ocean with estimates of daily ration. *Mar. Eco. Prog. Ser.*, **22**, 238–47.

Sugimoto, T. and H. Tameishi, 1992. Warm-core rings, streamers and their role on the fishing ground formation around Japan. *Deep-Sea Res.*, **39**, S183–201.
Sumich, J. L., 1999. *An Introduction to the Biology of Marine Life*. 2nd Edn. Boston: McGraw-Hill.
Sverdrup, H. U., M. W. Johnson, and R. H. Fleming, 1942. *The Oceans*. Englewood Cliffs, NJ: Prentice-Hall.
Taylor and Hellberg, 2003. Genetic evidence for local retention of pelagic larvae in a Caribbean reef fish. *Science*, **229**, 107–9.
Valiela, I., 1984. *Marine Ecological Processes*. New York: Springer-Verlag.
Van Winkle, W., K. A. Rose, and R. C. Cambers, 1993. Individual-based approach to fish population dynamics: an overview. *Trans. Amer. Fish. Soc.*, **122**, 397–403.
Venrick, E. L., 1982. Phytoplankton in an oligotrophic ocean: observation and questions. *Ecol. Monogr.*, **52**, 129–54.
Venrick, E. L., 1986. Patchiness and the paradox of the plankton. In *Pelagic Biogeography*, UNESCO Tech. Paper Mar. Sci. **49**, ed. A. C. Pierrot-Bults, S. van der Spoel, B. J. Zahuranec, and R. K. Johnson, 261–9.
Vogel, S., 1994. *Life in Moving Fluids: The Physical Biology of Flow*. Princeton, NJ: Princeton Press.
von Bertalanffy, L., 1968. *General System Theory*. New York: George Braziller.
Von Foerster, H., 1959. Some remarks on changing populations. In *The Kinetics of Cellular Proliferation*, ed. F. Stohlman. New York: Grune and Stratton.
Waring, G. T., C. P. Fairfield, C. M. Ruhsam, and M. Sano, 1993. Sperm whales associated with Gulf Stream features off the northeastern USA shelf. *Fish. Oceanogr.*, **2**, 101–5.
Washburn, L., D. C. Kadko, B. H. Jones, T. Hayward, P. M. Kosro, T. P. Stanton, S. Ramp, and T. Cowles, 1991. Water mass subduction and transport of phytoplankton in a coastal upwelling system. *J. Geophys. Res.*, **96**, 14,927–45.
Webb, G. F., 1985. *Theory of Nonlinear Age-Dependent Population Dynamics*. New York: Marcel Dekker.
Welch, J. M. and R. B. Forward Jr., 2001. Flood tide transport of blue crab, *Callinectes sapidus*, postlarvae: behavioral responses to salinity and turbulence. *Mar. Bio.*, **139**, 911–18.
Werner, F. E., R. H. Page, D. R. Lynch, J. W. Loder, R. G. Lough, R. I. Perry, D. A. Greenberg, and M. M. Sinclair, 1993. Influences of mean advection and simple behavior on the distribution of cod and haddock early life stages on Georges Bank. *Fish. Oceanogr.*, **2**, 43–64.
Wiebe, P. and G. Flierl, 1983. Euphausiid invasion/dispersal in Gulf Stream cold-core rings. *Aust. J. Mar. Freshw. Res.*, **34**, 625–52.
Wiebe, P., V. Barber, S. Boyd, C. Davis, and G. Flierl, 1985. Macrozooplankton biomass in Gulf Stream warm-core rings: spatial distribution and temporal changes. *J. Geophys. Res.*, **90**, 8885–901.
Wilson, W. D. and W. E. Johns, 1997. Velocity structure and transport in the Windward Islands Passages. *Deep-Sea Res.*, **44**, 487–520.
Wilson, E. O., 2000. *Sociobiology*. Boston: Harvard Press.
Withers, P. C., 1992. *Comparative Animal Physiology*. Fort Worth: Saunders College Pub.
Yamamoto, T. and S. Nishizawa, 1986. Small-scale zooplankton aggregations at the front of a Kuroshio warm-core ring. *Deep-Sea Res.*, **33**, 1729–40.
Yamazaki, H., D. L. Mackas, and K. L. Denman, 2002. Coupling small-scale physical processes with biology. In *The Sea*, Vol. 12, ed. A. R. Robinson, J. J. McCarthy, and B. J. Rothschild. New York: John Wiley and Sons, 51–112.

Yeung, C. and M. F. McGowan, 1991. Differences in inshore-offshore and vertical distribution of phyllosoma larvae of Panulirus, Scyllarus, and Scyllarides in the Florida Keys in May–June, 1989. *Bull. Mar. Sci.*, **49**, 699–714.

Yeung, C., J. T. Couillard, and M. F. McGowan, 1993. The relationship between vertical distribution of spiny lobster phyllosoma larvae (Crustacea: Palinuridae) and isolume depths generated by a computer model. *Revista de Biol. Tropical*, **41**, 63–7.

Yeung, C. and T. N. Lee, 2002. Larval transport and retention of the spiny lobster, Panulirus argus, in the coastal zone of the Florida Keys, U.S.A. *Fish. Oceanogr.*, **11**, 286–309.

11

Plankton: Lagrangian inhabitants of the sea

GARY L. HITCHCOCK AND ROBERT K. COWEN

Division of Marine Biology and Fisheries, Rosenstiel School of Marine and Atmospheric Science, University of Miami, Miami, Florida, USA

11.1 Introduction

Plankton have inhabited the Earth's oceans for hundreds of millions of years as evidenced by the fossil record. The exterior covering of identifiable dinoflagellates, for example, are well preserved in Mesozoic rock strata. Pelagic diatoms possess siliceous frustules with identifiable species dating from early Cretaceous sediments (see Falkowski *et al.*, 2004). Given the extent of fossil plankton, it is apparent that a drifting mode of life has been a successful means for survival in the sea for much of life's history.

With the importance of plankton in marine ecosystems, it is surprising that biological oceanographers have only recently begun to use drifting, or more formally Lagrangian, techniques. However, as with other aspects of biological oceanography, the Lagrangian 'tools' for studying plankton are relatively recent, and have often followed technique development by physical oceanographers and engineers. The main goal of this chapter is to summarize how biological oceanographers have applied Lagrangian and related methods to further our understanding of oceanic plankton distributions and dynamics, as well as biogeochemical processes. Our target audience is physical oceanographers and mathematicians who will hopefully gain some benefit from this exercise, while biological oceanographers may also be encouraged to further consider Lagrangian approaches in their field studies. We include studies on bacterio-, phyto-, zoo-, and ichthyoplankton and discuss the advances made in specific sub-disciplines of biological oceanography through the use of Lagrangian techniques. This review is timely in that new, low power sensors are now being adapted for deployments on a variety of Lagrangian platforms. A recent workshop addressed the utility of Autonomous and Lagrangian Platforms and Sensors (ALPS), which concluded with the recommendation that new technologies, such as instrumented Lagrangian platforms, will enable significant advances in understanding biogeochemical processes in the sea and the dynamics of plankton organisms (Rudnick and Perry, 2003).

Lagrangian Analysis and Prediction of Coastal and Ocean Dynamics, ed. A. Griffa, D. Kirwan, A. Mariano, T. Özgökmen, and T. Rossby. Published by Cambridge University Press.

In this chapter, Lagrangian methods are broadly defined as those techniques that provide information on water parcel or organism trajectories. We recognize that visual observations of larvae and 'mark/recapture' methods are not true Lagrangian techniques, but include them because they describe organism trajectories. Additionally, it is clear that few surface platforms truly follow the three-dimensional pathways of water parcels in the surface layer (see Geyer, 1989; D'Asaro *et al.*, 1996). Nevertheless, studies with surface drogues and drifters are included here since these platforms have a long history of use by biologists.

We begin with a brief historical view of the plankton and then consider the nature of drifting life modes in the sea. Next we survey three categories of Lagrangian studies utilized by biological oceanographers: direct observations of larval trajectories, tracers, and surface drifters and subsurface floats. The examples cited emphasize studies of larval dispersal, biogeochemical processes, and the definition of spatial and temporal distributions of marine plankton. We end with a view towards the future, briefly discussing some research areas where we believe Lagrangian techniques will make substantial contributions and discuss how new Lagrangian techniques could be used to better understand plankton dynamics. Our review is incomplete in the sense that we have not included all of the Lagrangian studies conducted by biological oceanographers, nor do we include the diverse Lagrangian models now being applied to plankton dynamics (e.g. Hofmann *et al.*, 1991; Siegel *et al.*, 2003), which are discussed by Olson in Chapter 10 of this volume. Instead, we have sought to illustrate how the increasing recognition and application of Lagrangian methods has advanced our understanding of plankton dynamics and biogeochemical cycling in the sea.

11.1.1 A historical view of plankton

In contrast to its current usage, 'plankton' was initially applied to all particulate matter in the ocean. Victor Hensen, a German professor of physiology, defined particulate matter "both living or dead" as plankton (1887; translation by F. J. R. Taylor, 1980). Hensen derived the term plankton from the Greek *planktos*, translated variously as "wanderer" or "drifting". As a founding member of the Kiel Commission, Hensen studied plankton distributions in an effort to explain the success or failure of North Sea fisheries. A very readable history of this era in biological oceanography is included in Schlee (1973). Hensen's definition effectively replaced the term 'Auftrieb' introduced by Johannes Muller for suspended matter collected in small mesh nets. Subsequently, Ernst Haeckel (1890) discussed plankton in the context of living organisms, providing our contemporary definition (Taylor, 1980). The terminology was well established in the early 1900s, as Gran (1912) stated in the first English text on marine science, *The Depths of the Ocean*, "The term plankton is now used for all floating organisms which are passively carried along by currents." Thus, in name, as well as by nature, marine ecologists have recognized the drifting existence of plankton for more than a century.

In the late 1890s and early 1900s, individual plankton species were interpreted as indicators of the physical properties of water masses. The Norwegian marine ecologist, Per Teodor Cleve, hypothesized that indicator plankton species exhibited an extremely high fidelity to individual currents in Scandinavian coastal waters. Subsequently Cleve collaborated with the physical oceanographers Vagn Walfrid Ekman and Otto Petersen on solving the problem of fisheries fluctuations in European coastal waters that motivated the formation of the International Council for the Exploration of the Sea, or ICES (Schlee, 1973). Ekman proposed that the spatial distributions of fish species reflected their ability to identify, and selectively stay within, waters with well-defined salinity and temperature conditions. Cleve then hypothesized that fish distributions indicated a preference for plankton species associations that resided within the specific water masses. As plankton surveys were conducted in northern European waters during the late nineteenth and early twentieth centuries, Cleve refined his hypothesis to state that characteristic species groups were viable indicators of the physical properties of surface waters and currents (e.g., Cleve, 1896). *Didymus-plankton* was characterized by the diatom *Chaeotoceros didymus* and associated diatoms. The water mass in which they resided was hypothesized to have originated in the southern North Sea, and when present, signaled the arrival of herring along the Scandinavian coast. Another distinct plankton association, the *Sira-plankton*, was dominated by a different diatom species assemblage, and its presence was interpreted as an indicator of waters that had originated in the Arctic Ocean. Cleve's hypothesis required an extraordinary fidelity of plankton species to specific water properties and currents. This association was assumed to be sufficiently strong such that plankton types were considered valid biological tracers of the presence, and source of, individual water masses.

Within a decade, Cleve's interpretation of plankton distributions was challenged by H. H. Gran (Gran, 1912). Gran argued that while various plankton species often did occur within identifiable salinity and temperature regimes, environmental conditions in water masses and currents invariably changed with time. Thus species occurred within a given region as environmental factors were favorable to enhance their growth. In other words, Cleve's plankton types serve as useful indices to associations of phytoplankton and zooplankton species, but they are not invariable indicators of water properties.

11.1.2 The planktonic life mode

The drifting nature of plankton is commonly attributed to their small size and limited swimming capability. In fact, the physical dimensions of marine plankton span several orders of magnitude. The classification scheme of Sieburth *et al.* (1978) is widely accepted and divides plankton into seven size-dependent categories with diameters that range from 0.02 µm at the lower size limit of the femtoplankton to an upper limit of 200 cm in megaplankton. At the smallest scales are sub-micron viruses, bacteria, and micron-sized photoautrotrophic picoplankton. The photoautotrophic biomass in oligotrophic seas is dominated by the photoautotrophic *Prochlorococcus*, with an average cell diameter of 0.5–0.7 µm (Chisholm *et al.*, 1988), and slightly larger *Synechococcus* (diameter *c.* 1 µm; see Morel *et al.*, 1993). Cell diameters of these plant cells approach the theoretical minimum cell diameter estimated for viable photoautotrophic organisms of *c.* 0.3 µm (Raven, 1986). Zooplankton include heterotrophs such as copepods and the ichthyoplankton, with typical body lengths of mm to cm. Zooplankton are the largest planktonic organisms, and the largest zooplankton likely is the Arctic Lion's mane jellyfish *Cyanea capillata* with a bell diameter of more than a meter.

Introductory marine science textbooks compare the small size of phytoplankton to terrestrial plants, yet few pose the question as to why plankton dominate lower trophic levels in the sea. The answer is that the physical properties of the sea, in particular the density and viscosity of seawater, dictate that primary producers in the ocean must be small to efficiently acquire limiting resources such as light and nutrients in a turbulent, aquatic environment. Photoautotrophs must remain in the sunlit, mixed layer of the ocean to photosynthesize. Stokes Law dictates that micron-size primary producers such as picoplankton are effectively retained within the euphotic zone. Larger phytoplankton with dense exterior cell coverings such as diatoms, dinoflagellates and coccolithophorids possess both physiological and morphological adaptations that minimize sinking from the euphotic zone (Smayda, 1970) while turbulence further aids in the suspension of large phytoplankton (Denman and Gargett, 1983; Ruiz *et al.*, 1996; Huisman *et al.*, 2002). The low ambient nutrient concentrations in the surface ocean also favor small photoautotrophs. Since the area/volume ratio of a cell increases as cell diameter decreases, small unicellular organisms have an inherently high capacity to acquire nutrients, per unit cell volume, at the low ambient nutrient concentrations that potentially limit phytoplankton growth in oligotrophic surface waters (Raven, 1986).

The size structure of marine phytoplankton communities dictates that grazers such as herbivorous and omnivorous zooplankton are also microscopic. Herbivorous zooplankton feed on their dilute prey by suspension or filter feeding. They range in size from heterotrophic unicellular protozoa to larger copepods, krill, as well as invertebrate and fish larvae. Some of the largest herbivorous zooplankton are pelagic tunicates that can harvest food ranging in size from colloids to netplankton (Bone, 1998). Although these herbivores attain lengths of several cm, and even longer dimensions in colonies, they are truly planktonic organisms with limited swimming capabilities.

In addition to their small size, the drifting life mode of plankton reflects their limited means of locomotion. Swimming speeds for large dinoflagellates, which are among the fastest phytoplankton (Raven and Richardson, 1984), are on the order of tens to hundreds of $\mu m\,s^{-1}$ (Kamykowski, 1995). Perhaps the fastest speed for a marine unicellular organism is that of *Mesodinium rubrum* Lohmann 1908 (= *Myrionecta rubra* Jankowski 1976), a protist with cryptophyte organelles that undergoes vertical migrations at speeds of 2 to $7\,m\,h^{-1}$ (Gustafson *et al.*, 2000). However, the maximum swimming speeds of photoautotrophic plankton are typically orders of magnitude less than the $cm\,s^{-1}$ speed of surface currents. Thus even the fastest swimming phytoplankton can be readily transported horizontally by surface currents.

Many planktonic species, including photoautotrophic dinoflagellates (Kamykowski *et al.*, 1992), some large diatoms (Villareal and Carpenter, 1994), as well as zooplankton such as copepods (Batchelder *et al.*, 2002) and invertebrate and fish larvae (Garland and Zimmer, 2002), undergo directed vertical migrations. These vertical motions occur over diel and ontogenetic periods. The ability to move vertically can confer an advantage to a photoautotroph with the ability to seek optimal light and nutrient levels, since these properties that have strong vertical gradients. Additionally, swimming can provide a mechanism for phytoplankton to maintain aggregated populations (e.g., Liu *et al.*, 2002).

Zooplankton, such as copepods, swim as a means to escape predation (Strickler *et al.*, 1997) as well as to maintain aggregated populations (Price, 1989; McGehee and Jaffe, 1996). They respond to stimuli in their immediate environment, as exemplified by rapid bursts of swimming in response to a nearby predator. Many species undertake extensive diurnal vertical migrations, swimming up at dusk and down at dawn, that extend over several tens of meters in the vertical. This migratory behavior affords protection from visual predators that might otherwise prey on zooplankton in surface waters during the day (e.g., Zaret and Suffern, 1976; Clark and Levy, 1988; Ohman, 1988). Vertical movements, especially by crustaceans and fish larvae, have also been implicated in enhancing horizontal dispersal in estuaries (Cronin and Forward, 1982) and coastal waters (Epifanio, 1988; Shanks *et al.*, 2000; Hare *et al.*, 1999). An extensive literature exists on species dispersal by benthic marine invertebrates and fish that release eggs, sperm and larvae in the sea (see Scheltema, 1986; Strathmann, 1990).

11.2 Lagrangian studies of plankton

Plankton distributions in the sea exhibit variability across a wide range of scales in both time and space (Haury *et al.*, 1977). This variability reflects the influence of physical and biological processes that concurrently act on the growth and distribution of marine organisms (Dickey, 1991). The space and time scales of processes studied in the sea must therefore be resolved by an appropriate observation method. Biological oceanographers have utilized Lagrangian techniques to observe plankton populations in time, and thereby constrain variability that otherwise could be attributed to their movement through space. In practice, biological oceanographers have relied upon these techniques to observe the temporal distribution of organisms as they evolve in "tagged" water parcels.

There are three major methods that have been used to describe plankton trajectories and biogeochemical processes from a Lagrangian frame of reference. These include direct observations of plankton trajectories, tracers (primarily fluorescent dyes and more recently inert gases), and surface drifters and subsurface floats. These tools have provided biologists with the means to define variability in plankton distribution and growth over a range of scales in both time and space as broadly defined in Figure 11.1. Each technique has an inherent "window" in time and space at which it can be applied to study processes of interest. Visual observations of individual larvae by SCUBA divers are obviously limited to short dispersal scales (meters) and short time periods (minutes to hour). Fluorescent tracers, in contrast, provide useful information on larval dispersal and property fluxes over periods of days, while gas tracers permit measurements over time scales of several months to a year. Surface drifters have been deployed for periods of days to months, and autonomous subsurface floats are now providing observations over months to years. Investigators need to carefully consider these characteristic space and time scales when Lagrangian methods are selected to study specific processes.

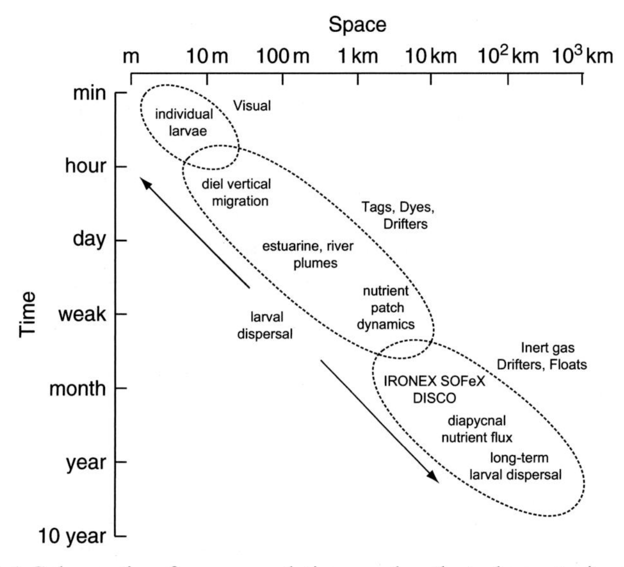

Figure 11.1 Schematic of space and time scales that characterize biological and physical processes examined with Lagrangian techniques. The figure includes scales at which processes are typically studied with methods as described in the text. The reader should consult Dickey (1991) and Bidigare *et al.* (1992) for figures with relevant space and time scales of biophysical processes and related methodologies.

11.2.1 Fate of individual larvae

Direct observations of larval dispersal

A Lagrangian method to trace organism dispersal would ideally provide marine ecologists with the capability to describe the temporal history of the spatial location of each individual within a "target" population. Given the patchy nature of plankton, as well as their size, individual plankton can rarely be followed, although individual larvae of sessile adult invertebrate species have been followed visually for brief periods. These studies have been mainly conducted to assess rates of larval mortality, with an objective to determine if "pre-settlement" mortality (of planktonic larvae) or, alternatively, "post-settlement" mortality (of settled juveniles) regulates the size of adult populations. On one hand, the size of adult benthic invertebrate populations may be a function of processes that influence larval survival prior to settlement on a suitable substrate. Alternatively, a population might produce sufficient larvae that survive to settle on a suitable substrate, but the subsequent mortality of the benthic juvenile stage determines the abundance of the adults. Direct measures of larval mortality, however, are notoriously difficult (Rumrill, 1991; Morgan, 1995).

The fate of individual planktonic larvae has been determined for several ascidians, the benthic tunicates typified by the "sea squirt." Free-swimming ascidian larvae are tadpole-shaped, and several species produce larvae of sufficient size (cm) such that individuals can be visually followed by divers. Ascidian larvae are often colored with enhanced visibility due to pigments that originate in their symbiotic photoautotrophs (e.g., Olson, 1985; Stoner, 1990). A defining characteristic of ascidian larvae is that they remain pelagic for a few minutes to at most an hour before they settle on the bottom (Olson, 1985). Individual larvae are released sporadically between mid-morning to mid-afternoon, and while in the water column remain within a few meters of the bottom. These attributes have permitted divers to describe trajectories and the fate of individual larvae. When sufficient larvae are followed, the fate of the dispersed sample of individuals can be assessed. At Coconut Island, Hawaii, for example, Stoner (1990) reported that 66% of 259 total larvae were successfully followed to their fate, with 29% of these captured or eaten within a few minutes of release. Those that settled did so within a few minutes, settling at a mean distance from the parent of only 2.21 m (SD $\pm$ 1.4 m). Although this short dispersal distance suggests a high proportion of larvae will settle successfully, all of the post-settlement individuals died within a month.

In St. Vincents Gulf and Investigator Strait, South Australia, Davis and Butler (1989) report a similar result for larvae of the colonial ascidian *Podoclavella moluccensis* (Sluiter). About 100 of 270 larvae were successfully followed to their fate, with planktonic mortality <20% in each of two consecutive years of observations. At Davies Reef on the Great Barrier Reef of Australia, in contrast, the planktonic larvae of another ascidian, *Lissolinum patella* Gottschaldt, suffered 87% mortality during the brief pelagic phase. The principal losses were to fish, mainly pomacentrids, with grazing by corals and other benthic organisms of secondary importance. In summary, these "quasi-Lagrangian" observations suggest that larval mortality rates are variable even at short intervals of only few minutes.

Mark/recapture techniques and geochemical signatures

Since few species are amenable to direct observations, "mark/recapture" methods have been developed to study individual-scale larval dispersal. These studies indicate both an origin and, when successful, a settlement site for larval populations. The approach is analogous to drift bottle studies in which positions are recorded for the release and recovery of drifting platforms (e.g. Wyatt *et al.*, 1972). The main goal of "mark/recapture" experiments is to link larval sources to settlement without knowledge of the intermediate trajectories. Mark/recapture studies have relied upon stains, radioisotopes, trace elements, a strontium substitution of calcium in $CaCO_3$, and genetic markers to label invertebrate and fish larvae prior to their release (Thorrold *et al.*, 2002; Thorrold and Hare, 2002). Levin (1990) provides a comprehensive review and critique of several of these methods.

One difficulty with this approach is that high larval mortality rates, coupled with wide dispersal, often result in the recovery of few labeled larvae. Cowen *et al.* (2000) examined this problem in a modeling exercise. Despite the limitations of relatively few recovered larvae, Jones *et al.* (1999) successfully marked approximately 10 million damselfish embryos and in the subsequent examination of 5000 individuals at the end of their planktonic stage, recovered 15 marked individuals. Using mark/recapture models, they estimated that 15.6% of the juveniles may have returned to their natal reef at the end of the planktonic stage. To bypass the problem of recapturing marked larvae, recent efforts have been directed at identifying unique geochemical signatures of larval source populations. These signatures are derived from the trace metal composition of invertebrate larvae (e.g. DiBacco and Levin, 2000) and the chemical composition of larval fish otoliths (e.g., Campana and Thorrold, 2001; Thorrold *et al.*, 2001, 2002). Natural geochemical signatures offer the advantage that the entire larval population is naturally marked as the elemental signature is incorporated in the larvae. A variety of recent applications have successfully identified source locations for migrating fish (Thorrold *et al.*, 2001) and dispersing larvae (Swearer *et al.*, 1999; Gillanders and Kingsford, 2003) based on element composition. It is clear that tagging and natural markers will continue to be developed as a unique experimental approach to study larval dispersal.

11.2.2 Tracers

Fluorescent dyes

Natural and anthropogenic tracers have had an integral role in the study of both oceanic turbulence and larval dispersal. Oceanographers have employed a wide variety of objects to quantify dispersion in the sea, including dyes, radioisotopes, drifters and drogues, pieces of paper, vegetables, and even bath tub toys and sneakers. Here we concentrate on fluorescent dyes, one of the most common tracers applied to parameterize turbulent mixing, and inert gases (Table 11.1). Relationships between rates of dye diffusion and space and time scales were developed by, among others, Bowden (1965) and Okubo (1971). Their work formed the basis for understanding diffusive processes in plankton dispersal and nutrient diffusion in the sea.

Table 11.1 *Studies of organism dispersal and biogeochemical fluxes that have incorporated tracers. These examples demonstrate the utility of tracers in field experiments at various time and spatial scales. See text for detail.*

Location	Scale	Purpose	Reference
Fluorescent dyes			
Davies Reef, Australia	min – hour	Flow rates to estimate zooplankton grazing by planktivorous reef fish	Hamner *et al.* (1988)
Conception Bay, Newfoundland		Assess advection, diffusion to model larval capelin mortality	Taggart and Leggett (1987)
North Sea		Diffusion of plaice egg and larval patches to estimate mortality	Talbot and Talbot (1974) Talbot (1977)
James River, Virginia USA	multiple tidal cycles	Dye tests to estimate nontidal circulation and retention of oyster larvae	Ruzecki and Hargis (1989)
Chesapeake Bay Estuaries	hours-days	Flow variability affects of oyster settlement	Kennedy and Boicourt (1981)
Long Island Coastal waters	18 days	Hard calm transplant site retention	Becker (1978)
Seto Inland Sea, Japan	days	Advection and diffusion of fish eggs and larvae	Nakata and Hirano (1978)
Sulfur hexafluoride			
Equatorial Pacific	days/10s km	Productivity response to iron enrichment in high nutrient, low chlorophyll area.	Martin *et al.* (1994)[1] Coale *et al.* (1996)[2]
Southern Ocean		Southern ocean phytoplankton response to iron enrichment (SOIREE)	Boyd *et al.* (2000) Boyd and Law (2001)
		Nitrate flux through the pycnocline	Law *et al.* (2003)
Southern Ocean		Phytoplankton response to iron enrichment in high and low silicate regions. (SOFeX).	Coale *et al.* (2004)
Northeast Atlantic	weeks	Biogeochemical evolution in an anticyclonic eddy in the Northeastern Atlantic	Martin *et al.* (1998)
		Eddy mixed layer dynamics	Law *et al.* (2000)
		Zooplankton time series within the Eddy tracer patch.	Irigoien *et al.* (2000)

North Sea[3]	days-week	Plankton community metabolism of dimethyl sulphide during a phytoplankton bloom	Burkill *et al* (2002)
Subarctic Pacific	weeks	Phytoplankton response to iron enrichment	Tsuda *et al*. (2003)
		Carbon dynamics in an iron-stimulated diatom bloom	Boyd *et al*. (2004)
Georges Bank	10 days	Air-sea transfer experiment with gaseous tracers	Wanninkhof *et al*. (1993)
		Carbon cycling within the tracer patch	Sambrotto and Langdon (1994)
West Florida Shelf[4]	11 days	Air-sea transfer on oligotrophic shelf	Wanninkhof *et al*. (1997)
		Carbon cycling in the mixed layer	Hitchcock *et al*. (2000a)
Indian River, Florida	2 weeks	Diffusion and advection of *Mercenaria* larvae	Arnold *et al*. (in press)

[1] IRONEX I;

[2] IRONEX II;

[3] DISCO

[4] FSLE

Fluorescent dyes, in particular Rhodamine-B and Rhodamine-WT, offered the first practical means to follow the spread of "tagged" water parcels. Colored dyes were first used in the 1920s to observe mixing in oceanic surface waters (Carter and Okubo, 1978). These qualitative diffusion experiments relied upon visual or colorimetric methods for detection and required large quantities of released material. As an example, Seligman (1955), as cited by Carter and Okubo (1978), released 10 tons of a dilute fluorescein solution in the Irish Sea in each of eleven experiments to assess the capacity of surface waters about the Windscale nuclear facility to dilute radioisotopes. Samples were collected at night, presumably to minimize photodegradation, and compared to standards under ultraviolet light to achieve a sensitivity of 1 ppb. The main contribution of this study, and prior colorimetric experiments, was principally in technique development.

In the late 1950s, D. W. Pritchard and J. H. Carpenter began using the fluorescent dye Rhodamine-B as an artificial tracer of water motion. Okubo *et al.* (1957) previously selected this tracer for diffusion studies off the coast of Japan, relying on spectrophotometric measurements to quantify concentrations. The higher sensitivity inherent in fluorometric measurements, compared to spectrophotometric detection, coupled with the availability of the stable and relatively inexpensive Turner Designs Model 110 fluorometer, led to field studies with a routine sensitivity of *c.* 0.01 ppb. Additionally, it was possible to continuously record dye fluorescence on a strip chart recorder to define the longitudinal and lateral "spread" of the tracer with time.

These technological advances encouraged studies of turbulent diffusion in a variety of coastal regions, resulting in Okubo's (1971) classic paper "Oceanic diffusion diagrams." Relationships of K_h, the apparent diffusivity, with length scale and the variance of a dye patch (e.g., σ_{rc}) with time (Fig. 11.2), indicate that tracers, and by inference plankton, are successively "spread" by turbulent eddies at increasingly larger length scales at a power of 2 to 3. A key concept in the relationships between diffusion in space and time is that advection is not the sole determinant of larval dispersal (Okubo, 1994). Some ecologists have neglected this, drawing conclusions that oversimplify larval dispersal (e.g., Roberts, 1997).

Throughout the 1970s, marine ecologists conducted tracer release experiments to better understand the role of turbulence in plankton dispersal. Fluorescent dyes, notably Rhodamine-WT, have minimal toxicity (e.g., Parker, 1973), an obviously necessary trait for a successful tracer experiment with live eggs and larvae. At the very short spatial scales of a few meters Rhodamine-WT defined the role of vertical motion in larval dispersal in wave-swept rocky shorelines (Koehl and Powell, 1994). Large-scale dye studies have characterized dispersal properties of invertebrate larval populations in many

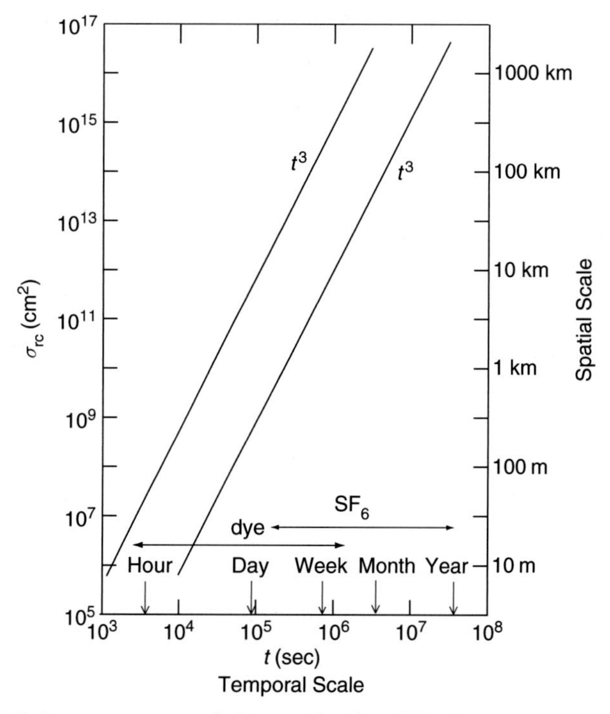

Figure 11.2 Relevant space and time scales for diffusion rates as measured by fluorescent dye and sulfur hexafluoride (SF_6) tracers. The figure summarizes observations in marine waters with the t^3 boundaries constraining the majority of observations. Modified from Okubo (1971).

environments (Table 11.1), including Chesapeake Bay tributaries (Kennedy and Boicourt, 1981), Long Island Sound (Becker, 1978), the coastal waters of the Seto Sea (Nakata and Hiroano, 1978) and United Kingdom estuaries and coastal waters (Talbot and Talbot, 1974).

Some of the first studies of the advective and diffusive processes affecting larval dispersal were conducted with dye tracers in the North Sea. Throughout the 1960s the United Kingdom Ministry of Fisheries conducted field studies that assessed the distribution and abundance of the eggs and larvae of plaice, a commercially important flatfish (Talbot and Talbot, 1974; Talbot, 1977). Mortality in plaice eggs and larvae is high during the interval between spawning and larval metamorphosis, and estimates of larval mortality are essential to judge year-class strength. A series of cruises mapped the extent of egg and larval patches annually in the Southern Bight of the North Sea where plaice routinely spawn. Egg and larval distributions were mapped as isolines defining areas with equal concentrations.

Talbot (1977) approached the problem of quantifying the magnitude of diffusive processes in larval dispersal with an approach proposed by Joseph and Sender (1958). In this analysis it is assumed that larval, or egg, densities in a patch decrease as it spreads due to turbulent diffusion. Note that the method does not include a term for mortality which would also potentially cause organism concentrations to decrease with time. As concentrations decrease with time (t) from an initial value $N(0, t)$, the 'equivalent radius' (r) of a circular patch increases. The data can therefore be expressed as a time-dependent change in organism densities, $N(r, t)$, as a function of the patch characteristic length scale (A) by the expression:

$$N(r,t) = N(0,t)\exp(-r/A)$$

In 1962, 1963, and 1969 patches of plaice eggs and larvae in the Southern Bight of the North Sea were observed to reach dimensions equivalent to a radius of several tens of km. After patches expanded to this size, however, there was little change in the length scale, A, over the next few weeks. The observations suggest that the variability found in patch size reflected the fact that egg patches were spawned at different times. The distribution of dye patches released in the Southern Bight was fairly uniform with depth, and after a day the horizontal diffusion coefficients (K_x, K_y) remained fairly constant (Fig. 11.3). Constant patch dimensions were interpreted as an indication that vertical shear associated with tides, in contrast to turbulent diffusion, dominated the horizontal dispersion of larvae. However, entrainment into the strong salinity frontal zones of the Southern Bight (e.g., Munk *et al.*, 2002), or eddies (Olson, Chapter 10 this volume), could also account for aggregated plaice larvae, and the relatively low rates estimated for turbulent diffusion. Additionally there was no assessment of the effect of mortality on larval abundance in this approach.

Despite the importance of larval fish mortality in structuring the strength of year classes in commercial fisheries, direct measurements of mortality in the sea are difficult and rare. Taggart and Leggett (1987a,b) approached the problem of constraining mortality rates for capelin (*Mallotus villosus*), an important prey item for cod and seabirds throughout the North Atlantic, through a series of dye tracer experiments. Field experiments were conducted in the 1 km^2 embayment of Bryants Cove, Conception Bay, Newfoundland. The limited spawning region made it feasible to monitor larval emergence from the spawning grounds, as well as quantify larval immigration and emigration from the Cove. The role of advective transport and diffusion in the horizontal dispersion of larvae was evaluated by dye studies and current meters. After Rhodamine-B was released at the spawning location, the dye patch rapidly

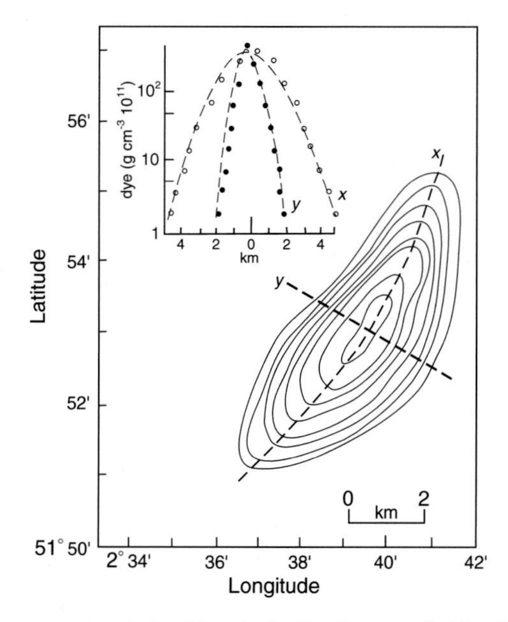

Figure 11.3 An example of the dispersal of a dye patch (rhodamine-B) in the North Sea modified from Talbot (1974). The dye patch has evolved about 2.5 days following release, and spread several kilometers in both the longitudinal (x) and lateral (y) axes. Inset shows the spread across the two orthogonal axes.

dispersed in a manner predicted by a third power law. The dye patches flowed seaward, in agreement with the circulation pattern described by current meters (Taggart and Leggett, 1987a). Advection dominated over horizontal diffusion in the cove, as reported for other coastal environments (e.g., Gagnon and Lacroix, 1981; Fortier and Leggett, 1982). The average daily mortality rates were estimated at 60% day^{-1}, with a range of 3 to 100% day^{-1}. Much of the daily variability in larval abundance could be attributed to larval immigration, which influenced the magnitude of apparent mortality.

SF_6

Improvements in both the extraction of fluorescent dyes (Laane *et al.*, 1984) and their detection, largely through high performance liquid chromatography and improved fluorescence methods, have lengthened the duration of rhodamine tracer experiments to several months in freshwater lakes (Suijlen and Buyse, 1994). In the sea, however, the maximum duration of a dye experiment is typically several days to a few weeks (e.g., Upstill-Goddard *et al.*, 2001) principally due to the reduction of dye concentrations by photolysis and mesoscale turbulence. A need to measure rates of mixing in the ocean at time scales of weeks to months has motivated development of tracers with longer useful lifetimes. Inert tracers such as the anthropogenic gas sulfur hexafluoride (SF_6) can be detected at concentrations well below those of fluorescent dyes, typically at levels of 1 fmol kg^{-1}. These tracers are therefore detectable at extremely low concentrations, a characteristic that enables them to be followed in experiments lasting several months. Watson and Ledwell (2000) provide a comprehensive review of the development of inert gases as oceanic tracers.

The anthropogenic gas sulfur hexafluoride is inert, nontoxic, and has a low background level, making it an optimal tracer for oceanic mixing studies. Initial releases of SF_6 were made in freshwater lakes to quantify air-water gas exchange rates (e.g., Wanninkhof *et al.*, 1985). These early experiments proved that gas exchange could be measured over tens of km and for periods that extend beyond a few weeks (e.g., Watson *et al.*, 1991; Wanninkhof *et al.*, 1993). Longer-term releases of SF_6 have provided estimates of diapycnal diffusivity in the main pycnocline of the North Atlantic on the order of $0.10\,cm^2\,s^{-1}$ (e.g., Ledwell *et al.*, 1998). These field studies show that tracer distributions remain "streaky" over long periods of time (see Garrett, 1983), and reveal that lateral mixing rates in the North Atlantic pycnocline are higher than those for diapycnal mixing (see Watson and Ledwell, 2000 for a summary).

The inert and nontoxic properties of SF_6 make it well suited for *in situ* experiments with natural plankton populations. The addition of a nontoxic tracer with a nutrient enrichment has provided ecologists with the means to follow the response of phytoplankton to the addition of a hypothesized "limiting" nutrient. The first of several large-scale enrichment experiments that have utilized SF_6 for this purpose were conducted in the Equatorial Pacific to test the "iron hypothesis" developed by the late John Martin and colleagues (Martin, 1990). The history of our understanding of iron utilization in the sea is long and complex (Chisholm and Morel, 1991). Martin's iron hypothesis specifically addresses the apparent discrepancy in "high nutrient low chlorophyll" (HNLC) regions of the surface ocean where the standing stock of primary producers is low, yet inorganic nutrient (e.g., nitrate) concentrations remain relatively high. Elevated nitrate concentrations in HNLC regions have been attributed to an insufficient input of available iron required to meet physiological requirements for enzymes involved in nitrate utilization. Refining and testing the iron hypothesis has been an active area of research, particularly since it is relevant to understanding variability in the Earth's climate.

The two largest HNLC regions of the global ocean are the Equatorial Pacific and Southern Ocean. To test the hypothesis, field experiments were first conducted with iron enrichments and SF_6 releases in the HNLC surface waters of the Equatorial Pacific (Table 11.1). The initial experiment, IRONEX I, was conducted in 1993 in surface waters southwest of the Galapagos Islands. The phytoplankton biomass response following iron enrichment was rather modest, but the surface ocean inorganic carbon fugacity (fCO_2, the tendency of the gas to escape from the ocean), declined in the tracer patch as the iron enrichment rapidly disappeared (Martin *et al.*, 1994). However, the observed decrease in inorganic carbon was less than expected if all available iron had been utilized to support the observed rates of primary production (Watson *et al.*, 1994). Subsequently, IRONEX II was conducted in 1995 with a larger enrichment, and as in IRONEX I, surface *f*CO2 rapidly declined in the tracer patch (Coale *et al.*, 1996). In contrast to the modest response seen in plankton biomass in IRONEX I, a diatom bloom developed within a week as iron concentrations declined in the tracer patch. Linear relationships between fCO_2 and surface SF_6 concentrations in both experiments reinforce the conclusion that iron availability is a key component to biogenic utilization of inorganic carbon in HNLC regions of the surface ocean (see Fig. 5 in Watson and Ledwell, 2000). Subsequent tracer studies have been conducted in the Southern Ocean (Boyd *et al.*, 2000; Coale *et al.*, 2004) and Subarctic Pacific (Tsuda *et al.*, 2003; Boyd *et al.*, 2004) to test the efficacy of iron enrichments in stimulating productivity in HNLC regions (Table 11.1). These studies have relied upon SF_6 for sequential sampling of an iron-enriched "patch," and their results have supported the hypothesis that iron availability limits phytoplankton productivity in large HNLC regions of the ocean.

Since the early 1990s, sulfur hexafluoride has also been increasingly deployed as a tracer to identify water parcels in biogeochemical studies within coastal and shelf waters. Nutrient-productivity relationships were defined in a tracer patch on Georges Bank in a very spatially heterogeneous environment (Wanninkhof *et al.*, 1993; Sambrotto and Langdon, 1994). The repeated sampling of the tracer patch provided estimates of carbon dynamics as primary production varied in response to declining nutrient concentrations and net community respiration was high, exceeding $1\,g\,C\,m^{-2}\,day^{-1}$. Within the tracer patch, plankton biomass changed from a dominantly autotrophic to heterotrophic population, consistent with the high rates of net community respiration. Given the inherent spatial variability in coastal waters, it would have been very difficult to isolate temporal variability in carbon and nutrient dynamics from spatial variability as the plankton population underwent a rapid change in composition over several days to weeks.

A subsequent tracer study on Georges Bank utilized the fluorescent dye Fluorescein to quantify the diapycnal motion of bottom waters in the vicinity of a strong tidal front (Houghton and Ho, 2001). By measuring the warming of the tagged water mass over time the investigators derived rates of cross-frontal exchange on the Bank and demonstrated that the heat input to the tagged water mass must have arisen from vertical mixing in the front. While the study did not directly address the biological implications of cross-frontal exchange, it demonstrates that not only gases but also dyes can yield information on exchange processes of interest to biological oceanographers.

A tracer experiment analogous to Wanninkhof *et al.* (1993) was conducted with SF_6 in the oligotrophic, subtropical waters of the West Florida shelf (Wanninkhof *et al.*, 1997). This experiment was one of the first to directly measure the magnitude of air–sea exchange concurrently with a study of biogenic carbon dynamics (Hitchcock *et al.*, 2000a). The ratio of two tracers, SF_6 and ^3He, in surface waters provided an estimate of an average gas transfer velocity of $8.4\,cm\,hr^{-1}$ over two weeks (normalized to 20° C). Dissolved inorganic carbon (DIC) increased at $1\,\mu mol\,kg^{-1}\,day^{-1}$ in the tracer patch, a rate that suggested air–sea exchange accounted for *c.* 20% of the net change in inorganic carbon (Wanninkhof *et al.*, 1997). The majority of DIC increase was attributed to biogenic remineralization, which was consistent with the heterotrophic nature of the plankton community where net community production rates ($6 \pm 6\,mmol\,C\,m^{-2}\,day^{-1}$) were almost five-fold lower than community dark respiration rates ($-40 \pm 3\,mmol\,C\,m^{-2}\,day^{-1}$). Conventional light-dark O_2 bottle measurements were combined with $H^{18}O_2$ incubations to derive light respiration rates ($-51 \pm 8\,mmol\,C\,m^{-2}\,day^{-1}$) that, unexpectedly, exceeded dark respiration rates. In summary, plankton community respiration was high enough to account for observed carbon remineralization and the increase in DIC within the SF_6 patch. As on Georges Bank, spatial variability would have precluded a consistent picture of carbon dynamics on the West Florida shelf in the absence of a Lagrangian frame of reference.

Inorganic nutrient dynamics and plankton community structure have also been addressed with a SF_6 release in the oceanic waters of the Northeast Atlantic during the Plankton Reactivity in the Marine Environment (PRIME) program. The major objective of the PRIME study was to interpret nutrient and carbon cycling within planktonic ecosystems in the context of the size structure and trophic structure of pelagic ecosystem models (Savidge and Williams, 2001). The size structure approach to plankton models is formulated as in Moloney and Field (1991), with the trophic structure modeling exemplified by Taylor *et al.* (1993). Each class of model requires a characteristic data suite, with data for size structure models provided by time series data from a Lagrangian sampling scheme, with the trophic structure model requiring input of several environmental parameters at specified locations through time.

The PRIME cruises included a SF_6 release in a cold core anticyclonic eddy near 59° N, 20° W to meet the needs of size structure modeling efforts. The anomalous anticyclonic circulation associated with a cold core feature was the result of a "lens" of cool water residing above a nearly stationary, deeper anticyclonic circulation (Martin *et al.*, 1998). A central objective was to constrain vertical nutrient transport rates in near surface waters. The vertical tracer distributions of SF_6, corrected for background tracer levels, gave vertical eddy diffusivities, K_z, on the order of 1.5–1.9 $cm^2\ s^{-1}$ in the eddy (Law *et al.*, 2000). These values were considerably higher than found in previous tracer releases, or from most microstructure-based estimates. The high K_z values were evaluated in terms of various error sources, but they appear realistic. When combined with vertical nutrient gradients, the K_z values suggest that vertical nitrate (1.8 $mmol\ m^{-2}\ d^{-1}$) and phosphate (1.25 $mmol\ m^{-2}\ d^{-1}$) fluxes could support less than half the daily "new" nutrient demands of primary producers (Law *et al.*, 2000). In contrast to other oceanic regimes, vertical turbulence had less of a contribution in supplying new nutrients in the eddy than did advection.

A process study of dimethyl sulphide (DMS) was conducted in the North Sea within a SF_6 tracer patch to assess the role of the plankton community, particularly the coccolithophore *Emiliania huxleyi*, in the biogeochemical cycling of DMS and its precursor dimethyl sulphoniopropionate (DMSP). Burkill *et al.* (2002) provides a summary of conclusions reached by the DISCO (dimethyl sulphide biogeochemistry within a coccolithophore bloom) Program. The composition, productivity and respiration of individual components of the plankton community revealed that *E. huxleyi* contributed a relatively small percentage to particulate DMSP; other flagellate phytoplankton were the dominant particulate source. Dissolved DMSP in the mixed layer mainly resulted from microzooplankton grazing, while one species of proteobacteria accounted for much of the subsequent DMS production. The study was sufficiently comprehensive that a DMS budget was derived for the mixed layer in the tracer patch. As in the PRIME, Georges Bank and FSLE studies, SF_6 provided the means for temporal sampling of an identifiable community within the mixed layer.

In summary, tracer experiments have been utilized to address larval dispersion and biogeochemical cycling in the sea. Both fluorescent dyes and inert gases have capability to "tag" specific water parcels, with dyes frequently utilized in short-term studies in coastal waters, and SF_6 well suited for longer-term biogeochemical studies in oceanic waters. The nontoxic nature of SF_6 makes it ideal for tracer studies with oceanic organisms. Pilot studies with intentional larval releases in coastal waters (Arnold *et al.*, 2005) have also demonstrated that SF_6 is also useful in short-term larval dispersion studies in shallow embayments.

11.2.3 Surface drifters

Applications of surface drifters by biologists

Biological oceanographers have studied plankton distributions and processes with precursors of contemporary surface drifters for more than four decades. Many early drifters had "shade" or parachute drogues attached to surface floats and were tracked by radar or by aerial observations from aircraft in coastal waters. These platforms had limited means for tracking since a ship had to stay within radar range or line of sight if the platform had a vhf transmitter. These designs are also compromised since their motion is biased by windage (effects of wind stress), wave effects, and drag (Kirwan *et al.*, 1975; Geyer, 1989; Davis, 1991). Nevertheless, studies conducted before the development of the contemporary TriStar, WOCE holey sock (Niiler *et al.*, 1987; Mackas *et al.*, 1989) and surface CODE drifters (Davis, 1985) demonstrated that a Lagrangian reference frame provides considerable information on the temporal change in "tagged" plankton populations.

Limitations in following surface platforms by radar or visually restricted early studies to nearshore waters. Biological oceanographers began to deploy surface platforms to describe water motion and plankton populations in the early 1960s. A central objective was to document plankton populations and surface properties with time, particularly in coastal waters (Denman and Powell, 1984). As an example, Alexander and Corcoran (1963) followed a drogue deployed at 100 m in the oligotrophic Florida Current off Miami to define the spatial variation in the chlorophyll field. Temporal changes in particulate, dissolved organic, and inorganic phosphorus were also sampled within a Lagrangian observational framework in the Gulf of Maine by Ketchum and Corwin (1965). With time, inorganic dissolved phosphorus declined as organic constituents increased and particulate matter sank below the pycnocline as the drogued surface platform drifted south through the Gulf. Temporal changes were extrapolated to release rates for dissolved phosphorus fractions in the surface layer, and remineralization in deeper waters. Since the pycnocline (nutricline) shoaled along the drogue trajectory, the changes in property fields were interpreted with respect to density surfaces, rather than depth. This ambitious study was one of the first to define biogeochemical fluxes in a Lagrangian observation scheme. These and others studies in Table 11.2 demonstrated that marine ecologists have recognized advantages of Lagrangian sampling schemes for several decades.

Zooplankton ecologists began to define the spatial variability in coastal mesozooplankton distributions with surface platforms in the 1970s. Spatial

Table 11.2 *Studies of organism dispersal and distributions, biogeochemical properties and fluxes utilizing Lagrangian platforms. Observations include use of platforms for guiding sampling, as well as sensors deployed on platforms. Platforms include parachute and windowshade drogues, surface drifters, and subsurface floats and profilers. These examples are intended to convey the diverse application of Lagrangian methods by marine ecologists. Time scale refers to the duration of deployment.*

Location	Scale	Purpose	Reference
Drogues			
Florida Current	days	Temporal/diel variation in pigment distributions	Alexander and Corcoran (1963)
Peru upwelling	days	Carbon/chlorophyll ratio variability	Lorenzen (1968)
Gulf of Maine	days	Phosphorus cycling	Ketchum and Corwin (1965)
Baja California	days	Diel zooplankton composition following a vertically migrating drogue	Miller (1970)
Peru Upwelling	days	Carbon dynamics in upwelled waters	Ryther *et al.* (1971)
California Current, North Pacific Gyre	hours to diurnal	Zooplankton horizontal spatial distribution 100s to 1000s meters	Haury (1976a)
California Current	hours to diurnal	Zooplankton horizontal spatial variability at 10s to 1000s meters	Haury (1976b)
Norwegian Sea	days	Feeding preference (gut content) of young herring	Bjoerke (1978)
Mauritania Upwelling	days	Ammonium source/fate in recently upwelled surface waters	Le Borgne (1978)
Sea of Japan	hours	Short-term variability in zooplankton composition, abundance	Itoh *et al.* (1979)
Yonge Reef, Australia	hours	Bacteria, organic matter flux over reef (included fluorescein tracer)	Moriarty (1979)
Equatorial Pacific	days	Phytoplankton biomass/production in recently upwelled waters	Oudot *et al.* (1979)

Table 11.2 *(cont.)*

Location	Scale	Purpose	Reference
Peru Upwelling	days	Physical/plankton evolution in recently upwelled waters	Brink *et al.* (1981)
Benguela Upwelling	days	Nutrient, phytoplankton, distributions in upwelled surface waters	Brown and Hutchings (1987)
Peru Upwelling	days	Evolution of nitrogen uptake in recently upwelled waters	MacIssac *et al.* (1985)
Gulf of Mexico	diel	Nitrite dynamics in subsurface chlorophyll maximum	French *et al.* (1983)
Campeche Bank, Mexico	days	Flow over Campeche Bank and nitrate influx	Furnas and Smayda (1987)
Sargasso Sea	days	Horizontal diffusion of atmospheric nitrogen inputs to surface waters	Hitchcock *et al.* (1990)
Kuroshio Current offshore waters	days	Growth/survival of Japanese sardine	Zanitani *et al.* (1996)
Surface drifters:			
Chesapeake Bay outflow plume[1]	day/season	Physical, chemical and biological properties in estuarine plume	Boicourt *et al.* (1987)
	day/season	Nitrogen and carbon dynamics in an estuarine outflow plume	Malone and Ducklow (1990); Jones *et al.* (1990); Glibert *et al.* (1991); Glibert and Garside (1992)
	day/season	Physical processes and larval distributions	Roman and Boicourt (1990, 1999); Johnson (1995)
German Bight	days	Nutrient plankton dynamics in fronts	Raabe *et al.* (1997)
Delaware Bay plume	days	Physical regulation of larval crab distribution	Natunewicz *et al.* (2001)
Rhone River plume	days	Plankton community evolution	Rick (1999)

Mississippi River plume	days	Nutrient, pigment distributions Surface CDOM distributions	Hitchcock *et al.* (1997); Hitchcock *et al.* (2004)
Newfoundland Shelf	weeks	Atlantic cod eggs and larval transport	Pepin and Helbig (1997)
Northwest Indian Ocean	weeks	Pigment fields about the Great Whirl	Hitchcock *et al.* (2000b)
Gulf Stream/Slope Sea	weeks	Dispersal pathways of larval fishes	Hare *et al.* (2001)
California Current	days/weeks	Phytoplantkon/property distributions in an intense eddy current field	Abbott *et al.* (1990, 1995, 1998); Abbott and Letelier (1997)
Southern Ocean	days/weeks	Seasonal variation in phytoplankton distributions	Letelier *et al.* (1995, 1997)
Antarctic Polar Front	days/weeks	Phytoplankton dynamics and physical processes	Abbott *et al.* (2000); Landry *et al.* (2001)
Subsurface floats and Profilers			
Gulf Stream	days	Resolve property distributions along surface front (Isopyncal Swallow float)	Hitchcock *et al.* (1994)
Gulf Stream meanders	days	Pigment fields in mesoscale meander (POGO floats)	Hitchcock *et al.* (1993)
Coastal waters, Duck, NC	minutes	Photosynthesis response to vertical light field (SUPA)	Kirkpatrick *et al.* (1990)
Sea of Japan	day/weeks	Variability in vertical extinction coefficient	Mitchell *et al.* (2000)
North Pacific	days/months	Plankton response to aeolian (dust) input	Bishop *et al.* (2002)
Southern Ocean	months	Carbon biomass and flux	Bishop *et al.* (2004)
Labrador Sea	week/months	Vertical temperature, salinity, oxygen profiles	Körtzinger *et al.* (in press)

[1] Microbial Exchanges and Coupling in Coastal Atlantic Systems (MECCAS) Project.

variability in mesozooplankton species composition was assessed with radar tracked parachute drogues in the California Current, Baja California, and North Pacific Central Gyre at scales of 100–1000 meters by Haury (1976 a, b). Plankton samples were collected with a Hardy Longhurst Plankton recorder that continuously preserved the collected material on a long roll of mesh. The length scales that could be resolved by this technique were comparable to the distances at which individual zooplankton interact, or aggregate. Although the duration of the drogue trajectories was relatively short at 24 to 48 hours, the observed scales of aggregation in coastal waters corresponded to a few km. Mesozooplankton aggregations in the North Pacific central gyre, in contrast, were longer than in coastal waters. Although day–night differences were noted in the community structure in both environments, no day–night differences were observed in the scales of aggregation in either coastal or oceanic waters. Parachute drogues were similarly deployed to assess spatial variability in zooplankton community structure in the Sea of Japan (Itoh *et al.*, 1979). Over the course of three days copepod densities of various species increased in surface waters at night due to diel vertical migration. Additionally, *Calanus* and *Paracalanus* copepodite stages increased over the three day study, although no causative factors were identified for the observed change.

A novel vertically migrating drogue was designed by Miller (1970) to mimic trajectories of zooplankton undergoing diel vertical migration. An automated winch on a surface platform raised and lowered a parachute drogue between depths of 10 m at night and 100 m during the day to simulate the path of vertically migrating organisms. Changes in species composition off Baja California were assessed along the drogue trajectory by conducting net tows at 2 hour intervals during a 53 hour experiment. The nocturnal zooplankton species composition was similar during three consecutive nights, and it was more uniform than in samples collected at fixed positions. Daytime zooplankton communities, in contrast, exhibited less similarity from day-to-day, suggesting night populations in the surface layer were fairly coherent. More recently, an autonomous profiler has been successfully developed as a vertically migrating zooplankton mimic by De Robertis and Ohlmann (1999). This profiling vehicle can be programmed to hover or passively drift at prescribed depths during predetermined times of the day. Its vertical speed through the water column is comparable to those of migrating zooplankton. Test deployments in California coastal waters have proven that it reliably mimics the vertical motion of zooplankton populations over diel cycles.

The horizontal and vertical motion of herring larvae was similarly documented over a 13 day study in the northeastern Atlantic by Heath and Rankine (1988). High speed nets were deployed from ships as they tracked buoys with radar reflectors and cone-shaped drogues (see Dooley, 1974), as well as an Argos-tracked buoy. Several larval cohorts were identified through size analysis, with the Coastal and Fair Island Currents identified as the primary mechanism by which larvae were advected from spawning grounds into the North Sea. Length distributions of the larvae indicated that two of the cohorts ranged in age from 9 to 13 days, with a third cohort hatching during the study. The drifter trajectories deviated from the center of the larval patch with time. After evaluating factors such as vertical shear and the effects of differences between the vertical distribution of larvae and the drogue depth, the investigators concluded that windage (drag imposed by wind on the buoy) was likely responsible for the observed deviation between the surface buoys and larval trajectories. They recommended that when following tagged larval patches with buoys, the deployments must be limited to a few days and that future buoy designs should minimize the effects of windage.

Upwelling zones

Surface drogues have also characterized temporal changes in plankton communities in coastal upwelling zones. In the spring of 1966 John Ryther and colleagues followed parachute drogues set at 10 and 100 meters depth for several days in the Peru coastal upwelling region (Ryther *et al.*, 1971). The objective of this field experiment was to constrain primary productivity and carbon cycling rates after waters had upwelled. As a ship tracked shallow drogues, phytoplankton biomass and ^{14}C-based productivity rates indicated that peak productivity occurred *c.* three days after nutrient-rich subsurface waters had upwelled into the euphotic zone. Phytoplankton biomass and productivity declined as upwelled waters were advected further offshore, presumably due to enhanced grazing pressure. Subsequently, the Coastal Upwelling Ecosystem Analysis (CUEA) Program included Lagrangian observations in upwelling regions off Mexico, Oregon, Peru and northwest Africa. The ultimate goal of the CUEA Program was to gain sufficient knowledge of coastal upwelling ecosystems such that fisheries resource managers might develop a capability to predict fishery stocks in these rich oceanic regions (Jennings, 1981). In the latter stages of the program physical, chemical, geological and biological oceanographers from several countries worked with engineers and meteorologists in interdisciplinary field experiments that predated satellite oceanography and the recognition of the microbial loop. In several respects, the CUEA Program set a precedent for the Joint Global Ocean Flux (JGOFS) Program.

In early CUEA field studies the spatial distributions of surface properties suggested that plankton communities respond to upwelling of "new" nutrients after a time lag of a few days (Jones *et al.*, 1983). Surface temperature, nutrient, and plankton distributions led to the hypothesis that the upwelling process could be conceptually described as four sequential stages. In stage I, newly upwelled waters arrived at the surface close to the coastal boundary; as the cool, high nutrient waters flowed offshore in stage II, surface temperatures increased and phytoplankton biomass increased as nutrients were utilized to support primary production. Stage III corresponds to the maximum phytoplankton biomass; in this stage nutrient concentrations had declined and mesozooplankton levels might also be at maximum concentrations. As the upwelled waters continued to age, stage IV waters were found furthest offshore where nutrient depletion limited primary production; here the nitrogen requirements were met by regenerated ammonium. However, coastal upwelling zones are dynamically complex, with coastally trapped waves and day-to-day variations in winds resulting in alterations in surface currents and property distributions in less than a day. Ship-based surveys of surface properties often yield patterns that are heterogeneous and cannot be easily interpreted in terms of a sequential "stage" of upwelled waters.

The hypothesis of sequential upwelling stages was, however, supported by drogue studies in the highly productive Peru upwelling region. Two ship operations near 15° S permitted simultaneous mapping of the upwelling regime on one vessel and process studies on a second vessel as it followed drogues as they were advected offshore. Two types of Lagrangian platforms, parachute drogues and near surface platforms with 1 m^2 vanes (configured similarly to the future CODE surface drifters), were tracked after they were deployed in cool, nearshore waters (Brink *et al.*, 1981). Maximum upwelling was evident within a few km of the shore where surface nitrate and silicate exceeded 20 $\mu mol\,kg^{-1}$ (Brink *et al.*, 1981). Maximum phytoplankton biomass, in contrast, developed *c.* 15 km further offshore along the drogue trajectories as nutrient concentrations declined. Primary productivity, euphotic zone depths, nitrate and silicate uptake rates, plankton biomass, and enzymatic-based rates of nitrate assimilation contributed to a coherent interpretation of the time course of upwelling from stage I through IV as surface drogues were followed across the shelf (Brink *et al.*, 1981; MacIssac *et al.*, 1985). The temporal sequence from stage I to IV was essentially completed within ten days, with maximum nitrate uptake occurring three days after subsurface waters had first upwelled into the euphotic zone and were subsequently retained offshore within a strongly stratified surface layer. Thus the "lag" between upwelling and maximum nutrient utilization is rapid, on the order of only two days in these dynamic regions. Highly variable drogue trajectories illustrated that although the origin of upwelling can be readily identified from surface property distributions, day-to-day variations occur in upwelling plume orientation. Rapid variations in the offshore flow make it nearly impossible to predict where maximum productivity and biomass develop in an Eulerian space (MacIssac *et al.*, 1985). Thus the Lagrangian observational scheme was essential to interpret the sequence and duration of stages.

Estuarine and river plumes

The development of the CODE surface drifter (Davis, 1985) has facilitated Lagrangian studies in dynamic coastal waters. One of the first truly interdisciplinary studies to utilize these platforms was the Microbial Exchanges and Coupling in Coastal Atlantic Systems (MECCAS) Program. This field experiment examined the flow field and pelagic microbial ecosystem in the buoyant Chesapeake Bay outflow plume. Estuarine and river plumes are highly dynamic structures that rapidly couple freshwater outflow to coastal marine ecosystems. The central objectives of the MECCAS Program included characterization of the role of wind forcing and ambient coastal flows on plume dynamics, and carbon and nitrogen cycling within the buoyant surface layer. Temporal variability in nitrogen uptake, utilization and regeneration rates were quantified in repeated surface drifter deployments during different seasons (Boicourt *et al.*, 1987).

Drifter trajectories revealed that the time scale for surface water parcels in the plume was short, between one to three days. Orientation of the plume on the shelf was typically to the south, in agreement with ambient coastal currents and Coriolis forcing. Both seasonal and diel variations in nutrient cycling were observed as water parcels were advected through the plume. Carbon turnover in winter was relatively low, with a decline in photoautotrophs (phytoplankton) occurring in the plume as heterotrophs (bacteria) and heterotrophic microflagellates increased. In contrast, during summer the particulate carbon cycling rates were enhanced and bacteria increased relative to phytoplankton, largely due to copepod grazing of the phytoplankton.

Nitrogen uptake and regeneration rates of various nitrogen species, including NH_4^+, NO_3^-, NO_2^-, urea and dissolved free amino acids, were assessed while following drifters (Glibert *et al.*, 1991; Glibert and Garside, 1992). Although there was little evidence of seasonal signals in nitrogen uptake rates, NH_4^+ regeneration rates were highest in late summer, with maximum amino acid cycling in early summer. Release rates for organic nitrogen were also seasonal, with winter release potentially attributed to nutrient-limited photoautotrophs, and grazers the dominant source in summer.

Lagrangian experiments have now been conducted in several river plumes. As an example, surface drifters were an integral component of spring bloom studies of nutrient fluxes and biogeochemical transformations in the Elbe River plume (Raabe *et al.*, 1997; Rick, 1999). In combination with ship-based sampling and microcosm experiments, a surface drifter was followed through the surface plume in spring 1995. As surface waters flowed into the German Bight, diatoms rapidly took up silica as the entire photoautotrophic community assimilated inorganic nitrogen and phosphorus. Surface chlorophyll values attained high levels during the ten day drift, exceeding 80 $\mu g\ l^{-1}$. Biogenic nutrient consumption rates were estimated in the plume, although the potential effect of mixing on nutrient concentrations was not assessed.

Seasonal deployments of CODE drifters in the Mississippi River plume indicate that mixing has a major influence on the distribution of surface nutrient concentrations within a few days as water parcels transit through the high velocity jet within the "core" of the plume (Hitchcock *et al.*, 1997). Transit times for water parcels through the plume, from the mouth of the Mississippi River to the plume perimeter, are only a few days. As riverine source waters exit the Mississippi delta, the freshwater plume rapidly shoals and spreads laterally. Low-nutrient, high-salinity subsurface waters are mixed across a sharp pycnocline into the fresher surface layer, diluting the riverine surface waters. Mixing is evident in increased surface salinities, with linear relationships observed between surface nutrients and surface salinity in the plume axis from samples collected along drifter trajectories. These relationships were interpreted as evidence that mixing has a major influence on surface nutrient distributions. The one to two day transit through the highly turbid plume was too short to permit a major phytoplankton response. Nonlinear relationships were found between salinity and nutrient concentrations at the plume edge, and exterior to the plume, where phytoplankton biomass is high and nutrients are rapidly utilized.

Instrumented surface drifters

Throughout the past two decades there have been major advances in the development of sensors with low power consumption. These sensors measure physical (temperature, conductivity) and optical properties that estimate a suite of particulate and dissolved constituents. The fluorescent fraction of chromophoric dissolved organic matter (CDOM), for example, can be detected at concentrations of ppb by fluorometers that have exceptionally low power requirements. A Lagrangian experiment was conducted to define the time scale for transport of CDOM through the Mississippi River plume with surface drifters instrumented with CDOM fluorometers and a towed, profiling vehicle (Hitchcock *et al.*, 2004). CDOM concentrations decreased in a linear manner along the plume axis with increasing surface salinity in a manner analogous to that observed in surface nutrient concentrations. During the short transit through the turbid plume (<two days), the riverine surface waters are mixed with higher-salinity, low-CDOM subsurface waters. Transformations of CDOM likely occur mainly at the plume perimeter, where water residence times may be longer, and surface CDOM distributions are very patchy. Photodegradation was not apparent in the plume axis, since the plume transit time was *c.* one to two photoperiods, and high turbidity reduces the potential for photodegradation by ultraviolet light.

Bio-optical sensors can reliably measure particulate matter concentrations (beam transmittance), and indices of phytoplankton biomass and productivity. Phytoplankton distributions can be assessed with "active" strobed and *in situ* fluorometers and "passive" sensors for upwelling radiance plus downwelling irradiance. Recent field studies in coastal and oceanic waters have demonstrated that surface drifters instrumented with bio-optical sensors and System Argos capability can describe temporal changes in phytoplankton populations over weeks to months. The description of the temporal changes in drifting populations at these time scales cannot be easily assessed by ship-based sampling or, obviously, by moored instrumentation.

A unique and comprehensive suite of bio-optical instrumentation and automated water samplers described the temporal changes in near-surface property distributions within cool surface filaments of the eddy-rich California Current System (Abbott *et al.*, 1990, 1995). The drifter package was deployed in 1987 and 1988 near the origin of cool surface filaments that often are generated in the intense nearshore eddy field. Since the data was recorded onboard the package, it had to be followed by a ship and retrieved at the end of the deployment. As the drifter package moved offshore, surface nutrient concentrations declined as phytoplankton biomass increased over a few days. This pattern is analogous to that in upwelling regions at Point Conception, California (Jones *et al.*, 1983) and the Peruvian upwelling zone (MacIsaac *et al.*, 1985). Abbott *et al.* (1990, 1995) observed short term variations in nutrient levels that likely resulted from mixing of "non-filament" waters into the plume during the offshore transition. Bio-optical indices of pigment concentration reached maximum values within a few days. Preserved phytoplankton samples were examined post-cruise by microscopy, and they indicated that the increased primary producer biomass corresponded to changes in species composition. Although the length of observations were relatively short in these deployments, about a week, the spectra of bio-optical signals from a strobed fluorometer and beam transmissometer contained distinct semi-diurnal and diurnal peaks. These variations were attributed to the diel variation in solar radiation, as well as internal tides. Downwelling occurred in the filament at rates comparable to nearshore upwelling, and exerted a strong influence on the distribution of phytoplankton as the filament flowed offshore.

An analogous study of an eddy-filament system was conducted in the Great Whirl, a mesoscale eddy that forms annually during the southwest monsoon in the Somali Current (Hitchcock *et al.*, 2000b). A limited number of surface drifters equipped with expendable strobed fluorometers were launched during the summer monsoon in August 1995. They were deployed at the seaward edge of the coastal upwelling region adjacent to the eddy. The platforms were rapidly exported from the upwelling zone, transiting the northern perimeter of the 200 km diameter eddy in less than two days at velocities $>100\,\mathrm{cm\,s^{-1}}$. The drifters were ejected from the cool surface filament and entrained in the Socotra Gyre, and thereafter meandered through the northern Arabian Sea for several months. The Lagrangian data yield decorrelation scales for velocities, temperature, and surface pigment fluorescence. Decorrelation times correspond to the time at which a temporal data record no longer correlates with itself. Thus it is an index to the time required for a parameter to effectively "lose any memory" of previous values. Decorrelation scales for drifter u and v velocity components, and the surface chlorophyll concentrations, ranged from four to seven days, commensurate with the timescale at which the drifters exited the domain of the Great Whirl. As in the California Current System and Peru upwelling region, time scales for plankton advection through the offshore filament were fairly rapid, on the order of only a few days.

The recent development of a low-power spectroradiometer, the Satlantic OCR-100, has provided long-term observations of bio-optical properties on surface and subsurface platforms. Spectroradiometers measure upwelling radiance at several wavelengths and yield estimates of chlorophyll *a* concentrations as well as a time scale for fouling (Abbott and Letelier, 1997). Downwelling solar irradiance is typically measured at 490 nm with sea surface temperature (SST). In comparison to strobed fluorometers, spectroradiometers are passive, require less power, and provide an index to *in situ* productivity through measurements of upwelling radiance at 683 nm (Chamberlin *et al.*, 1990), albeit with some constraints (Olaizola *et al.*, 1994).

Satlantic spectroradiometers deployed in the California Current System on 24 WOCE surface drifters revealed distinct pigment patterns developed in response to physical processes at anticyclonic and cyclonic eddies (Abbott and Letelier, 1998). The length of valid optical records averaged *c.* 70 days, so decorrelation scales could be calculated for optical properties and SST in nearshore (0–200 km), transition (200–400 km), and offshore (>400 km) regimes. Decorrelation scales for near-surface SST and pigment were 2 days in nearshore waters, while in the transition zone SST time scales (6 days) were longer than that for chlorophyll *a* (4 days). In the offshore region there was further separation in time, with decorrelation scales for SST at 7 days and *c*. 5 days for chlorophyll *a*. These trends were interpreted as evidence that similar processes influence the physical (SST) and surface pigment fields in nearshore waters, while SST and chlorophyll fields are likely regulated by different processes further offshore. Furthermore, offshore phytoplankton communities appear close to an "equilibrium" state where picoplankton dominate. The nearshore environment, in contrast, is characterized by rapidly changing conditions (light, nutrient and grazing) that sustain a diverse phytoplankton community. The time scales derived from Lagrangian drifters provide insight into ecosystem functioning in the California Current System where mesoscale eddies dominate the flow regime and plankton transport (Hofmann *et al.*, 1991). These conclusions would otherwise be difficult to derive solely from ship-based surveys.

A subsequent drifter program in the Southern Ocean utilized bio-optical WOCE drifters to examine phytoplankton biomass and growth in relation to the physical environment near the Antarctic Polar Front. Long-term increases in natural fluorescence were observed from a drifter that circulated around a cyclonic eddy for 50 days. The trend in fluorescence suggested chlorophyll *a* varied in response to enhanced iron inputs to the mixed layer (Letelier *et al.*, 1995, 1997), a response consistent with the high inorganic nutrient (nitrate, phosphate) environment in the Southern Ocean. A series of drifters were deployed south of the Polar Front near 61° S, 170° W where the phytoplankton community is distinct from that north of the front (Landry *et al.*, 2001). Net chlorophyll *a* increased over periods of tens of days, yielding a net community growth rate estimate for phytoplankton of 0.03 day^{-1}, a value that agreed well with rates from on-deck bottle incubations (0.04 day^{-1}). Two of the drifters released south of the front, and late in the season, suggested lower net growth rates (*c*. 0.01 day^{-1}). The drifter estimates of growth rates were, however, less than those derived from SeaSoar surveys, at 0.05 day^{-1}. The apparent discrepancy was attributed to the effects of photoadaptation, which potentially reduced pigment levels in the near-surface layers sampled by the drifter sensors.

The physical and biological structure of the Polar Front, and their seasonal aspects, were further examined with bio-optical drifters deployed as lines and clusters during the austral spring bloom (Abbott *et al.*, 2000). The temporal pattern in surface pigments reflected a vernal bloom that was detected by all drifters, attaining maximum values in December through January or February. Productivity, based on an apparent quantum yield of fluorescence, was also initially high and then decreased after peak chlorophyll levels developed. This temporal pattern is consistent with a nutrient-limited phytoplankton community that is unable to fully utilize the increased light energy during summer. The Antarctic Polar Front is composed of meandering jets with extensive vertical motions, and vorticity budgets were derived from the motion of drifter clusters that yield maximum vertical rates on the order of a meter day^{-1} with lower rates typical in most meanders. Although there was no consistent relationship between variables such as chlorophyll concentration, daily estimated accumulation rates, fluorescence per unit chlorophyll (F/C), and vertical water velocity, there was a period where fluorescence per unit chlorophyll decreased as vertical velocities increased. This relationship may reflect a physiological response, as lowered F/C values, associated with enhanced nutrient delivery during upwelling. Alternatively, an environmental shift may have occurred such as the development of a more favorable light environment. Irrespective of the physiological basis for the observed pattern, it is clear that long-term deployments of bio-optical drifters provide information on temporal patterns in phytoplankton responses to environmental forcings that are otherwise difficult to obtain in dynamic regions.

Larval transport

Several studies have used long deployment drift bottles and cards to infer larval transport pathways (e.g., Tegner and Butler, 1985; Domeier, 2004). For example, Domeier (2004) released *c.* 1000 drift bottles (ballasted vials) during each of two years from a known fish spawning location, to assess likely dispersal paths of young fish to potential nursery habitats along the Florida Keys and east coast. Following each release, up to 11% of the vials were retrieved at distances up to 730 km. There was some interannual variation in the mean distance of dispersal ranging from ~300 km in 1999 to 400 km in 2000, with no retrieval of drifter vials between the release site and Key West (a distance of *c.* 120 km). This low-cost approach sheds some light on the potential dispersal, although it requires simplifying assumptions about larval propagules, notably that they are passive and share the same vertical distribution as the drifter (in this case, at the surface). Additionally, larval dispersal studies that have used drift bottles and cards to describe transport routes generally have had to rely on the public to find the bottles or cards along the shore and report them back to the investigators. Of course, the success of this approach will depend on the location of release, with higher success of retrieval in locations with high population density (i.e., locations where more people walk the beach). When a survey of potential retrieval locations is nonconstant, it is difficult to determine the cause for low return. In the above study, the lack of retrieval between the release site at Riley's Hump and Key

West could be due to lack of onshore currents, lack of suitable coastal habitat, and/or lack of human observation due to limited human habitation in that region. The end result of this sort of study is knowledge of the beginning (or source site) and various end points; obviously transport routes cannot be determined, nor can transit times since return dates will vary and depend on the time when the drifter bottle is retrieved.

Recent work has suggested that the swimming behavior and capabilities of some larvae negate the simplifying assumption of "passivity" (Leis, 1991a,b; Cowen and Castro, 1994; Paris and Cowen, 2004). However, passive drifter tracks represent a *possible* outcome of the dispersal process, but probably one that resembles the extreme case since vertical movement by larvae, interacting with the vertical shear of the water column, usually limits transport distances and directions (Cowen and Castro, 1994; Paris and Cowen, 2004; Hare and Govoni, in press). Even in the passive scenario, when coupled with typical larval mortality rates, the extreme transport distances obtained by a few drifters are likely rarely realized (Cowen *et al.*, 2000).

Nonetheless, drifters have been useful in identifying potential transport routes, especially when strong boundary currents or eddies exist, thus limiting diffusion and highlighting the role of advection (*sensu* Largier 2003). Hare *et al.* (2001) used several sets of drifter releases and hydrographic datasets to estimate probable transport routes of, and transit times for, larval fish along the US east coast between Cape Hatteras and the New York Bight. Drifter trajectories matched indices of larval transport from Cape Hatteras northward in association with the Gulf Stream. Another dataset allowed an examination of potential cross-slope transport from the edge of the Gulf Stream to the shelf/slope break. With estimated transit times (plus variance in those estimates based on the different drifters) for each segment along the route, the investigators calculated the expected larval age structures of larvae assumed to have been spawned near Cape Hatteras and transported into the New York Bight. The predicted age structures were very similar to the observed age structures for several species of larval fish collected at the end-point of the transport route, corroborating the predicted transport route.

Dispersion and recruitment of blue crab (*Callinectes sapidus*) larvae at the mouth of Chesapeake Bay has been evaluated on the basis of drifter trajectories from the MECCAS Program (Roman and Boicourt, 1990, 1999; Johnson, 1995). The Chesapeake Bay plume is a major source of blue crab larvae for the adjacent continental shelf. Shelf larval distributions are dependent, in part, on estuarine discharge and the local wind field. During downwelling (northwesterly) favorable wind conditions, larval concentrations are highest along the coast as the coastal plume turns to the south. Downwelling winds also favor transport of older larvae back towards the Bay mouth, where they enter to complete their life cycles as adults. During upwelling-favorable wind events (south, southwesterly), in contrast, Ekman circulation forces the plume and crab larvae offshore. The interpretation of the larval distributions in the highly dynamic surface waters of the continental shelf has critically relied upon surface drifters.

Dispersal of blue crab larvae has similarly been examined from a Lagrangian study near the mouth of Delaware Bay, a major estuary north of Chesapeake Bay on the eastern coast of North America. On the shelf, crab larvae remain in the plankton for 3 to 6 weeks, and are mainly concentrated in surface patches a few meters thick and several 100 meters wide (Natunewicz *et al.*, 2001). Larval transport in the estuarine coastal plume, and several patches on the shelf, were successfully followed with ARGOS-tracked surface platforms that permitted sequential larval sampling over 1 to 11 days. During the initial spawning season in early summer, larval patches were mainly found in the coastal plume and subsequently transported south under buoyancy-driven flows. Later, in August, larval patches were often found seaward of the low salinity plume on the shelf. The larvae in these patches were likely retained near the mouth of Delaware Bay. A two-dimensional advection-diffusion model based on Garvine *et al.* (1997) successfully reproduced the major features of larval transport, verifying the dominant role of buoyancy and wind-driven flows.

Larval transport into an estuary is a critical stage in the life history of several commercially important fish and invertebrates that rely upon estuaries for nursery grounds. The larvae of these and many other marine organisms can exhibit selective tidal stream transport as the larvae adjust their vertical position to optimize horizontal motion in relation to the tides. Surface drifters released near the mouth of Beaufort Inlet on the Outer Banks of North Carolina have defined the major inflow paths of surface waters during the flood tide. These trajectories describe pathways by which larvae can enter the estuary (Churchill *et al.*, 1999a). Estuarine inflow during flood tide, and therefore the path of larvae entering the estuary, occurs in three principal trajectories. Surface waters entering near the eastern edge of the Inlet flow as a narrow jet that enters the estuary and flows towards the east, while drifters entering the center and western edge of the Inlet follow pathways to the central and western interior channels of the estuary, respectively. The drifter paths were confirmed in a model analysis, and the results interpreted with respect to the distribution of fish larvae at Beaufort Inlet (Churchill *et al.*, 1999b). A review of the physical processes that influence the transport of larval crustacean and fish on the northeastern US continental shelf by Epifanio and Garvine (2001) further describes the application of drifter studies to larval transport studies. The major physical forcings that influence larval transport include wind stress, tides, estuarine outflow, heat fluxes, and the physical dynamics regulating exchange at the shelf edge.

Drifters have also been used to confirm the presence of hypothesized circulation features that may act to transport and/or retain larvae within other regions (e.g. Yeung and Lee, 2002; Dalgren *et al.*, 2001). Lee *et al.* (1994) utilized drifter tracks to verify the existence of the Tortugas Eddy, a recurrent gyre off the Florida Keys with a typical lifetime of 60 to 90 days. This feature potentially introduces shrimp larvae into the nursery grounds of western Florida Bay. In the same region Yeung and Lee (2002) used trajectories of ARGOS-tracked drifters along the Florida Keys to demonstrate how a Tortugas Gyre could influence the influx of post-larval lobsters from offshore waters to their inshore juvenile habitat in Florida Bay. During their biophysical study, two drifter deployments of 24 days and 45 days demonstrated a coastal counter-current flowing inshore of the Florida Current when the cyclonic, mesoscale eddy was present. The current reversal generated by the eddy was located near channels that lead into Florida Bay. These drifter trajectories support the hypothesis that these large cyclonic eddies influence onshore larval transport in the Florida Keys ecosystem.

Pathways of larval drift have also been extrapolated from drifter releases in the Gulf of Papua that identify the connections between lobster spawning and nursery sites in the Coral Sea. Dennis *et al.* (2001) deployed six drifters in paired releases with one drogued at 15 m, and the second at 80 m. The two depths were intended to describe the effects of diel vertical migration in different lobster stages; however, the results were not reported according to depth. The trajectories suggest that potential larval transport occurs as a loop circling around the Coral Sea gyre. Larval stage-specific distributions were consistent with the inferred pathways. However, the plankton sampling did not coincide with the drifter releases, since the two studies were ten years apart. Drifters were also deployed with an objective of identifying potential sources for recruitment of fish larvae along the Leeuwin Current near the western coast of Australia (Hutchins and Pearce, 1994). Examples of other field programs utilizing surface drifters and drogue trajectories to study transport of fish and invertebrate eggs and larvae on continental shelves include Zenitani *et al.* (1996), who studied transport along the shoreward edge of the Kuroshio, and Pepin and Helbig (1997) in the shelf waters off Newfoundland. While not all of these studies released drifters concurrently with biological sampling, they have further demonstrated that surface platforms are useful tools for defining potential transport routes and the presence of circulation features (e.g. eddies/fronts) that have an important role in larval dispersal and recruitment.

The use of surface drifters in studies of larval transport is not without problems. One of the major difficulties in drifter studies is the problem of under-sampling trajectories when only a few drifters can be released. Lagrangian analyses require a full range of variation in dispersal trajectories to define estimates of dispersal kernels, that is, the 2-D probability density function of dispersal from a given release site. To this end, models that incorporate Lagrangian techniques to properly account for diffusion may serve as a valuable tool when the models are coupled with drifter studies for validation. Siegel *et al.* (2003) describes the modeling approach in detail and provides a comparison of model results with drifter trajectories from the California coast. An alternative approach to identify larval transport routes, and the associated variance in those routes, is to incorporate or "seed" Lagrangian-based circulation models with "virtual" drifters. This approach must be ideally validated with real drifter tracks as is illustrated in Figure 11.4 (see also Cowen *et al.*, 2003; Paris *et al.*, 2002) before it can be used to define accurate larval pathways in the sea. However, once a model has been validated, it has the distinct advantage of being able to track 1000s of virtual drifters, and thereby establish probability density functions of the direction of larval transport, as well as seasonal and inter-annual variation (see Hare *et al.*, 1999; Werner *et al.*, 1999). Furthermore, a well-designed, individual-based model can also incorporate vertical larval movement to improve an assessment of the role of vertical shear in larval transport. Griffin *et al.* (2001) used this latter approach in modeling the transport and coastal return of lobster larvae in Western Australia coastal waters. Their simulations, which included vertical migrations that were stage-specific, demonstrated that vertical migration behavior was critical for successful transport into suitable nursery areas.

Drifter studies that encompass the season and duration of the pelagic larval stage can serve to provide managers with potential options in spatial management. The value of this approach is maximized when the drifter studies are interpreted with regard to the bio-physical interactions that regulate larval transport of fish and invertebrates (e.g. James *et al.*, 2002). Specifically, these drifter studies may elucidate potential dispersal pathways from known spawning sites. The resulting linkage, or connectivity, of populations via larval dispersal can be established and provides a spatial framework in which to manage economically or ecologically important species (Cowen *et al.*, 2003).

11.2.4 Subsurface Platforms

Subsurface floats were initially developed to aid the study of circulation in the deep sea. The first such platform was the quasi-Lagrangian Swallow float

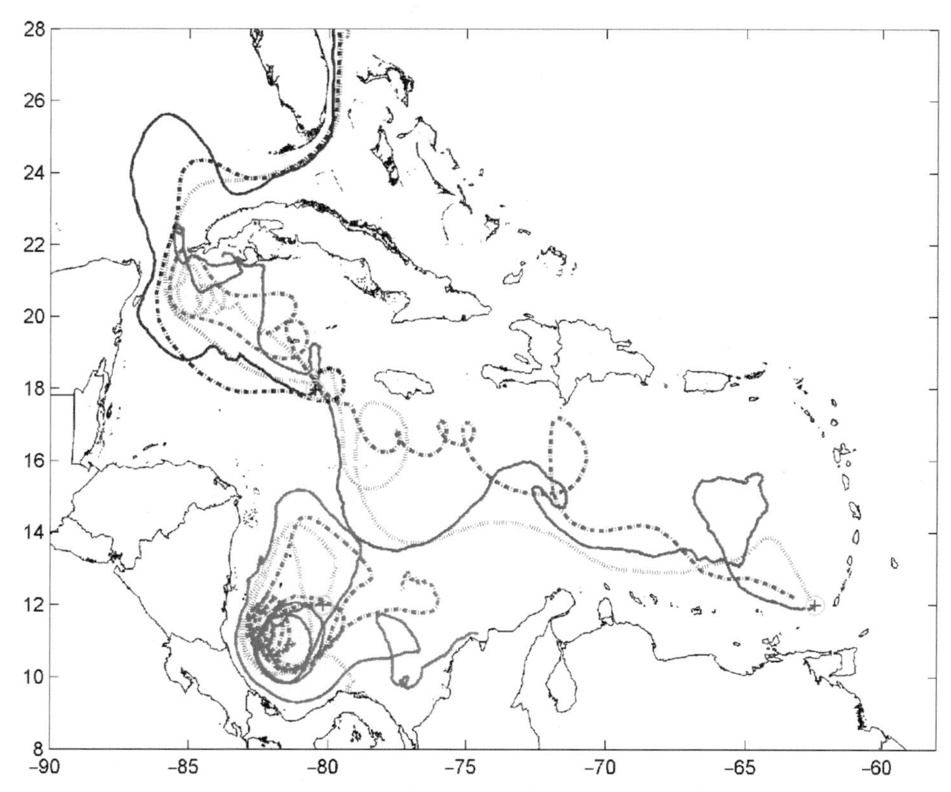

Figure 11.4 Comparison of drifter floats (solid lines) and simulated drifter tracks (dot and dashed lines) released in three different regions of the Caribbean Sea. Each color corresponds to an individual drifter release and each track is days duration. Simulations used output from the MICOM with wind forcing from ECMWF. Data for the real floats (solid lines) is from the Global Lagrangian Drifter Database available at: http://db.aoml.noaa.gov/cgibin/db/Bin/init_applet.x?gld + GLDKRIGGUI.class. See Plate 27 for color version.

(Swallow, 1955); Davis (1991) gives a history of the development of this technology. One of the major problems in the development of subsurface platforms was in determining the float's position. The nearly transparent acoustical nature of the ocean led to tracking systems that have relied upon sound as a means to locate platforms. Two basic systems have evolved. The first long-range tracking capability was provided by the SOFAR system (Rossby and Webb, 1970) that uses the float as a sound source and records the arrival time of a transmitted signal at a receiver in a fixed location. Advances in technology led to the RAFOS System (Rossby *et al.*, 1986) in which a fixed source transmits timing signals that are recorded at the float. The data needed to determine the history of the float's position is relayed ashore by the Argos System after the float surfaces at the end of its deployment. Low-power systems to control ballast have been added to floats that allow the platforms to rise and sink at predetermined times. This capability has resulted in a variety of instrumented subsurface floats in use by physical oceanographers (Davis, 2003). These platforms can remain at depth for predetermined periods, surface, transmit their position and data ashore via Argos or other telemetry systems, and then return to their predetermined working depth for another cycle.

Biological oceanographers began to deploy subsurface platforms in support of biological studies two decades ago. As an example, subsurface floats have provided direct measurements of upwelling and downwelling rates in current meanders. Acoustically tracked RAFOS floats that remain on a "target" density surface have been developed by ballasting the float to the compressibility of the target isopycnal surface (Rossby *et al.*, 1985). Many of these floats have been deployed in the upper and mid-thermocline of the Gulf Stream (Rossby *et al.*, 1986), where their trajectories describe pathways of fluid parcels in the "nutrient-bearing" isopycnal layers of the current (Pelegrí and Csanady, 1990). The vertical motion of neutrally buoyant floats demonstrates that water parcels upwell at rates exceeding 50 m day^{-1} as they enter anticyclonic meanders, and conversely sink at comparable rates when the current turns seaward. These motions are reflected in the vertical distribution of the plankton in the Gulf Stream front, with deep pigment maxima at shallower depths in anticyclonic meanders than in cyclonically-turning regions (Hitchcock *et al.*, 1993; Lohrenz *et al.*, 1993). Isopycnal floats have also been deployed in studies of smaller-scale features within the Gulf Stream surface front where thin "bands" of shelf and slope waters are actively stirred and mixed (Hitchcock *et al.*, 1994). A ship-tracked isopycnal Swallow float with a solid-state fluorometer for measuring chlorophyll *a* also described the fine-scale property distribution in the Gulf Stream front (Hitchcock *et al.*, 1989).

Low-power bio-optical sensors have been incorporated into autonomous subsurface floats to record vertical profiles of optical properties. Downwelling irradiance measurements recorded by sinking profiling floats have yielded values of K_d, the diffuse attenuation coefficient, for periods of weeks to months in the western Pacific (Mitchell *et al.*, 2000). The vertical gradient in the ambient light field is an important factor in validating oceanic productivity models. However, phytoplankton can alter their photosynthetic response to changes in light and temperature as their vertical position changes in the water column. The Self-contained Photosynthesis Apparatus (SUPA) is an autonomous platform that contains a phytoplankton culture within an upward-oriented quartz dome (Kirkpatrick *et al.*, 1990). The vertical motion of the

SUPA is linked to a companion instrument that measures vertical water velocities. As the SUPA traverses the water column, light intensity is measured with photosynthetic parameters of cultures derived from changes in oxygen concentrations and pH. The records are stored onboard in a microcomputer, and can be extrapolated to rates of carbon uptake along the trajectory of the platform. The SUPA data show that phytoplankton exhibit a rapid photosynthetic response to changes in light over diurnal cycles. These patterns could not be derived from "static" incubations in which cultures are held at specific depths throughout a diurnal cycle.

Lagrangian Carbon Explorers are profiling ARGOS floats equipped with transmissometers. The instruments measure attenuation coefficients (m^{-1}) which can be extrapolated to particulate organic carbon profiles (Bishop *et al.*, 2002, 2004). In the North Pacific, daily and diurnal variability in POC profiles illustrate the response of the plankton community to storms and wind-born dust (Bishop *et al.*, 2002). In the Southern Ocean these profilers also have described the response of plankton carbon biomass to iron enrichment during the Southern Ocean Iron Enrichment Experiment, SOFeX (Bishop *et al.*, 2004). Although the number of instrumented profiling platforms is now limited, further advances in instrumentation will undoubtedly result in novel means to conduct long-term measurements of vertical profiles of various properties, and ultimately rates of biogeochemical processes, in the sea.

11.3 Future directions

Future studies of larval transport will benefit from both more realistic, finer resolution Lagrangian-based models (e.g., Siegel *et al.*, 2003) and from biologically realistic drifters that behave as larval mimics. Critical regions of study for larval transport include shallow coastal waters, where water depths are <15 m, and regions where deep and shallow waters exchange. Lagrangian platforms will be essential to define distances of larval dispersal that are critical to the design of Marine Protected Areas (MPA). Marine protected areas are "no harvest" zones that are intended to sustain fish and invertebrate populations in the face of increased fishing pressure (e.g., Botsford *et al.*, 2001; Palumbi, 2001). Protected populations within MPAs also serve to replenish surrounding waters with propagules through larval dispersal. There are, however, few marine species for which the larval dispersal distances are well known (Shanks *et al.*, 2003). Thus the scale-dependent properties of dispersal (advection and diffusion) are essential to design a MPA (Largier, 2003). Larval mimics can contribute needed data to this process, especially if they include diel vertical migration and other behavioral attributes of the larvae.

Lagrangian mimics are also being developed to better understand the factors that regulate the development and fate of harmful algal blooms. An autonomous mimic is being deployed that imitates the diel behavior of *Karenia brevis*, a dinoflagellate that produces harmful algal blooms on the west Florida Shelf (Steidinger *et al.*, 2001). T. Wolcott, D. Kamykowski and colleagues have recently instrumented a vertically migrating platform with sensors that are capable of monitoring several environmental parameters that are hypothesized to regulate the diel motion of *K. brevis*. The platform is based on migrating "larval mimics" that have been developed by Wolcott (Wolcott and Wolcott, 1998). This platform will be able to test hypotheses of the factors that regulate *K. brevis* vertical distributions with bio-physical models (e.g. Walsh *et al.*, 2003; Liu *et al.*, 2002).

Instrumented Lagrangian platforms can also provide observations to improve our understanding of the role of the ocean in global climate change. Autonomous drifters equipped with sensors that measure key biogeochemical properties, such as oxygen, dissolved nutrients, and pH, can map the global ocean when ship access is limited (Johnson, 2003). Sounding Oceanographic Lagrangian Observers (SOLOs) instrumented with thermistors and spectroradiometers (Mitchell *et al.*, 2000), and the Carbon Explorer (Bishop *et al.*, 2002; 2004) are two examples of Lagrangian platforms that presently measure parameters relevant to the carbon cycle. The need for future sensor and platforms has been described in detail within the recent ALPS Workshop Report (Rudnick and Perry, 2003). It is clear that as instrumented Lagrangian platforms are further developed, they will make valuable contributions to our understanding of life in the sea.

Acknowledgments

The authors gratefully acknowledge the efforts of Dr. C. Paris in the development of the model, and the results, illustrated in Figure 11.4. We are in debt to the reviews provided by John Hare and Daniel Kamykowski. Their comments vastly improved an earlier version. We acknowledge our editors, A. Griffa, A. D. Kirwan, A. J. Mariano, T. Özgökmen, and T. Rossby for their exceptional patience and guidance. The authors have received support for Lagrangian studies from the Office of Naval Research including awards N00014-95-1-0131, N00014-97-1-0131 and N00014-98-1-0625 to G. Hitchcock, the National Oceanic and Atmospheric Administration to G. Hitchcock, and the National Science Foundation to R. Cowen.

References

Abbott, M. R., K. H. Brink, C. R. Booth, D. Blasco, L. A. Codispoti, P. P. Niiler, and S. R. Ramp, 1990. Observations of phytoplankton and nutrients from a Lagrangian drifter off northern California. *J. Geophys. Res.*, **95**, 9393–409.

Abbott, M. R., K. H. Brink, C. R. Booth, D. Blasco, M. S. Swenson, C. O. Davis, and L. A. Codispoti, 1995. Scales of variability of bio-optical properties as observed from near-surface drifters. *J. Geophys. Res.*, **100**, 13,345–67.

Abbott, M. R. and R. M. Letelier, 1997. Going with the flow – The use of optical drifters to study phytoplankton dynamics. In *Monitoring Algal Blooms: New Techniques for Detecting Large-Scale Environmental Change*, ed. M. Kahru and C. Brown. New York: R. G. Landes Co., 145–70.

Abbott, M. R. and R. M. Letelier, 1998. Decorrelation scales of chlorophyll as observed from bio-optical drifters in the California Current. *Deep-Sea Res. II. Top. Stud. Oceanogr.*, **45** (8–9), 1639–67.

Abbott, M. R., J. G. Richman, R. M. Letelier, and J. S. Bartlett, 2000. The spring bloom in the Antarctic Polar Frontal Zone as observed from a mesoscale array of bio-optical sensors. *Deep-Sea Res. II. Top. Stud. Oceanogr.*, **47** (15–16), 3285–314.

Alexander, J. E. and E. F. Corcoran, 1963. Distribution of chlorophyll in the Straits of Florida. *Limnol. Oceanogr.*, **8**(2), 294–6.

Arnold, W. S., G. L. Hitchcock, M. E. Frischer, R. Wanninkhof, and Y. P. Sheng, 2005. Dispersal of an introduced larval cohort in a coastal lagoon. *Limnol. Oceanogr.*, **50**(2), 587–97.

Batchelder, H. P., C. A. Edwards, and T. M. Powell, 2002. Individual-based models of copepod populations in coastal upwelling regions: implications of physiologically and environmentally influenced diel vertical migration on demographic success and nearshore retention. *Deep-sea Res.*, **53** (2–4), 307–33.

Becker, D. S., 1978. *Evaluation of a hard clam spawner transplant site using a dye tracer technique*. Marine Sciences Research Center. State Univ. of New York, Stony Brook (USA).

Bidigare, R. R., B. B. Prézelin, and R. C. Smith, 1992. Bio-optical models and the problems of scaling. In *Primary Productivity and Biogeochemical Scales in the Sea*, ed. P. G. Falkowski and A. D. Woodhead. New York: Plenum Press, 175–212.

Bishop, J. K. B., R. E. Davis, and J. T. Sherman, 2002. Robotic observations of dust storm enhancement of carbon biomass in the North Pacific. *Science*, **298**, 817–21.

Bishop, J. K. B., T. J. Wood, R. E. Davis, and J. T. Sherman, 2004. Robotic observations of enhanced carbon biomass and export at 55 degree S during SOFeX. *Science*, **304**, 417–20.

Bjoerke, H., 1978. Food and feeding of young herring larvae of Norwegian spring spawners. *Fiskeridir. Skr. (Havunders.)*, **16**(11), 405–22.

Boicourt, W. C., S.-Y. Chao, H. W. Ducklow, P. M. Glibert, T. C. Malone, M. R. Roman, L. P. Sanford, J. A. Fuhrman, C. Garside, and R. W. Garvine, 1987. Physics and microbial ecology of a buoyant estuarine plume on the continental shelf. *Eos.*, **68** (31), 666–8.

Bone, Q., 1998. *The Biology of Pelagic Tunicates*. Oxford: Oxford University Press.

Botsford, L. W., A. Hastings, and S. D. Gaines, 2001. Dependence of sustainability on the configuration of marine reserves and larval dispersal distance. *Ecol. Lett.*, **4**, 144–50.

Bowden, K. F., 1965. Horizontal mixing in the sea due to a shearing current. *J. Fluid. Mech.*, **21**, 83–95.

Boyd, P. W., A. J. Watson, C. S. Law *et al.*, 2000. A mesoscale phytoplankton bloom in the polar Southern Ocean stimulated by iron fertilization. *Nature*, **407**, 695–702.

Boyd, P. W. and C. S. Law, 2001. The Southern Ocean Iron RElease Experiment (SOIREE) – introduction and summary. *Deep-Sea Res. II Top. Stud. Oceanogr.*, **48** (11–12), 2425–38.

Boyd, P. W. *et al.*, 2004. The decline and fate of an iron-induced subarctic phytoplankton bloom. *Nature*, **428**, 549–53.

Brink, K. H., B. H. Jones, J. C. Van Leer, C. N. K. Mooers, D. W. Stuart, M. R. Stevenson, R. C. Dugdale, and G. W. Heburn, 1981. Physical and biological structure and variability in an upwelling center off Peru near 15° S during March 1977. In *Coastal Upwelling*, ed. F. A. Richards. Washington DC: American Geophysical Union, 473–95.

Brown, P. C. and L. Hutchings, 1987.The development and decline of phytoplankton blooms in the southern Benguela upwelling system. 1. Drogue movements, hydrography and bloom development. *S. Afr. J. Mar. Sci./S.-Afr. Tydskr. Seewet.*, **5**, 357–91.

Burkill, P. H., S. D. Archer, C. Robinson, P. D. Nightingale, S. B. Groom, G. A. Tarran, and M. V. Zubkov, 2002. Dimethyl sulphide biogeochemistry within a coccolithophore bloom (DISCO): an overview. *Deep-Sea Res. II.*, **49**, 2863–85.

Campana, S. E. and S. R. Thorrold, 2001. Otoliths, increments, and elements: keys to a comprehensive understanding of fish populations? *Can. J. Fish. Aquat. Sci.*, **58**, 30–38.

Carter, H. H. and A. Okubo, 1978. A study of turbulent diffusion by dye tracers: a review. In *Estuarine Transport Processes*, ed. B. Kjerfve. Columbia: Univ. South Carolina Press, 95–111.

Chisholm, S. W. and F. M. M. Morel, 1991. What controls phytoplankton production in nutrient-rich areas of the open ocean? *Limnol. Oceanogr.*, **36**, U1501–17.

Chisholm, S. W., R. J. Olson, E. R. Zettler, J. Waterbury, R. Goericke, and N. Welschmeyer, 1988. A novel free-living prochlorophyte occurs at high cell concentrations in the oceanic euphotic zone. *Nature*, **334**, 340–3.

Churchill, J. H., J. L. Hench, R. A. Luettich, J. O. Blanton, and F. E. Werner, 1999a. Flood tide circulation near Beaufort Inlet, North Carolina: Implications for larval recruitment. *Estuaries*, **22** (4), 1057–70.

Churchill, J. H., R. B. Forward, R. A. Luettich, J. L. Hench, W. F. Hettler, L. B. Crowder, and J. O. Blanton, 1999b. Circulation and larval fish transport within a tidally dominated estuary. *Fish. Oceanogr.*, **8** (suppl. 2), 173–89.

Clark, C. W. and D. A. Levy, 1988. Diel vertical migrations by juvenile sockeye salmon and the anti-predation window. *Am. Nat.*, **131**, 271–90.

Cleve, P. T., 1896. Microscopic marine diatoms in the service of hydrography. *Nature*, **55**, no. 1413, 89–90.

Coale, K. H., K. S. Johnson, S. E. Fitzwater, R. M. Gordon, S. Tanner, F. P. Chavez, L. Ferioli, C. P. Sakamoto, P. Rogers, F. Millero, P. Steinberg, P. Nightingale, D. Cooper, W. P. Cochlan, and R. Kudela, 1996. A massive phytoplankton bloom induced by an ecosystem-scale iron fertilization experiment in the Equatorial Pacific Ocean. *Nature*, **383**, 495–501.

Coale, K. H. *et al.*, 2004. Southern Ocean Iron Enrichment Experiment: Carbon Cycling in High-and Low-Si Waters. *Science*, **304**, 408–14.

Cowen, R. and L. R. Castro, 1994. Relation of coral reef fish larval distributions to island scale circulation around Barbados, West Indies. *Bull. Mar. Sci.*, **54** (1), 228–44.

Cowen, R. K., K. M. M. Lwiza, S. Sponaugle, C. B. Paris, and D. B. Olson, 2000. Connectivity of marine populations: open or closed? *Science*, **287** (5454), 857–9.
Cowen, R. K., C. B. Paris, D. B. Olson, and J. L. Fortuna, 2003. The role of long distance dispersal versus local retention in replenishing marine populations. *Gulf and Caribbean Res.*, **14**, 129–37.
Cronin, T. W. and R. B. Forward, 1982. Tidally timed behavior: effects on larval distributions in estuaries. In *Estuarine Comparisons*, ed. V. S. Kennedy. New York: Academic Press, 505–20.
Dahlgren, C. P., J. A. Sobel, and D. E. Harper, 2001. Assessment of the reef fish community, habitat, and potential for larval dispersal from the proposed Tortugas South Ecological Reserve. *Proc. Gulf Carib. Fish. Inst.*, **52**, 700–12.
D'Asaro, E. A., D. M. Farmer, J. T. Osse, and G. F. Dairiki, 1996. A Lagrangian float. *J. Atmosph. Ocean. Tech.*, **13**, 1230–46.
Davis, A. R. and A. J. Butler, 1989. Direct observations of larval dispersal in the colonial ascidian *Podoclavella moluccensis* Sluiter: Evidence for closed populations. *J. Exper. Mar. Biol. Ecol.*, **127**(2), 189–203.
Davis, R. E., 1985. Drifter observations of coastal surface currents during CODE: The statistical and descriptive view. *J. Geophys. Res.*, **90**, 4741–55.
Davis, R. E., 1991. Lagrangian ocean studies. *Annu. Rev. Fluid Mech.*, **23**, 43–64.
Davis, R. E., 2003. Neutrally buoyant floats. In *ALPS: Autonomous and Lagrangian Platforms and Sensors, Workshop Report*, ed. D. L. Rudnick and M. J. Perry.
Denman, K. L. and A. E. Gargett, 1983. Time and space scales of vertical mixing and advection of phytoplankton in the upper ocean. *Limnol. Oceanogr.*, **28** (5), 801–15.
Denman, K. L. and T. M. Powell, 1984. Effects of physical processes on planktonic ecosystems in the coastal ocean. *Oceanogr. Mar. Biol. Ann. Rev.*, **22**, 125–68.
Dennis, D. M., C. R. Pitcher, and T. D. Skewes, 2001. Distribution and transport pathways of *Panulirus ornatus* (Fabricius, 1776) and *Panulirus* spp. larvae in the Coral Sea, Australia. *Mar. Freshwat. Res.*, **52** (8), 1175–85.
De Robertis, A. and M. D. Ohman, 1999. A free-drifting mimic of vertically migrating zooplankton. *J. Plankton Res.*, **21**(10), 1865–75.
DiBacco, C. and L. A. Levin, 2000. Development and application of elemental fingerprinting to track the dispersal of marine invertebrate larvae. *Limnol. Oceanogr.*, **45** (4), 871–80.
Dickey, T. D., 1991. The emergence of concurrent high-resolution physical and bio-optical measurements in the upper ocean and their application. *Rev. Geophys.*, **29**, 383–413.
Domeier, M., 2004. A potential larval recruitment pathway originating from a Florida marine protected area. *Fisheries Oceanogr.*, **13**, 287–94.
Dooley, H. D., 1974. A comparison of drogue and current meter measurements in shallow waters. *Rapports et Procès-verbaux des Réunions Conseil International pour l'Eploration de la Mer*, **167**, 225–30.
Epifanio, C. E., 1988. Transport of invertebrate larvae between estuaries and the continental shelf. *American Fisheries Society Symposium*, **3**, 104–14.
Epifanio, C. E., and R. W. Garvine, 2001. Larval transport on the Atlantic Continental Shelf of North America: a Review. *Estuar., Coast. and Shelf Sci.*, **52**(1), 51–77.
Falkowski, P. G., M. E. Katz, A. H. Knoll, A. Quigg, J. A. Raven, O. Schofield, and F. J. R. Taylor, 2004. The evolution of modern eukaryotic phytoplankton. *Science*, **305**, 354–60.

Fortier, L. and W. C. Leggett, 1982. Fickian transport and the dispersal of fish larvae in estuaries. *Can. J. Fish. Aquat. Sci.*, **39** (8), 1150–63.
French, D. P., M. J. Furnas, and T. J. Smayda, 1983. Diel changes in nitrite concentration in the chlorophyll maximum in the Gulf of Mexico. *Deep-Sea Res.*, **30** (7A), 707–21.
Furnas, M. J. and T. J. Smayda, 1987. Inputs of subthermocline waters and nitrate onto the Campeche bank. *Cont. Shelf Res.*, **7** (2), 161–75.
Gagnon, M. and G. Lacroix, 1981. The effects of tidal advection and mixing on the statistical dispersion of zooplankton. *J. Exp. Mar. Biol. Ecol.*, **56**(1), 9–22.
Garland, E. D. and C. A. Zimmer, 2002. Hourly variations in planktonic larval concentrations on the inner shelf: Emerging patterns and processes. *J. Mar. Res.*, **60**(2), 311–25.
Garrett, C., 1983. On the initial streakiness of a dispersing tracer in two- and three-dimensional turbulence. *Dyn. Atmos. Oceans*, **7** (4), 265–77.
Garvine, R. W., C. E. Epifanio, C. C. Epifanio, and K.-C. Wong, 1997. Transport and recruitment of blue crab larvae: A model with advection and mortality. *Estuar. Coast. Shelf Sci.*, **45** (1), 99–111.
Geyer, W., 1989. Field calibration of mixed layer drifters. *J. Atmosph. Ocean. Tech.*, **6**, 333–42.
Gillanders, B. M. and M. J. Kingsford, 2003. Spatial variation in elemental composition of otoliths of three species of fish (family Sparidae). *Estuar. Coast. Shelf Sci.*, **57** (5–6), 1049–64.
Glibert, P. M., C. Garside, J. A. Fuhrman, and M. R. Roman, 1991. Time-dependent coupling of inorganic and organic nitrogen uptake and regeneration in the plume of the Chesapeake Bay estuary and its regulation by large heterotrophs. *Limnol. Oceanogr.*, **36** (3), 895–909.
Glibert, P. M. and C. Garside, 1992. Diel variability in nitrogenous nutrient uptake by phytoplankton in the Chesapeake Bay plume. *J. Plankton Res.*, **14** (2), 271–88.
Gran, H. H., 1912. Pelagic plant life. In *The Depths of the Ocean*, ed. J. Murray and J. Hjort. London: Macmillan, Chapter VI.
Griffin, D. A., J. L. Wilkin, C. F. Chubb, A. F. Pearce, and N. Caputi, 2001. Ocean currents and the larval phase of Australian western rock lobster, *Panulirus cygnus*. *Mar. Freshwater Res.*, **52**, 1187–99.
Gustafson, D. E. Jr., D. K. Stoecker, M. D. Johnson, W. F. Van Hukelem, and K. Sneider, 2000. Cryptophyte algae are robbed of their organelles by the marine ciliate Mesodinium rubrum. *Nature*, **405**, 1049–52.
Haeckel, E. 1890. Planktonstuden. Jena.
Hare, J. A., J. H. Churchill, R. K. Cowen, T. J. Berger, P. C. Cornillon, P. Dragos, S. M. Glenn, J. J. Govoni, and T. N. Lee, 2001. Routes and rates of larval fish transport from the southeast to the northeast United States continental shelf. *Limnol. Oceanogr.*, **47**(6), 1774–89.
Hare, J., J. Quinlan, F. Werner, B. Blanton, J. Govoni, R. Forward, C. Settle, and D. Hoss, 1999. Larval transport during winter in the SABRE study area: results of a coupled vertical larval behavior three-dimensional circulation model. *Fish. Oceanogr.*, **8**(S2), 57–76.
Haury, L. R., 1976a. A comparison of zooplankton patterns in the California Current and North Pacific Central Gyre. *Mar. Biol.*, **37**, 159–67.
Haury, L. R., 1976b. Small-scale pattern of a California current zooplankton assemblage. *Mar. Biol.*, **37**(2), 137–57.

Haury, L. A., J. A. McGowan, and P. H. Wiebe, 1977. Patterns and processes in the time-space of plankton distributions. In *Spatial Pattern in Plankton Communities*, ed. J. H. Steele. New York: Plenum Press, 277–327.
Heath, M. and P. Rankine, 1988. Growth and advection of larval herring (*Clupea harengus* L.) in the vicinity of the Orkney Isles. *Est. Coastal Shelf Sci.*, **27**, 547–65.
Hensen, V., 1887. Über die Bestimmung des Plankton oder des im Meer triebenden Materials an Pflanzen und Thieren. *Fünfter Ber. Komm. Wiss. Unters. Deut. Meer Kiel.*, 1–108.
Hitchcock, G. L., E. J. Lessard, D. Dorson, J. Fontaine, and T. Rossby, 1989. The IFF: The isopycnal float fluorometer. *J. Atmosph. Ocean. Tech.*, **6**, 17–26.
Hitchcock, G. L., D. B. Olson, G. A. Knauer, A. A. P. Pszenny, and J. N. Galloway, 1990. Horizontal diffusion and new production in the Sargasso Sea. *Global Biogeochem. Cycles*, **4**, 253–65.
Hitchcock, G. L., A. Mariano, and T. Rossby, 1993. Mesoscale pigment fields in the Gulf Stream: Observations in a meander crest and trough. *J. Geophys. Res.*, **98** (C5), 8425–45.
Hitchcock, G. L., T. Rossby, J. L. Lillibridge, III, E. J. Lessard, E. R. Levine, D. N. Connors, K. Y. Børsheim, and M. Mork, 1994. Signatures of stirring and mixing near the Gulf Stream front. *J. Mar. Res.*, **52** (5), 797–836.
Hitchcock, G. L., W. J. Wiseman, Jr., W. C. Boicourt, A. J. Mariano, N. Walker, T. A. Nelson, and E. Ryan, 1997. Property fields in an effluent plume of the Mississippi River. *J. Mar. Sys.*, **12**(2), 109–26.
Hitchcock, G. L., G. A. Vargo, and M. L. Dickson, 2000a. Plankton community composition, production, and respiration in relation to dissolved inorganic carbon on the west Florida Shelf, April, 1996. *J. Geophys. Res.*, **105** (C3), 6579–89.
Hitchcock, G. L., E. Key, and J. Masters, 2000b. The fate of upwelled waters in the Great Whirl, August, 1995. *Deep-Sea Res. II.*, **47** (7–8), 1605–21.
Hitchcock, G. L., R. F. Chen, G. B. Gardner, and W. J. Wiseman, Jr., 2004. A Lagrangian view of fluorescent chromophoric dissolved organic matter distributions in the Mississippi River plume. *Marine Chemistry*, **89**, 225–39.
Hofmann, E. E., K. S. Hedström, J. R. Moisan, D. B. Haidvogel, and D. L. Mackas, 1991. Use of simulated drifter tracks to investigate general transport patterns and residence times in the coastal transition zone. *J. Geophys. Res.*, **96** (C8), 15,041–52.
Houghton, R. W. and C. Ho, 2001. Diapycnal flow through the Georges Bank tidal front: A dye tracer study. *Geophys. Res. Lett.*, **28**(1), 33–6.
Huisman, J., M. Arrayás, U. Ebert, and B. Sommeijer, 2002. How do sinking phytoplankton species manage to persist? *Amer. Natural.*, **159**(3), 245–54.
Hutchins, J. B. and A. F. Pearce, 1994. Influence of the Leeuwin Current on recruitment of tropical reef fishes at Rottnest Island, Western Australia. *Bull. Mar. Science*, **54**, 245–55.
Itoh, H., T. C. Hirata, M. Tanaka, N. Inoue, and H. Irie, 1979. Short-term fluctuations of zooplankton community in company with movement of a curtain drogue-I. Fluctuations of copepod community neighboring a current drogue followed by YŌKŌ MARU. *Bull. Seikai Reg. Fish. Res. Lab.*, **53**, 113–23.
James, M. K., P. R. Armstrong, L. B. Mason, and L. Bode, 2002. The structure of reef fish metapopulations: modeling larval dispersal and retention patterns. *Proc. R. Soc. Lond. B*, **269**, 2079–86.
Jennings, F. D., 1981. The Coastal Upwelling Ecosystems Analysis program. Epilogue. IDOE Int. Symp. on Coastal Upwelling, Los Angeles, CA. Texas A&M Univ., Sea Grant Cent. for Marine Research, College Station, TX, 13–15.

Johnson, D. R., 1995. Wind forced surface currents at the entrance to Chesapeake Bay: Their effect on blue crab larval dispersion and post-larval recruitment. *Bull. Mar. Sci.*, **57**(3), 726–38.

Johnson, K., 2003. Biogecochemcial cycles. In *ALPS: Autonomous and Lagrangian Platforms and Sensors, Workshop Report*, ed. D. L. Rudnick and M. J. Perry. 22–5.

Jones, B. H., K. H. Brink, R. C. Dugdale, D. W. Stuart, J. C. Van Leer, D. Blasco, and J. C. Kelly, 1983. Observations of a persistent upwelling center off Point Conception, California. In *Coastal Upwelling: Its Sediment Record*, ed. S. Suess and J. Thiede. New York: Plenum Press, 37–60.

Jones, G. P., M. J. Milicich, M. J. Emslie, and C. Lunow, 1999. Self-recruitment in a coral reef fish population. *Nature*, **402**, 802–4.

Jones, T. W., T. C. Malone, and S. Pike, 1990. Seasonal contrasts in diurnal carbon incorporation by phytoplankton size classes of the coastal plume of Chesapeake Bay. *Mar. Ecol. Prog. Ser.*, **68**, 1–21.

Joseph, J. and H. Sender, 1958. Uber die horizontale Diffusion im Meere. *Dt. Hydrogr. Z.*, **11**, 49–77.

Kamykowski, D., 1995. Trajectories of autotrophic marine dinoflagellates. *J. Phycol.*, **31**, 200–8.

Kamykowski, D., R. E. Reed, and G. J. Kirkpatrick, 1992. Comparison of sinking velocity, swimming velocity, rotation, and path characteristics among six marine dinoflagellate species. *Mar. Biol.*, **113** (2), 319–28.

Kennedy, V. S. and W. C. Boicourt, 1981. Water circulation and oyster spat settlement in two adjacent tributaries of the Choptank River, Maryland. *J. Shellfish Res.*, **1** (1)118.

Ketchum, B. H. and N. Corwin, 1965. The cycle of phosphorus in a plankton bloom in the Gulf of Maine. *Limnol. Oceanogr.*, **10** (suppl.), R148–61.

Kirkpatrick, G. J., T. B. Curtin, D. Kamykowski, M. D. Freezor, M. D. Sartin, and R. E. Reed, 1990. Measurement of photosynthetic response to euphotic zone physical forcing. *Oceanography*, **3**(1), 18–22.

Kirwan, A. D., G. McNally, M.-S. Chang, and R. Molinari, 1975. The effect of wind and surface currents on drifters. *J. Phys. Oceanogr.*, **5**, 361–8.

Koehl, M. A. R. and T. M. Powell, 1994. Turbulent transport of larvae near wave-swept rocky shores: Does water motion overwhelm larval sinking? In *Reproduction and Development of Marine Invertebrates*, ed. W. H. Wilson, Jr., S. A. Stricker, and G. L. Shinn. Baltimore: The Johns Hopkins University Press, 261–74.

Laane, R. W. P. M., M. W. Manuels, and W. Staal, 1984. A procedure for enriching and cleaning up Rhodamine B and Rhodamine WT in natural waters. *Water Res.*, **18** (2), 163–5.

Landry, M. R., S. L. Brown, K. E. Selph, M. R. Abbott, R. M. Letelier, S. Christensen, R. R. Bidigare, and K. Casciotti, 2001. Initiation of the spring phytoplankton increase in the Antarctic Polar Front Zone at 170 degree W. *J. Geophys. Res.*, **106** (C7), 13,903–15.

Largier, J. L., 2003. Considerations in estimating larval dispersal distances from oceanographic data. *Ecol. Appl.*, **13**(1), S71–89.

Law, C. S., E. R. Abraham, A. J. Watson, and M. I. Liddicoat, 2003. Vertical eddy diffusion and nutrient supply to the surface mixed layer of the Antarctic Circumpolar Current. *J. Geophys. Res.*, **108** (C8).

Law, C. S., M. I. Liddicoat, A. P. Martin, K. J. Richards, and E. M. S. Woodward, 2000. A Lagrangian SF_6 tracer study of an anticyclonic eddy in the North

Atlantic: patch evolution, vertical mixing and nitrate supply to the mixed layer. *Deep-Sea Res. II.*, **48** (4–5), 705–24.
Le Borgne, R. P., 1978. Ammonium formation in Cape Timiris (Mauritania) upwelling. *J. Exp. Mar. Biol. Ecol.*, **31**(3), 253–65.
Ledwell, J. R., A. J. Watson, and C. S. Law, 1998. Mixing of a tracer in the pycnocline. *J. Geophys. Res.*, **103** (C10): 21,499–529.
Lee, T. N., M. E. Clarke, E. W. Williams, A. F. Szmant, and T. Berger, 1994. Evolution of a Tortugas gyre and its influence on recruitment in the Florida Keys. *Bull. Mar. Sci.*, **54**, 621–46.
Leis, J. M., 1991a. The pelagic stage of reef fishes: The larval biology of coral reef fishes. *In The ecology of fishes on coral reefs*, ed. P. F. Sale. San Diego: Academic Press, 183–230.
Leis, J. M., 1991b. Vertical distribution of fish larvae in the Great Barrier Reef Lagoon, Australia. *Mar. Biol.*, **109**, 157–66.
Letelier, R. M., M. R. Abbott, and D. M. Karl, 1995. Southern ocean optical drifter experiment. *Antarct. J. U.S.*, **30**(5), 108–10.
Letelier, R. M., M. R. Abbott, and D. M. Karl, 1997. Chlorophyll natural fluorescence response to upwelling events in the Southern Ocean. *Geophys Res. Letts.*, **24**(4), 409–12.
Levin, L. A., 1990. A review of methods for labeling and tracking marine invertebrate larvae. *Ophelia.*, **32** (1–2), 115–44.
Liu, G, G. S. Janowitz, and D. Kamykowski, 2002. Influence of current shear on *Gymnodinium breve* (Dinophyceae) population dynamics: a numerical study. *Mar. Ecol. Prog. Ser.*, **231**, 47–66.
Lohrenz, S. E., J. J. Cullen, D. A. Phinney, D. B. Olson, and C. S. Yentsch, 1993. Distributions of pigments and primary production in a Gulf Stream meander. *J. Geophys. Res.*, **98** (C8), 14,545–55.
Lorenzen, C. J., 1968. Carbon/Chlorophyll relationships in an upwelling area. *Limnol. Oceanogr.*, **13**, 202–4.
MacIsaac, J. J., R. C. Dugdale, R. T. Barber, D. Blasco, and T. T. Packard, 1985. Primary production cycle in an upwelling center. *Deep-Sea Res.*, **32** (5), 503–29.
Malone, T. C. and H. W. Ducklow, 1990. Microbial biomass in the coastal plume of Chesapeake Bay: Phytoplankton-bacterioplankton relationships. *Limnol. Oceanogr.*, **35** (2), 296–312.
Martin, J. H., 1990. Glacial-interglacial CO_2 change: The iron hypothesis. *Paleoceanography*, **3**, 1–13.
Martin, J. H., K. H. Coale, K. S. Johnson, S. E. Fitzwater, R. M. Gordon, S. J. Tanner, C. N. Hunter, V. A. Elrod, J. L. Nowicki, T. L. Coley, R. T. Barber, S. Lindley, A. J. Watson, K. Van Scoy, and C. S. Law, 1994. Testing the iron hypothesis in ecosystems of the Equatorial Pacific Ocean. *Nature*, **371**, 123–9.
Martin, A. P., I. P. Wade, K. J. Richards, and K. J. Heywood, 1998. The PRIME eddy. *J. Mar. Res.*, **56**, 439–62.
McGehee, D. and J. S. Jaffe, 1996. Three-dimensional swimming behaviour of individual zooplankters: observations using the acoustical imaging system FishTV. *ICES J. Mar. Sci.*, **53** (2), 363–9.
Miller, C. B., 1970. Some environmental consequences of vertical migration in marine zooplankton. *Limnol. Oceanogr.*, **15** (5), 727–41.
Mitchell, B. G., M. Kahru, and J. Sherman, 2000. Autonomous temperature-irradiance profiler resolves the spring bloom in the Sea of Japan. Proceedings Ocean Optics XV, Monaco, Oct 2000.

Moloney, C. L. and J. G. Field, 1991. The size-based dynamics of plankton food webs. 1. A simulation model of carbon and nitrogen flows. *J. Plankton Res.*, **13** (5), 1003–38.

Morel, A., Y.-W. Ahn, F. Partensky, D. Vaulot, and H. Claustre, 1993. *Prochlorococcus* and *Synechococcus*: A comparative study of their optical properties in relation to their size and pigmentation. *J. Mar. Res.*, **51** (3), 617–49.

Morgan, S. G., 1995. Life and death in the plankton: larval mortality and adaptation. In *Ecology of Marine Invertebrate Larvae*, ed. L. McEdward. Boca Raton: CRC Press, 279–321.

Moriarty, D. J. W., 1979. Biomass of suspended bacteria over coral reefs. *Mar. Biol.*, **53**(2), 193–200.

Munk, P., P. J. Wright, and N. J. Pihl, 2002. Distribution of the early larval stages of cod, plaice and lesser sandeel across haline fronts in the North Sea. *Estuar. Coast. Shelf Sci.*, **55** (1), 139–49.

Nakata, H. and T. Hirano, 1978. Dye-diffusion experiments in a narrow passage and approaches. *Bull. Jap. Soc. Fish. Oceanogr.*, **32**, 1–14.

Natunewicz, C. C., C. E. Epifanio, and R. W. Garvine, 2001. Transport of crab larval patches in the coastal ocean. *Mar. Ecol. Prog. Ser.*, **222**, 143–54.

Niiler, P. P., R. E. Davis, and H. J. White, 1987. Water-following characteristics of a mixed layer drifter. *Deep-Sea Res. I.*, **34**(11), 1867–81.

Ohman, M. D., 1988. Behavioral responses of zooplankton to predation. *Bull. Mar. Sci.*, **43**, 530–50.

Okubo, A., S. Hasegawa, M. Amano, and I. Takeda, 1957. Report of the observation concerning the diffusion of dye patch in the sea off the coast of Tokai-mura. *Research papers Japan Atomic Energy res. Inst.*, **2**, 17–21.

Okubo, A., 1971. Oceanic diffusion diagrams. *Deep-Sea Res.*, **18** (8), 789–802.

Okubo, A., 1994. The role of diffusion and related physical processes in dispersal and recruitment of marine populations. In *The Biophysics of Marine Larval Dispersal*, ed. P. Sammarco and M. L. Heron. Coastal Estuarine Studies, 45. New York: Springer Verlag, 5–31.

Olaizola, M., J. LaRoche, Z. Kolber, and P. G. Falkowski, 1994. Non-photochemical fluorescence quenching and the diadinoxanthin cycle in a marine diatom. *Photosynth. Res.*, **41**, 357–70.

Olson, R. R., 1985. The consequences of short-distance larval dispersal in a sessile marine invertebrate. *Ecology*, **66**(1), 30–9.

Oudot, C., P. Raul, and B. Wauthy, 1979. Western Pacific equatorial upwelling: physical and chemical distributions and standing crop following a drifting drogue. *Cahiers Indo-Pac.*, **1**(1), 39–81.

Palumbi, S. R., 2001. The ecology of marine protected areas. In *Marine Ecology: the New Synthesis*, ed. M. Bertness, S. D. Gaines, and M. E. Hay. Sunderland, MA: Sinauer, 509–30.

Paris, C. B. and R. K. Cowen, 2004. Direct evidence of a biophysical retention mechanism for coral reef fish larvae. *Limnol. Oceanogr.*, **49**(6), 1964–79.

Paris, C. B., R. K. Cowen, K. M. M. Lwiza, D.-P. Wang, and D. B. Olson, 2002. Multivariate objective analysis of the coastal circulation of Barbados, West Indies: implication for larval transport. *Deep-Sea Res. I.*, **49**, 1363–86.

Parker, G. G., Jr., 1973. Tests of rhodamine WT dye for toxicity to oysters and fish. *J. Res. U. S. Geol. Surv.*, **1** (4), 499.

Pelegrí, J. L. and G. T. Csanady, 1990. Nutrient transport and mixing in the Gulf Stream. *J. Geophys. Res.*, **96** (C2), 2577–83.

Pepin, P. and J. A. Helbig, 1997. Distribution and drift of Atlantic cod (*Gadus morhua*) eggs and larvae on the Northeast Newfoundland Shelf. *Can. J. Fish. Aquat. Sci.*, **54** (03), 670–85.

Price, H. J., 1989. Swimming behavior of krill in response to algal patches: A mesocosm study. *Limnol. Oceanogr.*, **34** (4), 649–59.

Raabe, T. U., U. H. Brockmann, C.-D. Duerselen, M. Krause, and H. J. Rick, 1997. Nutrient and plankton dynamics during a spring drift experiment in the German Bight. *Mar. Ecol. Prog. Ser.*, **156**, 275–88.

Raven, J. A., 1986. Physiological consequences of extremely small size for autotrophic organisms in the sea. In *Photosynthetic picoplankton*. (*Can. Bull. Fish. Aquat. Sci. 214*), ed. T. Platt and W. K. W. Li. Canadian Government Publishing Centre, 1–70.

Raven, J. A. and K. Richardson, 1984. Dinophyte flagella: A cost-benefit analysis. *New Phytol.*, **98**(2), 259–76.

Rick, S., 1999. The spring bloom in the German Bight: Effects of high inorganic N:P ratios on the phytoplankton development. 305. Berichte aus dem Institut für Meereskunde an der Christian-Albrechts-Universität Kiel.

Roberts, C. M., 1997. Connectivity and management of Caribbean coral reefs. *Science*, **278**, 1454–7.

Roman, M. R. and W. C. Boicourt, 1990. Temporal and spatial variations in the abundance of blue crab larvae in the Chesapeake Bay plume and surrounding shelf waters. *Bull. Mar. Sci.*, **46** (1), 249–50.

Roman, M. R. and W. C. Boicourt, 1999. Dispersion and recruitment of crab larvae in the Chesapeake Bay plume: physical and biological controls. *Estuaries*, **22** (3A), 563–74.

Rossby, T., A. S. Bower, and P.-T. Shaw, 1985. Particle pathways in the Gulf Stream. *Bull. Am. Meteorol. Soc.*, **66** (9), 1106–10.

Rossby, T., D. Dorson, and J. Fontaine, 1986. The RAFOS system. *J. Atmos. Oceanic Techn.*, **3**, 672–9.

Rossby, T. and D. Webb, 1970. Observing abyssal motion by tracking Swallow floats in the SOFAR channel. *Deep-Sea Res.*, **17**, 359–65.

Rudnick, D. L., and M. J. Perry, eds, 2003. *ALPS: Autonomous and Lagrangian Platforms and Sensors, Workshop Report*. www.geo-prose.com/ALPS

Ruiz, J., C. M. Garcia, and J. Rodriguez, 1996. Sedimentation loss of phytoplankton cells from the mixed layer: Effects of turbulence levels. *J. Plankton Res.*, **18**(9), 1727–34.

Rumrill, S. S., 1991. Natural mortality of marine invertebrate larvae. *Ophelia*, **32**, 163–98.

Ryther, J. H., D. W. Menzel, E. M. Hulburt, *et al.* 1971. *Production and utilization of organic matter in the Peru coastal current*. Anton Bruun Report No. 4. Texas A&M University. College Station, TX.

Sambrotto, R. N. and C. Langdon, 1994. Water column dynamics of dissolved inorganic carbon (DIC), nitrogen and O_2 on Georges Bank during April, 1990. *Continental Shelf Research*, **14** (7/8), 765–89.

Savidge, G. and P. J. le B Williams, 2001. The PRIME 1996 cruise: an overview. *Deep-Sea Res. II. Top. Stud. Oceanogr.*, **48** (4–5), 687–704.

Scheltema, R. S., 1986. On dispersal and planktonic larvae of benthic invertebrates: An eclectic overview and summary of problems. *Bull. Mar. Sci.*, **39**, 290–322.

Schlee, S., 1973. *The Edge of an Unfamiliar World: A History of Oceanography*. New York: E. P. Dutton & Co., Inc.

Seligman, H., 1955. The discharge of radioactive waste products into the Irish Sea. Proceedings of the International Conference of Peaceful Uses of Atomic Energy. Geneva. 701–11.

Shanks, A. L., J. Largier, L. Brink, J. Brubaker, and R. Hooff, 2000. Demonstration of the onshore transport of larval invertebrates by the shoreward movement of an upwelling front. *Limnol. Oceanogr.*, **45**(1), 230–6.

Shanks, A. L., B. A. Grantham, and M. H. Carr, 2003. Propagule dispersal distance and the size and spacing of marine reserves. *Ecol. Applications*, **13**(1), S159–69.

Sieburth, J. McN., V. Smetacek, and J. Lenz, 1978. Pelagic ecosystem structure: Heterotrophic components of the plankton and their relationship to plankton size-fractions. *Limnol. Oceanogr.*, **23**, 1256–63.

Siegel, D. A., B. P. Kinlan, B. Gaylord, and S. D. Gaines, 2003. Lagrangian descriptions of marine larval dispersion. *Mar. Ecol. Prog. Ser.*, **260**, 83–96.

Smayda, T. J., 1970. The suspension and sinking of phytoplankton in the sea. *Oceanogr. Mar. Biol. Ann. Rev.*, **8**, 353–414.

Stoner, D. S., 1990. Recruitment of a tropical colonial ascidian: Relative importance of pre-settlement vs. post-settlement processes. *Ecology*, **71**(5), 1682–90.

Strathmann, R. R., 1990. Why life histories evolve differently in the sea. *Am. Zool.*, **30** (1), 197–207.

Strickler, J. R., K. D. Squires, H. Yamakazi, and A. H. Abib, 1997. Combining analog turbulence with digital turbulence. *Sci. Mar. (Barc.).*, **61** (1), 197–204.

Suijlen, J. M. and J. J. Buyse, 1994. Potentials of photolytic rhodamine WT as a large-scale water tracer assessed in a long-term experiment in the Loosdrecht Lakes. *Limnol. Oceanogr.*, **39**, 1411–23.

Swallow, J. C., 1955. A neutral-buoyancy float for measuring deep currents. *Deep-Sea Res.*, **3**, 74–81.

Swearer, S. E., J. E. Caselle, D. W. Lea, and R. R. Warner, 1999. Larval retention and recruitment in an island population of a coral-reef fish. *Nature*, **402** (6763), 799–802.

Taggart, C. T. and W. C. Leggett, 1987a. Short-term mortality in post-emergent larval capelin *Mallotus villosus*. 1. Analysis of multiple *in situ* estimates. *Mar. Ecol. Prog. Ser.*, **41** (3), 205–17.

Taggart, C. T. and W. C. Leggett, 1987b. Short-term mortality in post-emergent larval capelin *Mallotus villosus*. 2. Importance of food and predator density, and density-dependence. *Mar. Ecol. Prog. Ser.*, **41** (3), 219–29.

Talbot, J. W. and G. A. Talbot, 1974. Diffusion in shallow seas and in English coastal and estuarine waters. *Rapp. P.-v. Réun. Cons. Int. Explor. Mer.*, **167**, 93–110.

Talbot, J. W., 1977. The dispersal of plaice eggs and larvae in the Southern Bight of the North Sea. *J. Cons. Int. Explor. Mer.*, **37**(3), 221–48.

Taylor, A. H., D. S. Harbour, R. P. Harris, P. H. Burkill, and E. S. Edwards, 1993. Seasonal succession in the pelagic ecosystem of the North Atlantic and the utilization of nitrogen. *J. Plankton Res.*, **15** (8), 875–91.

Taylor, F. J. R., 1980. Phytoplankton ecology before 1900: Supplementary notes to the "Depths of the Ocean". In *Oceanography: The Past*. ed. M. Sears and D. Merriman. New York: Springer-Verlag, 509–21.

Tegner, M. J. and R. A. Butler, 1985. Drift-tube study of the dispersal potential of green abalone (*Haliotus fulgens*) larvae in the Southern California Bight – Implications for recovery of depleted populations. *Mar. Ecol. Prog. Ser.*, **26**, 73–84.

Thorrold, S. R. and J. A. Hare, 2002. Otolith applications in reef fish ecology. In *Coral Reef Fishes: Dynamics and Diversity in a Complex Ecosystem*, ed. P. F. Sale. New York: Academic Press, 243–64.

Thorrold, S. R., C. Latkoczy, P. K. Swart, and C. M. Jones, 2001. Natal homing in a marine fish metapopulation. *Science*, **291** (5502), 297–9.

Thorrold, S. R., G. P. Jones, M. E. Hellberg, R. S. Burton, S. E. Swearer, J. E. Neigel, S. G. Morgan, and R. R. Warner, 2002. Quantifying larval retention and connectivity in marine populations with artificial and natural markers. *Bull. Mar. Sci.*, **70** (1), 291–308.

Tsuda, A. *et al.*, 2003. A mesoscale iron enrichment in the western Subarctic Pacific induces a large centric diatom bloom. *Science*, **300**, 958–61.

Upstill-Goddard, R. C., J. M. Suijlen, G. Malin, and P. D. Nightingale, 2001. The use of photolytic Rhodamines WT and sulpho G as conservative tracers of dispersion in surface waters. *Limnol. Oceanogr.*, **46**(4), 927–34.

Villareal, T. A. and E. J. Carpenter, 1994. Chemical composition and photosynthetic characteristics of *Ethmodiscus rex* (Bacillariophyceae): evidence for vertical migration. *J. Phycology*, **30**, 1–8.

Walsh, J. J., R. H. Weisberg, D. A. Dieterle, R. He, B. P. Darrow, J. K. Jolliff, K. M. Lester, G. A. Vargo, G. J. Kirkpatrick, K. A. Fanning, T. T. Sutton, A. E. Jochens, D. C. Biggs, B. Nababan, C. Hu, and F. E. Muller-Karger, 2003. Phytoplankton response to intrusions of slope water on the West Florida Shelf: Models and observations. *J. Geophys. Res.*, **108** (C6), 3190 10.1029/2002JC001406.

Wanninkhof, R., J. R. Ledwell, and W. S. Broecker, 1985. Gas-exchange wind-speed relation measured with sulfur-hexafluoride on a lake. *Science*, **227**, 1224–26.

Wanninkhof, R., W. Asher, R. Weppernig, H. Chen, P. Schlosser, C. Langdon, and R. Sambrotto, 1993. Gas transfer experiment on Georges Bank using two volatile deliberate tracers. *J. Geophys. Res.*, **98** (C11), 20,237–48.

Wanninkhof, R., G. L. Hitchcock, W. J. Wiseman, G. Vargo, P. B. Ortner, W. Asher, D. T. Ho, P. Schlosser, M.-L. Dickson, R. Masserini, K. Fanning, and J.-Z. Zhang, 1997. Gas exchange, dispersion and biological productivity on the west Florida Shelf: results from a Lagrangian tracer study. *Geophys. Res. Letts.*, **24** (14), 1767–70.

Watson, A. J. and J. R. Ledwell, 2000. Oceanographic tracer release experiments using sulphur hexafluoride. *J. Geophys. Res.*, **105** (C6), 14,325–37.

Watson, A. J., J. R. Ledwell, and S. C. Sutherland, 1991. The Santa Monica Basin tracer experiment: Comparison of release methods and performance of perfluorodecalin and sulfur hexafluoride. *J. Geophys. Res.*, **96** (C5), 8719–25.

Watson, A. J., C. S. Law, K. A. Van Scoy, F. J. Millero, W. Yao, G. E. Friedderich, M. I. Liddicoat, R. H. Wanninkhof, R. T. Barber, and K. H. Coale, 1994. Minimal effect of iron fertilization on sea-surface carbon dioxide concentrations. *Nature*, **371**, 143–5.

Werner, F. E., B. O. Blanton, J. A. Quinlan, and R. A. Luettich, Jr., 1999. Physical oceanography of the North Carolina continental shelf during the fall and winter seasons: implications for the transport of larval menhaden. *Fish. Oceanogr.*, **8**(S2), 7–21.

Wolcott, T. G. and D. L. Wolcott, 1998. Estuarine export and egress of larvae: test with larval mimics. Abstracts of the Annual Meeting of the Society for Integrative and Comparative Biology. Boston MA.

Wyatt, B., W. V. Burt, and J. G. Pattullo, 1972. Surface currents off Oregon as determined from drift bottle returns. *J. Phys. Oceanogr.*, **2**, 286–93.

Yeung, C. and T. N. Lee, 2002. Larval transport and retention of the spiny lobster, *Panulirus argus*, in the coastal zone of the Florida Keys, USA. *Fish. Oceanogr.*, **11**(5), 286–309.

Zaret, T. M. and J. S Suffern, 1976. Vertical migration in zooplankton as a predator avoidance mechanism. *Limnol. Oceanogr.*, **21**(6), 804–13.

Zenitani, H., K. Nakata, and R. Kimura, 1996. Survival and growth of sardine larvae in the offshore side of the Kuroshio. *Fish. Oceanogr.*, **5** (1), 56–62.

12

A Lagrangian stochastic model for the dynamics of a stage structured population. Application to a copepod population

GIUSEPPE BUFFONI
ENEA, La Spezia, Italy
MARIA GRAZIA MAZZOCCHI
Stazione Zoologica A. Dohrn, Napoli, Italy
AND
SARA PASQUALI
CNR-IMATI, Milano, Italy

12.1 Introduction

We produce the population dynamics of a stage structured population, where the stages are defined by sharp biological events (egg hatching, molt, adult emergence, beginning and end of oviposition, death), by means of a stochastic individual-based model that simulates the life histories of its individuals (Judson, 1994; Berec, 2002; Buffoni *et al.*, 2002; Buffoni *et al.*, 2004). Aspects of the life history of an individual, such as survival probabilities, development rates and egg production, depend on its "status," on the population size, and on external factors such as the environmental conditions (e.g. physical factors, food availability). In general, the status of an individual can be identified by means of a number of physiological variables or biometric descriptors, which describe the behavior of an individual in a given situation, and define its physiological age. The physiological age of an individual is generally described only by a variable. Here the status of an individual is individuated by its stage and its physiological age in the stage. The physiological age is defined as the percentage of development for non-reproductive individuals, and as the percentage of the potential reproductive effort for an adult female. The life history is obtained by the time evolution of the status of an individual, from birth to death, following its development and, when the individual is an adult female, the production of eggs. The dynamics of the overall population, i.e. the time evolution of the stage and physiological age distributions of the individuals, is then obtained by performing numerical simulations of the life histories of the individuals of the initial population, and those yielded by recruitment over time, generation after generation.

Lagrangian Analysis and Prediction of Coastal and Ocean Dynamics, ed. A. Griffa, D. Kirwan, A. Mariano, T. Özgökmen, and T. Rossby. Published by Cambridge University Press.

Both deterministic and stochastic models for the population dynamics can be formulated in Eulerian and Lagrangian approachs (Buffoni and Pasquali, 2003). In the Eulerian formalism, balance evolution equations are written for the stage and physiological age distributions of the individuals: generalized von Foerster hyperbolic equations in the deterministic case (Hoppensteadt, 1975), and Fokker–Planck parabolic equations in the stochastic case (Gardiner, 1994, p. 117; Carpi and Di Cola, 1988), together with suitable initial and boundary conditions. In the Lagrangian formalism (random flights, Monte Carlo formulations) the dynamics is obtained by means of a stochastic model describing the time evolution of the life history of an individual (Judson, 1994; Buffoni *et al.*, 2003, 2004). Stochastic difference equations are written for development and production processes; moreover, choice processes to determine when an individual survives or dies, and when an individual is a female or a male, are implemented. The dynamics of the overall population, i.e. the time evolution of the physiological age distribution of the individuals, is obtained by performing numerical simulations of the life histories of the individuals of the population.

Here the formulation of a Lagrangian or individual-based stochastic model is presented, trying to formalize, as far as possible, the biological processes in terms of equations. O. P. Judson (1994), introducing Monte Carlo methods for individual-based models, writes "Equations, unlike words, leave little room for different interpretation. As the descriptions of most individual-based models are verbal, one way to combat the confusion and suspicion that words may cause is to be careful to define terms that may mean different things to different people."

The model has been applied so far to insect populations (Buffoni *et al.*, 2002, 2004). It can also be applied to copepods, small aquatic crustaceans that have stage structured populations, and we present here, as a case study, the modeling of a planktonic species that is abundant in the Mediterranean Sea.

The chapter is organized as follows. In Section 12.2 the basic assumptions of the modeling approach are stated. In Section 12.3 the equations of development, mortality, reproduction processes are presented; by means of these equations the life history of an individual is realized. In Section 12.4 the qualitative behaviour of the dynamics of the overall population is illustrated, depending on the characteristics of the model (linear, density-dependent, with low and high levels of uncertainties). In Section 12.5 an application to a copepod population is shown and in Section 12.6 some concluding remarks are presented.

12.2 Basic assumptions of the modeling approach

It is assumed that the life history of an individual is completely determined by the biological processes of development, mortality and reproduction (Figure 12.1).

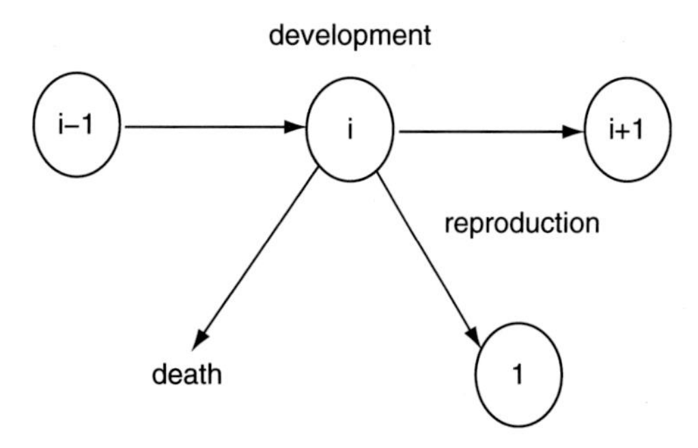

Figure 12.1 Main biological processes governing the life history of an individual. i is the stage index, 1 is the egg stage.

In general, the average values and the standard deviations of the rates of the processes of development, mortality and reproduction depend on time through the environmental variables, in particular temperature, and food; furthermore, some of them may depend on the overall population size or on some individual characteristics, which gives rise to a feedback on the population growth.

We shall consider $n+1$ stages; the first $n-1$ stages include pre-reproductive individuals, the stage n adult reproductive females and males, the stage $n+1$ post-reproductive individuals. At any time the status of an individual is individuated by its stage and its physiological age in the stage, which is defined as the percentage of development in a stage for a non-reproductive individual, and as the percentage of the potential reproductive effort realized for an adult female (Curry and Feldman, 1987; Munholland and Dennis, 1992). Thus, the model equations describe the time evolution of the status of an individual, from birth to death, following its development and, when the individual is an adult female, the production of eggs. The dynamics of the overall population, in terms of its density (number of individuals per spatial unit), is obtained by performing numerical simulations of the life histories of the individuals of the initial population, and those yielded by recruitment over time.

12.3 Life history of an individual

The equations governing the development, mortality and reproduction processes are here described in some details, by considering the noise as a diffusion process. At the end of the section, a different formulation of the dispersion process is summarized.

To simplify the notation, the dependence of the rates, and of other quantities introduced in what follows, on temperature and population size is not generally explicitly expressed; for clarity, it is introduced in the equations for the reproduction process.

12.3.1 Development and mortality processes

The development process of an individual is viewed as an accumulation of small increments of development over time, and it is described in terms of the random variable dependent on time

$$X_t^i \;=\; \textit{percentage of development of individual in stage } i \textit{ at time } t.$$

Let t^i, $i = 1, \ldots, n+1$, be the initial time instant of the stay of an individual in stage i. We have $X_{t^i}^i = 0$; the time instant t^{i+1} will be determined by the equation $X_{t^{i+1}}^i = 1$.

An average development rate (Curry and Feldman, 1987, p. 38) $v^i = 1/D^i$, where D^i = *average duration in stage i* = *average development time in stage i*, and a mortality rate μ^i, together with their standard deviations σ^i, are given for individuals in the pre-reproductive stages $i = 1, 2, \ldots, n-1$. Assuming the status of an individual is known at a generic time t, i.e. its stage i and its physiological age X_t^i, we compute the status at age $t + \Delta t$ by means of the stochastic equation

$$X_{t+\Delta t}^i \;=\; X_t^i \;+ \max\{0;\; v^i\,\Delta t \;+\; g^i\,\Delta W_t^i\},\quad t > t^i;\;\; X_{t^i}^i = 0, \qquad (12.1)$$

where $g^i = \sigma^i\sqrt{\Delta t}$, is the level of the noise, $\Delta W_t^i = \alpha_t^i\sqrt{\Delta t}$, with α_t^i standard normal random numbers. The increments ΔW_t^i are independent increments of a Wiener process (Gardiner, 1994, p. 69) satisfying

$$E[\Delta W_t^i] \;=\; 0,\; E[(\Delta W_t^i)^2] \;=\; \Delta t,$$

with $E[\cdot]$ indicating the expected value of the random variable in argument.

The computation of the time evolution of the development in the stage i ends when either at a time t^{i+1} we register that $X_{t^{i+1}}^i \geq 1$ or the individual is eliminated due to its death. Then, when $i < n$ and the individual is not dead, the computation of the development in the stage $i+1$ begins; otherwise, the life history of a new individual begins.

The ageing of a male is described by equation (12.1) for X_t^n, where $v^n = 1/L_{\mathrm{m}}$, whith L_{m} = *male average life span*. It is assumed that the individuals in the post-reproductive stage can only die; thus $v^{n+1} = 0$, $g^{n+1} = 0$.

The choice process of the events survival or death of an individual in the time interval $(t, t + \Delta t)$ is carried out by considering the mortality process as a

Poisson process (Feller, 1957, p. 400). Let β_t^i be a uniform random number in the interval [0,1]. We assume that:

$$\text{if } \beta_t^i \leq s^i, \text{ then the event is survival}$$
$$\text{if } \beta_t^i > s^i, \text{ then the event is death}$$

where $s^i = \exp(-\mu^i \Delta t)$ is the survival probability in the time interval $(t, t+\Delta t)$.

When the standard deviations of the mortality rates are high, the mortality rate has to be considered as a $N(\mu^i, \sigma^i)$ random variable. Thus, the actual mortality rate is given by $\mu_t^i = \max(0, \mu^i + \hat{\beta}_t^i\sigma^i)$, with $\hat{\beta}_t^i$ standard random normal process, and the survival probability is $s_t^i = \exp(-\mu_t^i\Delta t)$.

12.3.2 Reproduction process

The sex of an adult entering in the reproductive stage n is attributed as follows. Let the sex ratio ρ be defined as

$$\rho = \frac{\textit{number of females}}{\textit{females} + \textit{males}}.$$

Let γ_t be a uniform random number in the interval [0,1]. We assume that:

$$\text{if } \rho \geq \gamma_t, \text{ then the adult is a female}$$
$$\text{if } \rho < \gamma_t, \text{ then the adult is a male.}$$

The ageing of a female is assumed dependent on the reproduction process, so that it is possible to define the female physiological age as the ratio between realized and potential fecundity (Roff, 1992; Buffoni *et al.*, 2002). Let $\tau = t - t^n$, $t \geq t^n$, be the reproductive age of a female and let the reproductive profile $f(\tau, T)$ (Curry and Feldman, 1987, p. 107) be defined as

$$f(\tau, T) = \begin{array}{l}\textit{average number of eggs produced per unit time by a} \\ \textit{female with reproductive age } \tau, \textit{ at temperature } T.\end{array} \tag{12.2}$$

We consider the case of either constant or unimodal reproductive profiles (Roff, 1992, p. 122) with $f(0, T) = 0$ and $f(\tau, T) = 0$ for all sufficiently large τ. An unimodal reproductive profile (Figure 12.2) can be approximated by a function of the type

$$f(\tau, T) = \begin{cases} f_0 \, \frac{\tau}{\tau_0} \, [\mathrm{e}^{-\frac{\tau}{\tau_0}} - \mathrm{e}^{-\frac{L_f}{\tau_0}}] & 0 \leq \tau \leq L_f \\ 0 & \tau > L_f \end{cases} \tag{12.3}$$

where f_0 = *eggs/female/unit time*, L_f = *female average life span*, and τ_0 is a characteristic time determining the shape of the profile (in particular the position of the maximum). The quantities f_0, τ_0, L_f depend on T, and are estimated by fitting experimental profiles (Buffoni *et al.*, 2002).

Let

$$q_t \;=\; \textit{number of eggs produced by a reproductive female at time } t.$$

The evolution equation for q_t is

$$q_{t+\Delta t} \;=\; q_t \;+\; \psi(t), \quad t > t^n; \qquad q_{t^n} = 0 \tag{12.4}$$

where $\psi(t) = \max\{0, f(t - t^n, T)\,\Delta t \;+\; g^0\,\Delta W_t^0\}$, $g^0 = \sigma^0\sqrt{\Delta t}$ and $\Delta W_t^0 = \delta_t\sqrt{\Delta t}$, with δ_t a standard normal random number. Let

$$\mathbf{\Psi}_0(t) \;=\; \int_0^t \mathrm{d}\xi\, f(\xi, T(t^n + \xi)), \;\; \mathbf{\Psi}_1(t) \;=\; \int_t^\infty \mathrm{d}\xi\, f(\xi, T(t)). \tag{12.5}$$

Then, the development of a reproductive female is described in terms of a physiological age X_t^n defined by

$$X_t^n \;=\; \frac{q_t}{\mathbf{\Psi}_0(t) + \mathbf{\Psi}_1(t)}, \;\; t > t^n; \qquad X_{t^n}^n = 0. \tag{12.6}$$

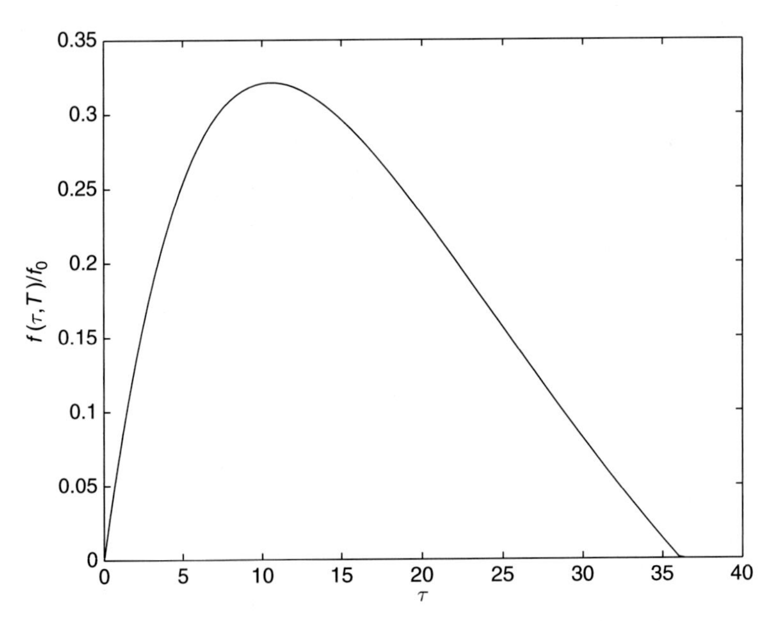

Figure 12.2 Example of unimodal reproductive profile $f(\tau, T)/f_0$ versus τ, for fixed T.

Furthermore we need to update the recruitment in terms of the total number of eggs produced

$$Q_t = \textit{total number of eggs at time } t \textit{ at the beginning of the egg stage.}$$

12.3.3 Formalization of choice processes

By introducing auxiliary processes, the evolution equation (12.1) of the development process, together with the choice process of survival or death of an individual, may be formulated as follows. Conventionally, we attribute the value $X_t^i = -1$ at the death of the individual. Thus we may write

$$X_{t+\Delta t}^i = \frac{1+\zeta_t^i}{2}\hat{X}_{t+\Delta t}^i - \frac{1-\zeta_t^i}{2} \tag{12.7}$$

where

$$\hat{X}_{t+\Delta t}^i = \hat{X}_t^i + \max\{0;\ v^i\,\Delta t + g^i\,\Delta W_t^i\}, \qquad t > t^i; \qquad \hat{X}_{t^i}^i = 0$$

and

$$\zeta_t^i = \frac{s^i - \beta_t^i}{|s^i - \beta_t^i|} = \begin{cases} 1 & \text{if } s^i - \beta_t^i \geq 0 \\ -1 & \text{if } s^i - \beta_t^i < 0 \end{cases}$$

Analogously, we may formalize the female–male choice process by introducing auxiliary processes in the reproduction process.

12.3.4 Infinitesimal and finite time scale of noise

Equation (12.1) describes the development process as a stochastic Markovian process in terms of X_t^i. As $E[(\Delta W_t^i)^2] = \Delta t$, the variance of the development rate

$$V_t^i = \frac{\Delta X_t^i - v^i \Delta t}{\Delta t}$$

where $\Delta X_t^i = X_{t+\Delta t}^i - X_t^i$, is given by

$$E[(V_t^i)^2] = \frac{(g^i)^2}{\Delta t}. \tag{12.8}$$

If the variance of V_t^i is $(\sigma^i)^2$, it follows that $(g^i)^2 = (\sigma^i)^2 \Delta t$, and the noise time scale is Δt; thus, as $\Delta t \to 0$, g^i becomes infinitesimal. Let

$$R(\tau) = \frac{E(V_t^i\, V_{t+\tau}^i)}{(\sigma^i)^2} = \frac{E(\Delta W_t^i\, \Delta W_{t+\tau}^i)}{\Delta t}$$

be the autocorrelation function of V_t^i. As W_t^i are independent increments of a Wiener process, it follows that

$$R(0) = 1; \qquad R(\tau) = 0 \text{ for } \tau > 0.$$

Other formulations of the development process are suggested by models of particle dispersion in water (Thomson, 1987; Zambianchi and Griffa, 1994). Let us consider the stochastic processes

$$X_{t+\Delta t}^i = X_t^i + (v^i + V_t^i)\Delta t, \tag{12.9}$$

$$V_{t+\Delta t}^i = V_t^i - v^i \frac{\Delta t}{\theta} + g^i\, \Delta W_t^i, \tag{12.10}$$

where θ is a characteristic time constant. Equations (12.9) and (12.10), describing the development process X_t^i, represent again a stochastic Markovian process, but in the velocity V_t^i instead of in X_t^i. The noise V_t^i is not uncorrelated from one time step to another; the individual conserves the memory of its initial development rate during a finite time of order θ. The autocorrelation function of V_t^i is now given by

$$R(\tau) = e^{-\frac{\tau}{\theta}}. \tag{12.11}$$

The parameter θ is a measure of the finite time scale of the noise. Taking into account this process of noise correlation could be significant for some populations; this approach is actually under way.

12.4 Dynamics of the overall population

Let an initial population be assigned at time t_0 (for example a given number of eggs and/or females). Then, according to the processes previously described, we perform numerical simulations of the life histories of the individuals of the initial population and of those from the recruitment yielded over time. The dynamics of the overall population is determined by the time evolution of the status of all its individuals. Thus, we can compute at any given time $t > t_0$ the number of individuals in stage i with physiological age in $(x - \Delta x/2, x + \Delta x/2)$. This number is clearly a random variable, and its average can be estimated by carring out a large number of realizations of life histories. Let

$$\phi^i(t,x)\Delta x = \textit{average number of individuals at time } t \textit{ in stage } i$$
$$\textit{with physiological age in } \left(x - \frac{\Delta x}{2}, x + \frac{\Delta x}{2}\right)$$

and

$$N_k^i(t) \;=\; \int_{x_{k-1}}^{x_k} \mathrm{d}x\, \phi^i(t,x), \quad N^i(t) \;=\; \int_0^1 \mathrm{d}x\, \phi^i(t,x). \tag{12.12}$$

We record the numbers of individuals $N_k^i(t_{jp})$, $N^i(t_{jp})$ at specified times $t_{jp} = jp\Delta t$, $j = 0, 1, 2, \ldots$, with p a given integer. Here $x_k = k\Delta x$, $k = 0, 1, \ldots, K$, with $\Delta x = 1/K$, define a partition of [0,1].

12.4.1 Linear model

Assume now that the population grows without any feedback dependent on the population size, i.e. the case of a linear model, and that the rates are constant. The dynamical response of the model, which is intuitively clear, is obtained from the analysis of the results of the numerical simulations and from a theoretical analysis of an equivalent Eulerian model briefly summarized in Section 12.4.4. The initial phase of the time evolution of the population is greatly influenced by the initial conditions. If either the production of eggs is low or the mortality rates are high, then the population goes to extinction. Otherwise, for time t long enough, the population attains an exponential growth; thus we have that

$$\phi^i(t,x) \simeq \mathrm{e}^{\lambda t}\varphi^i(x) \tag{12.13}$$

with $\lambda > 0$. It follows that during the asymptotic exponential growth, when $N^i(t) \simeq \mathrm{e}^{\lambda t} z^i$, with $z_i > 0$, a stable stage distribution is attained, according to

$$\frac{N^i(t)}{N(t)} \simeq \eta^i = \frac{z^i}{z},$$

where $N(t) = N^1(t) + N^2(t) + \cdots + N^n(t)$, $z = z^1 + z^2 + \cdots + z^n$, and

$$\eta^i \;=\; lim_{t\to\infty} \frac{N^i(t)}{N(t)}.$$

Furthermore, for g^i sufficiently small, we obtain that $\varphi^i(x)$ in (12.13) decreases exponentially:

$$\varphi^i(x) \simeq a^i \mathrm{e}^{\theta^i x} \tag{12.14}$$

with $a^i > 0$ and $\theta^i \simeq -(\lambda + \mu^i)/v^i < 0$. The distributions $\phi^i(t, x)$ in the physiological age x are discontinuous in the sense that $v^{i-1}\phi^{i-1}(t,1) = v^i\phi^i(t,0)$; in fact, $v^i\phi^i(t, v^i\tau)$ is the distribution in the chronological age τ, which is continuous in τ. A theoretical analysis of the Eulerian linear model can be found in Buffoni *et al.* (2004) and in Buffoni and Pasquali (submitted).

12.4.2 Feedback of the population size on the recruitment

When the system is either nonlinear, owing to a feedback of the population size on its growth, or some rates are explicitly time dependent, the response of the model depends strongly on the type of feedback and on the time-dependent driving forces.

A typical feedback is due to the control of the overall population density or of some linear functional of the population on the recruitment (Bergh and Getz, 1988; Abbiati *et al.*, 1992). In this class of models it is assumed that the mortality rate of the eggs is dependent on a linear functional $\chi(t)$ of the population:

$$\chi(t) = \sum_{i=1}^{n} \chi^i N^i(t),$$

with $\chi^i \geq 0$, while for stages other than the first the mortality is constant. Thus, the survival probability s^1 is given by: $s^1 = \exp[-\mu^1(\chi)\,\Delta t] = S(\chi, \Delta t)$. The control function $S(\chi, \Delta t)$ is non-negative, and it may be either a decreasing function of χ or an unimodal function, to take into account overcrowding and under-overcrowding effects, respectively. The results of numerical simulations, performed with a decreasing $S(\chi, \Delta t)$, show an asymptotic trend of the population toward either a steady state or oscillations around an average value, depending on the slope of $S(\chi, \Delta t)$ versus χ.

12.4.3 Effects of the uncertainty levels

Only few comments can be drawn about the dependence of the dynamics on the uncertainty levels. In general, when the population grows exponentially, the value of the Malthusian parameter λ increases with the uncertainties g^i. This can be explained by the fact that the uncertainty affecting development rates is expected to increase the variance of the individual distribution on physiological age; this makes the pattern of population growth less dependent on initial condition (a single cohort of individuals in $x = 0$). Furthermore, the ratio η^1 increases and η^i, with $i > 1$, decreases by increasing g^i.

12.4.4 Eulerian formulation

For comparison and completeness, we include a brief description of the equations for the distributions $\phi^i(t, x)$ in the Eulerian formulation (Buffoni *et al.*, 2004; Buffoni and Pasquali, submitted).

Under the assumption that the levels of the noise g^i are sufficiently small, the process (12.1) is a discrete diffusion process with positive drift $v^i\Delta t$. Thus, in a first approximation, the distributions $\phi^i(t, x)$ satisfy the balance equations (Fokker–Plank equations)

$$\frac{\partial\phi^i}{\partial t} + v^i\frac{\partial\phi^i}{\partial x} - k^i\frac{\partial^2\phi^i}{\partial x^2} + \mu^i\phi^i = 0 \qquad (t, x) \in (0, \infty) \times (0, 1), \tag{12.15}$$

where (Karlin and Taylor, 1981, p. 159)

$$k^i = \frac{1}{2}\frac{\mathrm{d}\sigma^2_{X^i_t}}{\mathrm{d}t} = \frac{(g^i)^2}{2}.$$

Moreover, $\phi^i(t, x)$ have to satisfy the initial conditions $\phi^i(0, x) = \tilde{\phi}^i(x), x \in (0, 1)$, and suitable boundary and regularity conditions, due to the discontinuous stage structure of the population.

The production of eggs is expressed in the form

$$v^1\phi^1(t, 0) - k^1\frac{\partial\phi^1}{\partial x}\bigg|_{x=0} = \rho\int_0^1 v^n b(t, x)\phi^n(t, x)\mathrm{d}x \tag{12.16}$$

where $b(t, x)\,\mathrm{d}x =$ *average number of eggs produced at time t by an individual with physiological age in* $(x, x + \mathrm{d}x)$. At the end of a stage, at $x = 1$, we assume that the advection process is the dominant process, i.e. the individuals change stage only due to advection, thus

$$-k^i\frac{\partial\phi^i}{\partial x}\bigg|_{x=1} = 0, \tag{12.17}$$

and we impose the continuity of the flux:

$$v^i\phi^i(t, 1) = v^{i+1}\phi^{i+1}(t, 0) - k^{i+1}\frac{\partial\phi^{i+1}}{\partial x}\bigg|_{x=0}, \tag{12.18}$$

for $i = 1, 2, \ldots, n - 1$. Moreover, at the end of their life the old and non-reproductive individuals are forced to leave the system, and to satisfy a condition of type (12.17).

A discrete form of the system (12.15)–(12.18) is obtained by integrating (12.15) along the characteristic lines, approximating the diffusion terms in the final points by means of the classical three points formulae, and taking into account the boundary conditions (Buffoni and Pasquali, submitted). Thus, the equivalent discrete Eulerian model is formulated in terms of a generalized Leslie matrix, which may be density dependent in the case considered in Section 12.4.2; the number and the stability of the equilibrium states of these models depend on the type of nonlinearities (Bergh and Getz, 1988; Buffoni and Cappelletti, 2000; Buffoni and Pasquali, submitted).

12.5 Application to a copepod population

Copepods are the most abundant metazoans in marine waters, where they dominate by numbers the mesozooplankton communities, particularly in the epipelagic domain. In terms of their size, diversity and abundance, planktonic copepods can be regarded as the insects of the seas. Most copepod species rely on phytoplankton as major food source and are, in turn, preyed by many larval and adult fishes. Copepods represent therefore, in the marine pelagic food webs, a very important link between the autotrophic compartment and the large consumers (Huys and Boxshall, 1991; Mauchline, 1998).

As a case study, we consider the temporal dynamics of the copepod *Temora stylifera*, a common and abundant species in coastal areas of the Western Mediterranean (Estrada *et al.*, 1985). The annual cycle of this species has been depicted in a long-term study conducted on the pelagic system in the Gulf of Naples (Mazzocchi and Ribera d'Alcalà, 1995). In this area, as in most Mediterranean regions, the annual cycle is characterized by a slow numerical increase of the population in early summer, a peak of abundance in late summer/autumn, and a rapid decline in winter. The peaks are generally recorded in October, when the fraction of the population represented by late copepodids (CIII-CV) and adults reaches 377 ind. m^{-3} (as mean value in the period 1984–1990; Mazzocchi and Ribera d'Alcalà, 1995). During the rest of the year, the species is present in the water column with only very few individuals (1 ind. m^{-3}), although it reproduces continuously (Ianora, 1998), and resting stages are not reported in its life cycle (Mauchline, 1998). On a pluriannual scale, the seasonal pattern of *T. stylifera* population can be variable in numerical amplitude, but is very recurrent in timing (Ribera d'Alcalà *et al.*, 2004). Because of its ecological relevance in the coastal zooplankton communities, this species has also been studied for its reproductive biology, and life history data of the populations in the Gulf of Naples have been published in recent years (e.g., Ianora and Poulet, 1993; Ianora, 1998; Carotenuto *et al.*, 2002). Data on egg production and viability, as well as development times of nauplii and copepodids have been used for reconstructing the population dynamics in the present work. Unfortunately, stage-specific mortality rates are still not available for larval stages and adults of *T. stylifera*, and data available for another pelagic copepod (Ohman and Wood, 1996; Asknes and Ohman, 1996) are used here for our modeling.

Plankton dynamics is subject to the action of the physical environment, which can interact with the development and the distribution of the organism through mechanical processes like advection, mixing upwelling and down-welling. Thus, the rates of the bioecological processes may change because of the different environmental conditions (in particular due to the temperature and food) experienced by the individuals. However, in this work we do not consider these dynamical factors to focus exclusively on the biological aspects, and we simulate the population evolution in a motionless medium at constant temperature.

In the present application, which is based on published data, we perform numerical simulations of

(i) the Malthusian growth of this copepod species during an annual cycle, in a generic situation at 20° C, starting from a reproductive female in a reference volume of 1 m^3, and
(ii) the limits to the exponential growth, i.e. the transition from the maximum level of abundance reached at 20° C to a low level at 15° C.

12.5.1 Malthusian growth

The individuals were grouped in six stages: stage 1 (eggs), stage 2 (NI), stage 3 (NII-NVI), stage 4 (CI), stage 5 (CII-CV), stage 6 (adult females and adult males). Nauplius NI and copepodid CI have been separately considered because they are generally reported as the most critical stages. The data used in the numerical simulation are reported in Table 12.1 (* from Carotenuto *et al.*, 2002; ** from Ianora, 1998; † from Ohman and Wood, 1996; ‡ from Asknes and Ohman, 1996).

Moreover, we have assumed a sex ratio $\rho = 0.5$, a constant reproductive profile f_0, $L_m = L_f/2$, $\mu^3 = \mu^2$, and that only late copepodids and adults were affected by predation; thus, μ^5 and μ^6 represent the rates due to physiological and predation mortality. The assumed levels of uncertainties, in terms of standard deviations, are in the range 0–20% of the average values.

Ianora and Poulet (1993) estimated a mean annual egg production rate of 50.9 ± 29.7 eggs female^{-1} d^{-1}. Taking $f_0 = 50$ eggs female^{-1} d^{-1}, together with the values in Table 12.1 for the other parameters, numerical simulations lead to unrealistic abundance of the population. This is supported by the

Table 12.1 *Values of the parameters.* D^i, L_f *in d;* μ^i *in* d^{-1}.

D^1	D^2	D^3	D^4	D^5	L_f	μ^1	μ^2	μ^3	μ^4	μ^5	μ^6
1 *	1 *	8 *	3 *	10 *	20 **	0.27 **	0.1 †	0.1	0.17 ‡	0.15 ‡	0.2

Table 12.2 f_0 *in eggs female*$^{-1}$ d^{-1}, $\hat{\lambda}$ *in* d^{-1}.

f_0	$f_0 L_f$	$f_0 e^{-\mu^1 D^1}$	$f_0 e^{-\mu^1 D^1 - \mu^2 D^2}$	$\hat{\lambda}$
25	500	19.1	17.3	0.034
50	1000	38.2	34.6	0.060

values of the quantities in Table 12.2, where $\hat{\lambda}$ in d^{-1} is the theoretical Malthusian parameter, i.e. the principal eigenvalue of the discrete Leslie matrix approximation associated to the Eulerian model described in Section 12.4.4.

Thus, in the numerical simulations of the exponential growth we are forced either to increase some mortality rates or to decrease the average production of eggs. For simplicity we have chosen the second alternative, and we have assumed $f_0 = 25$ eggs female^{-1} d^{-1}.

The Malthusian growth of a natural population describes the initial phase of the dynamics. Nevertheless, in order to correctly describe the asymptotic behavior, we perform numerical simulations of a long term population dynamics (180 days), considering population abundances probably exeeding the “carrying capacity” of the environment. Time steps $\Delta t = 0.1, 0.2$ d are used in the numerical simulations. At least twenty realizations of the time evolution of the population are performed, and the average outputs are reported; the convergence velocity depends on the levels of uncertainties (each realization consists of about 10 000 life histories of individuals). The convergence of the Lagrangian outcomes to the Eulerian one is illustrated by averaging $N^6(t)$ over different numbers of realizations (Figure 12.3).

The time evolution of the individuals $N^i(t)$ in each stage and of the total population $N(t)$ (Figures 12.4–12.6) reach an exponential growth in about 150 days. The oscillations reflect the succession of different generations. The value of the Malthusian parameter obtained by a best fit of $N(t)$ ($\lambda = 0.033$ d^{-1}) is in very good agreement with the theoretical value reported in Table 12.2. The asymptotic distribution $\Phi(t, x)$, $x \in [0, n]$, defined by

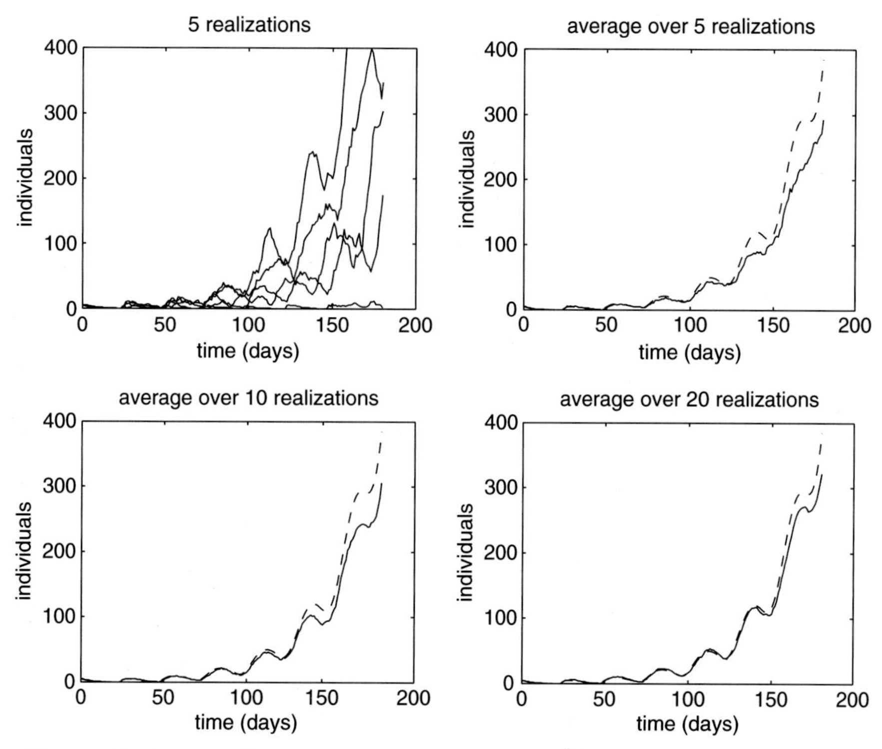

Figure 12.3 Lagrangian model: 5 realizations of $N^6(t)$ and averages over 5, 10, 20 realizations, continuous line; $N^6(t)$ from Eulerian model, dashed line.

$$\Phi(t, x) = v^i \phi(t, x) \qquad x \in (i-1, i],$$

is continuous, but with discontinuous derivatives, at the points $x = i$, $i = 2, \ldots, n-1$ (see Section 12.4.1); it can be approximated by $v^i N_k^i(t)$, $k \in (i-1, i]$, where $N_k^i(t)$ is defined in (12). The plot at 180 days (Figure 12.7) shows that a stable distribution is not yet reached in stages 3 (NII-NVI) and 4 (CI).

12.5.2 Limits to exponential growth

In the numerical simulations of the annual cycle we have to take into account the driving forces governing the population growth, and caused by the variability of the environmental conditions (for example, a decrease of the temperature produces a decrease in the rates of some biological processes, as the development process of an individual). Thus, a robust set of data and their dependence on environmental variables (mainly on temperature) should be available. Here we will consider only some processes which are able to limit the

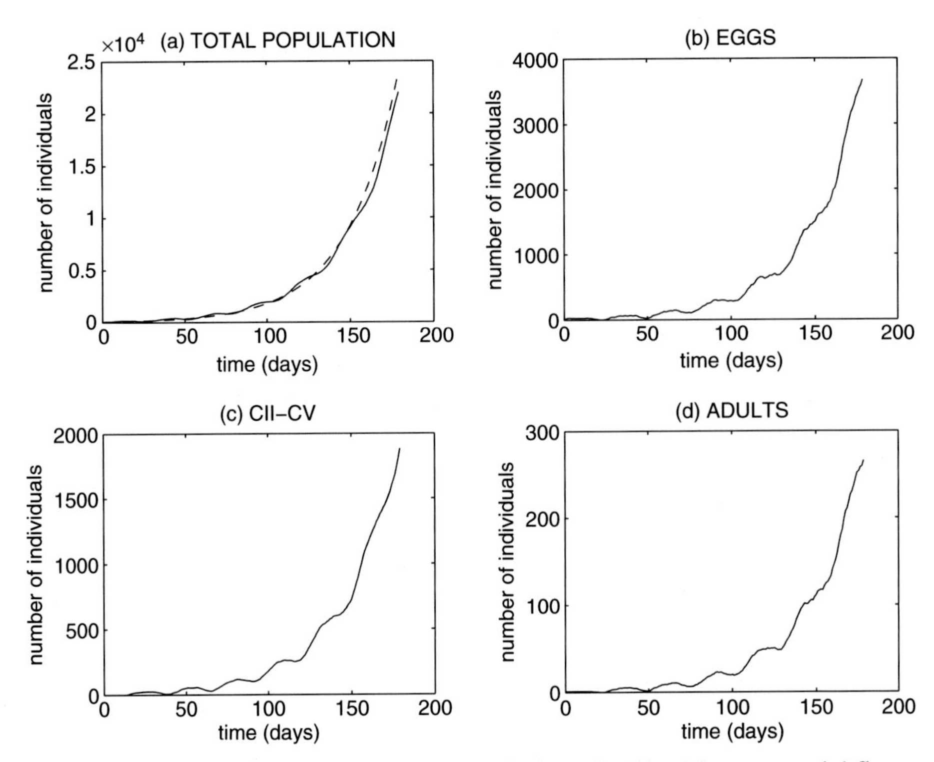

Figure 12.4 Malthusian growth: time evolution of $N(t)$ with exponential fit, $N^1(t)$, $N^5(t)$, $N^6(t)$

exponential growth, and simulate them by an abrupt change of some parameters after 180 days of Malthusian growth.

An attempt is to decrease the average development rates $\nu^i = 1/D^i$ by increasing the duration D^i as the temperature decreases. Ianora (1998) reports that $D^1 \simeq 1$ d at 20° C and $D^1 \simeq 2$ d at 15° C. Let us assume that for $t > 180$ days the values of the development times in Table 12.1 are increased by the factors $1 + \epsilon^i$. Assume $\epsilon^1 = \epsilon^2 = 1$, i.e. we double the development times of eggs and NI. In Table 12.3 are shown the theoretical values of $\hat{\lambda}$ by varying ϵ^i for $i > 2$.

From the values of $\hat{\lambda}$ in Table 12.3 it can be observed that there is no possibility for a decrease of the population size, unless large values of development times are used. This is not a general fact; it depends on the given set of parameters.

Assume that in the copepod population that we have considered, the egg cannibalism is practiced by juveniles (stage 5) and adults (stage 6) . Following Ohman and Hirche (2001), we express the eggs mortality rate as

$$\mu^1(\chi(t)) = \mu_0^1 + \mu_1^1 \frac{\chi(t)}{\chi^*},$$

Table 12.3 *Values of* $\hat{\lambda}$ *in* d^{-1} *for different values of* ϵ^i, $i > 2$.

ϵ^i, $i > 2$	$\hat{\lambda}$
0	0.02
0.1	0.009
0.2	$\simeq 0$
0.25	−0.004

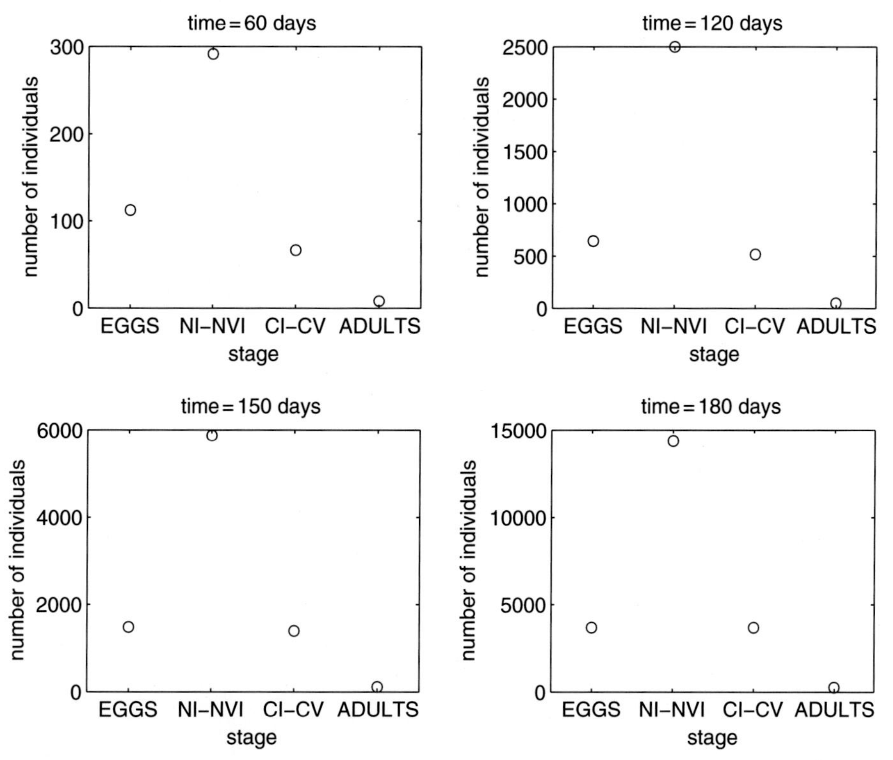

Figure 12.5 Malthusian growth: time evolution of the distribution $[N^1(t), N^2(t) + N^3(t), N^4(t) + N^5(t), N^6(t)]$.

where $\chi(t) = N^5(t) + N^6(t)$, χ^* is a reference value, and μ_0^1, $\mu_1^1 \chi(t)/\chi^*$ are the constant and density dependent mortality rates, respectively. Let $\chi^* = 100$ individuals m^{-3}, $\mu_0^1 = 0.15\ d^{-1}$, $\mu_1^1 = 0.1\ d^{-1}$, together the following values of ϵ^i: $\epsilon^1 = \epsilon^2 = 2$, $\epsilon^i = 0$ *for* $i > 2$. Then, the temporal trend of $N(t)$ (Figure 12.8)

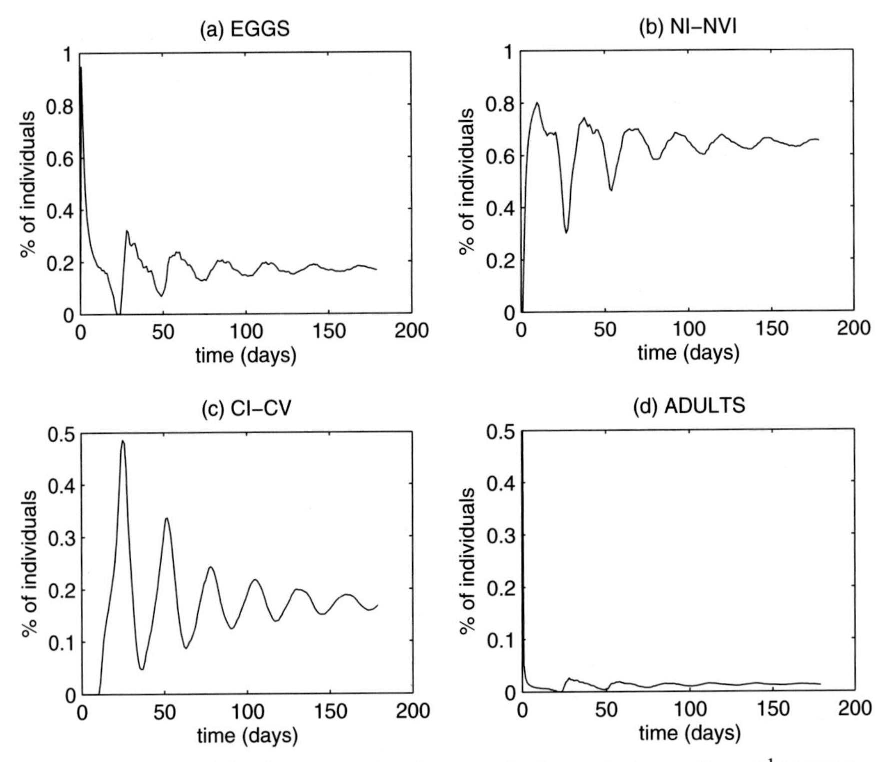

Figure 12.6 Malthusian growth: time evolution of the ratios $N^1(t)/N(t)$, $(N^2(t)+N^3(t))/N(t)$, $(N^4(t)+N^5(t))/N(t)$, $N^6(t)/N(t)$.

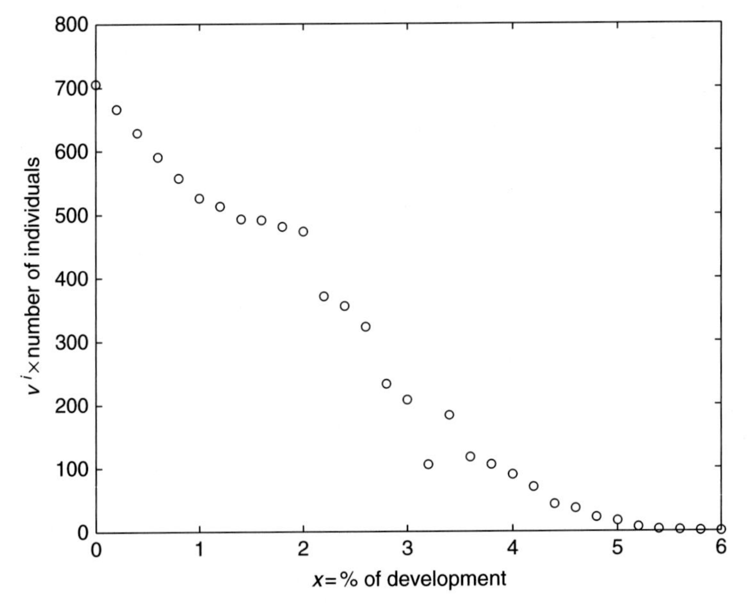

Figure 12.7 Malthusian growth: distribution $v^i N^i_k(t)$, $k \in (i-1,\ i]$, at time $t = 180$ days.

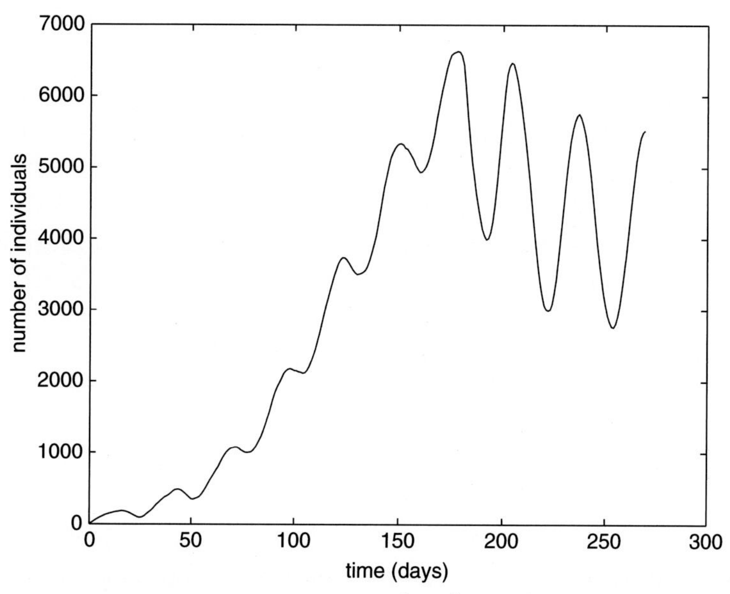

Figure 12.8 Time evolution of $N(t)$, for $\epsilon^1 = \epsilon^2 = 1$, $\epsilon^i = 0$ for $i > 2$, and with density dependent egg mortality.

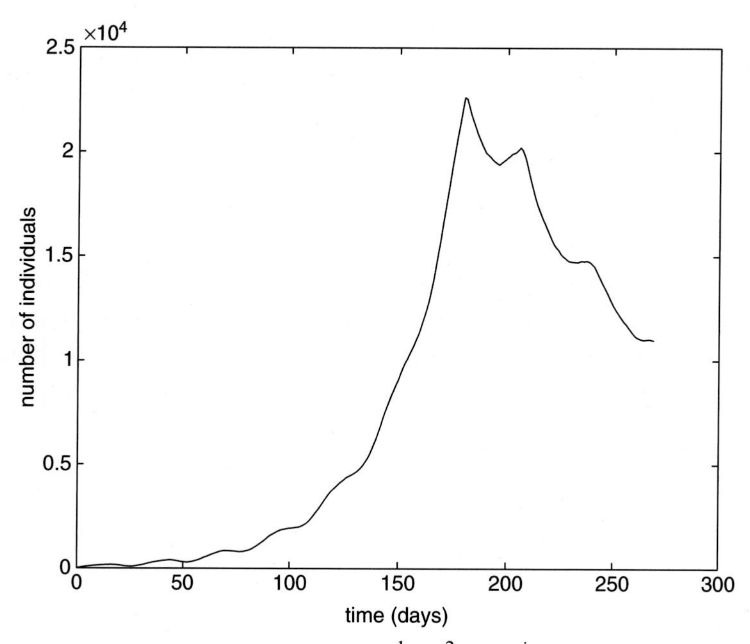

Figure 12.9 Time evolution of $N(t)$ for $\epsilon^1 = \epsilon^2 = 1$, $\epsilon^i = 0.2$ for $i > 2$, and with production of eggs given by $0.8 f_0$.

shows, after 180 days, a decrease of the mean population size with extensive oscillations.

Let us now assume that $\epsilon^1 = \epsilon^2 = 1$, $\epsilon^i = 0.2$ *for* $i > 2$, and that the production of eggs after 180 days is reduced by 20%, i.e. the actual production of eggs is $0.8 f_0$. In this situation $\hat{\lambda} = -0.007 \text{ d}^{-1}$, and the total population after 180 days is halved in 90 days (Figure 12.9).

The processes previously described can simultaneously occur, and other processes can influence the decrease of the population; therefore, the population size could decrease more considerably. At present we can only formulate suppositions until a complete and homogeneous set of biological data is available for the target species.

12.6 Concluding remarks

The development process of an individual may be described by equivalent advection–diffusion models: in either Eulerian or Lagrangian formulation. For some problems, Eulerian models allow the theoretical estimation of relevant parameters (such as time and age constants) which can be used to test best fits of results obtained by Lagrangian models. Moreover, the theoretical analysis of a Lagrangian model is often based on that of Eulerian equations. The numerical implementation of the Eulerian models needs some work of construction and analysis of algorithms, while that of the Lagrangian is immediate. In general, both of them can be successfully applied and produce results (temporal trends and age distributions) in good agreement. However, when the reproduction process of the population is very efficient, the computational cost of the numerical simulation of an exponential growth by a Lagrangian model may be very high. The Eulerian model equivalent to the Lagrangian model defined by equations (12.9)–(12.10) needs the introduction of distributions $\phi^i(t, x, v)$ in three independent variables: time, physiological age, development rate. In this case the numerical implementation of the Eulerian model is not quite reliable; moreover, its computational cost is high, while the application of (12.9)–(12.10) is again straightforward.

In the application to the dynamics of a natural population the availability of a "consistent" set of experimental data, able to describe the real time evolution of the population at least for a limited time period, is desirable. With "consistent" we mean that data should be relative to the same species, and to specified environmental conditions (e.g. temperature). In general the model stage structure does not correspond to the real structure of the population, which consists of many biological instars, characterized by different demographic, metabolic and trophic behavior; a simplified structure, to be used in modeling population dynamics, should be defined by grouping different instars in a few stages: individuals showing common behavior with respect to feeding, development, reproduction are grouped in the same stage. When some parameter takes a critical value, for example when the Malthusian parameter $\lambda \simeq 0$, the dynamical behaviour of the system may change abruptly for small variations of some data; in this case the knowledge of the uncertainties of the data is of crucial importance. Modeling the complete annual cycle of a population needs a set of parameters depending on the environmental conditions; furthermore, it makes a practice of regularizing the data by means of fitting procedures with smooth curves, in order to eliminate spurious fluctuations and to reproduce the observed main trends.

Acknowledgments

We would like to thank the referees for their useful suggestions. This work was partially supported by the Italian National Project Ambiente Mediterraneo – SINAPSI funded by MIUR.

References

Abbiati, M., G. Buffoni, G. Caforio, G. Di Cola, and G. Santangelo, 1992. Harvesting, predation and competition effects on a red coral population. *Netherlands J. Sea Res.*, **30**, 219–28.

Asknes, D. L. and M. D. Ohman, 1996. A vertical life table approach to zooplankton mortality estimation. *Limnol. Oceanogr.*, **41**, 1461–9.

Berec, L., 2002. Techniques of spatially explicit individual-based models: construction, simulation, and mean-field analysis. *Ecol. Model.*, **150** (2002), 55–81.

Bergh, M. O. and W. M. Getz, 1988. Stability of discrete age-structured and aggregated delay-difference population models. *J. Math. Biol.*, **26**, 551–81.

Buffoni, G. and A. Cappelletti, 2000. Size structured populations: dispersion effects due to stochastic variability of the individual growth rate. *Mathematical and Computer Modelling*, **31**, 27–34.

Buffoni, G., G. Gilioli, and S. Pasquali, 2002. *Population Dynamics: a Stochastic Model Based on Individual Life History Data.* IAMI Technical Report, 07–02.

Buffoni, G. and S. Pasquali, 2003. Structured population dynamics: Eulerian and Lagrangian approachs. In Proceedings of the Fourth International Conference "Tools for Mathematical Modelling", Saint Petersburg, June 23–28, 2003, ed. G. S. Osipeuko. *Electronic Journal "Differential Equations and Control Processes"*, **9**, 74–86.

Buffoni, G., S. Pasquali, and G. Gilioli, 2004. A stochastic model for the dynamics of a structured population. *Discrete and Continuous Dynamical Systems Series B*, **4**(3) (2004), 517–25.

Buffoni, G. and S. Pasquali. Structured population dynamics: continuous size and discontinuous stage structure. Submitted to *J. Math. Biol.*

Carotenuto, Y., A. Ianora, I. Buttino, G. Romano, and A. Miralto, 2002. Is postembryonic development in the copepod *Temora stylifera* negatively affected by diatom diets? *J. Exp. Mar. Biol. Ecol.*, **276**, 49–66.

Carpi, M. and G. Di Cola, 1988. Un modello stocastico della dinamica di una popolazione con struttura di età. *Quaderno del Dipartimento di Matematica dell'Università di Parma*, n. 31.

Curry, G. L. and R. M. Feldman, 1987. *Mathematical Foundations of Population Dynamics*. College Station, Texas: Texas AM University Press.

Estrada, M., F. Vives, and M. Alcaraz, 1985. Life and productivity of the open sea. In *Western Mediterranean*, ed. R. Margalef. Oxford: Pergamon Press, 148–97.

Feller, W., 1957. *An Introduction to Probability Theory and Its Applications*, Volume I. New York: J. Wiley.

Gardiner, C. W., 1994. *Handbook of Stochastic Methods*. Berlin: Springer-Verlag.

Hoppensteadt, F., 1975. *Mathematical Theories of Populations: Demographics, Genetics and Epidemics*. Philadelphia: Society for Industrial and Applied Mathematics.

Huys, R. and G. A. Boxshall, 1991. *Copepod Evolution*. London: The Ray Society.

Ianora, A., 1998. Copepod life history traits in subtemperate regions. *J. Mar. Sys.*, **15**, 337–49.

Ianora, A. and S. A. Poulet, 1993. Egg viability in the copepod Temora stylifera. *Limnol. Oceanogr.*, **38**, 1615–26.

Judson, O. P., 1994. The rise of the individual-based model in ecology. *Tree*, **9**, 9–14.

Karlin, S. and H. M. Taylor, 1981. *A Second Course in Stochastic Processes*. New York: Academic Press.

Mauchline, J., 1998. The biology of calanoid copepods. *Adv. Mar. Biol.*, **33**, 1–710.

Mazzocchi, M. G. and M. Ribera d'Alcalà, 1995. Recurrent patterns in zooplankton structure and succession in a variable coastal environment. *ICES J. Mar. Sci.*, **52**, 679–91.

Munholland, P. L. and B. Dennis, 1992. Biological aspects of a stochastic model for insect life history data. *Environ. Entomol.*, **21**(6), 1229–38.

Ohman, M. D. and H. J. Hirche, 2001. Density dependent mortality in an oceanic copepod population. *Nature*, **412**, 638–41.

Ohman, M. D. and S. N. Wood, 1996. Mortality estimation for planktonic copepods: *Pseudocalanus newmani* in a temperate fjord. *Limnol. Oceanogr.*, **41**, 126–35.

Ribera d'Alcalà, M., F. Conversano, F. Corato, P. Licandro, O. Mangoni, D. Marino, M. G. Mazzocchi, M. Modigh, M. Montresor, M. Nardella, V. Saggiomo, D. Sarno, and A. Zingone, 2004. Seasonal patterns in plankton communities in a pluriannual time series at a coastal Mediterranean site (Gulf of Naples): an attempt to discern recurrences and trends. *Sci. Mar.*, **67**(Suppl. 1), 65–83.

Roff, D. A., 1992. *The Evolution of Life Histories*. New York: Chapman and Hall.

Thomson, D. J., 1987. Criteria for the selection of stochastic models of particle trajectories in turbulent flows. *J. Fluid Mech.*, **180**, 529–56.

Zambianchi, E. and A. Griffa, 1994. Effects of finite scales of turbulence on dispersion estimates, *J. Mar. Res.*, **52**, 129–48.

13

Lagrangian analysis and prediction of coastal and ocean dynamics (LAPCOD)

ARTHUR J. MARIANO AND EDWARD H. RYAN
Rosenstiel School of Marine and Atmospheric Science, University of Miami, Miami, Florida, USA

13.1 Introduction

It was during the 1999 Liege Colloquium on, "Three-Dimensional Ocean Circulation: Lagrangian measurements and diagnostic analyses," that a number of researchers started to discuss the idea of having a meeting centered on studying the ocean, the atmosphere, and marine biology from a Lagrangian viewpoint. At the time of this writing, three Lagrangian Analysis and Prediction of Coastal and Ocean Dynamics (LAPCOD) meetings have been held in (i) Ischia, Italy from October 2–6, 2000, (ii) Key Largo, FL, USA from December 12–16, 2002, and (iii) Lerici, Italy from June 13–17, 2005. The LAPCOD meetings bring together a diverse group of scientists for the purpose of exchanging ideas on the collection, analysis, modeling, and assimilation of coastal and oceanic (quasi-)Lagrangian data. The purpose of this chapter is to provide both a tutorial for readers who are not specialists, and a summary of the material presented at the LAPCOD meetings and in this book. Since this chapter summarizes the material presented at LAPCOD meetings and because of space constraints, many important Lagrangian-based studies are not detailed here and the chapter topics, listed in the next paragraph, are those topics that have been central to the LAPCOD meetings and this book. There are a number of unpublished results presented at LAPCOD 2005 discussed here and referenced by personal communication, hereafter pers. com.

The material presented at LAPCOD meetings and in this book, as well as, Molinari and Kirwan (1975), Richardson (1976), Rossby *et al.* (1983), Kirwan *et al.* (1984), Davis (1985a), Brink *et al.* (1991, 2000), Davis (1991), Owens (1984, 1991), Bower and Lozier (1994), Hitchcock *et al.* (1997), Bauer *et al.* (1998), Boebel *et al.* (1999), Richardson *et al.* (2000), Niiler (2001), Davis and Zenk (2001), Fratantoni (2001), Zhang *et al.* (2001), Grodsky and Carton (2002), Mariano *et al.* (2002), Centurioni and Niiler (2003), Reverdin *et al.* (2003), Lumpkin (2003), Nùñez-Riboni *et al.* (2005), and Lavender *et al.* (2005),

Lagrangian Analysis and Prediction of Coastal and Ocean Dynamics, ed. A. Griffa, D. Kirwan, A. Mariano, T. Özgökmen, and T. Rossby. Published by Cambridge University Press.

clearly demonstrate that Lagrangian data are ideal for mapping the general circulation and variability of the oceans and marginal seas (Sections 13.2 and 13.3), for highlighting the role of topographical and potential vorticity constraints on geophysical and coastal flows (Section 13.4), for studying dispersion (Sections 13.5 and 13.7) and particle pathways (Sections 13.2 and 13.8), for providing insight into biological systems (Section 13.6), for assimilating into Eulerian and Lagrangian prediction models (Section 13.9), and for providing unique dynamical signatures of ocean and coastal processes (Section 13.10).

The fundamental observation provided by ocean drifters and subsurface floats is a set of positions following a "tagged fluid parcel" in time. This set of positions is a one-dimensional trajectory in a two-dimensional state space, usually defined by longitude and latitude or along-stream and cross-stream distance. The ability of drifters and floats to sample large, and in many cases, heterogeneous, two-dimensional slices (e.g. sea surface, isobaric, or isopycnal surface) of the ocean is one of the primary advantages of Lagrangian-based sampling. This ability is clearly evident in recent papers on the mid-depth circulation of the North East Atlantic (Bower *et al.*, 2002b), the general circulation of the North Atlantic surface currents (Reverdin *et al.*, 2003) and the Equatorial Atlantic (Grodsky and Carton, 2002), and in individual trajectories, such as those shown in Chapter 3. During the three years of its lifetime, Drifter 1611 (Grodsky and Carton, Chapter 3, this volume), for example, sampled the North Equatorial Countercurrent, the Guinea Current, the Angola Current, the South Equatorial Current from about 5° E to 35° W (>4000 km), the North Brazil Current, and was entering the Guiana Current when it stopped transmitting. The first comprehensive, nearly global map of the average near-surface circulation of the world's ocean from modern drifters was presented by Niiler (2001) and see Figure 13.1. In the North Atlantic, regional maps of the mean subsurface circulation are now being constructed with confidence (Bower *et al.*, 2002a,b) from RAFOS and SOFAR float data bases.

Other variables, especially temperature, salinity, and optical properties, are being measured by drifters and floats. These measurements are being used to calibrate satellite-based temperature and ocean color observations. Profiling floats have allowed oceanographers to make hydrographic measurements in parts of the world's oceans that have been extremely data sparse (see Section 13.11) and in weather conditions that would have kept ships in port. Autonomous gliders (Webb *et al.*, 2001) allow scientists to adaptively sample the ocean using a laptop to direct the gliders from thousands of miles away. As sensor and computational technology evolves, Lagrangian-based observations

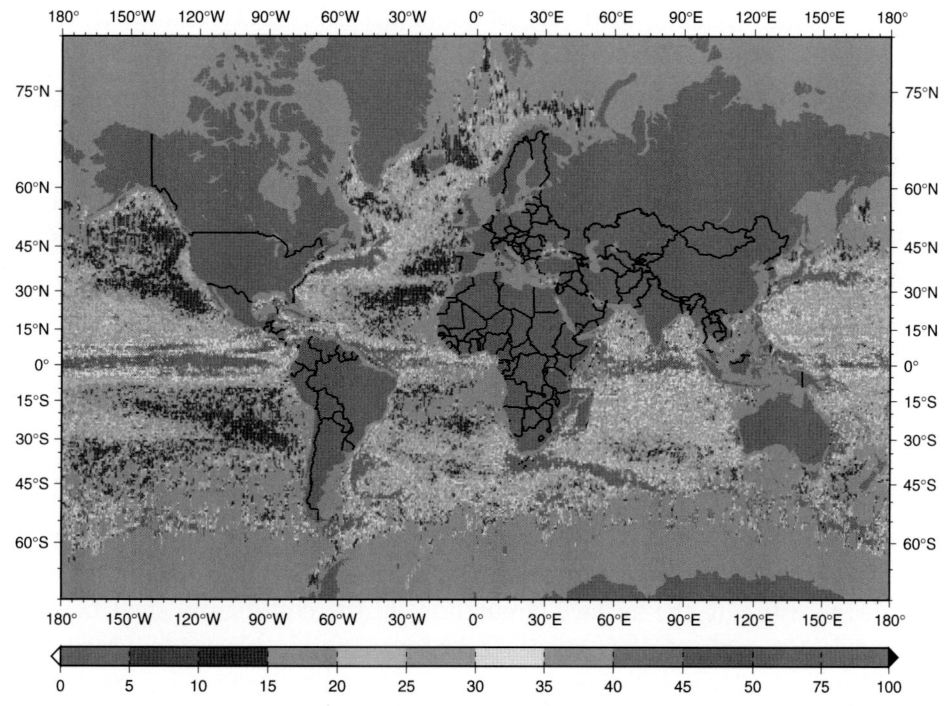

Figure 13.1 The average speed of sea surface currents, calculated from the climatological data base of the near-surface drifters, archived at AOML and detailed in Chapter 2 by Lumpkin and Pazos. The major ocean currents, along the western edge of ocean basins, and in both the equatorial and polar oceans are clearly visible in this global map. The average speed of the drifters is on the order of 10 cm/s in gyre interiors, 35 cm/s in regions close to the major currents and along the eastern boundaries of the ocean basins, and the average speed is on the order of 100 cm/s in the major ocean currents. See Plate 28 for color version.

of the ocean and its marginal seas have become more routine, more reliable, and more cost-effective.

13.2 The mean flow and flow variability

Tracked Lagrangian instruments are ideal for mapping the mean fluid flow on local to regional to global scales. The first comprehensive data set for determining the mean ocean circulation is the Maury Ship Drift Data, named after Lt. Matthew Fontaine Maury. Lt. Maury inaugurated the tabulation of ship drift in the 1840s and maintained the data for two decades. Sea surface velocity can be estimated from ship drift as the vector difference between dead-reckoning position and the actual position traveled over a 12 to 24 hour period.

Since velocities are computed over 24 hours, the effects of tides and other high frequency signals are effectively averaged out. The major error source is due to windage effects on ships, and that the data is spatially aliased along shipping routes and temporally aliased with much less data in winter (Wyrtki *et al.*, 1976; Richardson and McKee, 1984; Richardson, 1989).

Average fluid pathways, based on just the initial launch location and final recovery location, have been computed for drifting bottles and drift cards. Either the bottle or a plastic envelope contain a card that is mailed back, usually by a beachcomber. Drift cards are also made of wood or plastic with the return information imprinted on them. Many tens of thousands of drift cards have been launched over the last 60 years all over the world's oceans (Brodie, 1960; Ebbesmeyer and Coomes, 1993). A number of drifter bottles studies were conducted off of California starting in the 1930s as part of California Cooperative Oceanic Fisheries Investigation (CalCOFI) program. Burt and Wyatt (1964) used drift bottles to study the Davidson Current off of Oregon (Neuman, 1968). Bumpus and Lanzier (1965) used drift cards and bottles to estimate the surface circulation on the continental shelf between Florida and Canada. Cuban scientists deployed orange, plastic drift cards into the Florida Straits and Cuban coastal waters to study the energetic currents in this area, but unfortunately that data remains classified. Richardson and Stommel (1948) used parsnip to study fluid dispersion in Loch Long, Scotland (see Summers (2005)) for a new interpretation of this classical experiment). Oceanographers have also used sneakers, rubber ducks, toy animals, small model ships, large derelict ships, computer cards, water-filled balloons, copy paper, and other floating objects (e.g. Aliani *et al.* 2003) to study coastal and ocean flows from a Lagrangian perspective. Some great stories of long-distance floating objects can be found in a popular magazine for beachcombers called, *The Drifting Seed* (see, for example, September, 2003, Vol. 9, No. 2).

The first comprehensive mapping of the surface velocity field using tracked surface float was by Mosby, in 1954 (see Neumann, 1968) for Tromsö Sound. Reid *et al.* (1963) observed a coastal eddy with radar-tracked drifters. Davis (1985a, b) seeded the California Current in the early 1980s with surface drifters that were radio-tracked from shore and aircraft. Detailed tracking of surface floats required triangulation using fixed landmarks or ship-tracking and "high-precision" navigation like Loran. The advent of the satellite-based Service Argos system, starting in the 1970s, allowed drifters to be satellite-tracked over most of the ocean (Kirwan *et al.*, 1976, 1978; Cresswell, 1977; Cheney *et al.*, 1980; Richardson, 1980, 1981; McNally *et al.*, 1983; Davis, 1991). Drifters were primarily developed and tested at Scripps Institute of Oceanography (SIO),

MIT, and NOAA's Atlantic Ocean and Meteorological Laboratory (AOML) (Niiler *et al.*, 1987, 1995; Bitterman and Hansen, 1989, 1993). Chapter 2 by Lumpkin and Pazos details the development of (near-) surface drifters.

Work at SIO and WHOI on drogue design in the 1950s and 1960s improved the measurement of subsurface currents (Volkman *et al.*, 1956; Volkman, 1963; Pochapsky, 1961, 1963, 1966; Stalcup and Parker, 1965). As discussed in Chapter 1 by Rossby, Swallow (1955, 1957, 1971) revolutionized the field of oceanography with his neutrally buoyant float measurements starting in the 1950s, and starting in the late 1960s, Doug Webb, Tom Rossby, Russ Davis, and many others developed the SOFAR, RAFOS, and profiling float technology that is giving oceanographers a better understanding of the mean subsurface circulation and its variability.

After the success of the SOFAR float design, Webb and Davis designed a float that did not need land-based or autonomous listening stations. ALACE, the Autonomous LAgrangian Current Explorer, would float at depth and after a preprogrammed time, the float would rise to the surface and be tracked by Service Argos (Davis *et al.*, 1992). Davis first added a temperature sensor, and later on, a salinity sensor was added and Profiling-ALACE, or P-ALACE, floats started collecting hydrographic data from remote areas of the world's oceans. The newer APEX (Autonomous Profiling EXplorer) floats have become the workhorse of the international oceanographic community in the ARGO program, yielding a wealth of hydrographic profiles that would have been more costly and much more dangerous to collect by ship (Gould *et al.*, 2001; Roemmich *et al.*, 2001; Freeland and Cummins, 2005; Gould, 2005). Other ARGO float designs include IFREMER-designed Provor profiling float available from Metocean and Martec.

Lagrangian-based observations of coastal and ocean flows for more than fifty years, on local to global scales, have matured from prototype, one-of-a-kind instruments and solo experiments to "operational," buy-off-the-shelf instruments, and multi-national global experiments such as ARGO. A number of different sets of observations shown at the LAPCOD meetings are discussed next. In some situations, an investigator will not know if their Lagrangian observations sample local or global regions. Near-surface drifters were launched, bi-monthly, in the Shark River Plume (SW FL, 25.35° N, 81.23° W) in Florida Bay. These drifters, on average, travel through Florida Bay and the passages between the Florida Keys into the Dry Tortugas. Some drifters enter the Florida Current and travel thousands of km in the Gulf Stream System, while others remain local to the near-shore Florida coastal waters (Kourafalou *et al.*, Chapter 3). Seasonal shifts of the winds from northeast in fall/winter to southeast in the summer increase the pathways of drifters. They also found that about 75% of the subtidal motion of the drifters can be explained by the local wind forcing. These results and other results presented at the meetings indicate that seasonal wind forcing leads to nonstationary velocity statistics for near-surface drifters.

During the last decade and a half, a number of experiments deployed drifters in various basins of the Mediterranean Sea (Poulain, 2001; Falco *et al.*, 2000). Some of the major field programs have included the DOLCEVITA program that deployed surface drifters in the northern Adriatic. Many of the drifters had sampling rates of 30 to 60 minutes and were deployed in triplets within ten km of each other. In a time period of over one year that ended in the fall of 2003, 120 drifters were deployed. The drifters mapped out a clear picture of the surface flow field in the Adriatic with sub-basin scale gyres such as the northern cyclonic gyre, strong coastal jets such as the West Adriatic and the East Adriatic Currents, and energetic eddies, all seasonally modulated with minimum energy in the summer and maximum in the fall and winter. A quasi-stationary, stagnation point, defined by a small area of no significant mean flow and low eddy kinetic energy, was also identified near the Istria (Croatia) peninsula. With an average deployment rate of about ten drifters per month, all of the primary and energetic circulation features, as well as the quiescent, less energetic regions were identified and their variability quantified.

Thirty drifters were launched in the northern Aegean starting in March 2002 by Kourafalou, Johns, Kontoyiannis, Olson, and Zervakis. Trajectories from the Aegean Sea Pilot Drifter Program indicate that the flow is dominated by eddies trapped in deep sub-basins, there is a coastal jet on the western boundary, inflow to the Aegean Sea on the eastern margin, and outflow being focused by the Cyclades islands. The drifters were energetic over many scales; inertial, mesoscale, and basin-scale (Olson *et al.*, 2005). Thirty CODE drifters launched in the Tyrrhenian Sea exhibited strong eddy activity, particle trapping in the southeastern portion of the Tyrrhenian, and strong northward flow along the Italian coast. This data set verified the hypothesis of a basin-scale cyclonic gyre in the southern sector of the basin, a seasonally modulated, and wind-induced anticyclonic gyre in the northern sector. These Lagrangian measurements showed that the residence time is longer than estimates provided by simple theories, presumably due to observed particle trapping by basin-scale gyres.

Azol and Zambianchi (pers. com.) analyzed 50 CODE drifters that sampled the sub-basin scale gyres in the southern and northern Ligurian Sea. Trapping of particles in these gyres limits the connectivity between the two sub-basins. Testor and Gascard (2005) analyzed floats that were in the Levantine

Intermediate Water at 600 m and the floats clearly mapped out the cyclonic Algerian gyre, a boundary current along the slope south south of Sardinia, and an energetic, anticyclonic eddy. In the Black Sea, (Barbanti and Poulain, pers. com.) there are two cyclonic sub-basin gyres that occupy the western and eastern Black Sea and are part of the Rim Current. Drifter statistics are different in the Rim Current versus gyre interior and the velocity variance is very heterogeneous. All of these drifter data sets, as well as many other studies cited here, clearly illustrate the ability of drifters and floats to sample many different flow regimes and to give us insight into fluid dynamics at many different scales of motion.

The seasonal cycle of the near-surface velocity field of the North Equatorial Countercurrent (NECC) in the Eastern Tropical Pacific was calculated from over 200 drifters by Kennan *et al.* (pers. com.). Maximum eastward zonal flow occurred in June and July with maximum speeds of over 100 cm/s. The speed of the current decreases in the late fall and by early spring, it is difficult to track a continuous NECC. The seasonal cycle is very step-like, with large increases in surface velocity starting in May and then a gradual decline in the average current speed. The drifter data also revealed that the NECC is a fairly narrow current, in contrast to climatology, and that there are significant changes in the horizontal velocity shear of the current from south to north that needs to be accounted for in any large-scale vorticity balance. The ability of Lagrangian drifters to sample over large horizontal areas allows one not only to estimate the large-scale horizontal mean circulation, but also the large-scale velocity gradients needed for energy, momentum, and vorticity analyses.

Isern-Fontanet *et al.* (2006) analyzed eight years of altimetric sea surface height data, starting from October 1992, in order to build a census of Mediterranean eddies. The Algerian Basin and Levantine basin had the most mesoscale eddies, in agreement with an analysis based on AVHRR images. Eddies could be classified as strong (energetic mesoscale eddies) or weak, depending on the value of the Okubo–Weiss parameter (Okubo, 1970; Weiss, 1991). The Lagrangian velocity probability density function of the detected Mediterranean eddies is non-Gaussian and this property is due to intense vortices. The Okubo–Weiss parameter was also used by Pasquero *et al.* (2002, 2004) to identify vortices in numerical simulations of oceanic turbulence and they also found similar non-Gaussian behavior. Thus one would expect to find more non-Gaussian behavior near western boundaries and more Gaussian behavior in the tropics (LaCasce, pers. com.).

The census of eddies obtained from an Eulerian viewpoint, such as satellite images of the Mediterranean Sea and in global data sets (Yang *et al.*,

2004) that show many eddies of various sizes, basically agrees with what is seen in Lagrangian drifters. The ocean, its marginal seas, and coastal zones are populated with energetic eddy motion with a broad-spectrum of scales with a preference of scales centered about the baroclinic Rossby radius of deformation and scales set by basin geometry, bottom topography, and coastal morphology.

Twenty-five float years of North Atlantic RAFOS float data on the $\sigma_\theta = 27.5$ surface were analyzed for eddies. Richardson's looping criterion, that a float trajectory contains at least two consecutive loops in the same direction, was used to estimate that 16% of the floats were in mesoscale eddies. There was no significant difference in the number of clockwise and counter-clockwise rotating eddies in the whole data set, but there were some regional preferences. For instance, six anticyclonic eddies (clockwise rotating vortices) were found southwestward from the Goban Spur, a prominent bathymetric feature along the Iberian coast (Richardson *et al.*, 2000).

A number of different Lagrangian floats, drifters and profilers, as well as satellite altimetric data in the Northeast Atlantic, was combined by Le Cann *et al.* (2005) to estimate the circulation of the upper water column (500 m) and to document its mesoscale variability. The most energetic features in the south are the Meddies and to the north, "Northern Meddies." Northern Meddies are relatively warm, salty, ocean vortices, formed north of 40° N, containing central Atlantic and subpolar Mode water. Data suggests that these eddies merge by both horizontal interactions and by vertical alignment. The influence of topography on float trajectories, a recurring theme throughout the LAPCOD meetings (see Section 13.4) is clearly evident. Dramatic examples of Meddies colliding with seamounts and losing their identity were first documented by Richardson *et al.* (1989). These severe topographic interactions lead to an enhanced turbulent mixing of salinity and heat in the vicinity of the seamounts.

Many studies have demonstrated that combining drifter/float data with other data sets can be used to improve the space-time resolution of a Lagrangian data set, thus providing more detailed maps of the general circulation. Richardson and Reverdin (1987) combined surface drifters, current meters, and sea surface velocity estimated from ship drifts to analyze and document the seasonal cycle of the Atlantic North Equatorial Countercurrent surface velocity field. Kelly *et al.* (1998) combined altimetric, drifter, and current meter data to show that the dominant variability in the California Current had spatial scales of 240 to 370 km and a broad-band mesoscale signal with a dominant period of a few months. Uchida and Imawaki (2003) combined altimetric and drifter data to study the general circulation of the North Pacific. They found a decrease in the mean velocity of the Kuroshio Extension Current at 155° E is due to a bifurcation of the current, presumably induced by the Shatsky Rise. They confirm the existence of an average, eastward flowing Hawaii Lee Current and the southwestward flowing Alaska Stream. The ability of Lagrangian data to sample along the entire length of a current has greatly increased our knowledge of ocean currents all over the world.

The Atlantic cold tongue, an area of relatively cool, upwelled waters in the eastern Equatorial Atlantic Ocean was analyzed by Grodsky and Carton (2002). Their analysis of 55 near-surface drifters, plus ancillary data, finds four indirect, spatially dependent pathways for drifters to travel from the cold tongue to subtropical subduction zones, an component of the subtropical recirculation cells. These pathways are a: (i) fast (18 months) pathway from 15° W to the northern recirculation gyre via the South Equatorial Current (SEC) and the North Brazil current; (ii) slow (2 months) pathway from west of 15° W to the northern subtropical gyre via the SEC in the boreal springs; (iii) slow (24 months) pathway from south of the Equator and east of 20° W to the southern subtropical gyre via the southern branch of the SEC; and (iv) pathway from east of 20° W and north of the Equator to the Gulf of Guinea via the North Equatorial Counter Current.

Schmid *et al.* (2003) analyzed 50 profiling floats in the tropical Atlantic from June 1997 to February 2002 that drifted at a nominal depth of 800 to 1100 m, Miami Isopycnal Coordinate Ocean Model numerical simulations, and ship-based *in-situ* measurements. The measurements were dominated by annual and mesoscale (45 and 66 day spectral peaks) variability in the predominantly zonal current regime. Horizontal space scales were on the order of 500 to 1100 km. The dominant phase speed was 6 cm/s and to the west. The velocity variability can be parsimoniously modeled as a sum of two planetary waves. Perez-Brunius *et al.* (2004) combined RAFOS float data with historical hydrographic data in the North Atlantic Current (NAC)–SubPolar Front (SPF) current system. They found significant heat loss by cross-frontal exchange along the meandering NAC, the upper km of the NAC transports about 15 SV northward to the Northwest corner, and that the SPF consists of 2 branches from the bifurcation of the NAC at the Northwest corner to just west of the Charlie Gibbs fracture zone. These are just a few examples of many analyses that combine the horizontal and temporal resolution of floats with the vertical resolution of hydrography. These studies have yielded a wealth of information on ocean dynamics and are crucial for ocean modeling, in general, and for operational ocean forecasting, in particular (Chapter 9 by Chin *et al.*).

13.3 Methods for estimating mean flow and its variability from Lagrangian data

Almost all of the early maps of the mean circulation from Lagrangian data, e.g. Rossby *et al.* (1983) and Owens (1984), were constructed by averaging all the float velocities in some fixed space-time volume or bin. This technique is known as bin averaging and suffers from four problems: (1) slower floats in a bin yield more velocity averages in bins relative to faster floats that enter and leave a bin quickly (see Figure 13.2); (2) the resulting averages from bin to bin are not continuous; (3) there may be temporal aliasing due to inadequate sampling resolution (i.e., failure to resolve the Nyquist frequency whose only cure is higher sampling resolution), to seasonal sampling biases (Lumpkin, 2003) or the deployment of floats into a small number of energetic eddy features that dominates the mean flow statistics (Figs. 3 and 5 of Rossby *et al.*, 1983); and (4) spatial biasing caused by heterogeneous sampling in space and by heterogeneous diffusivity (Freeland *et al.*, 1975; Davis, 1991). These problems with bin averaging are compounded by sampling problems due to small sample sizes and the energetic, broad-band nature of ocean dynamics (Garraffo *et al.*, 2001a,b).

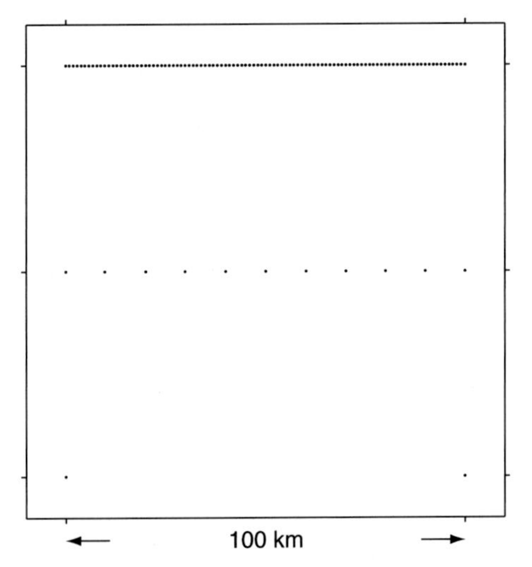

Figure 13.2 Three float trajectories in a simple, sheared flow whose velocity linearly changes from 1 cm/s to 10 cm/s over 100 km. Obviously, the number of velocity estimates in this bin contain more estimates from the northern trajectory and only a few from the southern trajectory. A simple arithmetic average using binned Lagrangian data can cause a low in the estimates of the mean flow. For example, the average speed is 5.5 cm/s in this bin but the arithmetic average from the plotted trajectories yields a velocity of 2.3 cm/s, a significantly lower, biased estimated of the average speed.

Table 13.1 *The first column is the number of float days, the second and third columns are the average horizontal velocity calculated by using a bin averaging technique and averaging along the trajectory technique, respectively.*

Total #	($u1$,$v1$)	($u2$,$v2$)
46	(−5.9, −2.5)	(−5.9, −2.2)
214	(−2.8, 0.1)	(−4.0, −0.4)
119	(−0.5, −4.0)	(−2.4, −5.1)
30	(4.5, 0.3)	(6.4, 5.0)
59	(−1.4, −1.0)	(−1.6, −1.2)
94	(−6.2, 2.9)	(−7.0, 4.5)
119	(−4.9, 0.2)	(−6.3, 0.0)
302	(−3.9, −3.0)	(−5.3, −5.6)
400	(−1.6, 0.1)	(−2.3, 0.1)
13	(−5.9, −6.7)	(−6.4, −6.5)
84	(−0.6, 3.4)	(−0.4, 4.8)
62	(1.9, 2.9)	(0.3, 2.6)
116	(−3.1, 1.3)	(−3.2, 0.4)
322	(−2.2, 0.2)	(−2.9, −0.9)
481	(−0.9, 1.1)	(−0.9, 1.4)
143	(0.1, −3.3)	(1.6, −4.6)
23	(0.1, 1.3)	(0.2, 0.9)
132	(1.1, 2.5)	(3.5, 5.8)
44	(3.5, 3.9)	(3.1, 4.6)
62	(0.5, 0.9)	(0.6, 1.4)
28	(1.7, −0.3)	(3.3, 1.7)
12	(4.9, −1.2)	(4.3, −1.4)
38	(−2.7, −1.4)	(−8.7, −2.7)
43	(−1.6, 1.9)	(−1.8, 1.4)
33	(1.1, 4.8)	(1.4, 4.9)

One can argue that problem (1) is counterbalanced by faster floats that enter more bins and offset the low bias of slower floats. However, given most float deployments in the ocean and the life expectancy of most floats, floats rarely revisit an averaging bin unless caught in an eddy on the boundary of a bin or a flow reversal is sampled.

One simple approach to avoid problem (1) is to perform an average along each float trajectory and count that as one velocity realization in a bin. So for the example shown in Figure 13.2, averages along each trajectory would yield values of 1 cm/s, 5.5 cm/s, and 10 cm/s, whose average is 5.5 cm/s. A set of 700 m SOFAR floats from the POLYMODE experiment was selected to test whether the low bias seen in this simple example is also significant when bin-averaging real floats. Table 13.1 shows that it is a significant bias and bin averaging should be avoided.

Simple pre-averaging along trajectories is not perfect, since short float trajectories are weighted as much as long trajectories. To circumvent this bias, float trajectories should be decomposed into segments whose lengths are the distance travelled by the float in a time period given by twice the integral scale (see Section 13.5). The velocities from each segment are averaged together and this counts as one velocity realization for averaging. It should also be noted if all the floats were launched in the northern domain of Figure 13.2, no matter what you did with only this data, any estimate of the mean flow for the whole domain would be low because of small sample size error and biased spatial sampling.

One solution to reduce bias problem 2 and the effects of spatial sampling biases, is to use spatial bins centered at each estimation location. Poulain presented results that used 30 km overlapping bins to increase the mean flow resolution, compared to the usual one degree by one degree horizontal bins, and for producing a mean flow field that is not as discontinuous as fixed bins. Note that overlapping bins allows data points that were in a corner of the old bin method to be used, not just once, but possibly in four different averages. Problem (2) can be avoided, at least in the spatial dimensions and reduced in the time, by the use of two-dimensional bicubic spline fits to the data in sliding, time windows. By construct, two-dimensional splines are continuous, and the first two derivatives of the resulting fields are also continuous, in latitude and latitude. Bauer *et al.* (1998) used this approach to calculate the mean flow for each of the four seasons from near-surface drifters in the tropical Pacific. There is a variance-bias trade-off, for example, monthly averages versus seasonal averages, given the limits of the present data set. To reduce estimation variance for high-resolution in time, e.g. daily estimates of the mean flow, a parsimonious model can be assumed to optimize the variance-bias trade-off.

One aspect of problem (3), seasonal sampling bias, can also be significantly reduced by fitting a parsimonious model of the form,

$$u = \bar{u} + \sum_{j=1}^{l} A_j \cos(\omega_j t) + B_j \sin(\omega_j t),$$

to each velocity component. There are l dominant periods assumed in the data set, usually based on an analysis of the Lagrangain velocity spectrum. Lumpkin (2003) used $l=2$ for the semi-annual and annual period in his analysis of the near-surface drifters in the tropical Atlantic. The coefficients A_j, B_j, and $\bar{u}$ are found by a simple least-squares procedure. For a velocity field dominated by an annual signal only, $l=1$ and 3 parameters are estimated;

the addition of each dominant period requires estimating 2 more parameters. Velocity estimates, that smoothly vary in time, are available for any of the 365 year days. Daily averages, based on arithmetic averaging data subsets, would not be continuous and have relatively, large estimation variance.

The incorporation of dynamical constraints can also be used to optimize information and provide spatio-temporal continuity. Faure and Speer (pers. com.) formulated an inverse model that had mass conservation and topographic steering constraints, and applied it to all available float data in the lower thermocline (1500 to 1800 m) in the North Atlantic. The model also included current meter velocities and boundary conditions to constrain the flow. The relative weighting of the constraints and the assumed level of allowed horizontal divergence are critical parameters that need to be estimated. The inverse model produced better estimates near the boundary than standard objective analysis techniques.

Spatial biases in estimates of the mean flow may result. In the absence of a mean flow, the center of mass of a float cluster can still have an observed mean flow due to a heterogeneous diffusivity tensor $\mathbf{K}, \mathbf{u} = \overline{\nabla \mathbf{K}}$. Thus particles can move up the diffusivity gradient. As discussed by Davis (1991) this is part of a more general tendency of particles to diffuse down the gradient of mean concentration. In addition, it is well known that drifters preferentially sample convergence zones and that drifter data densities are lower in divergence zones. These convergence zones can result from large-scale, wind-forced Ekman transport or mesoscale divergences associated with energetic features (Garraffo, 2001b). Particle trapping by coherent vortices can also introduce a bias in mean flow estimation. All these biases effect the estimation of the mean flow, which in turns, effects the calculation of velocity covariance functions.

Four-dimensional velocity covariance functions, either Eulerian or Lagrangian contain the second-order statistical information of the stochastic velocity field in physical space, while its Fourier transform pair, the power spectral density function summarizes this information in frequency–wavenumber space. In order to analyze typical, oceanic and coastal Lagrangian data sets, a mean field is usually constructed to capture the large-scale heterogeneous signal, usually in a limited space-time window of the data. This is done so that a simple covariance model, with a small number of parameters, can be fit to the data. The residual flow, about the constructed mean, is usually assumed to be stationary in time and valid over some depth range and in some ocean area, so that the Lagrangian velocity autocovariance functions are just a function of the temporal lag, τ,

$$C_{uu}(\tau) = \langle (u_o(t+\tau) - \langle u_o(t+\tau)\rangle)(u_o(t) - \langle u_o(t)\rangle)\rangle.$$

$$C_{uv}(\tau) = \langle (u_o(t+\tau) - \langle u_o(t+\tau)\rangle)(v_o(t) - \langle v_o(t)\rangle)\rangle.$$

$$C_{vv}(\tau) = \langle (v_o(t+\tau) - \langle v_o(t+\tau)\rangle)(v_o(t) - \langle v_o(t)\rangle)\rangle.$$

The subscript o denotes a tagged water particle at an initial location and time x_o, y_o, t_o. Subjective choices on how to replace the theoretical expectation operator, $\langle\rangle$, by a finite sum from usually, $O(10)$ to $O(100)$, float trajectories must be made. Of course, ergodicity is assumed because we don't have an infinite number of realizations of this stochastic process, but only a finite number of spatial realizations from limited float/drifter deployments and from one continuing, time realization. Therefore, as in almost all of signal processing, trajectory data are either segmented into m day time bins or are sorted either by time of the year (month), time of the day, or other physical parameter as wind forcing and these segments are averaged,

$$C_{uu}(\tau_k) = \frac{\sum_1^{n_k} (u_o(t_i) - \bar{u}_o(t_i))(u_o(t_j) - \bar{u}_o(t_j))}{n_k},$$

for all (n_k) of the $\tau_k = |(t_j - t_i)|$ segments in the data. As the size of the temporal lag, τ, increases so does its estimation variance. This is easily seen by considering one trajectory of length n, then n data points are available to estimate the variance (lag zero), $n-1$ product pairs for $\tau = \delta t$, $n-k$ estimates for $\tau = k\delta t$, and only two data points are used to calculate one estimate of the average covariance at lag $n-1$. This effect is compounded for a drifter cluster. Consequentially, rules exist to limit the number of lags calculated to be one-quarter of the total sampling time (Carter and Robinson, 1987).

What temporal lags to include in the integration of $C_{uu}(\tau)$ has been a topic of serious discussion in the literature. Of course, the answer will be a strong function of the sampling. Some oceanographers only integrate up to the temporal lag where $C_{uu}(\tau)$ is zero, the zero-crossing scale. However, this overestimates the integral time scale for particle realizations that have an oscillatory covariance function, e.g. floats in a vortex or in a planetary wave, that have significant negative lobes. As more and more values of $C_{uu}(\tau)$ are used from longer and longer temporal lags, those estimated covariance values become statistically noisy and can corrupt the estimation of the integral time scale. Other "stopping rules" have been used, the one-quarter rule referenced above, imposing integration limits of either twice the zero-crossing scale, twice the e-folding scale, twice the Lagrangian integral time scale of the velocity (see Section 13.5) or use only the estimated autocorrelation values that are significantly different than zero.

Another approach for estimating the Lagrangian velocity covariance function uses an assumed form for the covariance function (Griffa, 1996; Garraffo *et al.*, 2001a) that contains a few number of parameters, for example, $R(\tau) = C_1 e^{-\left(\frac{\tau^2}{C_2}\right)}$. These parameters are found by least-squares fitting the assumed functional form to the covariance estimates determined from your data. The advantage of this method is that the assumed covariance function is positive-definite and the method is robust with respect to highly variable covariance estimates and the cut-off problem that limit direct integration approaches.

In most oceanographic applications, the mean field used in calculating the covariance function is Eulerian-based, from either bin averaging or from a surface fit, such as the two-dimensional splines. Another approach is to calculate the mean over the trajectory either by simply averaging all velocity estimates from the trajectory or fitting a simple model like a Lagrangian mean that varies linearly in time (Garraffo *et al.*, 2001a) to each trajectory. An improper mean can cause the covariance function to be negative-definite, i.e. physically unrealizable, or the covariance function does not decay properly and the integral time scale, for example, would be biased too long. Since ocean and coastal environments are heterogeneous and nonstationary, data are autocorrelated and sparse, and that the transformation between Lagrangian and Eulerian coordinates is highly nonlinear, how to best estimate autocovariance from data is still an open question. One promising approach is to fit parsimonious models, based on Lagrangian Stochastic Models (LSMs), to observed and simulated Lagrangian data (see Section 13.7).

13.4 On the influence of topography on ocean motion

Riser *et al.* (1978), Rossby *et al.* (1983), Schmitz (1985), and Leaman and Vertes (1996) are some of the earlier studies that showed that topographic steering was an important mechanism for explaining the motion of (sub-) thermocline floats near the western boundary of the North Atlantic. Topographic steering is a direct consequence of potential vorticity conservation, see for example, LaCasce (2000), who showed a close relationship between float trajectories and f/h contours, where f is the local Coriolis frequency ($= 2\Omega\sin(\text{latitude})$, $\Omega = 1/24\,\text{hrs}$), and h is bottom depth. The quantity f/h is a good approximation for potential vorticity for barotropic, linear, large-scale ocean flows. The next few examples were presented at the LAPCOD meetings. Topographic steering can be seen in RAFOS meter floats deployed in the Mediterranean Undercurrent at target depths between 1000 and 1200 m, just south of Portugal. These isopycnal floats tightly hug the continental slope as they travel around Cape St. Vincent, northwards along the Iberian margin, and north of the Lisbon Canyon. At the Lisbon Canyon, f/h contours bifurcate and so does the flow (Bower *et al.*, 2002a). In areas of steep topographic slope, where effective beta ($f/h\nabla h$), is greater than planetary beta (the change of Coriolis force with latitude), the topography steers the flows. The strong steering of the flow by f/h contours would require a dense array of Eulerian observations to observe, but is easily seen in a few Lagrangian trajectories superimposed on topographic and/or potential vorticity maps.

Bower *et al.* (2002b) shows that at mid-depth in the Icelandic Basin, low-salinity Labrador Sea Water mixes with saltier Iceland-Scotland Overflow water from the Nordic Seas. Lagrangian data show that major topographic features, such as the Hatton Bank and the eastern flank of the Reykjanes Ridge, clearly control the mean cyclonic gyre circulation and mixing of these water masses, and that the Bight Fracture Zone at 57° N is an important pathway, like the Charlie Gibbs Fracture Zone at 52° N, for water mass exchange between the western and eastern North Atlantic. Rupolo *et al.* (2003) analyzed 500 000 particles released in an ocean circulation model of the Mediterranean Sea. The primary results from their Lagrangian analysis of the thermohaline flow is that most of the basins of the Mediterranean Sea have strong water mass formations but the sills and straits constrain the spreading and mixing of these intermediate and deep water so that the Levantine Intermediate Water is the dominant water mass that is exported out of Mediterranean Sea.

There was a large international effort to directly observe the circulation throughout the subpolar North Atlantic Ocean, where warm subtropical water is transported to high latitudes. A total of 223 acoustically tracked subsurface floats, representing 328 float-years of data, were combined to generate maps of mean absolute velocity and eddy kinetic energy at the main thermocline and Labrador Sea Water levels (Bower *et al.*, 2002b). Most of the mean flow transported northward by the North Atlantic Current at the thermocline level recirculated within the subpolar region, and relatively little entered Rockall Trough or the Nordic Seas. Saline Mediterranean water reached high latitudes not by continuous, broad-scale, mean advection along the eastern boundary as previously described, but by a combination of narrow slope currents and mixing processes. At the Labrador Sea Water level, a strong, topographically constrained current associated with the overflow of dense water from the Norwegian Sea flowed around the northwestern Iceland Basin along the continental slope and Reykjanes Ridge, and closed counterclockwise recirculations existed adjacent to this boundary current. At both levels, currents crossed the Mid-Atlantic Ridge, eastbound and westbound, preferentially over deep gaps in the ridge. The latter result demonstrates that seafloor topography can constrain even upper ocean circulation patterns, possibly limiting the oceans response to climate change (Bower *et al.*, 2002b).

The results of Olson *et al.* (2005) for near-surface drifters in the Aegean Sea also show the effects of topography with topographic steering by deep basins, islands, and, of course, continental boundaries. Coastal jets are formed along boundaries and by convergence of flow between islands. Striking examples of spatial bifurcation in a coastal flow field are evident in this data set. For example, a drifter would either continue in southward flowing coast jet or enter a small gulf and recirculate within the gulf. Flow bifurcations are seen in Lagrangian trajectories, satellite images, and numerical model simulations, in regions with complex coastal morphology, e.g. jagged coastlines, prominent points, capes, and steep bottom topography, e.g. seamounts, and features like the Charleston Bump that deflects the Gulf Stream offshore of the SE US coast (Bane and Dewar, 1988) and the Shatsky Rise, that is responsible for bifurcations of the Kuroshio (Hurlburt and Metzger, 1998).

Meddies are formed in distinct regions near major Portuguese capes, such as Cape St. Vincent, and follow distinct pathways determined by topographic steering (Bower *et al.*, 2002a). Richardson *et al.* (2000) in their census of twenty-seven Meddies, seeded with floats, and in some cases, hydrographic data (vertical profiles of temperature, salinity, and pressure), find that 90% of all Meddies collide with seamounts and are destroyed by this topographic interaction. The Meddies' anomalous salt and heat signals are preferentially mixed in the vicinity of seamounts.

Quantitative analysis of Lagrangian data has also revealed the strong constraints of topography and potential vorticity conservation. When numerous floats, at least three but preferably more to reduce estimation error, are deployed close together and stay together for awhile, the temporal evolution of differential properties of the flow field can be estimated. As with many nonlinear problems, the starting point for analysis is to use a first-order Taylor series expansion to expand the local Eulerian velocity about the centroid of a float cluster. This approach was first applied to Lagrangian data in the Caribbean Sea by Molinari and Kirwan (1975) and was further developed in Okubo and Ebbesmeyer (1976).

Let the ith float be at longitude x_i and y_i with zonal and meridional velocities, u_i and v_i, respectively. The Taylor series expansion yields the following relationship between the mean horizontal velocity, $(\bar{u}, \bar{v})$, the four local Eulerian velocity derivatives, and the measured Lagrangian velocities (u_i, v_i).

$$u_i = \bar{u} + \frac{\partial u}{\partial x}(x_i - \bar{x}) + \frac{\partial u}{\partial y}(y_i - \bar{y}) + \epsilon_i,$$

$$v_i = \bar{v} + \frac{\partial v}{\partial x}(x_i - \bar{x}) + \frac{\partial v}{\partial y}(y_i - \bar{y}) + \epsilon_i.$$

There are both measurement noise and truncation error for the expansion included in ϵ_i. The three unknowns, that are estimated separately for each velocity component using standard least-squares methods for n float observations, determine the average horizontal velocity and its gradient at the centroid of the float cluster $(\bar{x})$ and $(\bar{y})$. The horizontal divergence is estimated by $\partial u/\partial x + \partial v/\partial y$ and the relative vorticity by $\partial v/\partial y - \partial u/\partial x$.

Mariano and Rossby (1989) used this approach to examine the temporal evolution of the terms in the potential vorticity conservation equation calculated from selected clusters of 700 and 1300 m SOFAR floats from the POLYMODE experiment in the northwest Atlantic Ocean. Because of the Lagrangian nature of conservation equations and the Lagrangian-based velocity data, very accurate, to within a few percent, balance of terms was found. It was shown that the large-scale topographic beta term was as large or larger than the planetary beta term. One of the scientific results of this study is that the vortex stretching, induced by flow over topography, not by wind-forced Ekman pumping, controlled the mesoscale dynamics of thermocline waters. This result was not evident in the relatively dense current meter array of the Local Dynamics Experiment.

13.5 Dispersion and mixing

The foundation for studying dispersion in the ocean and coastal regions, as well as in the atmosphere, are the seminal papers by Taylor (1921) and Richardson (1926). At the heart of these papers is the assumption that fluid motion is not completely random but significantly autocorrelated in time for some finite amount of time and the velocity field is correlated over some spatial distance. They generalized the discontinuous Brownian particle displacement model to a more general stochastic process where particle movements are correlated. The dominance of energetic, turbulent, and coherent velocity fluctuations are readily observed in the atmosphere, the ocean, and in its marginal seas. Consequently, dispersion in geophysical fluids is not molecular diffusion and the standard Fickian model does not apply (Stommel, 1949). The interactions between mesoscale fluid stirring (Eckart, 1948; Welander, 1955; Rossby *et al.*, 1983; Aref, 1984), large-scale shear dispersion (Batchelor, 1952), and smaller-scale turbulent diffusion requires stochastic models for a spatio-temporal correlated process with multiple space and time scales.

Turbulent dispersion depends on a continuum of time scales and diffusion problems are usually studied by assuming just a few number of dominant time scales. There are advective time scales, turbulent time scales or slow and fast time scales, respectively, velocity integral time scales (defined below), the time scale of acceleration, dissipation time scale, forcing time scale, characteristic diffusive time scale, and rotational/eddy time scale, to name a few (Davis, 1983; Bennett, 1987; Cecconi *et al.*, 2005). When biology is included in a model, other important time scales are the growth rate and behavior time scales associated with diurnal migrations and metamorphism (Goodman, pers. com.).

Taylor (1921) showed the importance of a time scale defined by the integral of the Lagrangian velocity autocorrelation function. He related the Lagrangian velocity autocorrelation function for statistically stationary flows, $R_i(\tau) = C_{ii}/\langle u_i'^2 \rangle$, to mean-square eddy displacement in the ith direction, $\langle (x_i')^2 \rangle$, by

$$\langle x_i'^2 \rangle = \langle (u_i')^2 \rangle \int_0^t \mathrm{d}T \int_0^T R_i(\tau)\mathrm{d}\tau$$

For T much larger than $T_\mathrm{L} = \int_0^T R_i(\tau)\mathrm{d}\tau$, Taylor derived that

$$1/2 \frac{\mathrm{d}\langle x_i'^2 \rangle}{\mathrm{d}t} = \langle (u_i')^2 \rangle T_\mathrm{L} = \kappa_i$$

where T_L is integral time scale of the Lagrangian velocity, $\langle (u_i')^2 \rangle$ is the velocity variance and κ_i is the eddy diffusivity. For large times, so that particle velocities become uncorrelated, i.e. the Lagrangian velocity autocovariance function asymptotes to zero, the eddy diffusivity also asymptotes to a constant value, the particle displacements become more random and are no longer controlled by individual eddies but by the mean eddy kinetic energy. Rossby *et al.* (1983), based on a suggestion and significant input from Jim Price, showed that for depths between 700 and 2000 m in the western North Atlantic, the diffusivity κ_i is linearly related to the velocity variance implying a near constant value of T_L on the order of ten days. Zhang *et al.* (2001) also found good agreement between theory and a larger data base only if the ensemble average $\langle\rangle$ was properly defined. In practice, $\langle x_i \rangle$ and $\langle u_i \rangle$ used to form the residuals may be space- and time-dependent.

Taylor's analysis was the first to point out that "simple" diffusion depends only on the integral of velocity autocovariance function. Taylor's results imply that for short times, the diffusivity grows linearly in time since for small lags, $R(\tau)$ is approximately 1, $T_\mathrm{L} = t$, and the eddy diffusivity is linearly related to t

with the proportion constant equal to the variance of the Lagrangian velocity. There are a number of approximations involved in Taylor's theory including a linearization of statistical moments and the assumption that the observing period, T, is much larger than T_L. This latter assumption removes a τ/T term from the integral. For T on the order of $2T_L$ this leads to a substantial error in the estimation of T_L. For simple correlation functions, such as the exponential function, this error can be on the order of 10 to 40%.

In these classic formulations the value of the eddy diffusivity is decoupled from the mean advecting flow. Recent work by Mazzino *et al.* (2005) indicates that the classic approach is not the best because of multiple-scale interactions. Numerical simulations by Zambianchi (pers. com.) show as the average flow increases, the strength of mesoscale stirring increases, and particle diffusivity is much different as the mean flow varies over one order of magnitude, say 10 cm/s to 1 m/s. It should also be noted that the theory for the asymptotic values for diffusivity, for either small time after launch or large time after launch is fairly well developed. However, observations of diffusivity in the ocean are most reliable at intermediate times. At small times, statistics are a strong function of initial conditions (Flierl, 1981; Davis, 1983), while for large times, potential vorticity constraints, basin boundaries, and heterogeneous ocean and coastal dynamics, as well as forcing, lead to anisotropy of velocity statistics.

One-particle statistics are used to describe the evolution of the mean centroid of a cluster of particles. Multi-particles are needed to describe the spread of a cluster of particles about the cluster centroid. Because of economics, relative diffusion in the ocean has mostly been studied using two-particle statistics. Though hydrographic data and satellite AVHRR images are useful for studying cross-frontal mixing in current systems like the Gulf Stream, drifters and floats can provide quantitative estimates of the rate of that mixing. Mixing defined by relative, pairwise dispersion, was formulated by Richardson (1926) in a Lagrangian framework. Richardson derived his well-known "4/3 law" for isotropic, homogeneous turbulence in the inertial range where there are well-defined, coherent flow features with spatial scales much larger than the scales at which dissipation occurs. Richardson (1926), based on observing plumes from industrial smoke stacks, and releasing seeds and balloons into the wind (see the review by Hunt, 1998), assumed that the rate a pair of particles will spread apart increases as the separation increases between the particles. This assumption implies that the random particle motion is not Brownian, but depends on autocorrelations in space and time. These autocorrelations, in general, will also introduce non-Gaussian behavior in particle statistics (Sawford, 2001). Obukhov (1941) derived the 4/3 law using results of Kolmogorov's (1941) theory for isotropic turbulence. Batchelor (1952) extended Richardson's work, most notably to the study of the evolution of a passive scalar in a turbulent flow field.

Let $D_{ij} = ((x_i - x_j)^2 + (y_i - y_j)^2)^{1/2}$ be the usual Euclidean distance between a pair of particles i and j, the rate of relative dispersion is defined to be

$$1/2 \frac{\mathrm{d}D^2}{\mathrm{d}t} \propto D^{4/3}.$$

Kraichnan (1966, 1967) derived the same power law as Richardson for the large scales in two-dimensional turbulence, while on smaller scales Lin (1972) derived that the rate of relative dispersion is proportional to D^2. Scale separation between the two regimes is the energy injection scale. Richardson and Kraichnan's results indicate that $D(t) \propto t^{3/2}$, while Lin's result implies $D(t) \propto \exp(t)$. As separation between particles becomes much larger than the scales of the energy containing eddies, $D(t) = 2\sigma^2 T_L t$ and the relative diffusion is dominated by absolute dispersion because the motion of the particles at this scale are uncorrelated.

Stommel's (1949) observations of parsnips, copy paper, dye, "floats" at three different locations agreed with Richardson's 4/3 law on scales from 10 cm to 100 m. Okubo's (1971) analysis of different historical data sets extended this agreement out to scales of a few hundred km, but this requires constants that change with scale. Application of these results to coastal and oceanic regimes requires a definition of the energy injection scale. Is this scale associated with wind forcing, thermohaline forcing, the Rossby radius of deformation scale, or the Rhines scale? Okubo (1971), Kirwan *et al.* (1978), Bennett (1984), Davis (1985b), LaCasce and Bower (2000), LaCasce and Ohlman (2003), and Ollitrault *et al.* (2005) each present evidence that there does exist both $D^{4/3}$ and D^2 regimes with the "cut-off" scale being the Rossby radius of deformation scale for baroclinic eddies. However, this result is not so clear cut because of small sample size, errors in defining the mean, observation error, boundary effects, nonstationarity, and differences between regions, e.g. surface versus subsurface and coastal versus deep ocean. In coastal regimes, for example, submesoscale eddies (Shay *et al.*, 2000; Fraunie, pers. com.) of scales of a few km are also long-lived, trap particles, and are important for transport pathways and stirring in coastal waters. In marginal seas with semi-enclosed basins, the close proximity of boundaries introduces effects that severely limit the applicability of the classical dispersion theory.

Vulpiani and colleagues (Artale *et al.*, 1997; Aurell *et al.*, 1997; Lacorata *et al.*, 2001) introduced the concept of a Finite Scale Lyapunov Exponent (FSLE) to estimate relative diffusion in small basins. FSLE λ are a function of initial separation distance δ_0, longitude and latitude (x, y), and elapsed time τ,

$$\lambda(x, y, t, \delta_0, \delta_f) = \frac{1}{\tau}\log\left(\frac{\delta_f}{\delta_0}\right)$$

where τ is the time it takes a trajectory to move from a distance δ_0 from (x, y) to a separation δ_f from a reference trajectory that started at (x, y) at time t. The exponent λ is picked to be the maximum from 4 choices of δ_0 based on the $\pm$ separation in each of the two basis directions. FSLE λ represents an inverse time scale for mixing particles from length scale δ_0 to scale δ_f with $\lambda > 0$ implies divergence and $\lambda < 0$ implies convergence.

Now if λ is calculated for both positive time + and negative time − and denoted by, λ_+ and λ_-, respectively, where λ_+ and λ_- intersect is a good proxy for a distinguished hyperbolic trajectory. dÒvidio *et al.* (2004), using FSLEs, characterized the circulation and horizontal mixing of the Mediterranean from a numerical simulation. Their distribution of the FSLEs clearly shows enhanced horizontal mixing along the boundaries of mesoscale eddies with much lower values of horizontal mixing near the center of the eddies. Also evident in their analysis is a clear seasonal signal in mixing activity in Mediterranean surface waters.

FSLEs allow a more detailed analysis of scale-dependent dispersion than the classic analysis (Boffetta *et al.*, 2000; Lacorata *et al.*, 2003). FSLEs were calculated for surface drifters in the northern Gulf of Mexico by LaCasce and Ohlmann (2003). They found that for scales less than the deformation radius, the dispersion is consistent with a local enstrophy cascade and particles separate exponentially in time. For larger scales, the dispersion is non-exponential and is consistent with an inverse energy cascade. This result was also seen in an analysis of balloons in the lower stratosphere (Lacorata *et al.*, 2004). The integral time scale is arguably the important time scale for determining which dispersion regime particles are in. In a series of experiments in the early 1950s (Swallow, 1955, 1957; Rossby, Chapter 1), Swallow showed that the deep ocean was not "stagnant," but much more turbulent and energetic than the prevailing wisdom at that time. From this early beginning, Lagrangian exploration of the ocean has revealed us details of particle pathways dominated by turbulent mesoscale dynamics. Cheney *et al.* (1976) describe the motion of a SOFAR float (Rossby, Chapter 1) in a cold core ring and the clear trapping effect of being in an oceanic vortex. It is during severe events like ring formation, eddy merger, or filamentation, that particles can escape the trapping effects of the vortices. Rossby *et al.* (1983) and Richardson (1993) list the average statistics of floats that exhibited significant rotary motion in their trajectories and appear to be in coherent oceanic eddies. In these studies, the anisotropic dispersion of particles, due to the constraint of the large-scale potential vorticity field, is evident, as is the westward propagation of oceanic vortices, except where the floats are advected by major ocean currents like the Gulf Stream, Kuroshio, the Antarctic Circumpolar Current, and the equatorial counter-currents. In regions with highly nonlinear dynamics, e.g. the Brazil–Malvinas extension, characterized by large values of relative vorticity, the meridional planetary vorticity constraint is reduced and dispersion can be more isotropic (Figueroa and Olson, 1989). Observational evidence indicates that oceanic and coastal tracers are found in patches of high concentrations. One explanation for this non-Gaussian behavior is that vortices, such as Gulf Stream rings, Meddies, and submesoscale coastal eddies, trap particles and tracers. This phenomena is well documented, as referenced above, and see Dewar and Flierl (1985), Provenzale (1999), Mariano *et al.* (2002), and Chapter 4 by Pasquero *et al.*

In order to model tracer patchiness, Beron-Vera *et al.* (2004) simulated particle trajectories in two-dimensional (2-D), nonlinear incompressible flows consisting of a background flow and unsteady, spatially periodic perturbations on a beta-plane. Two background flows, a single Stommel gyre and a symmetric, basin-scale, clockwise gyre, are evaluated. Perturbations are calculated by 100 2-D spatial modes, with random phases, that are modulated by a cosine function with periods determined by the Rossby wave dispersion relationship for that spatial mode. The resulting phase space of the (x, y) positions of the simulated trajectories is mixed with "regular islands," where particle dispersion follows a classic power-law, and in a "background chaotic sea," where dispersion is exponential. As the amplitude of the perturbations increase, particle instability increases. They also demonstrated that the partitioning between these two states, i.e. the particle trajectory stability, is largely controlled by the shear of the background flow for the class of problems they considered. Their analysis exploits the action-angle formalism for a Hamiltonian system.

At times the superpositioning of different features, such as warm-core rings, cyclonic eddies, and fronts, leads to the formation of efficient pathways for long-distance fluid transport and efficient mixing of different water masses. Toner *et al.* (2003) analyzed simulated pathways from a data-assimilative numerical model of the Gulf of Mexico and observed chlorophyll plumes from satellite images. They showed that cyclonic–anticyclonic eddy interactions can form efficient, basin-wide, pathways for the transport of chlorophyll-rich water off of the shelf. Two examples, both with excellent agreement between the numerical simulations and satellite data, showed chlorophyll, from the South Florida shelf, was transported in a thin strip along the north wall of the Loop Current after an eddy-shedding event, and that shelf water from Louisiana, was advected by another event associated with a cyclone/anticyclone eddy pair. In both scenarios, shelf water from the northern Gulf of Mexico ended up along the Yucatan Peninsula. This transport event would not be seen in an analysis based on any climatological surface current data set but required a high-resolution numerical circulation model, forced with synoptic winds and surface fluxes, and assimilating altimetric heights and sea surface temperature data that resolved the location, size, and strength of the primary circulation features.

Hebert and Rossby launched over 90 floats south of The Cape Verde Islands as part of the Lagrangian Isopycnal Dispersion Experiment (LIDEX) in 2003. They deployed tight clusters of order ten floats on two density surfaces in the low-oxygen tongue off of the West African coast. The plan is to put the floats on specific isopycnal surfaces to reduce vertical dispersion, in order to enhance our ability to understand isopycnal dispersion and horizontal mixing on scales of km to basin-scale. One of the salient results is how dispersion characteristics change for some time after launch, but by year two, the anisotropic dispersion (stronger in zonal direction) remained relatively constant. The fact that more experiments and more planned Lagrangian-based experiments are initially deploying floats in large, tight clusters is a testimony to the increased dialogue between oceanographers and mathematicians, and in the specific case of LIDEX, the lead author of this chapter convincing his mentor (after 20 years!). Given the inherent, time-dependent chaotic dispersion, and dispersion by large-scale shear flow, floats should be deployed in tight clusters to provide an adequate amount of first- and second-order statistical information at small spatial lags.

Knowing efficient pathways can be used in determining the launch locations of drifters. For example, an optimal observing strategy goal is to launch floats in a limited region, and that they disperse as fast as possible to maximize coverage for mapping out the mean circulation, for Eulerian field reconstruction, or for data assimilation. Numerical results by Molcard *et al.* (2003) showed that floats deployed in high velocity regions reconstructed Eulerian fields with lower errors than floats randomly deployed. Numerical results (Poje *et al.*, 2002; Toner and Poje, 2004; Molcard *et al.*, Chapter 7) found that the optimal launch location based on identifying hyperbolic trajectories and deploying drifters along outflowing material curves/Lagrangian boundaries, in general, maximized Lagrangian data coverage leading to either more reliable Eulerian velocity field reconstruction or more reliable forecasts. These studies, presented at the LAPCOD meetings, confirmed that high velocity deployments were better than random deployments, but not as optimal as those based on knowing the out-flowing manifolds of the flow field.

Manifolds are useful for identifying flow features, such as ocean vortices and strong currents, and for identifying particle pathways. Stable and unstable manifolds are used to delineate invariant sets in phase planes. Given that dominance of horizontal flow in the ocean, most analyses are two-dimensional and there is a correspondence between the physical space of the fluid (the velocity field) and the phase space of the dynamical system. Grossly speaking, invariant sets are subregions of the plane that are generated from the initial conditions in that region. So a trajectory in an invariant set is confined to that invariant set or region. For example, in a simple, incompressible, kinematic flow field that consists of a vortex and a zonal, periodic jet, the invariants sets are well-defined. Particles in the vortex will stay in the vortex, particles in the jet will stay in the jet, and those in the surrounding background flow will not be entrained into any of the strong flow features. For this ideal case, the manifolds of the system are where the isotachs of the velocity field are zero, $(u, v) = (0,0)$. Given the inherent chaotic nature of particle trajectories, and the sensitivity of the phase plane trajectories to initial conditions, invariant sets can have complicated shapes and the invariant sets can be a strong function of the parameters of the dynamical system.

Samelson (1992), Duan and Wiggins (1996), Rogerson *et al.* (1999), Poje and Haller (1999), and others have studied cross-stream transport and mixing in time-dependent meandering jets using tools from the nonlinear analysis toolbox. When non-periodic time-dependence, strong nonlinearities and turbulent mixing phenomena are considered, one has to define a suitable method to delineate the strongly evolving flow features that are interacting across a multitude of scales. A fixed point of a system is not changed by the dynamics of the system and so the study of manifolds begins with identifying the fixed points of the differential equation. The stable manifold consists of all those points which approach the fixed points as time goes to infinity, while the unstable manifold consists of all those points that are repelled away from the fixed point as time goes to infinity. Ironically, the calculation of the stable manifold can be unstable near an instantaneous stagnation point because of its proximity to the unstable manifold. For linear differential equations, the eigenvectors of the matrix used to define the dynamical equation are calculated and the eigenvalues determine the phase plane possibilities; saddle point, stable node, unstable nodes, etc. For a nonlinear system, the eigensystem for the local Jacobian is solved and for time-varying systems, a number of algorithms are being developed and evaluated (Malhotra and Wiggins, 1998; Ide *et al.*, 2002a,b; Haller, 2000; Kuznetsov *et al.*, 2002; Mancho *et al.*, 2004) since the stationary points can quickly change their location in time. Kirwan *et al.* (2003) used these methods to analyze Lagrangian-based mixing and dispersion in oceanic flows and for determining flow predictability.

Another optimization problem is to determine the optimal release time for coastal pollutants, such as a sewage outflow. The goal would be to release large volumes of stored dirty water at times when the flow would take the water away from shore and when the water would be quickly diluted. Lekien *et al.* (2005) used estimates of the material lines and manifolds from near-shore velocity measurements by High Frequency Radar (Shay *et al.*, 2002) inshore of the northward flowing Florida Current to determine optimal release times. When repelling material lines, which are barriers to transport, are north of the release point, fluid should not be released, since the fluid would be locally recirculated. Large volumes of fluid should be released when the attachment point of the stable manifolds is south of the release point so that the water is quickly advected out of the release area and is diluted.

New theoretical methods are being developed and applied to numerical circulation model output and to coastal/ocean data sets for quantifying and analyzing turbulent fluid transport. Transport Induced by Mean-Eddy Interaction (TIME) theory, based on a hybrid Lagrangian–Eulerian approach was presented by Ide. This analysis suggests that the efficiency of large-scale transport, in the presence of oceanic eddies, can be parameterized by the ratio of the characteristic time scales of the dynamic variability to the length scales of the instantaneous particle flux. Small and Wiggins discussed the importance of modeling float trajectories as a finite time, nonstationary process. Such a process is necessary because Lagrangian observations clearly show particles leaving vortices and entering the background flow, significant flow into and out of jets, and flow bifurcations.

Lipphardt *et al.* (2000) used a method, based on fitting normal modes to HF measurements in Monterey Bay that produced a regular space-time gridded velocity set. Large numbers of float trajectories were simulated and analyzed to study transport and mixing in the bay. These trajectories revealed nonstationary, complex flow regimes with particle fates that were highly dependent on initial locations and times. Floats launched in one location would quickly leave the bay; floats launched hours later would recirculate in the bay for days. Maps of residence times are highly variable in both space and time, a signature of a chaotic or a highly nonstationary process. Besides the effects of seasonal forcing and episodic forcing, e.g. a hurricane, nonstationary Lagrangian statistics are also seen in nature due to turbulent mixing over a broad spectrum of scales that allow particles to escape from one flow feature into another flow feature.

Lagrangian observations are important for studying transport between different water masses (Bower and Rossby, 1989). Talley (2003) and Brambilla (pers. com.) are investigating why only a handful out of 273 near-surface drifters go from subtropics to subpolar; an amount much smaller than one would expect from estimates of the meridional overturning circulation. Lagrangian sampling at different depths, such as the Iceland Faroe Front experiment (Soiland and Rossby, pers. com.) with RAFOS floats at 200 m and 800 m, will be needed to find the pathways of intergyre exchange. As Rossby pointed out, the modification of potential vorticity required to cross the Iceland Faroe Ridge may be forced by the inflow to the Nordic Seas from the Atlantic required to close the thermohaline circulation. As the Lagrangian database grows, these and other important components of the climate system will be better documented.

The multi-disciplinary interactions between the different communities, mathematicians armed with their nonlinear theories of fluid flow and oceanographers with their Lagrangian data sets and simulations using simple kinematic models to high resolution numerical circulation models, was very evident at all of the LAPCOD meetings. Synergistic research between the different communities, if sufficiently funded, will accelerate our understanding of turbulent dispersion and mixing leading, for example, to more reliable biophysical models.

13.6 Biological Applications

As pointed out in Chapter 11 by Hitchcock and Cowen, plankton are the original Lagrangian platforms and their distribution has aided our understanding of turbulent mixing and dispersion. For example, Ashjian (1993) used the distribution of different copepod species from the Slope Water and the warmer Gulf Stream to infer entrainment and detrainment along and across the Gulf Stream North Wall. Temporal sequences of satellite images of ocean color have yielded a wealth of information on both the biology of and the dynamics of coastal zones and of the global ocean. Comparison of satellite data and Lagrangian data, for example in the California Current (e.g., Garfield *et al.*, 2001), have both indicated that energetic eddies entrain coastal waters and their communities and transport them offshore.

Hitchcock *et al.* (1997) conducted a series of observations in the Mississippi River plume in which the spatial distribution of nutrients and CDOM (Chromophoric Dissolved Organic Matter) were mapped while following surface drifters. In all cases, concentrations of dissolved materials in the river plume decrease in a quasi-linear manner with time. These observations argue for physical dilution as a primary mechanism controlling the initial distribution of dissolved materials in the observed plume. Convergent surface fronts at the edge of the plume appear to be the primary site where biological and other biogeochemical processes transform dissolved and particulate matter as river waters are mixed with ambient shelf water. In this experiment and in earlier experiments, reviewed in Hitchcock *et al.* (1997), Lagrangian-based sampling schemes reveal that surface concentrations of many water properties, such as temperature, salinity, nitrate, and silicates, in major river plumes (e.g. Mississippi River, Amazon River, Savannah River, and the Zaire River) are consistent with conservative mixing and that the mixing time scales are less than one day in near-shore coastal regions.

A major research question in marine fisheries is the importance of long-distance transport of fish larvae versus local retention (Cowen *et al.*, 2000). Early calculations based on just mean-flow and turbulent advection and very simple biology, would lead one to believe that many fish populations are "seeded" by larvae from noncompeting populations from distant locations. Recent results using either high-resolution, multi-disciplinary data sets (Paris and Cowen, 2004) or sophisticated circulation models coupled to much more complex biological models and forcing functions are painting a different picture. The modeling approach of Cowen *et al.* (2003) is based on a coupled, biophysical model incorporating a high-resolution ocean circulation model (MICOM), a Lagrangian scheme with larval sub-grid turbulent motion, and a biological model with functions for larval sensory capabilities and for settlement habitat availability. The eddy component of the Lagrangian flow field was modeled as a random flight model (see next section; Griffa, 1996), with parameters estimated from observations and realistic larval behavior, mortality estimates, and production variability was incorporated into their models. Their primary results are consistent with the hypothesis that marine populations must rely on mechanisms enhancing self-recruitment rather than depend on distant "source" populations. Cowen *et al.* (2003) used the same approach to look at the transport pathways of fish larvae and found local retention. Aliani and Griffa (pers. com.) used a similar approach to show that invasive species from ship ballast like goose barnacles may not invade distant areas in one continuous movement. They may settle in intermediate areas, grow, and resettle in steps.

Basin-scale gyres, set up by the geometry of the topography, that retain particles and narrow passages that inhibit flow can reduce connectivity between biological populations. For example, Cherubin (pers. com.) discussed the effects of passages in the Caribbean on the population of reef fish. If larvae are passively advected for a month or two and this is much less than the residence time of the gyres, they will more likely remain local. Mesoscale eddies may increase connectivity by providing efficient transport pathways. Bracco *et al.* (2000) found that eddy trapping can also increase the probability of having more coexisting plankton species.

Lobel and Robinson (1986) suggested that mesoscale eddies can entrain fish larvae, transport the larvae offshore into regions of lower predation, and then back to the reef after the larvae have grown. Upwelling in mesoscale eddies is also an important physical mechanism for maintaining plankton populations (Bauer *et al.*, 1991). McGillicuddy *et al.* (1997, 1998) used models and data to estimate that mesoscale eddies significantly influence the nutrient flux into the euphotic zone. Thus it is crucial that biophysical models are able to resolve or parameterize eddy dynamics because those dynamics are crucial in determining the distribution of marine organisms. In many applications, this would require accurate estimates of the parameters and the order of the Lagrangian Stochastic Model (Section 13.7). In Eulerian models, adequate model resolution and the ability to reproduce the energetic eddy field are important for realistic simulations. Of course, when the modeled specie has swimming speeds on the order of the mean and eddy velocities, that behavior needs to be adequately modeled. Pantalone and Ruddick (pers. com.) used Monte-Carlo chains and transition probability matrices to model swimming behavior of reef fish. Swimming rates average 20 cm/s and with peak speeds of 60 cm/s, the same order as the currents. They also include behavior-based swimming and this step is critical in real-world applications.

Accurate modeling of the vertical velocity is also very important in biological models such as those estimating plankton distributions and the sinking rates of inorganic and organic materials. Brandt showed the importance of incorporating a salinity-dependent vertical velocity component besides a vertical velocity for diurnal migration in explaining the distribution of mussels in coastal regions of the North Sea.

Olson (2006) used drifter observations to consider larval dispersal in both pelagic and meroplanktonic organisms. Applications range from understanding the distribution of zooplankton, dispersal of reef organisms across ocean basins, and the spawning of large pelagics such as squid and swordfish. Olson also favored a hybrid Lagrangian–Eulerian approach to modeling marine biological system since the forcing functions, based on data, are only estimated on an Eulerian grid. On the other hand, Monte-Carlo simulations using data-based advective velocities and data-fitted Lagrangian Stochastic Models (Section 13.7) produce realistic statistics of larval dispersal and different flow scenarios can easily be tested. See Chapter 10 by Olson for details of his analysis. The Lagrangian framework is the only feasible viewpoint to use in a biological model for understanding populations that are structured in terms of age and past history. The Lagrangian viewpoint introduces the need to evaluate forcing functions, like light levels, that have traditionally been observed and modeled in an Eulerian viewpoint. The resulting hybrid models usually require interpolating grids of temperature fields, for example, to the position trajectories of plankton.

Buffoni *et al.* in Chapter 12 use a stochastic model to describe the dynamics of a single specie in a stage structured population. Stage structured populations are required for species with well-defined events in their life history, e.g. egg stage, larval stage, moult stage, juvenile, reproductive adults, and death. Individuals' physiological age is an accumulation of the response to both deterministic and stochastic forcing that vary as a function of the stage. Population statistics are determined by Monte-Carlo simulations. This class of models require an average egg production and development rates, as well as its standard deviation, for each of the development stages. These rates can be a function of temperature and of population number depending on model complexity. A Poisson process is assumed for survival mortality. Two primary sets of simulations were performed with and without feedback from population size. If egg production is too low, a population quickly goes to extinction, otherwise there is exponential growth in the population without any feedback. For this case, it takes typical populations five to six months to explode. With feedback, the population either asymptotes to a steady-state if the survivorship rapidly decreases with the population size or oscillates around a fixed value that depends on the derivative of the survivorship with respect to population size. See Chapter 12 by Buffoni *et al.* for further details.

Idrisi *et al.* (2004) analyzed particle trajectories from a numerical simulation of the Arabian Sea, using MICOM and and a biological model for hibernating copepod, that are upwelled from deep to intermediate waters along the coasts of Oman, Somali, and the Gulf of Aden. These simulations had different behavior mechanisms to deal with different life histories of their respective modeled populations. Sabia and Zambianchi (pers. com.) coupled a light-based behavior model to a physical model to study how phytoplankton can coexist. In their simulations, deep mixed layers and high diffusivity limit the growth rate of each specie allowing coexistence.

Doglioli *et al.* (2004) modeled the dispersion characteristics of different particles (dissolved conserved tracer, decaying particle, sedimentary) for a fish farm in the Ligurian Sea in the western Mediterranean Sea. Their circulation model is based on a 2-D version of POM (Princeton Ocean Model) and their particle model is LAMP3D (Lagrangian Assessment for Marine Pollution Three-Dimensional Model). LAMP3D uses the 2-D field from POM to compute a high resolution theoretical Ekman profile for transporting particles and a random walk, zero-order stochastic model. Their primary experimental parameters were release schedule, continuous versus periodic, and particle behavior (conserved, exponential decay, sinking rate). Observations of fecal settling rates were usually between 2 and 4 cm/s with a mean rate of 3.2 cm/s. Uneaten food sinks at the rate of up to 12 cm/s. The simulated horizontal flow speeds ranged from 1 to 20 cm/s over the domain and were the same order of magnitude as the settling velocity for uneaten food pellets in the vicinity near the fish farms. During times of predominate NE winds, dissolved particles are effectively diluted and are transported offshore. However, as the winds turn from SE to SSW, more particles end up on the coast. Regardless of release schedule or whether the particles sank at 4 cm/s or 12 cm/s, there was a build up of sediments under the fish cages and the sedimentary particle concentrations had a Gaussian decay with distance away from the fish farm.

These are just a few examples, based on material presented at the LAPCOD meetings, from the growing field of biophysical modeling. There is a large number of important studies we unfortunately do not have the space to review. In many of the biological models, parameterization of the turbulent flow is crucial for any realistic modeling of population dispersal in the marine environment. More and more modeling studies are using turbulent flow parameterizations based on Lagrangian Stochastic Models.

13.7 Lagrangian stochastic models

Most of the published Lagrangian Stochastic Models (LSMs) are based on Auto-Regressive (AR) models for the ith particle position $(x_i(t),\ y_i(t))$ (Thomson, 1987, 1990; Griffa, 1996). The usual starting point for Lagrangian Stochastic Models (LSMs) is to assume isotropic, homogeneous, stationary, and incompressible turbulence with Gaussian velocity statistics of mean zero and variance σ^2. The horizontal velocity field is decomposed into large-scale, mean components, U and V, and a LSM is assumed for u and v, the turbulent velocity component.

$$\mathrm{d}x(t) = (U(x,y,t) + u(t))\mathrm{d}t$$

$$\mathrm{d}y(t) = (V(x,y,t) + v(t))\mathrm{d}t$$

Since it is known from observations that both u and v are correlated in time, a first-order $AR(1)$ or Markov models are often assumed, e.g.

$$\mathrm{d}u(t) = \frac{-u}{T_{\mathrm{L}}}\mathrm{d}t + \sqrt{\frac{2\sigma_u^2}{T_{\mathrm{L}}}}\mathrm{d}\eta(t),$$

T_{L} is the Lagrangian integral time scale, $T_{\mathrm{L}} = \int_0^\infty C(\tau)\mathrm{d}\tau$, and $\mathrm{d}\eta$ is a random process defined by the Ito differential equation (Castronovo and Kramer,

2004). Our Markov or $AR(1)$ model implies an exponential autocorrelation function (see Chapter 9 by Rupolo). The velocity autocovariance function, $C(\tau)$, exponentially decays to zero with $C(2T_{\mathrm{L}}) \approx 0$. Absolute particle dispersion, defined as the mean-square particle displacement from initial position, grows quadratically with time during the initial ballistic regime and then asymptotes to a linear function of time for times much greater than T_{L}. In the ballistic regime, the growth in particle variance is also proportional to the velocity variance. In the diffusive regime, the proportionality is twice the product of the integral time scale and the velocity variance.

The $AR(1)$ model generalizes to an $AR(2)$ model by also assuming that the turbulent particle acceleration, a, can be modeled as a Markov process,

$$\mathrm{d}u(t) = a(t)\mathrm{d}t$$

$$\mathrm{d}a(t) = \frac{\sigma_a^2}{\sigma_u^2} u(t)\mathrm{d}t - \frac{\sigma_a^2 T_{\mathrm{L}}}{\sigma_u^2} a(t)\mathrm{d}t + \sqrt{\frac{2\sigma_a^4 T_{\mathrm{L}}}{\sigma_u^2}}\mathrm{d}\eta(t).$$

The Lagrangian velocity autocorrelation function has two time scales for the $AR(2)$ model and the correlation can become negative. The two time scales can be considered as time scales associated with an oscillatory component and a decaying component. $AR(3)$ models are formulated in Castronovo and Kramer (2004) and in Berloff and McWilliams (2002). Higher-order models are being formulated to capture the complex correlation structure seen in observed trajectories and in simulated trajectories from high resolution circulation models. The statistical properties of auto-regressive stochastic process are reviewed in Chapter 9 by Rupolo and application to predicting Lagrangian motion in the ocean is reviewed in Chapter 6 by Piterbarg *et al.*

One-dimensional, first-order models for each component are accurate models for particle dispersion in quiet regions with little velocity shear and no coherent features (Bauer *et al.*, 1998). As the velocity shear increases and the number of coherent vortices increase, LSMs used to model Lagrangian trajectories and dispersion, must either increase in order to $AR(2)$ or $AR(3)$ models (Berloff *et al.*, 2002; Berloff and McWillaims, 2002, 2003; Pasquero *et al.*, 2001), incorporate nonlinear parameters, and/or become coupled. Castronovo and Kramer (2004) review the diffusivity properties of $AR(1)$, $AR(2)$, and $AR(3)$ models and how well they can capture the properties of a Lagrangian velocity field with a power law spectrum. Each model can qualitatively capture some aspects of the temporal evolution of the diffusivity but each model has its limitations in describing sub- and superdiffusion.

Besides increasing the order of the auto-regressive model, one can either couple the velocity components or assume a Moving-Average (MA) component, and use the more general $ARMA(p, q)$ model, where p is the order of the AR process and q is the order of the MA model. For a $MA(q)$ model, the signal at time t is the noise at time t plus the noise from the previous $t-1, \ldots, t-q$ times. Petrissans *et al.* (2002) assume an $ARMA(2,1)$ model to model particle motion when the particles are heavy, that is, the particle motion is significantly influenced by a nonlinear drag force. A simple one-dimensional LSM for coupling the velocity components is

$$\mathrm{d}u_i = -u_i T_{\mathrm{L}}^{-1}\mathrm{d}t + \epsilon_{ijk}\,\boldsymbol{\Omega}_j\,u_k\,\mathrm{d}t + b_{ij}\,\mathrm{d}\xi_i,$$

where ϵ_{ijk} is the antisymmetric unit tensor, $b_{ij}=\delta_{ij}(\ 2\sigma^2/T_{\mathrm{L}})^{1/2}$, and Ω_j are rotational frequencies. Coupled LSMs with a significant spin term are associated with spiraling particle trajectories, oscillatory velocity autocorrelation functions, crosscorrelation functions significantly different from zero, and suppressed rates of turbulent dispersion for given turbulent kinetic energies. These are the attributes of energetic coherent flow structures, such as eddies and vortices. The spin-term therefore provides a means of incorporating, into the LSM framework, structures seen in Lagrangian trajectories and satellite images of the ocean (Veneziani *et al.*, 2004).

Veneziani *et al.* (2004, 2005) used this model to simulate the statistical properties of SOFAR floats in the Northwest Atlantic thermocline (700 m). At minimum, a bimodal value of the spin parameter, Ω, is needed to accurately model the observed auto-covariance functions. If the float is in an energetic eddy, a significant spin value on the order of (ten days)$^{-1}$ should be used, while a zero spin value, that uncouples the equations, can be used for floats in the background flow. In other words, the Lagrangian velocity covariance function calculated in homogeneous region of the Northwest Atlantic from 700 m SOFAR float data can be modeled as a superpositioning of a background autocovariance function, accurately modeled as an uncoupled Markov process, and an eddy autocovariance function, that can be modeled as a coupled process with a significant spin parameter.

Autocovariance functions calculated for buoy data from the Equatorial Pacific Ocean can often be characterized by a non-oscillatory zonal Lagrangian velocity autocorrelation function having a “knee” (secondary maxima), and by an oscillatory meridional Lagrangian velocity autocorrelation function having pronounced negative lobes. The trajectories of these buoys will be very cycloidal in nature. Recent work by Reynolds on formulating both a LSM, based on a generalized wave structure, that could be used to model buoys in Equatorial Pacific Ocean, and a LSM for heterogeneous and nonstationary flows allows more realistic descriptions of real flows at the cost of larger data sets required to estimate a larger number of parameters. Due to the existence of coherent features, LSMs must adapt to spatial and temporal variations. For example, when a float leaves an eddy and enters the background flow field, the LSM that describes its behavior fundamentally changes and modeling this change is important. Recent work by Veneziani *et al.* (2004, 2005) has shown that Monte-Carlo simulations using vortex detrainment rates, based on climatological statistics, can be used to model the transition in LSM between vortex dynamics and the general turbulence of the background flow field. Eddies, for small times, cause ballistic diffusion, but at intermediate times, there is particle trapping and this decreases diffusion. At long times, coherent vortices transport particles great distances, mostly zonal, leading to anisotropic diffusion. Since there is no unique way to model nonstationary behavior in LSMs, this will be a fruitful area of research for decades.

LSMs can also be used to study mixing. In contrast to approaches that evolve statistical moments of a passive tracer, e.g. Fokker–Planck equation, LSMs produce an ensemble of tracer realizations. Piterbarg (2005) formulated a LSM to study the evolution of material lines in two different flows: a linearly distributed tracer in a Brownian stochastic flow in the limit of zero spatial correlation; and a flow with memory, characterized by a finite Lagrangian correlation time and a finite velocity spatial correlation scale. LSMs can be formulated to reproduce the classical results of Taylor (1921) and Batchelor (1952). Work in progress by Castronovo and Kramer (pers. com.) showed that different dispersion regimes exist depending on the interactions between a spatially dependent mean flow and turbulence flow. The effects of the nonlinear interaction increase as the shear of the mean flow increases. G. I. Taylor's seminal paper (1921) on turbulence is a Lagrangian-based analysis; it is refreshing to see the increasing use of Lagrangian stochastic models for the study of turbulence (Sawford, 1999, 2001; Reynolds, 2002a,b, 2003a,b).

13.8 Numerical simulations of ocean circulation and Lagrangian trajectories

As with observations, numerical models can be used to study circulation on scales ranging from local to regional to global. A few examples from LAPCOD are now presented. Drifter trajectories were simulated for the East Florida Shelf by Fiechter and Mooers using 25 sigma levels, O(4 km) horizontal resolution, Princeton Ocean Model (see Mooers and Fiechter (2003) for an EOF analysis of the Eulerian fields and see Fiechter and Mooers (2003) for an analysis of simulated frontal eddies). The presented Lagrangian results contained trajectories in well-known features such as meanders of the Florida

Current and small-scale frontal eddies, and regional differences in how Florida Current waters and coastal waters are detrained and entrained, respectively. This physical model is also being used to investigate larvae dispersion and dynamics in the coral reefs of the Florida Keys. Results presented at the LAPCOD meetings show a growing trend in both the number of coupled models and the complexity from simple Nitrogen–Phytoplankton–Zooplankton (NPZ) models to those coupling temperature-dependent, light-dependent, nutrient-dependent, salinity-dependent behavior, and multi-interacting species and trophic levels to high-resolution, data-assimilative models.

Lagrangian observations are often used for model validation. Garraffo *et al.* (2001a,b) compared numerical trajectories from a high-resolution simulation of the Northern Atlantic by the Miami Isopycnic Coordinate Ocean Model (MICOM) to *in-situ* near-surface drifters from the AOML/NOAA data set. They found that the sampling error in the available *in-situ* data base limits the comparison. Nevertheless, over most of the domain there was a good agreement in the mean velocity field, but the Lagrangian integral time scales were too large in the model trajectories by a factor of two. McClean *et al.* (2002), using a high-resolution eddy-resolving configuration of the Los Alamos National Laboratory (LANL) Parallel Ocean Program (POP) model, also compared simulated and *in-situ* data in the North Atlantic. Two of their results are that an eddy-resolving model produced much better Eulerian and Lagrangian velocity statistics than an eddy-permitting model simulation and that their simulated trajectories had excellent agreement with the time scales determined by the data.

Mooers *et al.* (2005) presented results from an eddy-resolving, numerical simulation of the circulation in the Japan (East) Sea. Model-based temperature and salinity profiles, as well as the overall circulation pattern are in good agreement with observations from the ARGO profiling floats. However, model velocities at 800 m, for example, were 50% larger than velocities computed from float displacements. Considerable discussion has taken place at the meetings about how to partition model–data differences between model error, sensor error, sampling error, and subgrid-scale variability.

Halliwell *et al.* (2003) released simulated floats in a numerical ocean circulation model (HYCOM) to study three-dimensional fluid pathways in the upper limb of the tropical/subtropical Atlantic thermohaline circulation. A vorticity analysis shows that upper limb water approaches the equator from the south within a predominantly inertial western boundary layer, and then requires equatorial processes, such as turbulent inertial boundary layer dynamics, upwelling, heating, to reset potential vorticity properties and permit the fluid to cross the equator. They found that it is necessary to use Lagrangian floats advected by all three velocity components to properly track upper-limb pathways. A similar conclusion of the importance of using three-dimensional velocities was reached by the TRACMASS-funded groups analyzing the annual cycle of the North Atlantic (Da Costa and Blanke, 2004), low-frequency (decades to centuries) circulation in the Mediterranean Sea (Rupolo *et al.*, 2003) and in global studies of interocean exchange (Blanke *et al.*, 2001).

13.9 Lagrangian data assimilation and prediction

Lagrangian observations are usually given as a set of positions or trajectory data. It is common practice to convert this data into Eulerian velocities before assimilation. This conversion throws out the time-integrated dynamical information hidden in the trajectories and is not optimal use of the information contained in Lagrangian trajectory data. The pioneering works by Kamachi and O'Brien (1995) with a variational method, by Ide and Ghil (1998a, 1998b) for vortex systems, and by Molcard *et al.* (2003) for wind-driven gyre flows in the ocean all show that assimilation of Lagrangian data by methods that take into account the Lagrangian nature of the observational data significantly reduces nowcast or forecast errors, relative to treating as moving Eulerian velocity data, provided that there is adequate sampling in time to resolve the dominant dynamics.

An assimilation method based on using the differences between trajectory positions and simulated model trajectories are used to correct the model's velocity field to reduce these differences is detailed in Chapter 7 by Molcard *et al.* and also see Molcard *et al.* (2003, 2005) and Özgökmen *et al.* (2003). This Lagrangian-based method minimizes the difference between simulated and observed position by adjusting the model's velocity field in an influential radius around the observation's location. This method has been applied to both quasi-geostrophic and primitive equation ocean models. Contemporaneous adjustment of layer thickness assuming geostrophy significantly reduces forecast errors and accelerates model convergence in twin experiments. The method has been recently improved using a variational approach (Taillandier *et al.*, 2006a). Lagrangian-based assimilation results are encouraging and point to the need to use the Lagrangian information in drifter and float data for oceanic and coastal data assimilation problems.

Assenbaum and Reverdin (2005) combined analyses from a data-assimilative quasi-geostrophic model that assimilates satellite-based altimetric measurements of sea surface heights and hydrographic data, as well as displacement data from a variety of Lagrangian instruments from the POMME (Programme d'Oceanographie Multi-disciplinaire Meso-Echelle) experiment including RAFOS, PROVOR profiling floats, and surface drifters. The displacement data was used with velocities calculated from the model's dynamic height maps by assuming geostrophy to estimate a misfit velocity vector. These misfits are blended in with hydrographic data using a multi-variate objective analysis (OA) approach popularized by Freeland and Gould (1976) that assumes an isotropic, nondivergent flow. In general, the OA maps were more energetic than the quasi-geostrophic maps produced by the data-asssimilative model. This work and other works presented at LAPCOD all noted that the outstanding research issue for assimilating position data from profiling floats is the coarse sampling in time, on the order of ten days, of the displacement measurements, and the effect of vertical velocity shear during ascent and descent of the profiling floats.

A new method that augments the position of trajectories to the state space consisting of the prognostic variables of the model is being developed by Ide *et al.* (2002a), Kuznetsov *et al.* (2003), and Salman *et al.* (2006). This method introduces the cross-correlation between the original model variables and the Lagrangian trajectories. Therefore, it provides a natural platform for direct assimilation of the Lagrangian observation in the framework of the extended Kalman filtering (*op. cit.*; Chapter 8 by Chin *et al.*). This approach can also be used for an ocean circulation model augmented with tracer advection equations. The augmented model state vector includes tracer coordinates and is updated through the correlations to the observed tracers. The technique works efficiently when the observations are accurate and frequent enough. Low quality data and large intervals between observations can lead to the divergence of this and other data assimilation schemes. Nonlinear effects, responsible for the failure of the extended Kalman filter, are triggered by the exponential separation rate of tracer trajectories in the neighborhood of the saddle points of the velocity field. Recent progress on assimilating data near saddle points was reported by Jones and Ide at the last LAPCOD meeting.

ARGO float displacement data were assimilated into a regional, open-boundary version of MICOM for the years 2000 and 2001. During this time, intensive hydrographic surveys and float deployments were made in order to initialize, supply boundary conditions, assimilate into, and evaluate the model. The float displacement velocities calculated from the typical ten day cycles, biased the model when standard assimilation techniques were used (Assenbaum, pers. com.). More advanced assimilation methods are being investigated to handle ARGO float displacement data. Taillandier *et al.* (2006b) generalized the method of Taillandier *et al.* (2006a) to include the difference in velocity as the float profiles, surfaces, and returns to its parking depth. They assimilated four ARGO floats in a high resolution model of the

Mediterranean Sea and found that their Lagrangian assimilation method improved model performance.

Nodet (2006) used a four-dimensional variational approach and the adjoint of a primitive equation model simulation of the classic, double-gyre circulation of mid-latitudes to test different stragies for assimilating float positions. In her twin-experiments, sampling time, number of floats, and depth of float observations were varied. In contrast to earlier studies, e.g. Molcard *et al.* (2003), Nodet's results were not sensitive to sampling period, the planned ARGO sampling density is not dense enough to constrain the velocity field, a result seen by others, the best vertical level is an intermediate model level that is energetic, and for most sampling periods, a method that takes into account the Lagrangian nature of the data produced forecasts with lower estimation error than treating the Lagrangian data as moving Eulerian measurements.

One approach for avoiding the Lagrangian to Eulerian transformation needed for assimilating drifter and float data into models is to use a circulation model based in Lagrangian coordinates (Bennett and Chua, 1999; Mead and Bennett, 2001; Mead, 2004). The big advantage of having the model state space and data space the same is offset by the technical challenges of using generalized Lagrangian coordinates. This coordinate system is sheared, stretched, and compressed in the Lagrangian formulation, as well as being efficiently stirred by chaotic advection (Aref, 1984). In other words, the Jacobian of the transformation from Eulerian to Lagrangian coordinates can be ill-conditioned. Nevertheless, there has been significant progress on Lagrangian-based modeling and assimilation on operational time scales and for regional models. The field of Lagrangian data assimilation has come a long way since Carter (1989) first assimilated Lagrangian data into a shallow water equation model using a Kalman filter.

Besides estimating Eulerian fields from Lagrangian data (see also Toner *et al.*, 2001a,b), the prediction of particle trajectories using different data sets was one of the main themes of all of the meetings. Most of the earlier attempts for predicting Lagrangian motion were motivated by search and rescue operation (Schneider, 1998) and used climatological mean flows, usually based on binned averages of all available velocity data, for open ocean applications and tidal current predictions for inshore applications. Search and rescue operations also require estimates of windage that depend on ship design or how much a person is floating out of the water. A thorough discussion of this and empirical coefficients to use in practical applications can be found in Allen and Plourde (1999).

As detailed in Chapter 6 by Piterbarg *et al.* using both near surface drifters in the tropical Pacific Ocean and in numerical simulations of wind-forced double gyre flows, predicting float trajectories, based on using climatological velocity fields to advect particles, is unreliable. In particular, climatological velocity fields usually underestimate the mean flow because of the inherent smearing of calculating an Eulerian-based average of a meandering jet (Mariano and Chin, 1996) and, as described above, may be biased low by construct. Either direct velocity measurements from nearby observations, velocities based on feature models and satellite data collected within an integral time scale of the time prediction (Mariano and Chin, 1996), velocity fields from a reliable, high-resolution circulation model (Schneider, 1998), or a high-resolution, near real-time wind fields and a slab model for oceanic motion (Paldor *et al.*, 2004), is needed for reliable prediction of Lagrangian motion, a critical component of search and rescue operations (Breivik and Allen, 2005), and for predicting the spread of pollutants (Korotenko *et al.*, 2002; Spaulding, 1988). Breivik reviewed recent developments in search and rescue algorithms and highlighted the cooperation of the US Coast Guard, in particular with Art Allen, the Norwegian Rescue Co-ordination Center, and other European centers.

Paldor *et al.* (2004) argue that a particle does not have to obey continuity so that velocities from a circulation model, besides the inherent model and forcing errors, may not be the right velocities to advect particles. They used a hybrid, wind-forced slab model with three adjustable parameters to predict trajectories in the tropical Pacific. The three adjustable parameters are the relative weight of climatological advection versus wind-driven advection from the slab model, the mixed-layer depth for the slab model, and the Rayleigh friction coefficient. A brute-force optimization technique was used to find the three optimal parameters for different subsets of near-surface drifters. In almost all cases, almost no weight was given to climatology velocities and the wind-driven model significantly improved trajectory forecasts, relative to using only climatological advective velocities.

Significant progress has been made in the forward prediction problem for particle trajectories (see Chapter 6 by Piterbarg *et al.*). Recent research into the inverse Lagrangian prediction problem has concentrated on the problem of finding the unknown initial "launch" location of an object given its location at a later time and some noisy, incomplete sea surface velocity and wind data. Chin and Mariano (pers. com.) formulate two solution methods for the inverse Lagrangian prediction problem. A unique and/or numerically stable solution for such an inversion problems is difficult to obtain computationally because of current and wind errors, the chaotic nature of the forward prediction problem, convergent and divergent flows, deployment constraints, and bi-modal, down-wind drag coefficients. An "ensemble" of solutions describing a set of possible deployment locations/times for each final location might be the best one can achieve.

One solution method is a brute-force optimization technique that minimizes the mean-square distances between predicted final locations from a Monte-Carlo based simulation of trajectories using mean currents and winds, and random flight models. The second solution method uses a particle filter. A particle filter is a generalization of the Kalman filter that incorporates higher order statistical information and predicts the distribution of possible outcomes of your state variables, not just the expected value and covariances. A particle filter and its variants for the assimilation of Lagrangian trajectory data are being developed by Thompson (pers. com.) and Jones (pers. com.) and their colleagues.

13.10 Lagrangian-based dynamics

Many dynamical processes are more easily observed in a Lagrangian framework. For example, SOFAR and RAFOS data in the Gulf Stream clearly illustrate that the Gulf Stream is not a transport barrier and that particles can have significant cross-stream velocities, be entrained and detrained into local, mesoscale recirculation cells that are associated with meanders or be trapped in warm and cold core rings and travel hundreds of km from the Gulf Stream (Owens, 1984; Bower and Rossby, 1989). Lozier *et al.* (1996) showed, with numerical simulations performed using a kinematic model that incorporates the observed dynamics, floats move onshore and upwell from a meander trough to a meander crest and move offshore and downwell from crest to trough, that it is the cross-stream, projected, relative-velocity vector difference between the particle velocity and the phase speed of the meander, that determines cross-stream particle exchange. As the relative-velocity vector difference increases, as it does for large-amplitude meanders, so does cross-stream particle exchange.

Bower (1991) calculated dynamical balances along RAFOS float trajectories and showed that on the anticyclonic side of the stream (the offshore side), particles diverge downstream of a trough and converge downstream of the crest, and that on the onshore, cyclonic side of the stream, particles converge downstream of a trough and diverge upstream of the crest. It would take a very dense Eulerian array to observe this pattern. Such a dense array is now practical for monitoring the coastal zones with larger arrays of more accurate High Frequency radar-based surface velocity mappers being deployed.

Direct, reliable estimates of important dynamical terms are not possible with an Eulerian-based approach to sampling the velocity field of the ocean because errors inherent in derivative calculations are amplified in the non-linear advection term in the Eulerian formulation of conservation equations. The fact that position errors in locating drifters and floats are highly correlated in time leads to accurate velocity estimates because the correlated errors are canceled when differencing, for example, position data for velocity estimates (Mariano and Rossby, 1989). Lagrangian position data are an integral of all energetic scales of motion that are resolved by the temporal sampling. The eight hour sampling rate of SOFAR float observations, for example, would require Eulerian sampling on the scale of 1–3 km to resolve typical, mid-ocean, mesoscale features to the same level of dynamical accuracy. Given that these high spatial resolutions must be met in latitude and longitudes leads to uneconomical sampling by subsurface Eulerian platforms for the purpose of calculating detailed dynamical balances of deep oceanic processes. HF radar (see Section 13.11) sensing of the coastal surface velocity field is one technology that has the adequate sensing resolution for Eulerian-based measurements.

Of course, there are situations where Eulerian-based observations may be more advantageous than Lagrangian-based observations. Two examples that come immediately to mind are flow through "choke" points and areas with strong horizontal divergence such as the equatorial upwelling regions. Leaman *et al.* (1987) survey of the Florida Current at 27° N, Sheinbaum *et al.* (2002) current meter measurements spanning the Yucatan Channel, and Bryden *et al.* (1994) analysis in the straits of Gibraltar are some examples of oceanographic studies in strongly constrained flows that would be much more expensive to get the required information from Lagrangian instruments. Likewise, the TOGA-TAO array in the Equatorial Pacific allows oceanographers to get near surface velocity measurements in areas where strong upwelling and off-equatorial transport of near surface drifters produced low data density areas on and near the equator.

As discussed in the previous sections, Lagrangian-based measurements have given us a wealth of information about the structure, kinematics, and dynamics of coherent vortices. Lagrangian-based sampling, for example, the tagged Meddies discussed in the last section, has made it possible to estimate the decay rate of ocean vortices and provide estimates of oceanic diffusion. Estimation of the decay rate of a moving, long-lived vortex would be near impossible using *in-situ* Eulerian-based sampling schemes with a fixed grid of measurements in space. Newer satellite-based observations, like wide-swath altimeters, will soon be available with better space-time resolution for studying ocean and coastal eddies.

The ability to make hydrographic profiles in the same Meddy over the course of a few years, for the purpose of quantifying ocean diffusivity and determining their lifetime, was made possible using floats to tag the Meddies. Of course, the fact that oceanic vortices trap fluid particles and provide long-distance pathways, is both a blessing, as just described, and is a curse. If temperature data, for example, from all the SOFAR floats in tagged Meddies were used in a climatology, would they introduce a warm bias? It is well known that repeated XBT profiles in cold core rings, south of the Gulf Stream, introduced a cold bias into the climatology of that region, and that averaging of nearby data to form super observations or other processing is necessary to reduce this bias (Festa and Molinari, 1992). Also, trapped particles do not visit all of the turbulent velocity field and may lead to a bias in the probability density function of velocity. This bias is an inherent in all of Lagrangian sampling (Pasquero *et al.*, 2002). Nevertheless, Lagrangian-based sampling of eddies has yielded a wealth of information on the distribution, formation, death, life history, size, and the dynamics of coherent oceanic vortices. For example, based on Richardson *et al.* (2000) census, the average lifetime of a Meddy is about two years and that about 29 Meddies co-exist in the Atlantic at any one time. Meddies may remain intact for around five years. Remarkably, 14 Meddies were sampled in February 1994. The ability to sample approximately 50% of the Meddies in the Atlantic Ocean at one time is a clear example of both the maturing technology of *in-situ* and satellite-based observations, and our increased understanding of the physical oceanography of the region. Twenty-five years ago no one even knew these eddies existed!

13.11 Future prospects

The technology revolution is leading to new Lagrangian instruments that allow more frequent temporal sampling, are more light-weight, contain new sensors, have better water-following characteristics, can adapt sampling rates and sampling depth, and/or whose designs are optimized for application for shallow water or to track the bottom. Ohlman *et al.* (2005b) used a recoverable, light-weight, shallow-water drifter, developed with Sybrandy and Niiler, that is GPS tracked to a spatial accuracy of 10 m (offshore of southern California), transmits and records its position up to every one minute, and has a slippage of up to 0.2% of the wind speed. This slippage is comparable to that of a WOCE drifter, CODE or Tristar drifter (Lumpkin and Pazos, Chapter 2; Poulain, pers com.). Poulain, Deponte, and Ursella have used GPS drifters equipped with Aquadopp acoustic velocity meters to measure velocity shear, tilt, and pressure that allows the estimation of wave parameters and parameters for an Ekman velocity spiral model of the upper ocean, for example.

Zervakis, Ktistakis, Georgopoulos, and Kantidakis, have used another GPS-tracked drifter design, the TELEPHOS, that uses either a Davis-type drifter or a surface float with a subsurface drogue. Users can call up and change sampling requirements and get a suite of Eulerian and Lagrangian statistics of the flow. Algorithms are being designed to guide underwater gliders to sample in interesting dynamical regions defined by Direct Lyapunov Exponent contours (Bhatta *et al.*, 2006). In his algorithm, repelling lines are computed forward in time, attracting material lines are computed backward in time and the glider is steered, by redistributing internal weights, to these lines that delineate fronts. The algorithm is motivated from results from nonlinear dynamics and is fundamentally different than traditional, oceanographic signal processing. The use of gliders (Webb *et al.*, 2001; Rudnick *et al.*, 2004), computer-controlled by scientists thousands of km away in the comfort of their office or lying on the couch at home with a laptop and wireless hook-up, is a testimony to the vision of Henry Stommel and advances in many technologies in the information age.

Studies by Davis (1998) and Lavender *et al.* (2005) estimated that sampling basin-scale regions of the open ocean with approximately 100 floats would produce a reliable estimate of the mean flow, but at least a thousand floats are needed to get reliable estimates of the time variability of subsurface ocean currents. As the Lagrangian data base grows, robust statistics of oceanic variability can be estimated by combining data from different experiments and from different observing platforms. About one-half of the planned six thousand profiling ARGO floats (Davis *et al.*, 1992; Roemmich *et al.*, 2001; Centurioni and Gould, 2004; Freeland and Cummins, 2005) are already in the ocean yielding a wealth of information on the hydrography of the upper ocean, especially in regions far from any home ports and during times of weather conditions that would not permit any ship-based profiling using CTDs. This data is extremely useful for constraining the salinity and temperature of ocean circulation models. Given the lack of global, high-resolution, unbiased, low-variance estimates of the forcing terms for models, evaporation, precipitation, and heat flux terms, and the state of mixed-layer models, the hydrographic data from profiling floats is of paramount importance for operational ocean predictions. However, there is a lot of work still needed on optimizing the float displacement observations, that are coarse in time, for assimilation into numerical circulation models (Mariano *et al.*, 2002).

Results of Paduan and Cook (1997), Paduan and Shulman (2004), Shay *et al.* (2002), Peters *et al.* (2002), and Lipphardt *et al.* (2000) indicate that HF radar sensing of the ocean surface is a technological revolution for the study of surface dynamics since it samples the velocity field with adequate spatial resolution of O(km). The combination of HF radar and surface drifters that sample on one to ten minute time scales, like the Microstar drifters (Ohlman *et al.*, 2005a), will compound the information content of both. The combination of tight clusters of Lagrangian observing platforms being deployed in coastal regimes, and surface velocity fields measured by an ever growing suite of High-Frequency radar sights along American and European coasts, will significantly increase our ability to model and predict the transport and dispersion of (near-) surface particles in our coastal zones. The higher spatial resolution and "near-" continuous monitoring of the surface velocity field by HF radars, O(100 obs/day), will allow the sampling of many different realizations of significant dynamical events like eddy formation and merging, strong upwelling, coastal jets, current bifurcations, and upper ocean response to storms. Combined with Lagrangian observations, collected at similar temporal sampling rate, these measurements will yield new insight into coastal processes.

More detailed numerical trajectories at higher spatial and temporal resolution, simulated from both ocean circulation models and from real data, are now possible because of increasing computer power. Lipphardt (pers com) released many circular rings of fluids or blobs in objectively analyzed maps of surface velocity and looked at mixing processes in Monterey Bay. Changes in blob area and perimeter boundary closely showed the transition from stirring to mixing to small-scale diffusion (Eckart, 1948). There are now simulations of particles in three dimensions on both regional and global scales from the European Union-funded TRACMASS program (Blanke *et al.*, 2001; Vries and Doos, 2001; Drijfhout *et al.*, 2003).

The ability to assimilate Lagrangian, satellite, and hydrographic data into numerical circulation models at ever-increasing spatio-temporal resolution has allowed new insight into the ocean dynamics. At LAPCOD 2005, Kirwan showed how the classic picture of the death of a Loop Current Eddy on the Texas shelf was wrong since years of data-assimilative model simulations only have a few strong eddies west of 93°.

There is an increased interest in explaining the shape and geometry of float trajectories. Rupolo (Chapter 9) classifies trajectories by the ratio of the velocity time scale and the acceleration time scale, defined by a second-order LSM. Floats with a similar ratio also had similar trajectory shapes, Lagrangian velocity autocovariance functions, and diffusion properties. Centurioni (pers. com.) presented cusp-like trajectories that look like a classic case of a drifter in an eddy translating with a similar rotational speed, but the analysis of attached thermistor chains and high frequency GPS-based sampling indicates tidally generated internal waves and solitons. Recent analysis of data along the east coast of Florida by Shay and colleagues has also seen similar trajectories and reached similar conclusions.

As discussed in Chapter 1 by Rossby, Lagrangian instruments are now being customized/designed to study parcel motion in different dynamical regimes, and their water-tagging abilities are improving (Harcourt *et al.*, 2002). Ocean and coastal measurements are now more Lagrangian and less quasi-Lagrangian. Instruments are being developed to track features such as the bottom (Prater and Rossby, 2005). New Lagrangian-based geophysical fluid dynamics theories are being proposed by Paldor and his collaborators (Dvorkin and Paldor, 1999; Paldor, 2000, 2001; Paldor *et al.*, 2003). Bennett (2006) has just published a marvelous textbook, *Lagrangian Fluid Dynamics*, that concentrates on Lagrangian-based dynamics and statistical analysis. Contemporaneous advancements in computational resources for Monte-Carlo simulations for stochastic Lagrangian motion and biological systems, and for numerical circulation modeling, in dynamical systems theory for three spatial-dimensions, and in modeling vertical velocities will only fuel the synergetic activity between observationalists, modeling, and theoreticians, and will produce a deeper understanding of ocean and coastal circulation, dynamics, and biology. The authors hope that future LAPCOD meetings will only accelerate the exchange of ideas between the different research communities as much as the first three meetings.

Acknowledgments

The authors appreciate the support of the Office of Naval Research and the encouragement of Dr. Manuel Fiadeiro during the writing of this review chapter. Comments by Drs. Annalisa Griffa and Tamay Özgökmen on an earlier version of the chapter greatly improved it. We also thank all the participants of the LAPCOD meetings for sharing all of their data, new methods, and theories with everyone. In particular, AJM acknowledges Dr. Tom Rossby for introducing me to Lagrangian data; it was love at first sight.

References

Aliani, S., A. Griffa, and A. Molcard, 2003. Floating debris in the Ligurian Sea, North-Western Mediterranean. *Marine Pollution Bulletin*, **46**, 1142–9.

Allen, A. and J. V. Plourde, 1999. *Review of Leeway: Field Experiments and Implementation.* Report CG-D-08-99, US Coast Guard Research and Development Center, 1082 Shennecossett Road, Groton, CT, USA.

Aref, H., 1984. Stirring by chaotic advection. *J. Fluid Mech.*, **143**, 1–21.

Artale, V., G. Boffetta, A. Celani, M. Cencini, and A. Vulpiani, 1997. Dispersion of passive tracers in closed basins: beyond the diffusion coefficient. *Phys. Fluids*, **9**, 3162.

Ashjian, C. J., 1993. Trends in copepod species abundances across and along a Gulf Stream Meander: evidence for entrainment and detrainment of fluid parcels from the Gulf Stream. *Deep-Sea Res.*, **40**, 461–82.

Assenbaum, M. and G. Reverdin, 2005. Near real-time analyses of the mesoscale circulation during the POMME experiment, *Deep-Sea Res. I*, **52**, 1345–73.

Aurell, E., G. Boffetta, A. Crisanti, G. Paladin, and A. Vulpiani, 1997. Predictability in the large: an extension of the concept of Lyapunov exponent. *J. Phys. A*, **30**, 1–26.

Bane, J. M. and W. K. Dewar, 1988. Gulf Stream bimodality and variability downstream of the Charleston bump. *J. of Geophys. Res.*, **93**(C6), 6695–710.

Batchelor, G. K., 1952. Diffusion in a field of homogeneous turbulence. II. The relative motion of particles. *Proc. Cambridge Philos. Soc.*, **48**, 345–62.

Bauer, S., G. L. Hitchcock, and D. B. Olson, 1991. Influence of monsoonally-forced Ekman dynamics upon surface layer depth and plankton biomass distribution in the Arabian Sea. *Deep-Sea Res.*, **38**(5A), 531–53.

Bauer, S., M. S. Swenson, A. Griffa, A. J. Mariano, and K. Owens, 1998. Eddy-mean flow decomposition and eddy-diffusivity estimates in the tropical Pacific Ocean. *J. Geophys. Res.*, **103**, 30855–71.

Bennett, A. F., 1984. Relative dispersion: local and nonlocal dynamics. *J. Atmos. Sci.*, **41**(11), 1881–6.

Bennett, A. F., 1987. A Lagrangian analysis of turbulent diffusion. *Rev. Geophys.*, **25**, 799–822.

Bennett, A. F. and B. S. Chua, 1999. Open boundary conditions for Lagrangian geophysical fluid dynamics. *J. Comput. Phys.*, **153**(2), 418–36.

Bennett, A., 2006. *Lagrangian Fluid Dynamics (Cambridge Monographs on Mechanics)*. Cambridge: Cambridge University Press.

Berloff, P., J. C. McWilliams, and A. Bracco, 2002. Material transport in oceanic gyres. Part I: Phenomelogy. *J. Phys. Oceanogr.*, **32**, 764–96.

Berloff, P. and J. C. McWilliams, 2002. Material transport in oceanic gyres. Part II: Hierarchy of stochastic models. *J. Phys. Oceanogr.*, **32**, 797–830.

Berloff, P. and J. C. McWilliams, 2003. Material transport in oceanic gyres. Part III: Randomized stochastic models. *J. Phys. Oceanogr.*, **33**, 1416–45.

Beron-Vera, F. J., M. J. Olascoaga, and M. G. Brown, 2004. Tracer patchiness and particle trajectory stability in incompressible two-dimensional flows. *Nonlin. Proc. Geophys.*, **11**, 67–74.

Bhatta, P., E. Fiorelli, F. Lekien, N. E. Leonard, D. A. Paley, F. Zhang, R. Bachmayer, R. E. Davis, D. Fratantoni, and R. Sepulchre, 2006. Coordination of an Underwater Glider Fleet for Adaptive Sampling, International Workshop on Underwater Robotics, (in press), 61–69.

Bitterman, D. S. and D. V. Hansen, 1989. Direct measurements of current shear in the Tropical Pacific Ocean and its effect on drift buoy performance. *J. Atmos. Ocean. Tech.*, **6**, 274–9.

Bitterman, D. B. and D. V. Hansen, 1993. Evaluation of sea surface temperature measurements from drifting buoys. *J. Atmos. Ocean. Tech.*, **10**(1), 88–96.

Blanke, B., S. Speich, G. Madec, and K. Doos, 2001. A global diagnostic of interocean mass transfers. *J. Phys. Oceanogr.*, **31**(6), 1623–32.

Boebel O., R. E. Davis, M. Ollitrault, R. G. Peterson, P. L. Richardson, C. Schmid, and W. Zenk, 1999. The Intermediate Depth Circulation of the Western South Atlantic. *Geophys. Res. Lett.*, **26**(21), 3329–32.

Boffetta, G., M. Cencini, S. Espa, and G. Querzoli, 2000. Chaotic advection and relative dispersion in an experimental convective flow. *Amer. Inst. of Physics*, **12/12**, 3160–7.

Bower, A. S., 1991. A simple kinematic mechanism for mixing fluid parcels across a meandering jet. *J. Phys. Oceanogr.*, **21**, 173–80.

Bower, A. S. and M. S. Lozier, 1994. A closer look at particle exchange in the Gulf Stream. *J. Phys. Oceanogr.*, **24**, 1399–418.

Bower, A. S. and H. T. Rossby, 1989. Evidence of cross-frontal exchange processes in the Gulf Stream based on isopycnal RAFOS float data. *J. Phys. Oceanogr.*, **19**, 1177–90.

Bower, A. S., B. Le Cann, T. Rossby, W. Zenk, J. Gould, K. Speer, P. L. Richardson, M. D. Prater, and H.-M. Zhang, 2002a. Directly measured mid-depth circulation in the northeastern North Atlantic Ocean, *Nature*, **419**, 603–7.

Bower, A. S., N. Serra, and I. Ambar, 2002b. Structure of the Mediterranean Undercurrent and Mediterranean Water spreading around the southwestern Iberian Peninsula. *J. Geophys. Res.*, **107**(C10), 3161, doi:10.1029/2001JC001007.

Bracco, A., J. H. LaCasce, C. Pasquero, and A. Provenzale, 2000. The velocity distribution of barotropic turbulence. *Phys. Fluids*, **12**, 2478–88.

Breivik, O. and A. Allen, 2005. An operational search and rescue model for the Norwegian Sea and the North Sea. *submitted to J. Mar. Syst.*

Brink, K. H., R. C. Beardsley, P. P. Niiler, M. R. Abbott, A. Huyer, S. R. Ramp, T. P. Stanton, and D. Stuart, 1991. Statistical properties of near-surface flow in the California coastal transition zone. *J. Geophys. Res.*, **96**(C8), 14693–706.

Brink, K. H., R. C. Beardsley, J. Paduan, R. Limeburner, M. Caruso, and J. Sires, 2000. A view of the 1993–1994 California current based on surface drifters, floats, and remotely sensed data. *J. Geophys. Res.*, **105**(C4), 8575–604.

Brodie, J. W., 1960. Coastal surface currents around New Zealand. *New Zealand Journal of Geology and Geophysics*, **3**(2), 235–52.

Bryden, H. L., J. C. Candela, and T. H. Kinder, 1994. Exchange through the Strait of Gibraltar. *Prog. Oceanogr.*, **33**, 201–48.

Bumpus, D. F. and L. M. Lauzier, 1965. Surface Circulation on the Continental Shelf off Eastern North America between Newfoundland and Florida. *Serial Atlas of the Marine Environment*. New York: American Geographical Society.

Burt, W. V. and B. Wyatt, 1964. Drift bottle observations of the Davidson Current off Oregon. In *Studies on Oceanography*, ed. K. Yoshida. Seattle: Univ. Washington Press, 156–65.

Carter, E. F., 1989. Assimilation of Lagrangian data into a numerical model. *Dyn. Atmos. Oceans*, **13**, 335–48.

Carter, E. F. and A. R. Robinson, 1987. Analysis models for the estimation of oceanic fields. *J. Atmos. Ocean. Tech.*, **4**, 49–74.

Castronovo, E. and P. R. Kramer, 2004. Subdiffusion and superdiffusion in Lagrangian stochastic models of oceanic transport, *Monte Carlo Methods and Appl.*, **10**(34), 245–56.

Cecconi, F., M. Cencini, M. Falcioni, and A. Vulpiani, 2005. Brownian motion and diffusion: from stochastic processes to chaos and beyond. *Chaos*, **15**, 26–102.

Centurioni, L. R. and P. P. Niiler, 2003. On the surface currents of the Caribbean Sea. *Geophys. Res. Lett.*, **30**(6), doi:10.1029/2002GL016231, 1279.

Centurioni, L. R. and W. J. Gould, 2004. Winter conditions in the Irminger Sea observed with profiling floats. *J. Mar. Res.*, **62**, 313–36.
Cheney, R. E., W. H. Gemmill, M. K. Shank, P. L. Richardson, and D. Webb, 1976. Tracking a Gulf Stream ring with SOFAR floats. *J. Phys. Oceanogr.*, **6**, 741–9.
Cheney, R. E., P. L. Richardson, and K. Nagasaka, 1980. Tracing a Kuroshio ring with a free-drifting surface buoy. *Deep-Sea Res.*, **27**, 641–54.
Cowen, R. K., K. M. M. Lwiza, S. Sponaugle, C. B. Paris, and D. B. Olson, 2000. Connectivity of marine populations: Open or closed? *Science*, **287**, 857–9.
Cowen, R. K., C. B. Paris, D. B. Olson, and J. L. Fortuna, 2003. The role of long distance dispersal versus local retention in replenishing marine populations. *J. Gulf Caribbean Res.*, **14**(2), 129–37.
Cresswell, G. R., 1977. The trapping of two drifting buoys by an ocean eddy. *Deep-Sea Res.*, **24**, 1203–9.
Da Costa, M. V. and B. Blanke, 2004. Lagrangian methods for flow climatologies and trajectory error assessment. *Ocean Modelling*, **6**, 335–58.
Davis, R. E., 1983. Oceanic property transport, Lagrangian particle statistics, and their prediction. *J. Mar. Res.*, **41**, 163–94.
Davis, R. E., 1985a. Drifter observations of coastal currents during CODE: The method and descriptive view. *J. Geophys. Res.*, **90**, 4741–55.
Davis, R. E., 1985b. Drifter observations of coastal surface currents during CODE: The statistical and dynamical views. *J. Geophys. Res.*, **90**(C7), 4756–72.
Davis, R. E., 1991. Observing the general circulation with floats. *Deep Sea Res.*, **38**(Suppl.1), 531–71.
Davis, R., D. C. Webb, L. A. Regier, and J. Dufour. 1992. The Autonomous Lagrangian Circulation Explorer (ALACE). *J. Atmos. Ocean. Tech.*, **9**, 264–85.
Davis, R. E., 1998. Preliminary results from directly measuring mid-depth circulation in the tropical and South Pacific. *J. Geophys. Res.*, **103**, 619–39.
Davis, R. E. and W. Zenk, 2001. Subsurface Lagrangian Observations during the 1990s. In *Ocean Circulation and Climate: Observing and Modeling the Global Ocean*, ed. J. Church, G. Siedler, and J. Gould. San Diego: Academic Press, Chapter 3.2.
Dewar, W. K. and G. R. Flierl, 1985. Particle trajectories and simple models of transport in coherent vortices. *Dyn. Atmos. Oceans*, **9**, 21–52.
Doglioli, A. M., M. G. Magaldi, L. Vezzulli, and S. Tucci, 2004. Development of a numerical model to study the dispersion of wastes coming from a marine fish farm in the Ligurian Sea (Western Mediterranean). *Aquaculture*, **231**(1), 215–35.
Drijfhout, S. S., P. de Vries, K. Doos, and A. C. Coward, 2003. Impact of eddy-induced transport on the Lagrangian structure of the upper branch of the thermohaline circulation. *J. Phys. Oceanogr.*, **33**, 2141–55.
Duan, J. Q. and S. Wiggins, 1996. Fluid exchange across a meandering jet with quasi-periodic time variability. *J. Phys. Oceanogr.*, **26**, 1176–88.
Dvorkin, Y. and N. Paldor, 1999. Analytical considerations of Lagrangian cross-equatorial flow. *J. Atmos. Sci.*, **56**(9), 1229–37.
Ebbesmeyer, C. C. and C. A. Coomes, 1993. Historical shoreline recoveries of drifting objects: an aid for future shoreline utilization. *Oceans 93 Proceedings*, **III**, 159–64.
Eckart, C., 1948. An analysis of the stirring and mixing processes in incompressible fluids. *J. Mar. Res.*, **7**, 265–75.
Falco, P., A. Griffa, P.-M. Poulain, and E. Zambianchi, 2000. Transport properties in the Adriatic Sea as deduced from drifter data. *J. Phys. Oceanogr.*, **30**, 2055–71.

Festa, J. F. and R. L. Molinari, 1992. An evaluation of the WOCE volunteer observing ship-XBT network in the Atlantic. *J. Atmos. Ocean. Tech.*, **9**, 305–17.

Fiechter, J. and C. N. K. Mooers, 2003. Simulation of frontal eddies on the East Florida Shelf. *Geophys. Res. Lett.*, **30**(22), doi:10.1029/2003GL018307.

Figueroa, H. A. and D. B. Olson, 1989. Lagrangian statistics in the South Atlantic as derived from SOS and FGGE drifters. *J. Mar. Res.*, **47**(3), 525–46.

Flierl, G. R., 1981. Particle motions in large amplitude wave fields. *Geophys. Astrophys. Fluid Dyn.*, **18**, 39–74.

Fratantoni, D. M., 2001. North Atlantic surface circulation during the 1990's observed with satellite-tracked drifters. *J. Geophys. Res.*, **106**(C10), 22067–93.

Freeland, H. J., P. B. Rhines, and T. Rossby, 1975. Statistical observations of the trajectories of neutrally buoyant floats in the North Atlantic. *J. Mar. Res.*, **33**, 383–404.

Freeland, H. J. and W. J Gould, 1976. Objective analysis of mesoscale ocean circulation features. *Deep-Sea Res.*, **23**, 915–24.

Freeland, H. J. and P. F. Cummins, 2005. Argo: A new tool for environmental monitoring and assessment of the world's oceans, an example from the NE Pacific. *Prog. Oceanog.*, **64**, 31–44.

Garfield, N., M. E. Maltrud, C. A. Collins, T. A. Rago, and R. G. Paquette, 2001. Lagrangian flow in the California Undercurrent, an observation and model comparison. *J. Mar. Syst.*, **29**, 201–20.

Garraffo, Z. D., A. J. Mariano, A. Griffa, C. Veneziani, and E. P. Chassignet, 2001a. Lagrangian data in a high resolution numerical simulation of the North Atlantic. I: Comparison with in-situ drifter data. *J. Mar. Syst.*, **29**/1–4, 157–76.

Garraffo, Z. D., A. Griffa, A. J. Mariano, and E. P. Chassignet, 2001b. Lagrangian data in a high resolution numerical simulation of the North Atlantic. II: On the pseudo-Eulerian averaging of Lagrangian data. *J. Mar. Syst.*, **29**/1–4, 177–200.

Gould, W. J., 2005. From Swallow floats to Argo – the development of neutrally buoyant floats, *Deep-Sea Res. II*, **52**/3–4, 529–43.

Gould, W. J., *et al.*, 2001. Hydrographic Observations. In *Observing the Oceans in the 21st Century*, ed. C. Koblinsky and N. Smith. Victoria, Australia: CSIRO Publishing.

Griffa, A., 1996. Applications of stochastic particle models to oceanographic problems. In *Stochastic Modeling in Physical Oceanography*, ed. R. Adler, P. Muller, and B. Rozovoskii. Cambridge, MA: Birkhäuser Boston.

Grodsky, S. A. and J. A. Carton, 2002. Surface drifter pathways originating in the equatorial Atlantic cold tongue. *Geophys. Res. Lett.*, **29**, 23, 2147, doi:10.1029/2002GL015788.

Haller, G., 2000. Finding finite-time invariant manifolds in two-dimensional velocity fields. *Chaos*, **10**, 99–108.

Halliwell, G. R., R. H. Weisberg, and D. Mayer, 2003. A synthetic float analysis of upper-limb meridional overturning circulation interior ocean pathways in the tropical/subtropical Atlantic. In *Interhemisphere Water Exchange in the Atlantic Ocean*, ed. G. Goni and P. Malanotte-Rizzoli. Amsterdam: Elsevier, 193–204.

Harcourt, R. R., S. L. Elizabeth, R. W. Garwood, and E. A. D'Asaro, 2002. Fully Lagrangian floats in Labrador Sea Deep Convection: comparison of numerical and experimental results. *J. Phys. Oceanogr.*, **32**, 493–510.

Hitchcock, G. L., W. L. Wiseman, Jr., W. C. Boicourt, A. J. Mariano, N. Walker, T. Nelsen, and E. H. Ryan, 1997. Property fields in the effluent plume of the Mississippi River. *J. Mar. Syst.*, **12**, 109–26.

Hunt, J. R. C., 1998. Lewis Fry Richardson and his contributions to mathematics, meteorology, and models of conflict. *Annu. Rev. Fluid Mech.*, **30**, xiii–xxvi.

Hurlburt, H. E. and E. J. Metzger, 1998. Bifurcation of the Kuroshio Extension at the Shatsky Rise. *J. Geophys. Res.*, **103**, 7549–66.

Ide, K. and M. Ghil, 1998a. The extended Kalman filtering for vortex Systems, Part I. Methodology and point vortices. *Dyn. Atmos. Oceans*, **27**(1–4), 301–32.

Ide, K. and M. Ghil, 1998b. The extended Kalman filtering for vortex Systems, Part II. Rankine vortex and observing-system design. *Dyn. Atmos. Oceans*, **27**(1–4), 333–50.

Ide, K., L. Kuznetsov, and C. K. R. T. Jones, 2002a. Lagrangian data assimilation for point-vortex system. *J. Turbulence*, **3**, 053.

Ide, K., D. Small, and S. Wiggins, 2002b. Distinguished hyperbolic trajectories in time-dependent fluid flows: analytical and computational approach for velocity fields defined as data sets. *Nonlinear Proc. Geoph.*, **9**, 237–63.

Idrisi, N., M. J. Olascoaga, Z. D. Garraffo, D. B. Olson, and S. L. Smith, 2004. Mechanisms for emergence from diapause of Calanoides carinatus in the Somali Current. *Limnol. Oceanogr.*, (Biocomplexity Special Issue) **49**(4, Part 2), 1262–8.

Isern-Fontanet, J., E. Garcia-Ladona, and J. Font, 2006. The vortices of the Mediterranean sea: an altimetric perspective. *J. Phys. Oceanogr.*, **36**(1), 87–103.

Kamachi, M. and J. J. O'Brien, 1995. Continuous assimilation of drifting buoy trajectories into an equatorial Pacific Ocean model. *J. Mar. Sys.*, **6**, 159–78.

Kelly, K. A., S. Singh, and R. X. Huang, 1998. Seasonal variations of sea surface height in the Gulf Stream region. *J. Phys. Oceanogr.*, **29**(3), 313–27.

Kirwan, A. D. Jr., G. McNally, and J. Coehlo, 1976. Gulf Stream kinematics inferred from a satellite tracked drifter. *J. Phys. Oceanogr.*, **6**(5), 750–5.

Kirwan, A. D. Jr., G. McNally, E. Reyna, and W. J. Merrell, Jr., 1978. The near-surface circulation of the eastern North Pacific. *J. Phys. Oceanogr.*, **8**(6), 937–45.

Kirwan, A. D. Jr., W. J. Merrell, Jr., J. K. Lewis, and R. E. Whitaker, 1984. Lagrangian observations of an anticyclonic ring in the western Gulf of Mexico. *J. Geophys. Res.*, **89**(NC3), 3417–24.

Kirwan, A. D., Jr., M. Toner, and L. Kantha, 2003. Predictability, uncertainty, and hyperbolicity in the ocean. *Int. J. Engin. Sci.*, **41**, 249–58.

Kolmogorov, A. N., 1941. The local structure of turbulence in incompressible viscous fluid for very large Reynolds numbers. *C. R. Acad. Sci. URSS*, **30**, 301.

Korotenko, K. A., R. M. Mamedov, and C. N. K. Mooers, 2002. Prediction of the transport and dispersal of oil in the South Caspian Sea resulting from blowouts. *Environmental Fluid Mechanics*, **1**, 383–414.

Kraichnan, R. H., 1966. Dispersion of particle pairs in homogeneous turbulence. *Phys. Fluids*, **9**, 1937–43.

Kraichnan, R. H., 1967. Inertial ranges in two-dimensional turbulence. *Phys. Fluids*, **10**(7), 1417–23.

Kuznetsov, L., M. Toner, A. D. Kirwan, C. K. R. T. Jones, L. H. Kantha, and J. Choi, 2002. The loop current and adjacent rings delineated by Lagrangian analysis of the near-surface flow. *J. Mar. Res.*, **60**(3), 405–29.

Kuznetsov, L., K. Ide, and C. K. R. T. Jones, 2003. A method for assimilation for Lagrangian data. *Mon. Wea. Rev.*, **131**, 2247–60.

LaCasce, J., 2000. Floats and f/H. *J. Mar. Res.*, **58**, 61–95.

LaCasce, J. H. and A. Bower, 2000. Relative dispersion in the subsurface North Atlantic. *J. Mar. Res.*, **58**(6), 863–94.

LaCasce, J. H. and C. Ohlman, 2003. Relative dispersion at the surface of the Gulf of Mexico. *J. Mar. Res.*, **61**, 285–312.

Lacorata, G., E. Aurell, and A. Vulpiani, 2001. Drifter dispersion in the Adriatic Sea: Lagrangian data and chaotic model. *Ann. Geophys.*, **19**, 121–9.

Lacorata, G., E. Aurell, B. Legras, A. Vulpiani, 2004. Evidence for a $k^{-5/3}$ spectrum from the EOLE Lagrangian balloons in the low stratosphere. *J. Atmos. Sci.*, **61**(23), 2936–42.

Lavender, K. L., W. B. Owens, and R. E. Davis, 2005. The mid-depth circulation of the subpolar North Atlantic Ocean as measured by subsurface floats. *Deep Sea Res. I*, **52**(5), 767–85.

Leaman, K. D., R. Molinari, and P. Vertes, 1987. Structure and variability of the Florida Current at 27N: April 1982–July 1984. *J. Phys. Oceanogr.*, **17**, 565–83.

Leaman, K. D. and P. Vertes, 1996. Topographic influences on recirculation in the Deep Western Boundary Current: Results from RAFOS float trajectories between the Blake-Bahama Outer Ridge and the San Salvador "Gate". *J. Phys. Oceanogr.*, **26**, 941–61.

Le Cann, B., M. Assenbaum, J.-C. Gascard, and G. Reverdin, 2005. Observed mean and mesoscale upper ocean circulation in the midlatitude northeast Atlantic. *J. Geophys. Res.*, **110**, C07S05, doi:10.1029/2004JC002768.

Lekien, F., C. Coulliette, A. J. Mariano, E. H. Ryan, L. Shay, G. Haller, and J. Marsden, 2005. Lagrangian structures in very high-frequency radar data along the coast of Florida and automated optimal pollution timing, *Physica D*, **210**(1–2), 1–20.

Lin, J. T., 1972. Relative dispersion in the enstrophy-cascading inertial range of homogeneous two-dimensional turbulence. *J. Atmos. Sci.*, **29**, 394–6.

Lipphardt, B. L., Jr., A. D. Kirwan, Jr., C. E. Grosch, J. D. Paduan, and J. K. Lewis, 2000. Blending HF radar and model velocities in Monterey Bay through normal mode analysis. *J. Geophys. Res.*, **105**(C2), 3425–50.

Lobel, P. S. and A. R. Robinson, 1986. Transport and entrapment of fish larvae by ocean mesoscale eddies and currents in Hawaiian waters. *Deep-Sea Res.*, **33**(4), 483–500.

Lozier, M. S., T. J. Bold, and A. S. Bower, 1996. The influence of propagating waves on cross-stream excursions. *J. Phys. Oceanogr.*, **26**(9), 1915–23.

Lumpkin, R., 2003. Decomposition of surface drifter observations in the Atlantic Ocean. *Geophys. Res. Lett.*, **30**(14), 1753, doi:10,1029/2003GL017519.

Malhotra, N., and S. Wiggins, 1998. Geometric structures, lobe dynamics, and Lagrangian transport in flows with aperiodic time-dependence with applications to Rossby wave flow. *J. Nonlin. Sci.*, **8**, 401–56.

Mancho, A., D. Small, and S. Wiggins, 2004. Computation of hyperbolic trajectories and their stable and unstable manifolds for oceanographic flows represented as data sets. *Nonlinear Process. Geophys.*, **11**(1), 17–33.

Mariano, A. J. and H. T. Rossby, 1989. The Lagrangian Potential Vorticity Balance during POLYMODE, *J. Phys. Oceanogr.*, **19**(7), 927–39.

Mariano, A. J. and T. M. Chin, 1996. Feature and contour based data analysis and assimilation in physical oceanography. In *Stochastic Modeling in Physical Oceanography*, ed. R. Adler, P. Muller, and B. Rozovskii. Cambridge, MA: Birkhäuser Boston, 311–42.

Mariano, A. J., A. Griffa, T. Ozgokmen, and E. Zambianchi, 2002. Lagrangian analysis and predictability of coastal and ocean dynamics. *J. Atmos. Ocean. Tech.*, **19**(7), 1114–26.

Mazzino, A., S. Musacchio, and A. Vulpiani, 2005. Multiple-scale analysis and renormalization for pre-asymptotic scalar transport. *Phys. Rev. E*, **71**, 111–13.

McClean, J. L., P.-M. Poulain, J. W. Pelton, and M. E. Maltrud, 2002. Eulerian and Lagrangian statistics from surface drifters and a high-resolution POP simulation in the North Atlantic. *J. Phys. Oceanogr.*, **32**(9), 2472–91.

McGillicuddy, D. J. and A. R. Robinson, 1997. Eddy-induced nutrient supply and new production in the Sargasso Sea. *Deep-Sea Res. I*, **44**(8), 1427–50.

McGillicuddy, D. J., A. R. Robinson, D. A. Siegel, H. W. Jannasch, R. Johnson, T. D. Dickey, J. McNeil, A. F. Michaels, and A. H. Knapp, 1998. Influence of mesoscale eddies on new production in the Sargasso Sea. *Nature*, **394**, 263–6.

McNally, G. J., W. C. Patzert, A. D. Kirwan, Jr., and A. C. Vastano, 1983. The near-surface circulation of the north Pacific using satellite tracked drifting buoys. *J. Geophys. Res.*, **88**(C9), 7507–18.

Mead, J. L., 2004. The shallow water equations in Lagrangian coordinates. *J. Comput. Phys.*, **200**, 654–69.

Mead, J. L. and A. F. Bennett, 2001. Towards regional assimilation of Lagrangian data: the Lagrangian form of the shallow water model and its inverse. *J. Mar. Syst.*, **29**, 365–84.

Molcard A., L. Piterbarg, A. Griffa, T. M. Özgökmen, and A. J. Mariano, 2003. Assimilation of drifter positions for the reconstruction of the Eulerian circulation field. *J. Geophys. Res.*, **108**(C3), doi:10.1029/200IJC001240.

Molcard, A., A. Griffa, and T. M. Özgökmen, 2005. Lagrangian data assimilation in multi-layer primitive equation ocean models. *J. Atmos. Ocean. Tech.*, **22**, 70–83.

Molinari, R. L. and A. D. Kirwan, 1975. Calculations of differential kinematic properties from Lagrangian observations in the western Caribbean Sea. *J. Phys. Oceanogr.*, **5**, 483–91.

Mooers, C. N. K. and J. Fiechter, 2003. Numerical simulations of mesoscale variability in the Straits of Florida. *Ocean Dynamics*, **55**(3–4), doi:10.1007/s10236-005-0019-0, 309–325.

Mooers, C. N. K., I. Bang, and F. J. Sandoval, 2005. Comparisons between observations and numerical simulations of Japan (East) Sea flow and mass fields in 1999 through 2001. *Deep-Sea Res. II*, **52**(11–13), 1639–61.

Neumann, G., 1968. *Ocean Currents*. Amsterdam: Elsevier.

Niiler, P., 2001. The world ocean surface circulation. In *Ocean Circulation and Climate: Observing and Modeling the Global Ocean*, ed. J. Church, G. Siedler, and J. Gould. San Diego: Academic Press, 193–204.

Niiler, P. P., R. E. Davis, and H. J. White, 1987. Water-following characteristics of a mixed-layer drifter. *Deep-Sea Res.*, **34**, 1867–81.

Niiler, P. P., A. S. Sybrandy, K. Bi, P. M. Poulain, and D. Bitterman, 1995. Measurements of the water-following capability of holey-sock and TRISTAR drifters. *Deep-Sea Res.*, **42**, 1951–64.

Nodet, M., 2006. Variational assimilation of Lagrangian data in oceanography. *Inverse Probl.*, **22**, 245–63 doi:10.1088/0266-5611/22/1/014.

Nùñez-Riboni, I., O. Boebel, M. Ollitrault, Y. You, P. Richardson, and R. Davis, 2005. Lagrangian circulation of Antarctic Intermediate Water in the subtropical South Atlantic. *Deep-Sea Res. II*, **52**, 545–64.

Oboukhov, A. M., 1941. Spectral energy distribution in a turbulent flow. *Izv. Acad. Nauk SSR Geogr. Geofiez.*, **5**, 453–66.

Ohlmann, J. C., P. F. White, A. L. Sybrandy, and P. P. Niller, 2005a. GPS-cellular drifter technology for coastal ocean observing systems, *J. Atmos. Ocean. Tech.*, in press.

Ohlmann, J. C., P. F. White, L. Washburn, E. Terrill, B. Emery, and M. Otero, 2005b. Interpretation of coastal HF radar derived surface currents with high resolution drifter data, *J. Atmos. Ocean. Tech.*, in preparation.
Okubo, A., 1970. Horizontal dispersion of floatable particles in the vicinity of velocity singularities such as convergences. *Deep-Sea Res.*, **17**, 445–54.
Okubo, A., 1971. Oceanic diffusion diagrams. *Deep-Sea Res.*, **18**, 789–802.
Okubo, A. and C. C. Ebbesmeyer, 1976. Determination of vorticity, divergence and deformation rates from analysis of drogue observations. *Deep-Sea Res.*, **23**, 345–52.
Ollitrault, M., C. Gabillet, and A. C. De Verdiere, 2005. Open ocean regimes of relative dispersion. *J. Fluid Mech.*, **533**, 381–407.
Olson, D. B., V. H. Kourafalou, W. H. Johns, G. Samuels, and M. Veneziani, 2005. Aegean surface circulation from a satellite-tracked drifter array. *submitted to J. Phys. Oceanogr.*
d'Ovidio, F., V. Fernàndez, E. Herǹandez-Garcìa, and C. Lòpez, 2004. Mixing structures in the Mediterranean Sea from finite-size Lyapunov exponents. *Geophys. Res. Lett.*, **31**, doi:10.1029/2004GL020328.
Owens, W. B., 1984. A synoptic and statistical description of the Gulf Stream and Subtropical Gyre using SOFAR floats. *J. Phys. Oceanogr.*, **14**(1), 104–13.
Owens, W. B., 1991. A statistical description of the mean circulation and eddy variability in the Northwest Atlantic using SOFAR floats. *Prog. Oceanog.*, **28**(3), 257–303.
Özgökmen, T. M., A. Molcard, T. M. Chin, L. I. Piterbarg, and A. Griffa, 2003. Assimilation of drifter positions in primitive equation models of midlatitude ocean circulation. *J. Geophys. Res.*, **108**(C7), 32–8.
Paduan, J. D. and M. S. Cook, 1997. Mapping surface currents in Monterey Bay with CODAR-type HF radar. *Oceanography*, **10**, 49–52.
Paduan, J. D. and I. Shulman, 2004. HF radar data assimilation in the Monterey Bay area. *J. Geophys. Res.*, **109**, C07S09, doi:10.1029/2003JC001949.
Paldor, N., 2000. The transport in the Ekman surface layer on the spherical Earth. *J. Mar. Res.*, **60**, 47–72.
Paldor, N., 2001. The zonal drift associated with time-dependent particle motion on the earth. *Q. J. Roy. Meteor. Soc.*, **127A**(577), 2435–50.
Paldor, N., A. Sigalov, and D. Nof, 2003. The mechanics of eddy transport from one hemisphere to the other. *Q. J. Roy. Meteor. Soc.*, **129B**(591), 2011–25.
Paldor, N., Y. Dvorkin, A. J. Mariano, T. M. Özgökmen, and E. Ryan, 2004. A practical, hybrid model for predicting the trajectories of near-surface ocean drifters. *J. Atmos. Ocean. Tech.*, **21**(8), 1246–58.
Paris, C. B. and R. K. Cowen, 2004. Direct evidence of a biophysical retention mechanism for coral reef fish larvae. *Limnol. Oceanogr.*, **49**, 1964–79.
Pasquero, C., A. Provenzale, and A. Babiano, 2001. Parameterization of dispersion in two-dimensional turbulence. (Under consideration for publication in *J. Fluid Mech.*, 1–30.)
Pasquero, C., A. Provenzale, and J. B. Weiss, 2002. Vortex statistics from Eulerian and Lagrangian time series. *Phys. Rev. Lett.*, **89**, 284–501.
Pasquero, C., A. Bracco, and A. Provenzale, 2004. Coherent vortices, Lagrangian particles and the marine ecosystem. In *Shallow Flows*, ed. G. H. Jirka and W. S. J. Uijttewaal. Leiden, NL: Balkema Publishers, 399–412.
Perez-Brunius, P., T. Rossby, and D. R. Watts, 2004. A method for obtaining the mean transports of ocean currents by combining isopycnal float data with historical hydrography. *J. Atmos. Ocean. Tech.*, **21**(2), 298–316.

Peters, H., L. K. Shay, A. J. Mariano, and T. M. Cook, 2002. Current variability on a narrow shelf with large ambient vorticity. *J. Geophys. Res.*, **107**(C8), doi:10.1029/2001JC000813.

Petrissans, A., A. Tanire, and B. Oesterl, 2002. Effects of nonlinear drag and negative loop correlations on heavy particle motion in isotropic stationary turbulence using a new Lagrangian stochastic model. *Aerosol Sci. Tech.*, **36**(9), 963–71.

Piterbarg, L. I., 2005. On predictability of particle clusters in a stochastic flow. *Stochastics and Dynamics*, **5**(1), 111–31.

Pochapsky, T. E., 1961. Exploring subsurface waves with neutrally buoyant floats, *Instrument Society of America Journal*, **8**, 34–7.

Pochapsky, T. E., 1963. Measurements of small scale oceanic motions with neutrally buoyant floats. *Tellus*, **4**, 5352–62.

Pochapsky, T. E., 1966. Measurements of deep water movements with instrumented neutrally buoyant floats. *J. Geophys. Res.*, **71**, 2491–504.

Poje, A. and G. Haller, 1999. Geometry of cross-stream Lagrangian mixing in a double gyre ocean model. *J. Phys. Oceanogr.*, **29**, 1649–65.

Poje, A. C., M. Toner, A. D. Kirwan Jr., and C. K. R. T. Jones, 2002. Drifter launch strategies based on Lagrangian templates. *J. Phys. Oceanogr.*, **32**, 1855–69.

Poulain, P.-M., 2001. Adriatic Sea surface circulation as derived from drifter data between 1990 and 1999. *J. Mar. Syst.*, **29**, 332.

Prater, M. D. and T. Rossby, 2005. Observations of the Faroe Bank Channel Overflow using bottom-following RAFOS floats. *Deep Sea Res. II*, **52**(3–4), 481–494. doi:10.1016/j.dsr2.2004.12.009.

Provenzale, A., 1999. Transport by coherent barotropic vortices. *Annu. Rev. Fluid Mech.*, **31**, 55–93.

Reid, J., R. A. Schwarzlose, and D. M. Brown, 1963. Direct measurements of a small surface eddy off northern Baja California. *J. Mar. Res.*, **21**, 205–18.

Reverdin, G., P. P. Niiler, and H. Valdimarsson, 2003. North Atlantic Ocean surface currents. *J. Geophys. Res.*, **108**(C1), doi:10.1029/2001JC001020.

Reynolds, A. M., 2002a. On Lagrangian stochastic modelling of material transport in oceanic gyres. *Physica D.*, **172**, 124–38.

Reynolds, A. M., 2002b. Lagrangian stochastic modeling of anomalous diffusion in two-dimensional turbulence. *Phys. Fluids.*, **14**, 1442–9.

Reynolds, A. M., 2003a. Third-order Lagrangian stochastic modeling. *Phys. Fluids.*, **15**, 2773–7.

Reynolds, A. M., 2003b. On the application of nonextensive statistics to Lagrangian turbulence. *Phys. Fluids.*, **115**, L1–4.

Richardson, L. F., 1926. Atmospheric diffusion shown on a distance-neighbour graph. *Proc. R. Soc.*, **A 110**, 709–37.

Richardson, P. L., 1976. Tracking a Gulf Stream Ring with SOFAR Floats. *J. Phys. Oceanogr.*, **6**(5), 741–9.

Richardson, P. L., 1980. Gulf Steam ring trajectories. *J. Phys. Oceangr.*, **10**, 90–104.

Richardson, P. L., 1981. North Atlantic subtropical gyre: SOFAR floats tracked by moored listening stations. *Science*, **213**(4506), 435–7.

Richardson, P. L., 1989. Worldwide ship drift distributions identify missing data. *J. Geophys. Res.*, **94**(C5), 6169–76.

Richardson, P. L., 1993. A census of eddies observed in North Atlantic SOFAR float data. *Prog. Oceanog.*, **31**, 1–50.

Richardson, P. L. and T. K. McKee, 1984. Average seasonal variation of the Atlantic equatorial currents from historical ship drifts. *J. Phys. Oceanogr.*, **14**(7), 1226–38.
Richardson, P. L. and G. Reverdin, 1987. Seasonal cycle of velocity in the Atlantic North Equatorial Countercurrent as measured by surface drifters, current meters, and ship drifts. *J. Geophys. Res.*, **92**(C4), 3691–708.
Richardson, P. L., D. Walsh, L. Armi, M. Schroeder, and J. F. Price, 1989. Tracking three meddies with SOFAR floats. *J. Phys. Oceanogr.*, **19**, 371–83.
Richardson, P. L., A. S. Bower, and W. Zenk, 2000. A census of meddies tracked by floats. *Prog. Oceanog.*, **45**, 209–50.
Richardson, R. L. and H. Stommel, 1948. Note on eddy diffusion in the sea. *J. Meteorol.*, **5**, 238–40.
Riser, S. C., H. J. Freeland, and H. T. Rossby, 1978. Mesoscale motions near the deep western boundary of the North Atlantic. *Deep-Sea Res.*, **25**, 1179–91.
Roemmich, D., O. Boebel, Y. Desaubies, H. Freeland, B. King, P.-Y. LeTraon, R. Molinari, W. B. Owens, S. Riser, U. Send, K. Takeuchi, and S. Wijffels, 2001. Argo: the global array of profiling floats. In *Observing the Oceans in the 21st Century*, ed. C. Koblinsky and N. Smith. Victoria, Australia: CSIRO Publishing.
Rogerson, A., P. D. Miller, L. J. Pratt, and C. K. R. T. J. Jones, 1999. Lagrangian motion and fluid exchange in a barotropic meandering jet. *J. Phys. Oceanogr.*, **29**(10), 2635–55.
Rossby, H. T., S. C. Riser, and A. J. Mariano, 1983. The western North Atlantic – a Lagrangian viewpoint. In *Eddies in Marine Science*, ed. A. R. Robinson. Heidelberg: Springer-Verlag, 66–91.
Rudnick, D. L., R. E. Davis, C. C. Eriksen, D. M. Fratantoni, and M. J. Perry, 2004. Underwater gliders for ocean research. *Mar. Technol. Soc. J.*, **38**(1), 48–59.
Rupolo, V., S. Marullo, and D. Iudicone, 2003. Eastern Mediterranean Transient studied with Lagrangian diagnostics applied to a Mediterranean OGCM forced by satellite SST and ECMWF wind stress for the years 1988–1993. *J. Geophys. Res.*, **108**(C9), doi:10.1029/2002JC001403.
Salman, H., L. Kuznetsov, C. K. R. T. Jones, and K. Ide, 2006. A method for assimilating Lagrangian data into a shallow-water equation ocean model. *submitted to Mon. Weather Rev.*
Samelson, R. M., 1992. Fluid exchange across a meandering jet. *J. Phys. Oceanogr.*, **22**, 431–40.
Sawford, B. L., 1999. Rotation of trajectories in Lagrangian stochastic models of turbulent dispersion. *Bound-Lay. Meteorol.*, **93**, 411–24.
Sawford, B. L., 2001. Turbulent relative dispersion. *Annu. Rev. Fluid Mech.*, **33**, 289–317.
Schmid, C., Z. Garraffo, E. Johns, and S. L. Garzoli, 2003. Pathways and variability at intermediate depths in the tropical Atlantic. In *Interhemispheric Water Exchange in the Atlantic Ocean*, ed. G. J. Goni and P. Malarotte-Rizzoli. Amsterdam: Elsevier, 233–68.
Schmitz, W. J., Jr., 1985. SOFAR float trajectories associated with the Newfoundland Basin. *J. Mar. Res.*, **43**, 761–78.
Schneider, T., 1998. Lagrangian drifter models as search and rescue tools. M. S. Thesis, Dept. of Meteorology and Physical Oceanography, University of Miami.
Shay, L. K., T. M. Cook, B. K. Haus, J. Martinez, H. Peters, A. J. Mariano, J. VanLeer, S. M. Smith, P. E. An, A. Soloview, R. Weisberg, and M. Luther, 2000. A submesoscale vortex detected by very high resolution radar. *Eos Trans. Amer. Geophys. Union*, **81**, 209–13.

Shay, L. K., T. M. Cook, H. Peters, A. J. Mariano, R. Weisberg, P. E. An, A. Soloview, and M. Luther, 2002. Very high frequency radar mapping of surface currents, *IEEE JOE*, **27**(2), 155–69.

Sheinbaum, J., J. Candela, A. Badan, and J. Ochoa, 2002. Flow structure and transport in the Yucatan Channel. *Geophys. Res. Lett.*, **29**(3), doi:10.1029/2001GL013990.

Spaulding, M. L., 1988. A state-of-the-art review of oil spill trajectory and fate modeling. *Oil Chem. Poll.*, **3**, 455–69.

Stalcup, M. C. and C. E. Parker, 1965. Drogue measurements of shallow currents on the equator in the Western Atlantic Ocean. *Deep-Sea Res.*, **12**, 535–6.

Stommel, H., 1949. Horizontal diffusion due to oceanic turbulence. *J. Mar. Res.*, **8**, 199–225.

Summers, D. M., 2005. Eddy diffusion in the sea: reinterpreting an early experiment. *Proc. R. Soc. Mat.*, **461**(2058), 1811–18.

Swallow, J. C., 1955. A neutral-buoyancy float for measuring deep currents. *Deep-Sea Res.*, **3**(1), 74–81.

Swallow, J. C., 1957. Some further deep current measurements using neutrally buoyant floats. *Deep-Sea Res.*, **4**, 93–104.

Swallow, J. C., 1971. The Aries current measurements in the Western North Atlantic. *Philos. Trans. Roy. Soc. Lond. A*, **270**, 451–60.

Taillandier, V., A. Griffa, and A. Molcard, 2006a. A variational approach for the reconstruction of regional scale Eulerian velocity fields from Lagrangian data. *Ocean Modelling*, **13**, 1–24.

Taillandier, V., A. Griffa, P.-M. Poulain, and K. Branger, 2006b. Assimilation of Argo float positions in the North Western Mediterranean Sea and impact on ocean circulation simulations. (Submitted to *Geophys. Res. Lett.*)

Talley, L. D., 2003. Shallow, intermediate, and deep overturning components of the global heat budget. *J. Phys. Oceanogr.*, **33**, 530–60.

Taylor, G. I., 1921. Diffusion by continuous movements. *Proc. Lond. Math. Soc.*, Ser. 2, **20**, 196–211.

Testor, P. and J. C. Gascard, 2005. Large scale flow separation and mesoscale eddy formation in the Algerian Basin. *Prog. Oceanog.*, **66**, 211–30.

Thomson, D. J., 1987. Criteria for the selection of stochastic models of particle trajectories in turbulent flows. *J. Fluid Mech.*, **180**, 529–56.

Thomson, D. J., 1990. A stochastic model for the motion of particle pairs in isotropic high Reynolds number, and its application to the problem of concentration variance. *J. Fluid Mech.*, **210**, 113–53.

Toner, M. and A. C. Poje, 2004. Lagrangian velocity statistics of directed launch strategies in a Gulf of Mexico model. *Nonlinear Proc. Geoph.*, **11**, 35–46.

Toner, M., A. Kirwan, L. Kantha, and J. Choi, 2001a. Can general circulation models be assessed and enhanced with drifter data? *J. Geophys. Res.*, **106**, 1366–83.

Toner, M., A. Poje, A. Kirwan, C. Jones, B. Liphardt, and C. Grosch, 2001b. Reconstructing basin-scale Eulerian velocity fields from simulated drifter data. *J. Phys. Oceanogr.*, **31**, 1361–76.

Toner, M., A. Kirwan, A. Poje, L. Kantha, F. E. Muller-Karger, and C. Jones, 2003. Chlorophyll dispersal by eddy-eddy interactions in the Gulf of Mexico. *J. Geophys. Res.*, **108**, 3105, doi:10.1029/2002JC001499.

Uchida, H. and S. Imawaki, 2003. Eulerian mean surface velocity field derived by combining drifter and satellite altimeter data. *Geophys. Res. Lett.*, **30**, 1229, doi:10.1029/2002GL016445.

Veneziani M., A. Griffa, A. M. Reynolds, and A. J. Mariano, 2004. Oceanic turbulence and stochastic models from subsurface Lagrangian data for the North-West Atlantic Ocean, *J. Phys. Oceanogr.*, **34**, 1884–906.

Veneziani, M., A. Griffa, Z. D. Garraffo, and E. P. Chassignet, 2005. Lagrangian spin parameter and coherent structures from trajectories released in a high-resolution ocean model. *J. Mar. Res.*, **63**(4), 753–88.

Volkman, G., J. Knauss, and A. Vine, 1956. The use of parachute drogues in the measurement of subsurface ocean currents. *Eos Trans. Amer. Geophys. Union*, **37**, 573–7.

Volkman, G., 1963. Deep-current measurements using neutrally buoyant floats. In *The Sea*, ed. M. N. Hill. New York: Interscience, 297–302.

Vries, P. de, and K. Doos, 2001. Calculating Lagrangian trajectories using time-dependent velocity fields *J. Atmos. Ocean. Tech.*, **18**, 1092–101.

Webb, D. C., P. J Simonetti, and C. P. Jones, 2001. SLOCUM: An underwater glider propelled by environmental energy. *IEEE J. Oceanic Eng.*, **26**(4), 447–52.

Weiss, J., 1991. The dynamics of enstrophy transfer in two-dimensional hydrodynamics. *Physica D*, **48**, 273–94.

Welander, P., 1955. Studies on the general development of motion in a two-dimensional, ideal fluid. *Tellus*, **7**, 141–56.

Wyrtki, K., L. Magaard, and J. Hager, 1976. Eddy energy in the oceans. *J. Geophys. Res.*, **81**(15), 2641–6.

Yang, Q., B. Parvin, A. J. Mariano, E. H. Ryan, R. Evans, and O. B. Brown, 2004. Seasonal and interannual studies of vortices in sea surface temperature data. An updated version of a paper originally presented at Oceans from Space 'Venice 2000' Symposium, Venice, Italy, 9–13 October 2000. *Int. J. Remote Sens.*, **25**(7–8), 1371–6.

Zhang, H.-M., M. D. Prater, and T. Rossby, 2001. Isopycnal Lagrangian statistics from the North Atlantic Current RAFOS floats observations. *J. Geophys. Res.*, **106**, 13817–36.

Index